ADVANCED TOPICS
IN SCIENCE AND TECHNOLOGY IN CHINA

ADVANCED TOPICS
IN SCIENCE AND TECHNOLOGY IN CHINA

Zhejiang University is one of the leading universities in China. In Advanced Topics in Science and Technology in China, Zhejiang University Press and Springer jointly publish monographs by Chinese scholars and professors, as well as invited authors and editors from abroad who are outstanding experts and scholars in their fields. This series will be of interest to researchers, lecturers, and graduate students alike.

Advanced Topics in Science and Technology in China aims to present the latest and most cutting-edge theories, techniques, and methodologies in various research areas in China. It covers all disciplines in the fields of natural science and technology, including but not limited to, computer science, materials science, life sciences, engineering, environmental sciences, mathematics, and physics.

Faxin Yu
Zheming Lu
Hao Luo
Pinghui Wang

Three-Dimensional Model Analysis and Processing

With 134 figures

ZHEJIANG UNIVERSITY PRESS
浙江大学出版社

Springer

Authors
Associate Prof. Faxin Yu
School of Aeronautics and Astronautics
Zhejiang University
Hangzhou 310027, China
E-mail: fxyu@zju.edu.cn

Prof. Zheming Lu
School of Aeronautics and Astronautics
Zhejiang University
Hangzhou 310027, China
E-mail: zheminglu@zju.edu.cn

Dr. Hao Luo
School of Aeronautics and Astronautics
Zhejiang University
Hangzhou 310027, China
E-mail: luohao@zju.edu.cn

Prof. Pinghui Wang
School of Aeronautics and Astronautics
Zhejiang University
Hangzhou 310027, China
E-mail: wangpinghui@tom.com

ISSN 1995-6819 e-ISSN 1995-6827
Advanced Topics in Science and Technology in China

ISBN 978-7-308-07412-4
Zhejiang University Press, Hangzhou

ISBN 978-3-642-12650-5 e-ISBN 978-3-642-12651-2
Springer Heidelberg Dordrecht London New York

Library of Congress Control Number: 2010924807

Cover design: Frido Steinen-Broo, EStudio Calamar, Spain

Printed on acid-free paper

Springer is a part of Springer Science+Business Media (www.springer.com)

Preface

With the increasing popularization of the Internet, together with the rapid development of 3D scanning technologies and modeling tools, 3D model databases have become more and more common in fields such as biology, chemistry, archaeology and geography. People can distribute their own 3D works over the Internet, search and download 3D model data, and also carry out electronic trade over the Internet. However, some serious issues are related to this as follows: (1) How to efficiently transmit and store huge 3D model data with limited bandwidth and storage capacity; (2) How to prevent 3D works from being pirated and tampered with; (3) How to search for the desired 3D models in huge multimedia databases. This book is devoted to partially solving the above issues.

Compression is useful because it helps reduce the consumption of expensive resources, such as hard disk space and transmission bandwidth. On the downside, compressed data must be decompressed to be used, and this extra processing may be detrimental to some applications. 3D polygonal mesh (with geometry, color, normal vector and texture coordinate information), as a common surface representation, is now heavily used in various multimedia applications such as computer games, animations and simulation applications. To maintain a convincing level of realism, many applications require highly detailed mesh models. However, such complex models demand broad network bandwidth and much storage capacity to transmit and store. To address these problems, 3D mesh compression is essential for reducing the size of 3D model representation.

Feature extraction is a special form of dimensionality reduction. When the input data to an algorithm is too large to be processed and is suspected to be notoriously redundant (much data, but not much information), the input data will be transformed into a reduced representation set of features (also named a feature vector). If the features extracted are carefully chosen, it is expected that the features set will extract the relevant information from the input data, in order to perform the desired task using this reduced representation instead of the full size input. Feature extraction is an essential step in content-based 3D model retrieval systems. In general, the shape of the 3D object is described by a feature vector that serves as a search key in the database. If an unsuitable feature extraction method has been used, the whole retrieval system will be unusable. We must realize that 3D objects can be saved in many representations, such as polyhedral meshes, volumetric data and parametric or implicit equations. The method of feature extraction should accept this fact and it should be independent of data representation. The method should also be invariant under transforms such as translation, rotation and scale of the 3D object. Perhaps this is the most important requirement, because the 3D objects are usually saved in various poses and on various scales. The 3D object can be obtained either from a 3D graphics program or from a 3D input device. The second way is more susceptible to some errors, therefore the feature extraction method should also be insensitive to noise. Perhaps the last requirement is that it has to be quick to compute and easy to index. The database may contain thousands of objects, so the agility of the system would also be one of the main requirements.

Content-based visual information retrieval (CBVIR) is the application of computer vision to the visual information retrieval problem, which solves the problem of searching for digital images/videos/3D models in large databases. "Content-based" means that the search will analyze the actual contents of the visual media. The term "content" in this context might refer to colors, shapes, textures, or any other information that can be derived from the visual media itself. Without the ability to examine visual media content, searches must rely on metadata such as captions and keywords, which may be laborious or expensive to produce. A common characteristic of all applications in multimedia databases (and in particular in 3D object databases) is that a query searches for similar objects instead of performing an exact search, as in traditional relational databases. Multimedia objects cannot be meaningfully queried in the classical sense (exact search), because the probability that two multimedia objects are identical is very low, unless they are digital copies from the same source. Instead, a query in a multimedia database system usually requests a number of objects most similar to a given query object or to a manually entered query specification. Therefore, one of the most important tasks in a multimedia retrieval system is to implement effective and efficient similarity search algorithms. Typically, the multimedia data are modeled as objects in a metric or vector space, where a distance function must be defined to compute the similarity between two objects. Thus, the similarity search problem is reduced to a search for close objects in the metric or vector space. The primary goal in a 3D similarity search is to design algorithms with the ability to effectively and efficiently execute similarity queries in 3D databases. Effectiveness is related to the ability to retrieve similar 3D objects while holding back non-similar ones, and efficiency is related to the cost of the search, measured e.g., in CPU or I/O time. But, first of all one should define how the similarity between 3D objects is computed.

Digital watermarking is a branch of data hiding (or information hiding). It is the process of embedding information into a digital signal. The signal may be audios, pictures, videos or 3D models. If the signal is copied, then the information is also carried in the copy. An important application of invisible watermarking is in copyright protection systems, which are intended to prevent or deter unauthorized copying of digital media. Another important application is to authenticate the content of multimedia works, where fragile watermarks are commonly used for tamper detection (integrity proof). Steganography is an application of digital watermarking, where two parties communicate a secret message embedded in the digital signal. Annotation of digital photographs with descriptive information is another application of invisible watermarking. While some file formats for digital media can contain additional information called metadata, digital watermarking is distinct in that the data is carried in the signal itself.

Reversible data hiding is a technique that enables images or 3D models to be authenticated and then restored to their original forms by removing the watermark and replacing the images or 3D data which had been overwritten. This would make the images or 3D models acceptable for legal purposes. Although reversible data hiding was first introduced for digital images, it has also wide application scenarios for hiding data in 3D models. For example, suppose there is a column on a 3D mechanical model obtained by CAD. The diameter of this column is changed with a given data hiding scheme. In some applications, it is not enough that the hidden content is accurately extracted, because the remaining watermarked model is still distorted. Even if the column diameter is increased or decreased by 1 mm, it may cause a severe effect for this mechanical model cannot be well assembled with other mechanical accessories. Therefore, it also has significance in the design of reversible data hiding methods for 3D models.

Based on the above background, this book is devoted to processing and analysis techniques for 3D models, i.e., compression techniques, feature extraction and retrieval techniques and watermarking techniques for 3D models. This book focuses on three main areas in 3D model processing and analysis, i.e., compression, content-based retrieval and data hiding, which are designed to reduce redundancy in 3D model representations, to extract the features from 3D models and retrieve similar models to the query model based on feature matching, to protect the copyright of 3D models and to authenticate the content of 3D models or hide information in 3D models. This book consists of six chapters. Chapter 1 introduces the background to three urgent issues confronting multimedia, i.e., storage and transmission, protection and authentication, and retrieval and recognition. Then the concepts, descriptions and research directions for the newly-developed digital media, 3D models, are presented. Based on three aspects of the technical requirements, the basic concepts and the commonly-used techniques for multimedia compression, multimedia watermarking, multimedia retrieval and multimedia perceptual hashing are then summarized. Chapter 2 introduces the background, basic concepts and algorithm classification of 3D mesh compression techniques. Then we discuss some typical methods used in connectivity compression and geometry compression for 3D meshes respectively. Chapter 3 focuses on the techniques of feature extraction from 3D models. First, the background, basic concepts and algorithm classification related to 3D model feature extraction are introduced. Then, typical 3D model feature extraction methods are classified into six categories and are, discussed in eight sections, respectively. Chapter 4 discusses the steps and techniques related to content-based 3D model retrieval systems. First, we introduce the background, performance evaluation criteria, the basic framework, challenges and several important issues related to content-based 3D model retrieval systems. Then we analyze and discuss several topics for content-based 3D model retrieval, including preprocessing, feature extraction, similarity matching and query interface. Chapter 5 starts with the description of general requirements for 3D watermarking, as well as the classification of 3D model watermarking algorithms. Then some typical spatial domain 3D mesh model watermarking schemes, typical transform-domain 3D mesh model watermarking schemes and watermarking algorithms for other types of 3D models are discussed respectively. Chapter 6 starts by introducing the background and performance evaluation metrics of 3D model reversible data hiding. Then some basic reversible data hiding schemes for digital images are briefly reviewed. Finally, three kinds of 3D model reversible data hiding techniques are extensively introduced, i.e., spatial domain based, compressed domain based and transform domain based methods.

This book embodies the following characteristics. Firstly, it has novelty. The content of this book covers the research hotspots and their recent progress in the field of 3D model processing and analysis. For example, in Chapter 6, reversible data hiding in 3D models is a very new research branch. Secondly it has completeness. Techniques for every research direction are comprehensively introduced. For example, in Chapter 3, feature extraction methods for 3D models are classified and introduced in detail. Thirdly it is theoretical. This book embodies many theories related to 3D models, such as topology, transform coding, data compression, multi-resolution analysis, neural networks, vector quantization, 3D modeling, statistics, machine learning, watermarking, data hiding, and so on. For example, in Chapter 2, several definitions related to 3D topology and geometry are introduced in detail in order to easily understand the content of later chapters. Fourthly it is practical. For each application, experimental results for typical methods are illustrated in detail. For example, in Chapter 6, three examples of typical reversible data hiding are illustrated with detailed steps and elaborate experiments.

In this book, Chapters 1, 4 and 5 were written by Prof. Zheming Lu, Chapters 2 and 3 were written by Prof. Faxin Yu, Chapter 6 was written by Dr. Hao Luo with the aid of student Hua Chen. The whole book was finalized by Prof. Faxin Yu. The research results of this book are based on the accumulated work of the authors over a long period of time. We would like to show our great appreciation for the assistance of other teachers and students in the Institute of Astronautics and Electronic Engineering of Zhejiang University. The work was partially supported by the National Natural Science Foundation of China, the foundation from the Ministry of Education in China for persons showing special ability in the new century, and the foundation from the Ministry of Education in China for the best national Ph.D dissertations. Due to our limited knowledge, it is inevitable that errors and defects will appear in this book and we invite our readers to comment.

The authors
Hangzhou, China
January, 2010

Contents

1

Introduction

The digitization of multimedia data, such as images, graphics, speech, text, audio, video and 3D models, has made the storage of multimedia more and more convenient, and has simultaneously improved the efficiency and accuracy of information representation. With the increasing popularization of the Internet, multimedia communication has reached an unprecedented level of depth and broadness, and multimedia distribution is becoming more and more manifold. People can distribute their own works over the Internet, search and download multimedia data, and also carry out electronic trade over the Internet. However, some serious issues accompany this as follows: (1) How can we efficiently transmit and store huge multimedia information with limited bandwidth and storage capacity? (2) How can we prevent multimedia works from being pirated and tampered with? (3) How can we search for the desired multimedia content in huge multimedia databases?

1.1 Background

We first introduce the background to three urgent issues for multimedia, i.e., (1) storage and transmission, (2) protection and authentication, (3) retrieval and recognition.

1.1.1 Technical Development Course of Multimedia

"Multimedia" [1] is a compound word composed of "multiple" and "media", which means "multiple media". Here, "media" is the plural form of the word "medium". In fact, the word "medium" has two kinds of meaning in the computer field: one stands for the entities for storing information, such as diskettes, CDs, magnetic tapes and semiconductor memorizers; the other stands for the carriers for transmitting information, such as digits, characters, audio clips, graphics and images. Here, the word "media" in multimedia technology means the latter. "Monomedia" is one (word) as opposed to "multimedia" and, literally, multimedia is composed of several "monomedia". People use various media during information communication, and multimedia is just the representation and transmission form for multiple information carriers. In other words, it is a technique to simultaneously acquire, process, edit, store and display more than two kinds of media, including text, audios, graphics, images, movies and videos, etc. In fact, it is the material development of computer and digital information processing technologies that enables people to process multimedia information and thus enables the realization of multimedia technology. Therefore, so-called "multimedia" stands no longer for multiple media themselves but for the whole series of techniques to deal with and apply them. In fact, "multimedia" has been viewed as a synonym of "multimedia technology". It is worth noting that multimedia technology nowadays is often associated with computer technology. The reason is that the computer's capability of digitization and interactive processing greatly promotes the development of multimedia technology. In general, people can view multimedia as the new technology or as product forming from the combination of advanced computer, video, audio and communication technologies.

The multimedia technique has been rapidly developed accompanied by the wide application of computer and network technologies, and computer network multimedia technology has become an area under rapid development and has gained research focus in the 21st century. As a rapidly developing all-round electronic information technology, multimedia technology has brought directional renovation to traditional computer systems and audio and video equipments, and will have a great effect on mass media. Since the mid to late 1980s, multimedia computer technology has become the focus of concern, and its definition is as follows: computers comprehensively process various kinds of multimedia information (text, graphics, images, audios and videos), which means various kinds of information is linked together to form a system with interactivity. Interactivity is one of the characteristics of multimedia computer technology, meaning the characteristic of interactive communication with users, which is the biggest difference from traditional media. Apart from providing users with solutions to problems on their own, such a change can help users learn and think with the aid of conversational communication and carry out systematical queries or statistical analysis in order to achieve the advancement of knowledge and the improvement of problem-solving ability. Multimedia computers will speed up the process of introducing computers to families and societies, and will bring a profound revolution to people's work, life and entertainment. Since the 1990s, the progress that the world has made towards an information society has been significantly expedited, in which the application of multimedia technology has been playing a vital role. Multimedia improves a human's information communication and shortens the communication path. The application of multimedia technology is a sign of the 1990s, and is a second revolution in the computer field.

On the whole, multimedia technology is nowadays developing in the following two directions.

One is networking, which means that, combined with wide-band network communication technology, multimedia technology enters areas such as scientific research, designing, enterprise management, office automation, remote education, telemedicine, retrieval, entertainment and automatic testing. In some recent films, we can often see a very personalized computer that can talk with humans and provide any information they want to know. It can play any music they want to listen to. If there is any accident anywhere in the world, it can report to them in time. It can monitor the status of all the apparatus at home, and can help to receive phone calls and remind humans what to do, and even transmit messages to their friends living far away. Today, because of the development of multimedia, all of the above dreams will come true.

The other direction is componentization together with intelligentization and embeddability of the multimedia terminal, which means improving the multimedia performance of computer systems to develop intelligent household appliances. The current household television system cannot be called a multimedia system, because although existing televisions also provide "sound, graphics, text" information, people can do nothing but select different channels, and people cannot interfere or change them but passively receive the programs from TV stations. This process is not two-way but one-way. However, we can forecast that, in the near future, the household television system will definitely be a multimedia system, which will combine many functions, such as entertainment, education, communication and consultation, all in one.

In summary, the birth of multimedia technology will definitely bring a revolution to the computer field once more. It indicates computers will not only be used in offices and laboratories but also be used in the household, in commerce, for travel, amusement, education and art, etc., i.e., in nearly all areas of daily life. At the same time, it means computers can be developed in the most ideal way for humans, i.e., with the integration of seeing and hearing, which completely plays down the human-computer interface.

1.1.2 Information Explosion

Real human civilization starts from the Internet. In fact, we are living with all kinds of networks, such as electrical networks, telephone networks, broadcast/television networks, commercial networks and traffic networks. However, all these networks are very different from the Internet, which has affected so many governments, enterprises and individuals in such a short time. Nowadays, the network has become a substitutable noun for the Internet. In the past few years, with the rapid development of computer and network techniques, the scale of the Internet has been suddenly expanded. The Internet technique breaks the traditional borderline, which makes the world smaller and smaller, while making the market larger and larger. The wide world is like a global village, where the global economy and information networking promote and depend on each other. The Internet makes the speed and scale of information acquisition and transmission reach an unprecedented level. In the era of information networking, the Internet should be considered for any product or technique. Network information systems are playing more and more important roles in politics, military affairs, finance, commerce, transportation, telecommunication, culture and education. Modern communication and transmission techniques have greatly improved the speed and extent of information transmission. The technical means include broadcasts, television, satellite communication and computer communication using microwave and optical fiber communication networks, which overcome traditional obstacles in space and time and further unite the whole world. However, the accompanying issues and side effects are as follows: A surge of information overwhelms people, and it is very hard to retrieve accurately and rapidly the information most needed from the tremendous amount of information. This phenomenon is called the information explosion [2], also called "information overload" or "knowledge bombing".

The information explosion describes the rapid development in the amount of information or human knowledge in recent years, whose speed is like a bomb engulfing all the world. With regard to the phrase "information explosion", it can date back to the 1980s. At that time, besides broadcasting, television, telephone, newspapers and various publications, new means of communication, i.e., computers and communication satellites emerged, making the amount of information increase suddenly like an explosion. Statistics show that over the past decade the amount of information all over the world doubled every 20 months. During the 1990s, the amount of information continued to increase dramatically. At the end of the 1990s, due to the emergence of the Internet, information distribution and transmission got out of control, and a great deal of false or useless information was generated, resulting in the pollution of information environments and the birth of "waste messages". Because everyone can freely air his opinion over the Internet, and the distribution cost can be ignored, in a sense everyone can become an information manufacturer on the global level, and thus information really starts to explode. As times go by, the information explosion manifests itself mainly in five aspects: (1) the rapid increase in the amount of news; (2) the dramatic increase in the amount of amusement information; (3) a barrage of advertisements; (4) the rapid increase in scientific and technical information; (5) the overloading of our personal receptiveness. However, faced with the inflated amount of information and the enormous pressure of "chaotic information space" and "information surplus", people out of the blue become hesitant in their urgent pursuit and expectation of information. Even if we take 24 hours every day to read information, we cannot take it all in, and besides, there is a great deal of useless or false information. Useful information can increase economic benefits and promote the development of human society, but if the information increases in a disorderly fashion and even runs out of control, it will bring about various social problems such as information crime and information pollution. People on the one hand are enjoying the convenience brought about by abundant information over the Internet; on the other hand they are suffering from annoyance due to the "information explosion". "Information explosion" has had a negative effect on the advance of the social economy. A recent survey of ten multinational corporations has revealed that, because they have to deal with a great deal of information that exceeds their ability to analyse it, their efficiency in decision-making is severely disturbed, even resulting in wrong decisions or difficulty in making the optimal decision. On detailed analysis, nowadays collecting information has cost us much more than the intrinsic value of that information. At present, besides an abundance of useful information, there is also a great deal of pornographic content, violent content and false advertising over the Internet. These junk messages have deluged us, to become a new public nuisance, just like the pollution produced by industrial waste, medical and other human refuse, and they have confused users in their rapid search for useful information.

The opposite of "information explosion" is "information shortage". On the one hand, from the quantitative angle, an information explosion refers to the phenomenon where web information increases exponentially because of the advance in transmission techniques and the openness of the transmission environment, while information shortage refers to a situation where the amount of information cannot satisfy the receiver's needs, because of congestion in the channels or a lack of information sources. In this sense, information shortage is a kind of absolute shortage. On the other hand, from the qualitative angle, accompanied by the information explosion, the really valuable information is submerged by a great deal of waste messages, and the receivers are thrown into great confusion because of numerous and jumbled items of information. In this sense, information shortage is a kind of relative shortage.

Nowadays people are devoting themselves to solving the "information explosion" problem from two aspects, i.e., technology and management. From the point of view of management, all governments have promulgated corresponding regulations and byelaws for network information. However, it is hard to have a unified worldwide standard due to the differences in constitutions, ideologies, conventions and moral values from country to country. Therefore, it is impractical to create a single regulation to control "waste messages" for worldwide webs. From such cognition, people try to seek technical solutions. Since the 1990s, every country has laid heavy stress on databases, data mining and information standardization technologies, resulting in the emergence of a new interdisciplinary field, knowledge discovery. Currently, the main technologies for obtaining information are retrieval technologies, e.g., search engines based on cataloguing, keywords-based search engines and content-based retrieval systems. In addition, some internet content providers (ICPs) push the special information to users through an intelligent proxy server according to users' customization, which is called the push service.

Based on the background to the information explosion era, this book focuses on applying retrieval technology to deal with the information explosion problem with regard to the new kind of media, 3D models, in Chapter 4. Apart from information retrieval, another effective technical solution to the information explosion is data compression technology. As is well known, the amount of digitalized information is huge, which brings extreme pressure to the storage capacity of memorizers, the transmission bandwidth of channels and the processing speed of computers. With regard to this problem, it is impractical to purely increase the storage capacity, the bandwidth or the CPU speed. If we adopt advanced compression algorithms to compress the digitalized audiovisual data, we can not only save the storage space but also make it possible for the computer to process and play the audiovisual information in a real-time manner. This book will focus on the 3D model compression problem in Chapter 2.

1.1.3 Network Information Security

People neglect the security problems of most modern computer networks at the beginning of construction and, even if they do not, they only base the security mechanism on the physical security. Therefore, with the enlargement of the networking scale, this physical security mechanism is but an empty shell in the network environment. In addition, the protocol in use nowadays, e.g., the TCP/IP protocol, does not take the security problem into account at the beginning. Thus, openness and resource sharing are the main rootstock of the computer networking security problem, and the security mainly depends on encryption, network user authentication and access control strategies. Facing such severe threats that harm network information systems and considering the importance of network security and secrecy, we must take effective measures in order to guarantee the security and secrecy of the network information. The network measures for security can be classified in the following three categories: logical-based, physical-based and policy-based. In the face of various threats that harm computer networking security more and more severely, only using physical-based or policy-based means cannot effectively keep away computer-based crime. People should therefore adopt logical-based measures, that is to research and develop effective techniques for network and information security. Even if we have very self-contained policies and rules for security and secrecy, very advanced techniques for security and secrecy and flawless physical security mechanisms, all efforts will go to waste if the above knowledge cannot be popularized.

People's understanding of information security is continually updated. In the era of host computers, people understand information security as the protection of confidentiality, integrality and availability of information, which is data-oriented. In the era of microcomputers and local networks in the 1980s, because of the simple structure of users and networks, information security was administrator-oriented and stipulation-oriented. In the era of the Internet in the 1990s, every user could access, use and control the connected computers everywhere, and thus information security over the Internet emphasizes connection-oriented and user-oriented security. Thus it can be seen that data-oriented security considers the confidentiality, integrality and availability of information, while user-oriented security considers authentication, authorization, access control, non-repudiation and serviceability, together with content-based individual privacy and copyright protection. Combining the above two aspects of security, we can obtain the generalized information security [3] concept, that is all theories and techniques related to information security, integrality, availability, authenticity and controllability, suming up physical security, network security, data security, information content security, information infrastructure security and public information security. On the other hand, information security in the narrow sense indicates information content security, which is the protection of the secrecy, authenticity and integrality of the information, avoiding attackers' wiretapping, imitating, beguilement and embezzlement and protecting the legal users' benefits and privacy. The secure service issues in the information security architecture rely on ciphers, digital signatures, authentication techniques, firewalls, secure audit, disaster recovery, anti-virus, preventing hacker intrusion, and so on. Among them, cryptographic techniques and management means are the core of information security, while the security standards and system evaluation methods are the bases of information security. Technically, information security is a marginal integrated subject involving computer science, network techniques, communication techniques, applied mathematics, number theory, information theory, and so on.

Network information security consists of four aspects, i.e., the security problems in information communication and storage, and the audit of network information content and authentication. To maintain the security of data transmission, it is necessary to apply data encryption and integrity identification techniques. To guarantee the security of information storage, it is necessary to guarantee the database security and terminal security. An information content audit checks the content of the input and output information from networks, so as to prevent or trace possible whistle-blowing. User identification is the process of verifying the principal part in the network. Usually there are three kinds of methods for verifying the principal part identity. One is that only the secret known by the principal part is available, e.g., passwords or keys. The second is that the objects carried by the principal part arc available, e.g., intelligent cards or token cards. The third is that only the principal part's unique characteristics or abilities are available, e.g., fingerprints, voices, retina, signatures, etc. The technical characteristics of network information security mainly embody the following five aspects: (1) Integrity. It means the network information cannot be altered without authority. It is against active attacks, guaranteeing data consistence and preventing data from being modified and destroyed by illegal users. (2) Confidentiality. It is the characteristic that the network information cannot be leaked to unauthorized users. It is against passive attacks so as to guarantee that the secret information cannot be leaked to illegal users. (3) Availability. It is the characteristic that the network information can be visited and used by legal users if needed. It is used to prevent information and resource usage by legal users from being rejected irrationally. (4) Non-repudiation. It means all participants in the network cannot deny or disavow the completed operations and promises. The sender cannot deny the already sent information, while the receiver also cannot deny the already received information. (5) Controllability. It is the ability to control the content of network information and its prevalence. Namely, it can monitor the security of network information.

The coming of the network information era also proposes a new challenge to copyright protection. Copyright is also called author's rights. It is a general designation of legal rights based on a special production and the economic rights which completely dominate this production and its interest. With the continuous enlargement of the network scope and the gradual maturation of digitalization techniques, the quantity of various digitalized books, magazines, pictures, photos, music, songs and video products has increased rapidly. These digitalized products and services can be transmitted by the network without the limitation of time or space, even without logistic transmission. After the trade and payment are completed, they can be efficiently and quickly provided for clients by the network. On the other hand, openness and resource sharing of the network will cause the problem of how to validly protect the digitalized network products' copyright. There must be some efficient techniques and approaches for the prevention of digitalized products from altering, counterfeiting, plagiarizing and embezzling, etc.

Information security protection methods are also called security mechanisms. All security mechanisms are designed for some types of security attack threats. They can be used individually or in combination according to different manners. Commonly used network security mechanisms are as follows. (1) Information encryption and hiding mechanism. Encryption makes an attacker unable to understand the message content and thus information is protected, while hiding conceals the useful information in other information, and thus the attacker cannot find it. It not only realizes information secrecy, but also protects the communication itself. So far, information encryption is still the most basic approach in information security protection, while information hiding is a new direction in information security areas. It draws more and more attention in the applications of digitalized productions' copyright protection. (2) Integrity protection. It is used for the prevention of illegal alteration based on cipher theory. Another purpose of integrity protection is to provide non-repudiation services. When information source's integrity can be verified but cannot be simulated, the information receiver can verify the information sender. Digital signatures can provide methods for us. (3) Authentication mechanism. This is the basic mechanism of network security, namely that network instruments should authenticate each other so as to guarantee the right operations and audit of a legal user. (4) Audit. It is the foundation for preventing inner criminal offenses and for taking evidence after accidents. Through the records of some important events, errors can be localized and reasons for successful attacks can be found when mistakes appear in the system or the system is attacked. Audit information should prevent illegal deletion and modification. (5) Power control and access control. It is the requisite security means of host computer systems. Namely, the system endows suitable operation power to a certain user according to the right authentication, and thus makes him not exceed his authority. Generally, this mechanism adopts the role management method. That is, aiming at system requirements, it defines various roles, e.g., manager, accountant, etc., and then endows them with different executive powers. (6) Traffic padding. It generates spurious communications or data units to disguise the amount of real data units being sent. Typically, useless random data are sent out in a vacancy and thus enhance the difficulty of obtaining information through the communication stream. Meanwhile, it also enhances the difficulty of deciphering the secret communications. The sent random data should have good simulation performance, and thus can mix the false with the genuine. This book focuses on applying digital watermarking techniques to solve copyright protection and content authentication problems for 3D models, involving the first three security mechanisms.

1.1.4 Technical Requirements of 3D Models

Before the emergence of 3D models, multimedia technology experienced three waves: digital sound in the 1970s, digital images in the 1980s and digital videos in the 1990s. Human visual perception possesses the 3D stereo property. 3D models and their corresponding 3D scenes can therefore afford more abundant visual perceptual details than 2D images. With the development of 3D data acquisition, 3D graphics modeling and graphics hardware technologies, people have generated more and more 3D object databases for virtual reality, 3D games and industrial solid CAD models, and so on. Here, CAD, i.e., Computer Aided Design, means that designers carry out the design work with the aid of computers and their graphics devices. With the increasing popularization of 3D scanning technologies and 3D modeling tools, 3D model databases have become more and more common in fields such as biology, chemistry, archaeology and geography. On the other hand, the dilatation of the Internet has enhanced the ability to retrieve 3D models that are dispersedly stored, and has created favorable conditions to efficiently transmit high-quality 3D models. Currently, 3D models have been applied to various fields: In the medical field, 3D models are used to accurately describe the organs; in the movie industry, 3D models are utilized to represent the characters, objects and scenes; in the video game industry, 3D models are adopted as the game sources in computers and video games; in the science field, 3D models can be used to show accurate structures of compounds; in the architecture industry, they are used to display the buildings and landscapes; in the engineering field, they are used to design new devices, vehicles, structures, and so on; in the geosciences, people start to construct 3D geologic models.

3D models have been the fourth generation of multimedia data type following audios, images and videos, and the increasingly developing Internet and function-enhanced computers have provided conditions for 3D model processing and sharing. Thus, in the near future people can freely use 3D models just like 2D images. The former problem of "how to acquire 3D models" has been changed into the current problem of "how to search for 3D models we need", which has resulted in the increasing need for 3D model retrieval technologies. For example, it is a long laborious process to carry out high-fidelity 3D modeling. If there are some former models that can be reused, the cost will be greatly reduced. At the same time, the research results of content-based 3D model retrieval techniques can be widely applied to fields such as virtual geographical environments, CAD, molecular biology, military affairs, medicine, chemistry, archaeology and industrial manufacturing, and one can also find applications in electronic business and web-based search engines. Therefore, how to rapidly search for the required 3D models has been a second popular topic following the retrieval techniques for texts, audios, images and videos. The 3D model retrieval technology involves several areas such as artificial intelligence, computer vision and pattern recognition. The underlying problem in content-based 3D model retrieval systems is to select appropriate features to distinguish dissimilar shapes and index 3D models. Based on these requirements, this book discusses 3D model feature extraction techniques in Chapter 3, and introduces 3D model retrieval techniques in Chapter 4.

On the other hand, with the ceaseless emergence of advanced modeling tools and the increasing maturation of 3D shape data scanning techniques, people have put forward greater requests for accuracy and details of 3D geometric data, which has at the same time brought about a rapid growth in the scale and complexity of 3D geometric data. Huge geometric data have enormously challenged the capacity and speed of current 3D graphics search engines. Furthermore, the development of the Internet makes the application of 3D geometric data broader and broader. However, the limitation of bandwidth has severely restricted the distribution of this kind of media. It is not sufficient to solve this problem merely based on the increase in the contribution of hardware devices, but we also need to research 3D model compression techniques. Thus, this book discusses 3D model compression techniques in Chapter 2.

More severely, with the development of computer technologies, CAD, virtual reality and network technologies have made considerable progress, and more and more 3D models have been created, distributed, downloaded and used. Because 3D models possess commercial value, visual value and economic benefits, the producers and copyright owners of these 3D products will inevitably have to face up to the practical issues of copyright (or intellectual property rights) protection and content authentication during the distribution of 3D models over the Internet. Thus, this book discusses the watermarking and reversible data hiding techniques of 3D models in Chapters 5 and 6.

Besides the above three technical requirements, there are some other technical requirements for 3D models including simplification, reconstruction, segmentation, interactive display, matching and recognition, and so on. For example, computer- aided geometric modeling techniques have been widely used during product development and manufacturing processes, but there are still many products not originally described by CAD models because the designers or manufacturers are faced with material objects. In order to utilize the advanced manufacturing technology, we should transform material objects into CAD models, and this has been a relatively independent research area in CAD or CAM (computer-aided manufacturing) systems, i.e., reverse engineering [4]. To take a second example, mesh segmentation [5] has become a hot research topic because it has become an important technical requirement to modify current models according to the new design goal by reusing previous models. Mesh segmentation stands for the technique of segmenting a closed mesh polyhedron or orientable 2D manifold, according to certain geometric or topological characteristics, into a certain number of sub-meshes with simple shapes, each sub-mesh self-connected. This work has been widely applied in research works on digital geometric processing such as mesh reconstruction based on 3D point cloud data, mesh simplification, levels of detail (LOD) modeling, geometric compression and transmission, interactive editor, texture mapping, mesh tessellation, geometry deformation, parameterization of local areas and spline surface reconstruction in reverse engineering.

1.2 Concepts and Descriptions of 3D Models

In the following, the concepts, descriptions and research directions for newly-developed digital media, 3D models, are presented. Based on three aspects of technical requirements, the basic concepts and the commonly-used techniques for multimedia compression, multimedia watermarking, multimedia retrieval and multimedia perceptual hashing are then summarized.

1.2.1 3D Models

A model is the abstract representation of an objective, including structures, attributes, variation laws and relationships among components. 3D models are the fourth generation of multimedia following sound, images and videos. A 3D model represents a 3D object using a collection of points in the 3D space, connected by various geometric entities such as triangles, lines, curved surfaces, etc. A typical example is shown in Fig. 1.1. Being a collection of data (points and other information), 3D models can be created by hand, algorithmically (procedural modeling), or scanned. 3D models have been widely used anywhere in 3D graphics. Actually, their use predates the widespread use of 3D graphics on personal computers. Many computer games use pre-rendered images of 3D models as sprites before computers can render them in real-time. Today, 3D models are used in a wide variety of fields. The medical industry uses detailed models of organs. The movie industry uses them as characters and objects for animated and real-life motion pictures. The video game industry uses them as assets for computer and video games. The science sector uses them as highly detailed models of chemical compounds. The architecture industry uses them to demonstrate proposed buildings and landscapes through software architectural models. The engineering community uses them as designs of new devices, vehicles and structures, as well as for a host of other uses. In recent decades, the earth science community has started to construct 3D geological models as a standard practice.

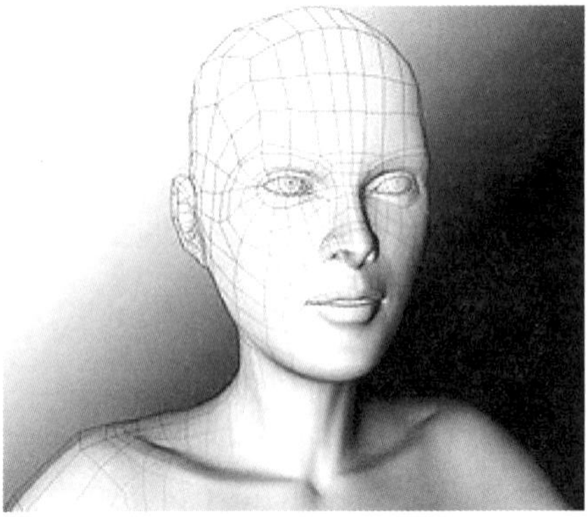

Fig. 1.1. A typical polygon mesh model

3D models can be roughly classified into two categories: (1) Solid models. These models define the volume of the object they represent (like a rock). These are more realistic, but more difficult to build. Solid models are mostly used for non-visual simulations such as medical and engineering simulations, and for CAD and specialized visual applications such as ray tracing and constructive solid geometry. (2) Shell/Boundary models. These models represent the surface, e.g., the boundary of the object, not its volume (like an infinitesimally thin eggshell). These are easier to work with than solid models. Almost all visual models used in games and films are shell models.

Because the appearance of an object depends largely on the exterior of the object, boundary representations are common in computer graphics. 2D surfaces are a good analogy for the objects used in graphics, though quite often these objects are non-manifold. Since surfaces are not finite, a discrete digital approximation is required: polygonal meshes are by far the most common representations, although point-based representations have been gaining some popularity in recent years. Level sets are a useful representation for deforming surfaces which undergo many topological changes, such as fluids.

The process of transforming representations of objects, such as the middle point coordinate of a sphere and a point on its circumference into a polygon representation of a sphere, is called tessellation. This step is used in polygon-based rendering, where objects are broken down from abstract representations ("primitives") such as spheres, cones, etc., to so-called meshes, which are nets of interconnected triangles. Meshes of triangles (instead of e.g. squares) are popular as they have proven to be easy to render using scan line rendering. Polygon representations are not used in all rendering techniques, and in these cases the tessellation step is not included in the transition from abstract representation to the rendered scene.

There are two types of information in a 3D model, geometrical information and topological information. Geometrical information generally represents shapes, locations and sizes in the Euclidean space, while topological information stands for the connectivity between different parts of the 3D model. The 3D model itself is invisible, but we can perform the rendering operation at different levels of detail based on simple wireframes or shading based on different methods. Here, rendering is the process of generating an image from a model by computer programs. The model is a description of 3D objects in a strictly defined language or data structure. It may contain geometry, viewpoint, texture, lighting and shading information. The generated image is a digital image or raster graphics image. This term may be analogous with an "artist's rendering" of a scene. Rendering is also used to describe the process of calculating effects in a video editing file to produce the final video output. Shading is a process in drawing for depicting levels of darkness on paper by applying media more densely or with a darker shade for darker areas, and less densely or with a lighter shade for lighter areas. In computer graphics, shading refers to the process of altering a color according to its angle to lights and its distance from lights to create a photorealistic effect. Shading is performed during the rendering process. However, a lot of 3D models are covered with texture, and we call this process texture mapping. It is a method for adding detail, surface texture, or color to a computer-generated graphic or 3D model. Its application to 3D graphics was pioneered by Dr. Edwin Catmull in his Ph.D thesis in 1974. A texture map is applied (mapped) to the surface of a shape or polygon. This process is akin to applying patterned paper to a plain white box. The way by which the resulting pixels on the screen are calculated from the texels (texture pixels) is governed by texture filtering. The fastest method is to use the nearest-neighbor interpolation technique, while bilinear interpolation and trilinear interpolation between mipmaps are two commonly used alternatives which reduce aliasing or jaggies. In the event of a texture coordinate being outside the texture, it is either clamped or wrapped.

1.2.2 3D Modeling Schemes

When we use computers to analyze and research objective things, it is essential to adopt suitable models to represent the actual objects or abstract phenomena. This process is called modeling. In 3D computer graphics, 3D modeling [6] is the process of developing a mathematical, wireframe representation of any 3D object (either inanimate or living) via specialized software. It can be displayed as a 2D image through a process called 3D rendering or used in a computer simulation of physical phenomena. The model can also be physically created using 3D printing devices. Models may be created automatically or manually. The manual modeling process of preparing geometric data for 3D computer graphics is similar to plastic arts such as sculpting. 3D modeling has played an important role in architecture, medical imaging, cultural relic preservation, 3D animation, 3D games, film's technical razzle-dazzle making, and so on.

3D scanners and image acquisition systems are rapidly becoming more affordable and allow the building of highly accurate models of real 3D objects in a cost- and time-effective manner. To construct 3D models for actual objects, we must first acquire related attributes of samples, such as geometrical shapes and surface textures. The data that record such information are called 3D data, and 3D data acquisition is the process by which the 3D information is acquired from samples and organized as the representation consistent with the samples' structures. The methods of acquiring 3D information from samples can be classified in the following five categories:

(1) Methods based on direct design or measurement. They are often used in early architecture 3D modeling. They utilize engineering drawing to obtain the three views of each model.

(2) Image-based methods. They construct 3D models based on pictures. They first obtain geometrical and texture information simultaneously by taking photos, and then construct 3D models based on obtained images.

(3) Mechanical-probe-based methods. They acquire the surface data by physical touch between the probe and the object. They require that the object hold a certain hardness.

(4) Methods based on volume data restoration. They adopt a series of slicing images of the object to restore the 3D shape of the object. They are often used in medical departments with X-ray slicing images, CT images and MRT images.

(5) Region-scanning-based methods. They obtain the position of each vertex in the space by estimating the distance between the measuring instrument and each point on the object surface. Two examples of the methods are optical triangulation and interferometry.

The main problem in 3D modeling is to render 3D models based on 3D data. To achieve a better visual effect, we should guarantee it has smooth surfaces, without burrs and holes, and make 3D models embody a third dimension and sense of reality. At the same time, we should organize the data in a better manner to reduce the storage space and speed up the displaying. Current modeling techniques can be mainly classified in three categories: geometric-modeling-based, 3D scanner-based and image-based, which can be described in detail as follows.

1.2.2.1 Geometric-Modeling-Based Techniques

Geometric modeling is a branch of applied mathematics and computational geometry that studies methods and algorithms for the mathematical description of shapes. The shapes studied in geometric modeling are mostly 2D or 3D, although many of its tools and principles can be applied to sets of any finite dimension. Today most geometric modeling processes are done with computers and for computer-based applications. 2D models are important in computer typography and technical drawing. 3D models are central to CAD/CAM, and widely used in many applied technical fields such as civil and mechanical engineering, architecture, geology and medical image processing. Geometric models are usually distinguished from procedural and object-oriented models, which define the shape implicitly by an opaque algorithm that generates its appearance. They are also contrasted with digital images and volumetric models which represent the shape as a subset of a fine regular partition of space, and with fractal models that give an infinitely recursive definition of the shape. However, these distinctions are often blurred. For instance, a digital image can be interpreted as a collection of colored squares, and geometric shapes such as circles are defined by implicit mathematical equations. Also, a fractal model yields a parametric or implicit model when its recursive definition is truncated to a finite depth. A geometric modeling technique involves the development from wireframe modeling through surface modeling to solid modeling, where the representation of geometric volume information becomes more and more accurate, and the range of "design" problems which we are able to solve is wider and wider. These three modeling techniques can be illustrated as follows.

(1) Wireframe modeling. A wireframe model is a visual presentation of a 3D or physical object used in 3D computer graphics. It is created by specifying each edge of the physical object where two mathematically continuous smooth surfaces meet, or by connecting an object's constituent vertices using straight lines or curves. The object is projected onto the computer screen by drawing lines at the location of each edge. Using a wireframe model allows visualization of the underlying design structure of a 3D model. Traditional 2D views and drawings can be created by appropriate rotation of the object and selection of hidden line removal via cutting planes. Since wireframe rendering is relatively simple and fast to calculate, it is often used in cases where a high screen frame rate is needed (for instance, when working with a particularly complex 3D model, or in real-time systems that model exterior phenomena). When greater graphical detail is desired, surface textures can be added automatically after completion of the initial rendering of the wireframe. This allows the designer to quickly review changes or rotate the object to new desired views without long delays associated with more realistic rendering. The wireframe format is also well suited and widely used in programming tool paths for direct numerical control (DNC) machine tools.

(2) Surface modeling. Unlike wireframe models, surface models introduce the concept of "surfaces". It is a mathematical technique for representing solid-appearing objects. Surface modeling is a more complex method for representing objects than wireframe modeling, but not as sophisticated as solid modeling. Surface modeling is widely used in CAD for illustrations and architectural renderings. It is also used in 3D animation for games and other presentations. Although surface and solid models appear the same on screen, they are quite different. Surface models cannot be sliced open as solid models. In addition, in surface modeling, the object can be geometrically incorrect, whereas, in solid modeling, it must be correct. Typical surface modeling techniques can be described as follows:

1) Polygonal modeling. In 3D computer graphics, polygonal modeling is an approach for modeling objects by representing or approximating their surfaces using polygons. Polygonal modeling is well suited to scan line rendering and is therefore the choice for real-time computer graphics. We will discuss this kind of model in detail in the next subsection.

2) NURBS modeling. Non-uniform rational B-spline (NURBS) is a mathematical model commonly used in computer graphics for generating and representing curves and surfaces which offers great flexibility and precision for handling both analytic and freeform shapes. The development of NURBS began in the 1950s by engineers who were in need of a mathematically precise

representation of freeform surfaces like those used for ship hulls, aerospace exterior surfaces and car bodies, which could be exactly reproduced whenever technically needed. Prior representations of this kind of surface only existed as a single physical model created by a designer. The pioneers of this development were Pierre Bézier who worked as an engineer at Renault, and Paul de Casteljau who worked at Citroën, both in France. Bézier worked almost in parallel to de Casteljau, neither knowing about the work of the other. But because Bézier published the results of his work, the average computer graphics user today recognizes splines — which are represented with control points lying off the curve itself — as Bézier splines, while de Casteljau's name is only known and used for the algorithms he developed to evaluate parametric surfaces. In the 1960s, it became clear that NURBSs are a generalization of Bézier splines, which can be regarded as uniform, NURBSs. At first, non-uniform rational B-splines were only used in the proprietary CAD packages of car companies. Later they became part of standard computer graphics packages. In 1985, the first interactive NURBS modeler for PCs, called Macsurf (later Maxsurf), was developed by Formation Design Systems, a small startup company based in Australia. Maxsurf is a marine hull design system intended for the creation of ships, workboats and yachts, whose designers have a need for highly accurate sculptured surfaces. Real-time, interactive rendering of NURBS curves and surfaces was first made available on Silicon Graphics workstations in 1989. Today, most professional computer graphics applications available for desktop use offer NURBS technology, which is most often realized by integrating a NURBS engine from a specialized company.

3) Subdivision surface modeling. Subdivision surface modeling, in the field of 3D computer graphics, is a method of representing a smooth surface via the specification of a coarser piecewise linear polygon mesh. The smooth surface can be calculated from the coarse mesh as the limit of a recursive process of subdividing each polygonal face into smaller faces that better approximate the smooth surface. The subdivision surfaces are defined recursively. The process starts with a given polygonal mesh. A refinement scheme is then applied to this mesh. This process takes that mesh and subdivides it, creating new vertices and new faces. The positions of the new vertices in the mesh are computed based on the positions of nearby old vertices. In some refinement schemes, the positions of old vertices might also be altered (possibly based on the positions of new vertices). This process produces a denser mesh than the original one, containing more polygonal faces. This resulting mesh can be passed through the same refinement scheme again. The limit subdivision surface is the surface produced from this process being iteratively applied infinitely many times. In practical use, however, this algorithm is only applied a limited number of times.

(3) Solid modeling. Solid modeling is the unambiguous representation of the solid parts of an object, which means models of solid objects suitable for computer processing. As we know, surface models are used extensively in automotive and consumer product design as well as entertainment animation, while wireframe models are ambiguous about solid volume. Primary uses of solid modeling are for CAD, engineering analysis, computer graphics and animation, rapid prototyping, medical testing, product visualization and visualization of scientific research.

1.2.2.2 3D Scanner-Based Techniques

A 3D scanner is a device that analyzes a real-world object or environment to collect data on its shape and possibly its appearance (e.g., color). The collected data can then be used to construct digital, 3D models useful for a wide variety of applications. These devices are used extensively by the entertainment industry in the production of movies and video games. Other common applications of this technology include industrial design, orthotics and prosthetics, reverse engineering and prototyping, quality control/inspection and documentation of cultural artifacts. Many different technologies can be used to build these 3D scanning devices, each coming with its own limitations, advantages and costs. It should be remembered that many limitations on the kind of object that can be digitized are still present: for example, optical technologies encounter many difficulties with shiny, mirroring or transparent objects. However, there are methods for scanning shiny objects, such as covering them with a thin layer of white powder that will help more light photons to reflect back to the scanner. Laser scanners can send trillions of light photons toward an object and only receive a small percentage of those photons back via the optics that they use. The reflectivity of an object is based upon the object's color or terrestrial albedo. A white surface will reflect lots of light and a black surface will reflect only a small amount of light. Transparent objects such as glass will only refract the light and thus give false 3D information.

The purpose of a 3D scanner is usually to create a point cloud of geometric samples on the surface of the subject. These points can then be used to extrapolate the shape of the subject (a process called reconstruction). If the color information is collected at each point, then the colors on the surface of the subject can also be determined. 3D scanners are very analogous to cameras. Like cameras, they have a cone-like field of view, and they can only collect information about surfaces that are not obscured. A camera collects color information about surfaces within its field of view, while a 3D scanner collects distance information about surfaces within its field of view. The "picture" produced by a 3D scanner describes the distance to a surface at each point in the picture. If a spherical coordinate system is defined, in which the scanner is the origin and the vector out from the front of the scanner is $\varphi = 0$ and $\theta = 0$, then each point in the picture is associated with a φ and a θ. Together with the distance, which corresponds to the r component, these spherical coordinates fully describe the 3D position of each point in the picture, in a local coordinate system relative to the scanner. For most situations, a single scan will not produce a complete model of the subject. Multiple scans, even hundreds, from many different directions are usually required to obtain information about all sides of the subject. These scans have to be brought into a common reference system, a process that is usually called alignment or registration, and then be merged to create a complete model. This whole process, going from the single range map to the whole model, is usually known as the 3D scanning pipeline.

There are two types of 3D scanners, i.e., contact and non-contact scanners. Non-contact 3D scanners can be further classified into two main categories, active scanners and passive scanners. There are a variety of technologies that fall under each of these categories.

(1) Contact. Contact 3D scanners probe the subject through physical touch. A coordinate measuring machine (CMM) is an example of a contact 3D scanner. It is used mostly in manufacturing and can be very precise. The disadvantage of CMMs is that they require contact with the object being scanned. Thus, the scanning operation might modify or damage the object. This fact is very significant when scanning delicate or valuable objects such as historical artifacts. The other disadvantage of CMMs is that they are relatively slow compared to the other scanning methods. Physically moving the arm that the probe is mounted on can be very slow and the fastest CMMs can only operate on a few hundred hertz. In contrast, an optical system like a laser scanner can operate from 10 to 500 kHz. Other examples are the hand-driven touch probes used to digitize clay models in the computer animation industry.

(2) Non-contact active. Active scanners emit some kind of radiation or light and detect its reflection in order to probe an object or environment. Possible types of emissions used include light, ultrasound or X-ray. For example, both time-of-flight and triangulation 3D laser scanners are active scanners that use laser lights to probe the subject or environment. The advantage of time-of-flight range finders is that they are capable of operating over very long distances, in the order of kilometers. These scanners are thus suitable for scanning large structures like buildings or geographic features. The disadvantage of time-of-flight range finders is their accuracy. Due to the high speed of light, timing the round-trip time is difficult and the accuracy of the distance measurement is relatively low, in the order of millimeters. Triangulation range finders are exactly the opposite. They have a limited range of some meters, but their accuracy is relatively high. The accuracy of triangulation range finders is in the order of tens of micrometers.

(3) Non-contact passive. Passive scanners do not emit any radiation themselves, but instead rely on detecting reflected ambient radiation. Most scanners of this type detect visible light because it is a readily available ambient radiation. Other types of radiation, such as infrared, could also be used. Passive methods can be very cheap, because in most cases they do not need particular hardware. For example, stereoscopic systems usually employ two video cameras, slightly apart, looking at the same scene. By analyzing the slight differences between the images seen by each camera, it is possible to determine the distance at each point in the images. This method is based on human stereoscopic vision. In contrast, photometric systems usually use a single camera, but take multiple images under varying lighting conditions. These techniques attempt to invert the image formation model in order to recover the surface orientation at each pixel. In addition, silhouette-based 3D scanners use outlines generated from a sequence of photographs around a 3D object against a well-contrasted background. These silhouettes are extruded and intersected to form the visual hull approximation of the object. However, some types of concavities in an object (like the interior of a bowl) cannot be detected by these techniques.

1.2.2.3 Image-Based Modeling Techniques

Recently, a trend in modeling is to reconstruct 3D models from photographs, i.e., IBM (image-based modeling). In computer graphics and computer vision, IBMR (image-based modeling and rendering) methods rely on a set of 2D images of a scene to generate a 3D model and then render some novel views of this scene. The traditional approach of computer graphics has been to create a geometric model in the 3D space and try to re-project it onto a 2D image. Computer vision, conversely, is mostly focused on detecting, grouping and extracting features (edges, faces, etc.) present in a given picture and then trying to interpret them as 3D clues. IBMR allows the use of multiple 2D images in order to generate directly novel 2D images, skipping the manual modeling stage. The main advantage of IBM is to create 3D photorealistic models by using textures directly extracted from the real world. Generally speaking, IBM refers to the reconstruction process of 3D geometries from images, which include real photographs, rendered images, video clips and range images, whereas the generalized-IBM techniques should also contain the reconstruction process of surface textures, reflectance characteristics, lighting conditions and kinematic properties. According to which image feature is used, this technique can be classified into the following categories.

(1) Texture based. This technique reconstructs the 3D feature point cloud by searching the similar texture area in multiple images. It can obtain models with high accuracy. However, the modeling effect for irregular objects is worse, and it is only suitable for regular objects such as buildings from which the texture is easily extracted.

(2) Contour based. This method obtains the 3D model of the object automatically by analyzing the object contour information in images. The robustness of this method is high, but because it is an ill-posed problem to restore the complete surface geometric information of the object from the contour, the accuracy will not be high, particularly for the depressed details on the object surface. We are unable to reflect them in the contour, and thus they will be lost in the 3D model.

(3) Color based. This method is based on Lambertian's diffuse reflection model; i.e., the colors under different view angles for the same point on the object's surface are basically similar. Based on the similar colors in multiple images, we can reconstruct the 3D model of the object. This method has higher accuracy, but because the colors on the object surface are very sensitive to the environment, it needs relatively harsh requirements for the illumination condition of the scanning environment, and thus the robustness is not high.

(4) Shadow based. This method performs the 3D modeling through analyzing the shadow of the object under lights. It can obtain 3D models with a relatively high accuracy, but the more requirements of light are not conducive to practical use.

(5) Light based. This approach illuminates the object with intense lights at close range. By analyzing the intensity distribution of the reflection of light on the object surface and applying the bidirectional reflectance distribution function, we can obtain the normal vectors of the surface and thus we can obtain the vertices

and faces of the object.

(6) Mixture information based. This method uses comprehensively the surface contours, colors, shadows and other information to improve the accuracy of modeling, but the comprehensive use of multiple kinds of information is difficult, and the problem of system robustness cannot be fundamentally resolved.

Although automatic IBM systems cannot reach the level of practical use, there have been some semi-automatic mature software tools. The IBM technique is not only the research hot spot of virtual reality modeling, but also the focus in the next few years, which can greatly reduce the threshold and cost of virtual reality modeling. Although there are still some technical thresholds to overcome, it is believed that in less than a few years, the use of the IBM technology can be achieved on the practical level. At that time, only using an ordinary digital camera, you will be able to "capture" a 3D model. Furthermore, we will be able to use our own 3D models to make a movie and play games.... Think about how exciting this thing will be! Generally speaking, virtual reality modeling technology is developing in the direction of high precision and high robustness.

1.2.3 Polygon Meshes

This book mainly focuses on 3D polygon meshes. A polygon mesh or unstructured grid is a collection of vertices, edges and faces that defines the shape of a polyhedral object in 3D computer graphics and solid modeling. The faces usually consist of triangles, quadrilaterals or other simple convex polygons, since this simplifies rendering, but may also be composed of more general concave polygons, or polygons with holes. A typical triangle mesh model is shown in Fig. 1.2.

Fig. 1.2. Example of a triangle mesh "dolphin"

The study of polygon meshes is a large sub-field of computer graphics and geometric modeling. Different representations of polygon meshes are used for different applications and goals. The variety of operations performed on meshes may include Boolean operators, smoothing, simplification, and so on. Network representations, "streaming" and "progressive" meshes, are used to transmit polygon meshes over a network. Volumetric meshes are distinct from polygon meshes in that they explicitly represent both the surface and volume of a structure, while polygon meshes only explicitly represent the surface (the volume is implicit). As polygonal meshes are extensively used in computer graphics, algorithms also exist for ray tracing, collision detection and rigid-body dynamics of polygon meshes.

Objects created with polygon meshes must store different types of elements, including vertices, edges, faces, polygons and surfaces. In many applications, only vertices, edges and either faces or polygons are stored as shown in Fig. 1.3. A renderer may support only 3-sided faces, so polygons must be composed of many of these. However, many renderers either support quadrangles and higher-sided polygons, or are able to triangulate polygons to triangles on the fly, making it unnecessary to store a mesh in a triangulated form. Also, in certain applications like head modeling, it is desirable to be able to create both 3- and 4-sided polygons.

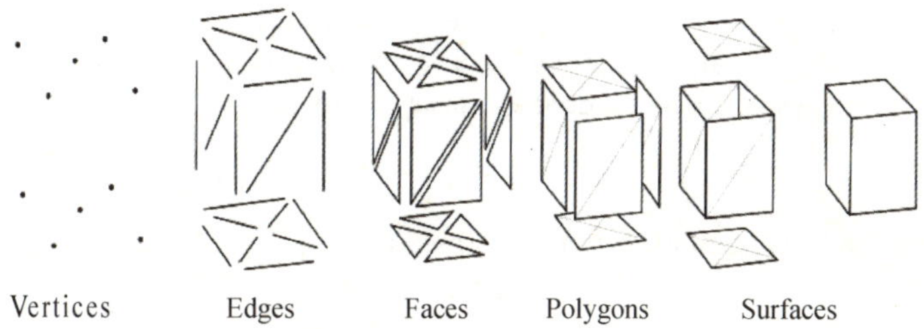

Fig. 1.3. Elements of polygonal mesh modeling

A vertex is a position along with other information such as colors, normal vectors and texture coordinates. An edge is a connection between two vertices. A face is a closed set of edges, in which a triangular face has three edges, and a quad face has four edges. A polygon is a set of faces. In systems that support multi-sided faces, polygons and faces are equivalent. However, most rendering hardware supports only 3- or 4-sided faces, so polygons are represented as multiple faces. Mathematically, a polygonal mesh may be considered an unstructured grid, or undirected graph, with additional properties of geometry, shape and topology.

Surfaces, more often called smoothing groups, are useful, but not required to group smooth regions. Consider a cylinder with caps, such as a soda can. For smooth shading of the sides, all surface normals must point horizontally away from the center, while the normals of the caps must point in the $(0, 0, \pm 1)$ directions. Rendered as a single, Phong shaded surface, the crease vertices would have incorrect normals. Thus, some way of determining where to cease smoothing is needed to group smooth parts of a mesh just as polygons group 3-sided faces. As an alternative to providing surfaces/smoothing groups, a mesh may contain other data for calculating the same data, such as a splitting angle (polygons with normals above this threshold are automatically treated as separate smoothing groups or some technique such as splitting or chamfering is automatically applied to the edge between them). Additionally, very high resolution meshes are less subject to issues that would require smoothing groups, as their polygons are so small as to make the need irrelevant. Furthermore, another alternative exists in the possibility of simply detaching the surfaces themselves from the rest of the mesh. Renders do not attempt to smooth edges across noncontiguous polygons.

Mesh format may or may not define other useful data. Groups may be defined, which define separate elements of the mesh and are useful for determining separate sub-objects for skeletal animation or separate actors for non-skeletal animation. Generally, materials will be defined, allowing different portions of the mesh to use different shaders when rendered. Most mesh formats also suppose some forms of UV coordinates, which are separate 2D representations of the mesh "unfolded" to show what portion of a 2D texture map to apply to different polygons of the mesh.

If there is no other special explanation, this book only involves the geometric data and their connection relationships in 3D mesh models. Thus, here we can define a 3D mesh model using mathematical symbols. A mesh model $\boldsymbol{M} = \{\boldsymbol{C}, \boldsymbol{G}\}$ is composed of the set of vertices $\boldsymbol{G}$ and the set of connections $\boldsymbol{C}$, where $\boldsymbol{G}$ includes N vertices $\boldsymbol{v}_i$, each one denoted as (x_i, y_i, z_i), i.e.,

$$\boldsymbol{G} = \{\boldsymbol{v}_i\}, i = 0, 1, \cdots, N-1, \quad \boldsymbol{v}_i = (x_i, y_i, z_i), \tag{1.1}$$

while the set of connections $\boldsymbol{C}$ can be defined as

$$\boldsymbol{C} = \{\{i_k, j_k\}\}_{k=0, \cdots, K-1}, \quad 0 \le i_k \le N-1, \ 0 \le j_k \le N-1, \tag{1.2}$$

where $\{i_k, j_k\}$ denotes the k-th edge that connects the i_k-th and j_k-th vertices.

1.2.4 3D Model File Formats and Processing Software

Currently, there are many types of software for 3D model generation, design and processing. The famous ones include AutoCAD, 3ds Max, Maya, Art of Illusion, ngPlant, Multigen, SketchUp, and so on. The most common ones are AutoCAD, 3DSMAX and MAYA, which will be introduced in detail below. 3D data can be stored in various formats, including 3DS, OBJ, ASE, MD2, MD3, MS3D, WRL, MDL, BSP, GEO, DXF, DWG, STL, NFF, RAW, POV, TTF, COB, VRML, OFF, and so on. Currently, the most common ones are 3DS, OBJ and DXF, and OFF and OBJ are the two most common formats used in academic research, which will be introduced in detail below. Before introducing these types of software and file formats, we must introduce OpenGL, the industrial standard for high-performance graphics.

1.2.4.1 OpenGL

OpenGL (Open Graphics Library) is a standard specification defining a cross-language, cross-platform application programming interface (API) for writing applications that produce 2D and 3D computer graphics. The interface consists of over 250 different function calls which can be used to draw complex 3D scenes from simple primitives. OpenGL was developed by Silicon Graphics Inc. (SGI) in 1992 and is widely used in CAD, virtual reality, scientific visualization, information visualization and flight simulation. It is also used in video games, where it competes with Direct3D on Microsoft Windows platforms. OpenGL is managed by the non-profit technology consortium, the Khronos Group. At its most basic level, OpenGL is a specification; i.e., it is simply a document that describes a set of functions and the precise behaviors that they must perform. From this specification, hardware vendors create implementations (libraries of functions) to match the functions stated in the OpenGL specification, making use of hardware acceleration where possible. Hardware vendors have to meet specific tests to be able to qualify their implementation as an OpenGL implementation. Efficient vendor-supplied implementations of OpenGL (making use of graphics acceleration hardware to a greater or lesser extent) exist for Mac OS, Microsoft Windows, Linux and many UNIX platforms.

OpenGL serves two main purposes: (1) to hide the complexities of interfacing with different 3D accelerators, by presenting the programmer with a single, uniform API; (2) to hide the different capabilities of hardware platforms, by requiring that all implementations support the full OpenGL feature set (using software emulation if necessary). The OpenGL's basic operation is to accept primitives such as points, lines and polygons, and convert them into pixels. This is done by a graphics pipeline known as the OpenGL State Machine. Most OpenGL commands either issue primitives to the graphics pipeline, or configure how the pipeline processes these primitives. Prior to the introduction of OpenGL 2.0, each stage of the pipeline performed a fixed function and was configurable only within tight limits. OpenGL 2.0 offers several stages that are fully programmable using the GLSL (OpenGL Shading Language). OpenGL is a low-level, procedural API, requiring the programmer to dictate the exact steps required to render a scene. This contrasts with descriptive APIs, where a programmer only needs to describe a scene and can let the library manage the details of rendering it. OpenGL's low-level design requires programmers to have a good knowledge of the graphics pipeline, but also gives a certain amount of freedom to implement novel rendering algorithms.

1.2.4.2 AutoCAD

AutoCAD is a CAD software for 2D and 3D design and drafting, developed by Autodesk, Inc. Initially released in late 1982, AutoCAD was one of the first CAD programs to run on personal computers, and notably the IBM PC. Most CAD software at the time must run on graphics terminals connected to mainframe computers or mini-computers. In early versions, AutoCAD used primitive entities (such as lines, poly-lines, circles, arcs and text) as the foundation for more complex objects. Since the mid-1990s, AutoCAD has supported custom objects through its C++ API. Modern AutoCAD includes a full set of basic solid modeling and 3D tools. With the release of AutoCAD 2007, it became easier to edit 3D models. AutoCAD 2010 has introduced parametric functionality and mesh modeling. Fig. 1.4 shows an example of 3D effects created by the AutoCAD software.

Fig. 1.4. 3D effects of outdoor buildings designed by AutoCAD

AutoCAD supports a number of APIs for customization and automation. These include AutoLISP, Visual LISP, VBA, .NET and ObjectARX. ObjectARX is a C++ class library, which was also the base for products extending AutoCAD functionality to specific fields, to create products such as AutoCAD Architecture, AutoCAD Electrical, AutoCAD Civil 3D, or third-party AutoCAD-based applications. AutoCAD currently runs exclusively on Microsoft Windows desktop operating systems. Versions for UNIX and Mac OS were released in the 1980s and 1990s respectively, but were later dropped. AutoCAD can run on an emulator or compatibility layer like VMware Workstation or Wine, albeit subject to various performance issues that can often arise when working with 3D objects or large drawings.

AutoCAD's native file format, DWG and, to a lesser extent, its interchange file format, DXF, have become de facto standards for CAD data interoperability. AutoCAD in recent years has included support for DWF, a format developed and promoted by Autodesk for publishing CAD data. In 2006, Autodesk estimated the number of active DWG files to be in excess of one billion. The current AutoCAD file format (.dwfx) is based on ISO/IEC 29500-2:2008 Open Packaging Convention. In the past, Autodesk has estimated the total number of DWG files in existence to be more than three billion.

1.2.4.3 3ds Max

Autodesk 3ds Max, formerly 3D Studio MAX, is a modeling, animation and rendering package developed by Autodesk Media and Entertainment. The original 3D Studio product was created for the DOS platform by the Yost Group and published by Autodesk. After 3D Studio Release 4, the product was rewritten for the Windows NT platform, and re-named "3D Studio MAX". This version was also originally created by the Yost Group. It was released by Kinetix, which was at that time Autodesk's division of media and entertainment. Autodesk purchased the product at the second release mark of the 3D Studio MAX version and internalized development entirely over the next two releases. Later, the product name was changed to "3ds max" (all lower case) to better comply with the naming conventions of Discreet, a Montreal-based software company which Autodesk had purchased. At release 8, the product was again branded with the Autodesk logo, and the name was again changed to "3ds Max" (upper and lower cases). At release 2009, the product name was changed to "Autodesk 3ds Max".

3ds Max is the third most widely-used off the shelf 3D animation program by content creation professionals. It has strong modeling capabilities, a flexible plug-in architecture and a long heritage on the Microsoft Windows platform. It is mostly used by video game developers, TV commercial studios and architectural visualization studios. It is also used for movie effects and movie pre-visualization.

In addition to its modeling and animation tools, the latest version of 3ds Max also features advanced shaders (such as ambient occlusion and subsurface scattering), dynamic simulation, particle systems, radiosity, normal map creation and rendering, global illumination, an intuitive and fully-customizable user interface and its own scripting language. A plethora of specialized third-party renderer plug-ins, such as V-Ray, Brazil r/s, Maxwell Render, and finalRender, may be purchased separately.

1.2.4.4 Maya

Autodesk Maya, or simply Maya, is a high-end 3D computer graphics and 3D modeling software package originally developed by Alias Systems Corporation, but now owned by Autodesk as part of the media and entertainment division. Autodesk acquired the software in October 2005 upon purchasing Alias. Maya is used in the film and TV industry, as well as for computer and video games, architectural visualization and design. In 2003, Maya (then owned by Alias/ Wavefront) won an Academy Award for "scientific and technical achievement", citing use on "nearly every feature using 3D computer-generated images".

Maya is a popular, integrated node-based 3D software suite, evolving from Wavefront Explorer and Alias PowerAnimator using technologies from both. The software is released in two versions: Maya Complete and Maya Unlimited. Maya Personal Learning Edition (PLE) was available (excluding the Linux version) at no cost for non-commercial use, with the resulting rendered image watermarked, but as of December 2, 2008, it was no longer made available. Maya was originally released for the IRIX operating system, and subsequently ported to the Microsoft Windows, Linux, and Mac OS X operating systems. IRIX support was discontinued after the release of Version 6.5. When Autodesk acquired Alias in October 2005, they continued the development of Maya. The latest version, 2009 (10.0), was released in October 2008. An important feature of Maya is its openness to third-party software, which can strip the software completely of its standard appearance and, using only the kernel, transform it into a highly customized version of the software. This feature in itself made Maya appealing to large studios, which tend to write custom codes for their productions using the provided software development kit. A Tcl-like cross-platform scripting language called Maya Embedded Language (MEL) is provided not only as a scripting language, but as a means to customize Maya's core functionality. Additionally, user interactions are implemented and recorded as MEL scripting codes which users can store on a toolbar, allowing animators to add functionality without experience in C or C++, though that option is provided with the software development kit. Support for Python scripting was added in Version 8.5. The core of Maya itself is written in C++. Project files, including all geometry and animation data, are stored as sequences of MEL operations which can be optionally saved as a human-readable file (.ma, for "Maya ASCII"), editable in any text editor outside of the Maya environment, thus allowing for a high level of flexibility when working with external tools. A marking menu is built into a larger menu system called Hotbox that provides instant access to a majority of features in Maya at the press of a key.

1.2.4.5 3DS File Format

The 3DS format is one of the file formats used by Discreet Software's 3D Studio Max. It is close to the most common format, and is supported by many applications. DirectX does not provide native support to load 3DS files, but you can find the code to convert a 3DS to the DirectX's internal format.

The 3DS file format is made up of chunks. They describe what information is to follow, what it is made up of, its ID and the location of the next block. If you do not understand a chunk you can quite simply skip it. The next chunk pointer is relative to the start of the current chunk and in bytes. The binary information in the 3Ds file is written in a special way. Namely, the least significant byte comes first in an integer. For example: 4A 5C (2 bytes in hex) would be 5C high byte and 4A low byte. In a long integer, it is 4A 5C 3B 8F where 5C 4A is the low word and 8F 3B is the high word. A chunk is defined as:

start	end	size	name
0	1	2	Chunk ID
2	5	4	Pointer to next chunk relative to the place where the Chunk ID is, in other words the length of the chunk

Chunks have a hierarchy imposed on them that is identified by its ID. A 3DS file has the primary chunk ID 4D4Dh. This is always the first chunk of the file. Within the primary chunk are the main chunks.

1.2.4.6 OBJ File Format

OBJ is a geometry definition file format first developed by Wavefront Technologies for its Advanced Visualizer animation package. The file format is open and has been adopted by other 3D graphics application vendors. For the most part, it is a universally accepted format. The OBJ file format is a simple data-format that represents 3D geometry alone, namely the position of each vertex, the UV position of each texture coordinate vertex, normals and the faces that make each polygon defined as a list of vertices, texture vertices and normals. A typical OBJ file looks like as follows:

```
# This is a comment
# Here is the first vertex, with (x,y,z) coordinates.
v 0.123 0.234 0.345
v ...
...
# Texture coordinates
vt ...
...
# Normals in (x,y,z) form; normals might not be unit.
vn ...
...
# Each face is given by a set of indices to the vertex/texture/normal
# coordinate array that precedes this.
# Hence f 1/1/1 2/2/2 3/3/3 is a triangle having texture coordinates and
# normals for those 3 vertices,
# and having the vertex 1 from the "v" list, texture coordinate 2 from
# the "vt" list, and the normal 3 from the "vn" list
f v0/vt0/vn0 v1/vt1/vn1 ...
f ...
...
# When there are named polygon groups or materials groups the following
# tags appear in the face section,
g [group name]
usemtl [material name]
# the latter matches the named material definitions in the external .mtl file.
# Each tag applies to all faces following, until another tag of the same type appears.
...
...
```

An OBJ file also supports smoothing parameters to allow for curved objects, and also the possibility to name groups of polygons. It also supports materials by referring to an external MTL material file. OBJ files, due to their list structure, are able to reference vertices, normals, etc., either by their absolute (1-indexed) list position, or relatively by using negative indices and counting backwards. However, not all software supports the latter approach, and conversely some software inherently writes only the latter form (due to the convenience of appending elements without the need to recalculate vertex offsets, etc.), leading to occasional incompatibilities.

Now let us see a practical case. We create a polygon cube using the Maya software as shown in Fig. 1.5. Select this cube, using the menu item "File→Export Selection..." to export as an OBJ file named "cube.obj". If OBJ is not found, please load "objExport.mll" in the Plug-in Manager. Using the notepad to open "cube.obj", we have the following codes:

```
# The units used in this file are centimeters.
g default
v -0.500000 -0.500000 0.500000\v 0.500000 -0.500000 0.500000
v -0.500000 0.500000 0.500000\v 0.500000 0.500000 0.500000
v -0.500000 0.500000 -0.500000\v 0.500000 0.500000 -0.500000
v -0.500000 -0.500000 -0.500000\v 0.500000 -0.500000 -0.500000
vt 0.000000 0.000000\vt 1.000000 0.000000
vt 0.000000 1.000000\vt 1.000000 1.000000
vt 0.000000 2.000000\vt 1.000000 2.000000
vt 0.000000 3.000000\vt 1.000000 3.000000
vt 0.000000 4.000000\vt 1.000000 4.000000
vt 2.000000 0.000000\vt 2.000000 1.000000
vt -1.000000 0.000000\vt -1.000000 1.000000
vn 0.000000 0.000000 1.000000\vn 0.000000 0.000000 1.000000
vn 0.000000 0.000000 1.000000\vn 0.000000 0.000000 1.000000
vn 0.000000 1.000000 0.000000\vn 0.000000 1.000000 0.000000
vn 0.000000 1.000000 0.000000\vn 0.000000 1.000000 0.000000
vn 0.000000 0.000000 -1.000000\vn 0.000000 0.000000 -1.000000
vn 0.000000 0.000000 -1.000000\vn 0.000000 0.000000 -1.000000
vn 0.000000 -1.000000 0.000000\vn 0.000000 -1.000000 0.000000
vn 0.000000 -1.000000 0.000000\vn 0.000000 -1.000000 0.000000
vn 1.000000 0.000000 0.000000\vn 1.000000 0.000000 0.000000
vn 1.000000 0.000000 0.000000\vn 1.000000 0.000000 0.000000
vn -1.000000 0.000000 0.000000\vn -1.000000 0.000000 0.000000
vn -1.000000 0.000000 0.000000\vn -1.000000 0.000000 0.000000
s off
g pCube1
usemtl initialShadingGroup
f 1/1/1 2/2/2 4/4/3 3/3/4
f 3/3/5 4/4/6 6/6/7 5/5/8
f 5/5/9 6/6/10 8/8/11 7/7/12
f 7/7/13 8/8/14 2/10/15 1/9/16
```

```
f 2/2/17 8/11/18 6/12/19 4/4/20
f 7/13/21 1/1/22 3/3/23 5/14/24
```

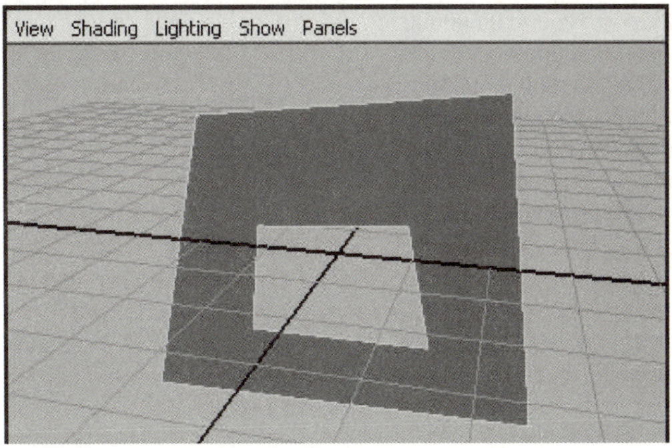

Fig. 1.5. The polygon with holes created by the Maya software

1.2.4.7 OFF File Format

Object file format (OFF) files are used to represent the geometry of a model by specifying the polygons of the model's surface. The polygons can have any number of vertices. The .off files in the Princeton Shape Benchmark conform to the following standard. OFF files are all ASCII files beginning with the keyword OFF. The next line states the number of vertices, the number of faces and the number of edges. The number of edges can be safely ignored. The vertices are listed with *x*, *y*, *z* coordinates, written one per line. After the list of vertices, the faces are listed, with one face per line. For each face, the number of vertices is specified, followed by indices into the list of vertices. Note that earlier versions of the model files had faces with −1 indices into the vertex list. That was due to an error in the conversion program and can be corrected now.

```
OFF numVertices numFaces numEdges
  x y z
  x y z
  ... numVertices like above
  NVertices v1 v2 v3 ... vN
  MVertices v1 v2 v3 ... vM
  ... numFaces like above
```

Note that vertices are numbered starting at 0 (not starting at 1), and that numEdges will always be zero. A simple example for a cube is as follows:

```
OFF
8 6 0
-0.500000 -0.500000 0.500000
0.500000 -0.500000 0.500000
-0.500000 0.500000 0.500000
0.500000 0.500000 0.500000
-0.500000 0.500000 -0.500000
0.500000 0.500000 -0.500000
-0.500000 -0.500000 -0.500000
0.500000 -0.500000 -0.500000
4 0 1 3 2
4 2 3 5 4
4 4 5 7 6
4 6 7 1 0
4 1 7 5 3
4 6 0 2 4
```

1.2.4.8 DXF File Format

The DXF format is a tagged data representation of all the information contained in an AutoCAD drawing file. Tagged data means that each data element in the file is preceded by an integer number that is called a group code. A group code's value indicates what type of data element follows. This value also indicates the meaning of a data element for a given object type. Virtually all user-specified information in a drawing file can be represented in the DXF format. The DXF reference presents the DXF group codes found in DXF files and encountered by AutoLISP and ObjectARXTM applications. This chapter describes the general DXF conventions. The remaining chapters list the group codes organized by the object type. The group codes are presented in the order they are found in a DXF file, and each chapter is named according to the associated section of a DXF file. In the DXF format, the definition of objects differs from entities: objects have no graphical representation but entities do. For example, dictionaries are objects without entities. Entities are also referred to as graphical objects, while objects are referred to as non-graphical objects. Entities appear in both the BLOCK and ENTITIES sections of the DXF file. The use of group codes in the two sections is identical. Some group codes that define an entity always appear; others are optional and appear only if their values differ from the defaults. The end of an entity is indicated by the next 0 group, which begins the next entity or indicates the end of the section. Group codes define the type of the associated value as an integer, a floating-point number, or a string, according to the table of group code ranges.

1.3 Overview of 3D Model Analysis and Processing

3D models are the fourth type of digital media following audio data, images and video data. Compared to the first three kinds of digital media, the 3D model has its own characteristics: (1) no data sequence; (2) no specific sampling rate; (3) non-unique description; (4) containing both the geometric information and topological information; (5) Both geometry and topology information can be modified easily. Therefore, the analysis and processing techniques for 3D models are very different from those for other media. Similar to other media, the analysis and processing techniques for 3D models include pre-processing, de-noising, coding and compression, copyright protection, content authentication, retrieval and identification, segmentation, feature extraction, reconstruction, matching and stitching, visualization, etc., but due to the specificity of 3D models, in the realization of these technologies or with the meaning, it is very different from traditional media. In addition, there are some special analysis and processing techniques for 3D models, including model simplification, model voxelization, texture mapping, speedup of the drawing, transformation of 2D graphics into 3D models, rendering techniques, reverse engineering, 2D projection of 3D models, contour line extraction algorithms, and so on. In the following subsections, we briefly introduce the concepts of 3D-model-related techniques in two aspects, i.e., 3D model processing techniques and 3D model analysis techniques. Detailed techniques will be discussed from Chapter 2 to Chapter 6.

1.3.1 Overview of 3D Model Processing Techniques

The so-called 3D model processing operations are those operations whose inputs and outputs are both 3D models or 3D objects. 3D model processing techniques comprise many aspects, including 3D model construction, format conversion, 3D model transmission and compression, 3D model management and retrieval.

1.3.1.1 Processing Techniques for 3D Model Construction

During the 3D object construction or 3D model reconstruction process, as well as in the 3D model format conversion process, we require processing techniques including 3D modeling, model simplification, model de-noising, voxelization, texture mapping, subdivision, splicing, and so on. The connotation of 3D modeling is relatively large, and this has already been described in the former section.

Model simplification [7] refers to representing a model with fewer geometric elements to obtain an approximate model to the original one. That is, during the rendering process, according to the number of covering pixels of the model on the screen, we select appropriate levels of detail, making the near objects rendered with relatively refined models and the far objects with relatively coarse models. The aim is to reduce the number of triangles representing the model as much as we can, while guaranteeing a good approximation in shape to the original model. We can describe this process as: (1) inputting the original triangle mesh data, including geometric data, surface data, color information, texture information, normal vectors, etc.; (2) generating automatically multiple levels of details through the model simplification method; (3) describing different parts of the model with different levels of detail during the rendering process, guaranteeing that the difference between the result image and the rendering result with the most refined model is within a predefined range.

Mesh de-noising [8] is used in the surface reconstruction procedure to reduce noise and output a higher quality triangle mesh which describes more precisely the geometry of the scanned object. 3D surface mesh de-noising has been an active research field for several years. Although much progress has been made, mesh de-noising technology is still not mature. The presence of intrinsic fine details and sharp features in a noisy mesh makes it hard to simultaneously de-noise the mesh and preserve the features. Mesh de-noising is usually posed as a problem of adjusting vertex positions while keeping the connectivity of the mesh unchanged. In the literature, mesh de-noising is often confused with surface smoothing or fairing, because all of them use vertex adjustment to make the mesh surface smooth. However, they have different purposes and different algorithms are needed to meet their specific requirements, and we should keep in mind the distinctions. The main goal of mesh fairing is related to aesthetics, while the goal of mesh de-noising has more to do with fidelity, and mesh smoothing generally attempts to remove small scale details. Another commonly used term, mesh filtering, is also often used in place of mesh fairing, smoothing or de-noising. Filtering, however, is a rather general term which simply refers to some black box which processes a signal to produce a new signal, and could, in principle, perform some quite different function such as feature enhancement.

Voxelization [9] refers to converting geometric objects from their continuous geometric representation into a set of voxels that best approximates the continuous object. As this process mimics the scan-conversion process that pixelizes (rasterizes) 2D geometric objects, it is also referred to as 3D scan conversion. In 2D rasterization, the pixels are directly drawn onto the screen to be visualized and filtering is applied to reduce the aliasing artifacts. However, the voxelization process does not render the voxels but merely generates a database of the discrete digitization of the continuous object.

Texture mapping [10] in computer graphics generally refers to the process of mapping a 2D image onto geometric primitives. The primitives are annotated with an extra set of 2D coordinates that orient the image on the primitive. The coordinate system axes of the image space are typically denoted as u and v for the horizontal and vertical axes, respectively. When the geometry is processed, the texture is applied to the geometry and appears draped over the geometry primitive like painting on cloth. The texture to be draped on the geometric primitive can be stored as an array of colors that will eventually be mapped onto the polygonal surface. The surface to be textured is specified with vertex coordinates and texture coordinates (u,v), the latter being used to map the color array on the polygon's surface. The u and v are interpolated across the span and then used as indices into the texture map to obtain the texture color. This color is combined with the primitive color (obtained by interpolating vertex colors across spans) or the colors specified by the application to obtain a final color value at the pixel location. Texture maps do not have to be color arrays but can be arrays of intensities used for color modulation. In this case, the application can specify two colors to modulate with the intensity, or it can take one of the colors from the primitive. The software takes the colors and uses the intensity in the texture map to determine how much of each color to be blended to produce the color of the pixel. This is useful for defining mottled textures found in landscape or cloth.

Subdivision surface refinement schemes [11] can be broadly classified into two categories: interpolating and approximating. Interpolating schemes are required to match the original position of vertices in the original mesh, while approximating schemes will adjust these positions as needed. In general, approximating schemes have greater smoothness, but editing applications that allow users to set exact surface constraints require an optimization step. This is analogous to spline surfaces and curves, where Bézier splines are required to interpolate certain control points, while B-splines are not. There is another classification of subdivision surface schemes as well, i.e., the type of polygon that they operate on. Some function for quadrilaterals (quads), while others operate on triangles. Approximating means that the limit surfaces approximate the initial meshes and that after subdivision, the newly generated control points are not in the limit surfaces. After interpolation-based subdivision, the control points of the original mesh and the newly generated control points are interpolated on the limit surface. Subdivision surfaces can be naturally edited at different levels of subdivision. Starting with basic shapes you can use binary operators to create the correct topology. You can edit the coarse mesh to create the basic shape and edit the offsets for the next subdivision step, and then repeat this at finer and finer levels. You can always see how your edit affects the limit surface via GPU (graphic processing unit) evaluation of the surface.

1.3.1.2 Processing Techniques for 3D Model Transmission and Storage

During the 3D model transmission or storage process, it usually involves compression, progressive transmission, encryption and information hiding techniques. To resolve the contradiction between the large amount of 3D data and the limited network bandwidth, it is of great significance to research the representation schemes of 3D models that are suitable for computer networks with small space requirements. Therefore, 3D model compression has become the research hot spot of computer graphics. Currently, most of the 3D models are approximated with meshes, and thus there are many research papers focusing on mesh model compression problems. The research work in this area can be roughly classified into two categories: one is the compression technology for connection relationships among vertices, edges and faces, which is called topological compression; the other is the compression method for the 3D vertex data and some other attribute data such as colors, texture and normal vectors, which is called geometric compression, among which vertex compression is the focus. In 1996, Hoppe presented a new representation scheme for 3D models, called progressive mesh [12]. It describes a dynamic data structure that is used to represent a given (usually quite complex) triangle mesh. At runtime, a progressive mesh provides a triangle mesh representation whose complexity is appropriate for the current view conditions. The purpose of progressive meshes is to speed up the rendering process by avoiding the rendering of details that are unimportant or completely invisible. This efficient, lossless, continuous-resolution representation addresses several practical problems in graphics: smooth geomorphing of level-of-detail approximations, progressive transmission, mesh compression and selective refinement. While conventional methods use a small set of discrete LODs, Schmalstieg et al. introduced a new class of polygonal simplification: Smooth LODs [13]. A very large number of small details encoded in a data stream allow a progressive refinement of the object from a very coarse approximation to the original high quality representation. Advantages of the new approach include progressive transmission and encoding suitable for networked applications, interactive selection of any desired quality, and compression of the data by incremental and redundancy-free encoding.

3D model encryption is the process of transforming 3D model data (referred to as plaintext) using an algorithm (called cipher) to make it unreadable to anyone except those possessing special knowledge, usually referred to as a key. The result of the process is the encrypted 3D model (in cryptography, referred to as ciphertext). In many contexts, the word encryption also implicitly refers to the reverse process, decryption (e.g. "software for encryption" can typically also perform decryption), to make the encrypted information readable again (i.e., to make it unencrypted).

3D model information hiding refers to the process of invisibly embedding the copyright information, the authentication information or other secret information into 3D models to fulfill the purpose of copyright protection, content authentication or covert communication. People usually embed information in 3D models with digital watermarking techniques, which will be discussed in Chapters 5 and 6 of this book.

1.3.1.3 Processing Techniques for 3D Model Management and Retrieval

In 3D model management and retrieval systems, it often involves 3D model pose normalization, content-based 3D model retrieval (which can fall into one direction in 3D model analysis techniques), volume visualization, and so on. 3D model pose normalization, also called pose estimation, is an important preprocessing step in 3D model retrieval systems. In the absence of prior knowledge, 3D models have arbitrary scales, orientations and positions in the 3D space. Because not all dissimilarity measures are invariant under scaling, translation, or rotation, one or more normalization procedures may be necessary. The normalization procedure depends on the center of mass, which is defined as the center of its surface points. To normalize a 3D model for scaling, the average distance of the points on its surface to the center of mass should be scaled to a constant. Note that normalizing a 3D model by scaling its bounding box is sensitive to outliers. To normalize for translation, the center of mass is translated to the origin. To normalize a 3D model for rotation, usually the principal component analysis (PCA) method is applied. It aligns the principal axes to the *x*-, *y*-, and *z*-axes of a canonical coordinate system by an affine transformation based on a set of surface points, e.g. the set of vertices of a 3D model. After translation of the center of mass to the origin, a rotation is applied so that the largest variance of the transformed points is along the *x*-axis. Then a rotation around the *x*-axis is carried out such that the maximal spread in the *yz*-plane occurs along the *y*-axis.

Content-based 3D model retrieval [14] has been an area of research in disciplines such as computer vision, mechanical engineering, artifact searching, molecular biology and chemistry. Recently, a lot of specific problems about content-based 3D shape retrieval have been investigated by researchers. At a conceptual level, a typical 3D shape retrieval framework consists of a database with an index structure created offline and an online query engine. Each 3D model has to be identified with a shape descriptor, providing a compact overall description of the shape. To efficiently search for a large collection online, an index of data structures and searching algorithms should be available. The online query engine computes the query descriptor, and models similar to the query model are retrieved by matching descriptors to the query descriptor from the index structure of the database. The similarity between two descriptors is quantified by a dissimilarity measure. Three approaches can be distinguished to provide a query object: (1) browsing to select a new query object from the obtained results; (2) handling a direct query by providing a query descriptor; (3) querying by example by providing an existing 3D model or by creating a 3D shape query from scratch using a 3D tool or sketching 2D projections of the 3D model. Finally, the retrieved models can be visualized. 3D model retrieval techniques will be discussed in Chapter 4.

Volume visualization is used to create images from scalar and vector datasets defined on multiple dimensional grids; i.e., it is the process of projecting a multidimensional (usually 3D) dataset onto a 2D image plane to gain an understanding of the structure contained within the data. Most techniques are applicable to 3D lattice structures. Techniques for higher dimensional systems are rare. It is a new but rapidly growing field in both computer graphics and data visualization. These techniques are used in medicine, geosciences, astrophysics, chemistry, microscopy, mechanical engineering, and so on.

1.3.2 Overview of 3D Model Analysis Techniques

So-called 3D model analysis operations are those operations whose inputs are 3D models or 3D objects while outputs are features, classification results, recognition results, matching results or semantics. 3D model analysis techniques comprise many aspects, such as feature extraction, perceptual hashing, segmentation, classification, matching, identification, retrieval, understanding, and so on.

3D model feature extraction is a necessary step in the identification, retrieval and classification techniques. Due to the overwhelming majority of 3D models being used for visualization, the documents representing 3D models often contain only the geometric properties of the model (vertex coordinates, normal vectors, topology connection, etc.) and appearance attributes (vertex color, texture, etc.); thus there are rarely descriptors suitable for automatic high-level description of semantic features. How to describe a 3D model (i.e., feature extraction) has become the problem to be solved first in the subject of 3D model retrieval, and it is also a difficult problem in 3D model retrieval. According to the different aspects of the content they represent, the features of a 3D model can be roughly categorized into two main types: (1) shape features, namely, geometry and topology features; (2) appearance features, which represent some important cognitive characteristics such as material colors, reflection coefficients and textures mapping. The characteristics of an ideal shape descriptor (SD) must satisfy the following conditions: (1) Both the expression and the calculation are easy; (2) It does not take up too much storage space; (3) It is suitable for similarity matching; (4) It is with geometric invariant, meaning invariance to the translation, rotation, scaling operations of 3D models; (5) It is with topological invariant, meaning when the same model embodies a number of topology descriptors, SD should be stable; (6) SD should be robust with regard to the vast majority of operations on 3D models, such as subdivision, simplification, adding noise and deformation; (7) SD must be unique, that is for different types of models, their features should be different. We will discuss the 3D model feature extraction techniques in Chapter 3.

Perceptual hashing is a one-way mapping from the multimedia dataset to the perceptual digest set [15], that is, to uniquely map the multimedia data with the same content to the same segment of digital digest, which satisfies the perceptual robustness and security. Perceptual hashing of multimedia content provides a safe and reliable technical support for identification, retrieval, authentication and other information services.

Model segmentation [16] has become an important and challenging problem in computer graphics, with applications in areas as diverse as modeling, metamorphosis, compression, simplification, 3D shape retrieval, collision detection, texture mapping and skeleton extraction. Mesh (and more generally shape) segmentation can be interpreted either in a purely geometric sense or in a more semantics-oriented manner. In the first case, the mesh is segmented into a number of patches that are uniform with respect to some property (e.g., curvature or distance to a fitting plane), while in the latter case the segmentation is aimed at identifying parts that correspond to relevant features of the shape. Methods that can be grouped under the first category may serve as a pre-processing for the recognition of meaningful features. Semantics-oriented approaches to shape segmentation have gained great interest recently in the research community, because they can support parameterization or re-meshing schemes, metamorphosis,

3D shape retrieval, skeleton extraction as well as the modeling by composition paradigm that is based on natural shape decompositions. It is rather difficult, however, to evaluate the performance of the different methods with respect to their ability to segment shapes into meaningful parts.

Pattern classification is the process of using a certain scheme in the feature space to classify the input pattern as a particular category, and it is the most basic and most important subject in the fields of pattern recognition and artificial intelligence. Things in the real world are complex, especially after the appearance of massive databases and the Internet, and the classification of 3D models will be essential research work.

3D model matching is the matching or shape comparison process in the space between the two models obtained from the same scene with different sensors, to confirm their similarity or the relative translation between them. It can be widely used in target tracking, resource analysis and medical diagnosis areas. In addition, how to perform the matching operation to search for in a 3D scene model similar to the input model is also a common technical problem.

Pattern recognition is a sub-topic in machine learning. It is "the act of taking in raw data and taking an action based on the category of the data". Most research in pattern recognition is about methods for supervised learning and unsupervised learning. Pattern recognition aims to classify data (patterns) based either on a priori knowledge or on statistical information extracted from the patterns. The patterns to be classified are usually groups of measurements or observations, defining points in an appropriate multidimensional space. This is in contrast to pattern matching, where the pattern is rigidly specified. 3D model recognition refers to the process of using mathematical techniques through computers to study the automatic processing and interpretation of the patterns of 3D models, and it needs the training and matching processes to finally identify the class of the input 3D model. 3D model retrieval is for calculating the similarity between the query model and the target model in the multi-dimensional feature space, and to realize the browsing and retrieval of 3D model databases. We will discuss the 3D model retrieval technique in Chapter 4.

3D model understanding should be one of the open problems in computer research, and its fundamental task is, from the semantics viewpoint, to make the computer correctly interpret the perceived 3D scenes and their content. The geometric and topology data are viewed as low-level data for 3D model understanding, and the corresponding theoretical starting point is computer vision and graphics. Knowledge information is viewed as high-level data for 3D model understanding, and the corresponding theoretical starting point is artificial intelligence. The key problems in 3D model understanding are the integration of knowledge and data, and the link between low-level processing and high-level analysis.

1.4 Overview of Multimedia Compression Techniques

Multimedia compression techniques include audio, images and video compression techniques.

1.4.1 Concepts of Data Compression

In computer science and information theory, data compression or source coding is the process of encoding information with fewer bits than an unencoded representation would use, based on specific encoding schemes. As with any communication, compressed data communication only works when both the sender and receiver of the information understand the encoding scheme. Similarly, compressed data can only be understood if the decoding method is known by the receiver. Compression is useful because it helps reduce the consumption of expensive resources, such as hard disk space or the transmission bandwidth. On the downside, compressed data must be decompressed to be used, and this extra processing may be detrimental to some applications. The design of data compression schemes therefore involves trade-offs among various factors, including the degree of compression, the amount of distortion introduced and the computational resources required.

Lossless compression algorithms usually exploit statistical redundancy in such a way as to represent the sender's data more concisely without error. Lossless compression is possible because most real-world data possess statistical redundancy. For example, in English text, the letter "e" is much more common than the letter "z", and the probability that the letter "q" will be followed by the letter "z" is very small. Another kind of compression, called lossy data compression, is possible if some loss of fidelity is acceptable. Generally, lossy data compression will be guided by research on how people perceive the data in question. For example, the human eye is more sensitive to subtle variations in luminance than it is to variations in color. JPEG image compression works in part by "rounding off" some of this less-important information. Lossy data compression provides a way to obtain the best fidelity for a given amount of compression. In some cases, transparent compression is desired, while in other cases fidelity is sacrificed to reduce the amount of data as much as possible. Lossless compression schemes are reversible so that the original data can be reconstructed, while lossy schemes accept some loss of data in order to achieve higher compression. However, lossless data compression algorithms will always fail to compress some files. For example, any compression algorithm will necessarily fail to compress any data containing no discernible patterns. An example of lossless vs. lossy compression is the following string: 25.888888888. This string can be compressed as: 25.[9]8, interpreted as "twenty five point 9 eights". The original string can thus be perfectly reconstructed, just written in a smaller form. In a lossy system, using 26 instead, the original data is lost, to the benefit of a smaller file size.

The theoretical background of compression is provided by information theory and rate-distortion theory. These fields of study were essentially created by Claude Shannon, who published fundamental papers on this topic in the late 1940s and early 1950s. Cryptography and coding theories are also closely related. The idea of data compression is deeply connected with statistical inference. Many lossless data compression systems can be viewed in terms of a four-stage model. Lossy data compression systems typically include even more stages, including prediction, frequency transformation and quantization. There is a close connection between machine learning and compression: a system that predicts the posterior probabilities of a sequence given its entire history can be used for optimal data compression, while an optimal compressor can be used for prediction. This equivalence has been used as justification for data compression and as a benchmark for "general intelligence".

1.4.2 Overview of Audio Compression Techniques

Audio compression [17] is a form of data compression designed to reduce the size of audio files. Audio compression algorithms are implemented in computer software as audio codecs. Generic data compression algorithms perform poorly with audio data, seldom reducing file sizes much below 87% of the original, and are not designed for use in real-time. Consequently, specific audio "lossless" and "lossy" algorithms have been designed. Lossy algorithms provide far greater compression ratios and are used in mainstream consumer audio devices. As with image compression, both lossy and lossless compression algorithms are used in audio compression, lossy being the most common for everyday use. In both lossy and lossless compression, information redundancy is reduced, using methods such as coding, pattern recognition and linear prediction to reduce the amount of information used to describe the data. The trade-off of slightly reduced audio quality is clearly outweighed for most practical audio applications, where users cannot perceive any difference and space requirements are substantially reduced. For example, on one CD, one can fit an hour of high fidelity music, less than two hours of music compressed losslessly, or seven hours of music compressed in MP3 format at medium bit rates.

1.4.2.1 Lossless Audio Compression

Lossless audio compression allows one to preserve an exact copy of one's audio files, in contrast to the irreversible changes from lossy compression techniques such as Vorbis and MP3. Compression ratios are similar to those for generic lossless data compression (around 50%–60% of original size), and substantially less than those for lossy compression (which typically yield 5%–20% of the original size).

The primary uses of lossless encoding are: (1) Archives. For archival purposes, one naturally wishes to maximize quality. (2) Editing. Editing lossily compressed data leads to digital generation loss, since the decoding and re-encoding introduce artifacts at each generation. Thus audio engineers use lossless compression. (3) Audio quality. Being lossless, these formats completely avoid compression artifacts. Audiophiles thus favor lossless compression. A specific application is to store lossless copies of audio, and then produce lossily compressed versions for a digital audio player. As formats and encoders are improved, one can produce updated lossily compressed files from the lossless master. As file storage space and communication bandwidth have become less expensive and more available, lossless audio compression has become more popular.

"Shorten" was an early lossless format, and newer ones include Free Lossless Audio Codec (FLAC), Apple's Apple Lossless, MPEG-4 ALS, Monkey's Audio and TTA. Some audio formats feature a combination of a lossy format and a lossless correction, which allows stripping the correction to easily obtain a lossy file. Such formats include MPEG-4 SLS (Scalable to Lossless), WavPack and OptimFROG DualStream. Some formats are associated with a technology, such as Direct Stream Transfer used in Super Audio CD, Meridian Lossless Packing used in DVD-Audio, Dolby TrueHD, Blu-ray and HD DVD.

It is difficult to maintain all the data in an audio stream and achieve substantial compression. First, the vast majority of sound recordings are highly complex, recorded from the real world. As one of the key methods of compression is to find patterns and repetition, more chaotic data such as audios cannot be compressed well. In a similar manner, photographs can be compressed less efficiently with lossless methods than simpler computer-generated images. But interestingly, even computer-generated sounds can contain very complicated waveforms that present a challenge to many compression algorithms. This is due to the nature of audio waveforms, which are generally difficult to simplify without a conversion to frequency information, as performed by the human ear. The second reason is that values of audio samples change very quickly, so generic data compression algorithms do not work well for audios, and strings of consecutive bytes do not generally appear very often. However, convolution with the filter [−1 1] tends to slightly whiten the spectrum, thereby allowing traditional lossless compression at the encoder to do its job, while integration at the decoder restores the original signal. Codecs such as FLAC, "Shorten" and TTA use linear prediction to estimate the spectrum of the signal. At the encoder, the inverse of the estimator is used to whiten the signal by removing spectral peaks, while the estimator is used to reconstruct the original signal at the decoder.

Lossless audio codecs have no quality issues, so the usability can be estimated by: (1) speed of compression and decompression; (2) degree of compression; (3) software and hardware support; (4) robustness and error correction.

1.4.2.2 Lossy Audio Compression

Lossy audio compression is used in an extremely wide range of applications. In addition to the direct applications, digitally compressed audio streams are used in most video DVDs, digital television, streaming media on the Internet, satellite and cable radio and increasingly in terrestrial radio broadcasts. Lossy compression typically achieves far greater compression than lossless compression by discarding less-critical data.

The innovation of lossy audio compression was to use psychoacoustics to recognize that not all data in an audio stream can be perceived by the human auditory system. Most lossy compression reduces perceptual redundancy by first identifying sounds which are considered perceptually irrelevant, i.e., sounds that are very hard to hear. Typical examples include high frequencies, or sounds that occur at the same time as louder sounds. Those sounds are coded with decreased accuracy or not coded at all.

While removing or reducing these "unhearable" sounds may account for a small percentage of bits saved in lossy compression, the real reduction comes from a complementary phenomenon: noise shaping. Reducing the number of bits used to code a signal increases the amount of noise in that signal. In psychoacoustics-based lossy compression, the real key is to "hide" the noise generated by the bit savings in areas of the audio stream that cannot be perceived. This is done by, for instance, using very small numbers of bits to code the high frequencies of most signals (not because the signal has little high frequency information, but rather because the human ear can only perceive very loud signals in this region), so that softer sounds "hidden" there simply are not heard.

If reducing perceptual redundancy does not achieve sufficient compression for a particular application, it may require further lossy compression. Depending on the audio source, this still may not produce perceptible differences. Speech, for example, can be compressed far more than music. Most lossy compression schemes allow compression parameters to be adjusted to achieve a target rate of data, usually expressed as a bit rate. Again, the data reduction will be guided by some model of how important the sound is as perceived by the human ear, with the goal of efficiency and optimized quality for the target data rate. Hence, depending on the bandwidth and storage requirements, the use of lossy compression may result in a perceived reduction of the audio quality that ranges from none to severe, but generally an obviously audible reduction in quality is unacceptable to listeners.

Because data is removed during lossy compression and cannot be recovered by decompression, some people may not prefer lossy compression for archival storage. Hence, as noted, even those who use lossy compression may wish to keep a losslessly compressed archive for other applications. In addition, the compression technology continues to advance, and achieving state-of-the-art lossy compression would require one to begin again with the lossless, original audio data and compress with the new lossy codec. The nature of lossy compression results in increasing degradation of quality if data are decompressed and then recompressed with lossy compression.

1.4.2.3 Coding Methods

There are two kinds of coding methods: transform dromain methods and time domain methods.

(1) Transform domain methods. To determine what information in an audio signal is perceptually irrelevant, most lossy compression algorithms use transforms such as the modified discrete cosine transform (MDCT) to convert time domain sampled waveforms into a transform domain. Once transformed, typically into the frequency domain, component frequencies can be allocated bits according to how audible they are. The audibility of spectral components is determined by first calculating a masking threshold, below which it is estimated that sounds will be beyond the limits of human perception.

The masking threshold is calculated with the absolute threshold of hearing and the principles of simultaneous masking (the phenomenon wherein a signal is masked by another signal separated by frequency) and, in some cases, temporal masking (where a signal is masked by another signal separated by time). Equal-loudness contours may also be used to weigh the perceptual importance of different components. Models of the human ear-brain combination incorporating such effects are often called psychoacoustic models.

(2) Time domain methods. Other types of lossy compressors, such as linear predictive coding (LPC) used for speech signals, are source-based coders. These coders use a model of the sound's generator to whiten the audio signal prior to quantization. LPC may also be thought of as a basic perceptual coding technique, where reconstruction of an audio signal using a linear predictor shapes the coder's quantization noise into the spectrum of the target signal, partially masking it.

1.4.3 Overview of Image Compression Techniques

Image compression [18] is the application of data compression on digital images. The objective is to reduce redundancy of the image data in order to be able to store or transmit data in an efficient form. Image compression can be lossy or lossless. Lossless compression is sometimes preferred for artificial images such as technical drawings, icons or comics. This is because lossy compression methods, especially when used at low bit rates, introduce compression artifacts. Lossless compression methods may also be preferred for high value content, such as medical imagery or image scans made for archival purposes. Lossy methods are especially suitable for natural images such as photos in applications where minor loss of fidelity is acceptable to achieve a substantial reduction in bit rate. The lossy compression that produces imperceptible differences can be called visually lossless.

1.4.3.1 Lossless Image Compression

Typical methods for lossless image compression are as follows.

(1) Run-length encoding (RLE). RLE is used as a default method in PCX and as one possible method in BMP, TGA and TIFF. RLE is a very simple form of data compression in which runs of data are stored as a single data value and its count, rather than as the original run. This is most useful in data that contains many such runs, for example, relatively simple graphic images such as icons, line drawings and animations. It is not recommended for use with files that do not have many runs as it could potentially double the file size.

(2) DPCM and predictive coding. DPCM was invented by C. Chapin Cutler at Bell Labs in 1950, and his patent includes both methods. DPCM or differential pulse-code modulation is a signal encoder that uses the baseline of PCM but adds some functionality based on the prediction of the samples of the signal. The input can be an analog signal or a digital signal. If the input is a continuous-time analog signal, it needs to be sampled first so that a discrete-time signal is the input to the DPCM encoder. There are two options. The first one is to take the values of two consecutive samples (if they are analog samples, quantize them). The difference between the first value and the next is calculated and the difference is further entropy coded. The other option is, instead of taking a difference relative to a previous input sample, to take the difference relative to the output of a local model of the decoder process, and in this option the difference can be quantized, which allows a good way of incorporating a controlled loss in the encoding. Applying one of these two processes, short-term redundancy of the signal is eliminated, and the compression ratios of the order of 2 to 4 can be achieved if differences are subsequently entropy coded, because the entropy of the difference signal is much smaller than that of the original discrete signal treated as independent samples.

(3) Entropy encoding. In information theory an entropy encoding is a lossless data compression scheme that is independent of the specific characteristics of the medium. One of the main types of entropy coding creates and assigns a unique prefix code to each unique symbol that occurs in the input. These entropy encoders then compress data by replacing each fixed-length input symbol by the corresponding variable-length prefix codeword. The length of each codeword is approximately proportional to the negative logarithm of the probability. Therefore, the most common symbols use the shortest codes. According to Shannon's source coding theorem, the optimal code length for a symbol is $\log_b P$, where b is the number of symbols used to make output codes and P is the probability of the input symbol. Two most commonly-used entropy encoding techniques are Huffman coding and arithmetic coding. If the approximate entropy characteristics of a data stream are known in advance, a simpler static code may be useful.

(4) Adaptive dictionary algorithms. They are used in GIF and TIFF. A typical one is the LZW algorithm, a universal lossless data compression algorithm created by Lempel, Ziv and Welch. It was published by Welch in 1984 as an improved implementation of the LZ78 algorithm published by Lempel and Ziv in 1978. The algorithm is designed to be fast to implement but is not usually optimal because it performs only limited analysis of the data.

(5) Deflation. Deflation is used in PNG, MNG and TIFF. It is a lossless data compression algorithm that uses a combination of the LZ77 algorithm and Huffman coding. It was originally defined by Phil Katz for Version 2 of his PKZIP archiving tool, and was later specified in RFC 1951. Deflation is widely thought to be free of any subsisting patents and, for a time before the patent on LZW (which is used in the GIF file format) expired, this led to its use in gzip compressed files and PNG image files, in addition to the ZIP file format for which Katz originally designed it.

1.4.3.2 Lossy Image Compression

Typical methods for lossy image compression are as follows.

(1) Color space reduction. The main idea is to reduce the color space to the most common colors in the image. The selected colors are specified in the color palette in the header of the compressed image. Each pixel just references the index of a color in the color palette. This method can be combined with dithering to avoid posterization.

(2) Chroma subsampling. This takes advantage of the fact that the eye perceives spatial changes in brightness more sharply than those in color, by averaging or dropping some of the chrominance information in the image. It is used in many video encoding schemes, both analog and digital, and also in JPEG encoding. Because the human visual system is less sensitive to the position and motion of color than luminance, bandwidth can be optimized by storing more luminance detail than color detail. At normal viewing distances, there is no perceptible loss incurred by sampling the color detail at a lower rate. In video systems, this is achieved through the use of color difference components. The signal is divided into a luma (Y′) component and two color difference components. Chroma subsampling deviates from color science in that the luma and chroma components are formed as a weighted sum of gamma-corrected R′G′B′ components instead of linear RGB components. As a result, luminance detail and color detail are not completely independent of one another. The error is greatest for highly-saturated colors. This engineering approximation allows color subsampling to be more easily implemented.

(3) Transform coding. This is the most commonly-used method. Transform coding is a type of data compression for "natural" data like audio signals or photographic images. The transformation is typically lossy, resulting in a lower quality copy of the original input. A Fourier-related transform such as DCT or the wavelet transform is applied, followed by quantization and entropy coding. In transform coding, knowledge of the application is used to choose information to be discarded, thereby lowering its bandwidth. The remaining information can then be compressed via a variety of methods. When the output is decoded, the result may not be identical to the original input, but is expected to be close enough for the purpose of the application. The JPEG format is an example of transform coding, one that examines small blocks of the image and "averages out" the color using a discrete cosine transform to form an image with far fewer colors in total.

(4) Fractal compression. Fractal compression is a lossy image compression

method using fractals to achieve high compression ratios. The method is best suited for photographs of natural scenes such as trees, mountains, ferns and clouds. The fractal compression technique relies on the fact that in certain images, parts of the image resemble other parts of the same image. Fractal algorithms convert these parts or, more precisely, geometric shapes into mathematical data called "fractal codes" which are used to recreate the encoded image. Fractal compression differs from pixel-based compression schemes such as JPEG, GIF and MPEG since no pixels are saved. Once an image has been converted into fractal code, its relationship to a specific resolution has been lost, and it becomes resolution independent. The image can be recreated to fill any screen size without the introduction of image artifacts or loss of sharpness that occurs in pixel-based compression schemes. With fractal compression, encoding is very computationally expensive because of the search used to find the self-similarities. However, decoding is quite fast. At common compression ratios, up to about 50:1, fractal compression provides similar results to DCT-based algorithms such as JPEG. At high compression ratios, fractal compression may offer superior quality. For satellite imagery, ratios of over 170:1 have been achieved with acceptable results. Fractal video compression ratios of 25:1–244:1 have been achieved in reasonable compression time (2.4 to 66 s/frame).

The quality of a compression method is often measured by the peak signal-to-noise ratio. It measures the amount of noise introduced through a lossy compression of the image. However, the subjective judgment of the viewer is also regarded as an important measure, perhaps the most important one. The best image quality at a given bit-rate is the main goal of image compression. However, there are other important requirements in image compression as follows:

(1) Scalability. It generally refers to a quality reduction achieved by manipulation of the bitstream or file. Other names for scalability are progressive coding or embedded bitstreams. Despite its contrary nature, scalability can also be found in lossless codecs, usually in the form of coarse-to-fine pixel scans. Scalability is especially useful for previewing images while downloading them or for providing variable quality access to image databases. There are several types of scalability: 1) Quality progressive or layer progressive: the bitstream successively refines the reconstructed image; 2) Resolution progressive: to first encode a lower image resolution and then encode the difference to higher resolutions; 3) Component progressive: to first encode the grey component and then color components.

(2) Region-of-interest coding. Certain parts of the image are encoded with a higher quality than others. This can be combined with scalability, i.e., to encode these parts first, others later.

(3) Meta information. Compressed data can contain information about the image which can be used to categorize, search or browse images. Such information can include color and texture statistics, small preview images and author/copyright information.

(4) Processing power. Compression algorithms require different amounts of processing power to encode and decode. Some compression algorithms with high compression ratios require high processing power.

1.4.4 Overview of Video Compression Techniques

Video compression [18] refers to reducing the quantity of data used to represent digital video frames, and is a combination of spatial image compression and temporal motion compensation. Compressed video can effectively reduce the bandwidth required to transmit video via terrestrial broadcast, cable TV or satellite TV services. Most video compression is lossy, for it operates on the premise that much of the data present before compression is not necessary for achieving good perceptual quality. For example, DVDs use a video coding standard called MPEG-2 that can compress around two hours of video data by 15 to 30 times, while still producing a picture quality that is generally considered high-quality for a standard-definition video. Video compression is a tradeoff between disk space, video quality, and the cost of hardware required to decompress the video in a reasonable time. However, if the video is overcompressed in a lossy manner, visible artifacts may appear. Video compression typically operates on square-shaped groups of neighboring pixels, often called macroblocks. These pixel groups or blocks of pixels are compared from one frame to the next and the video compression codec sends only the differences within those blocks. This works extremely well if the video has no motion. A still frame of text, for example, can be repeated with very little transmitted data. In areas of the video with more motion, more pixels change from one frame to the next. When more pixels change, the video compression scheme must send more data to keep up with the larger number of pixels that are changing. If the video content includes an explosion, flames, a flock of thousands of birds, or any other image with a great deal of high-frequency detail, the quality will decrease, or the variable bit rate must be increased to render this added information with the same level of detail.

The programming providers have control over the amount of video compression applied to their video programming before it is sent to their distribution system. DVDs, Blu-ray discs, and HD DVDs have video compression applied during their mastering process, though Blu-ray and HD DVD have enough disc capacity so that most compression applied in these formats is light, when compared to such examples as most of the video streamed over the Internet, or taken on a cellphone. Software used for storing videos on hard drives or various optical disc formats will often have a lower image quality, although not in all cases. High-bitrate video codecs, with little or no compression, exist for video post-production work, but create very large files and are thus almost never used for the distribution of finished videos. Once excessive lossy video compression compromises image quality, it is impossible to restore the image to its original quality.

A video is basically a 3D array of color pixels. Two dimensions serve as spatial directions of the moving pictures, and one dimension represents the time domain. A data frame is a set of all pixels that correspond to a single time moment. Basically, a frame is the same as a still picture. Video data contains spatial and temporal redundancy. Similarities can thus be encoded by merely registering differences within a frame (spatial), and/or between frames (temporal). Spatial encoding is performed by taking advantage of the fact that the human eye is unable to distinguish small differences in color as easily as it can perceive changes in brightness, so that very similar areas of color can be "averaged out" in a similar way to JPEG images. With temporal compression, only the changes from one frame to the next are encoded, as often a large number of the pixels will be the same on a series of frames.

Some forms of data compression are lossless. This means that when the data is decompressed, the result is a bit-for-bit perfect match with the original. While lossless compression of video is possible, it is rarely used, as lossy compression results in far higher compression ratios at an acceptable level of quality.

One of the most powerful techniques for compressing videos is interframe compression. Interframe compression uses one or more earlier or later frames in a sequence to compress the current frame. Intraframe compression is applied only to the current frame, where we can just adopt effective image compression methods. The most commonly-used method works by comparing each frame in the video with the previous one. If the frame contains areas where nothing has moved, the system simply issues a short command that copies that part of the previous frame, bit-for-bit, into the next one. If sections of the frame move in a simple manner, the compressor emits a command that tells the decompresser to shift, rotate, lighten, or darken the copy. This is a longer command, but still much shorter than intraframe compression. Interframe compression works well for programs that will simply be played back by the viewer, but can cause problems if the video sequence needs to be edited. Since interframe compression copies data from one frame to another, if the original frame is simply cut out, the following frames cannot be reconstructed properly. Some video formats, such as DV, compress each frame independently through intraframe compression. Making "cuts" in the intraframe-compressed video is almost as easy as editing the uncompressed video, i.e., one finds the beginning and end of each frame, and simply copies bit-for-bit each frame that one wants to keep, and discards the frames one does not want. Another difference between intraframe and interframe compression is that with intraframe systems, each frame uses a similar amount of data. In most interframe systems, certain frames are not allowed to copy data from other frames, and thus they require much more data than other frames nearby. It is possible to build a computer-based video editor that spots problems caused when frames are edited out (i.e., deleted) while other frames need them. This has allowed newer formats like HDV to be used for editing. However, this process demands much more computing power than editing intraframe-compressed videos with the same picture quality.

Today, nearly all video compression methods in common use, e.g., those in standards approved by the ITU-T or ISO, apply a discrete cosine transform for spatial redundancy reduction. Other methods, such as fractal compression, matching pursuit and the use of a discrete wavelet transform (DWT), have been the subjects of some research, but are typically not used in practical products. The interest in fractal compression seems to be waning, due to recent theoretical analysis showing a comparative lack of effectiveness of such methods.

1.5 Overview of Digital Watermarking Techniques

Digital watermarking [19] is a fast developing focus technique, which has been already of high interest to the international academic and business communities. The watermarking technique is a rising interdisciplinary technique, which refers to ideas and theories from different scientific and academic fields, such as signal processing, image processing, information theory, coding theory, cryptography, detection theory, probability theory, random theory, digital communication, game theory, computer science, network technique, algorithm design, etc., but also including public strategy and law. Therefore, whether from the point of theories or applications, carrying out research on digital watermarking techniques is not only a matter of great academic significance, but also a matter of great economic significance.

1.5.1 Requirement Background

The sudden increase in interest in the digital watermarking technique probably originates from people's concern about copyright protection. In recent years, with the abrupt development of the computer multimedia technique, people can use digital equipments to produce and process and restore information media, such as images, audios, texts and videos. In the meanwhile, the digital network communication is developing quickly, which means the release and transmission of information becomes digitized and networked. In the analog era, people used tapes as recording equipments, so the quality of pirate copies is usually lower than that of original copies. However, in the digital age, there is no quality loss in the digital copying process of songs and movies. Since the emergence of Marc Andreessen's Mosaic web browser in November 1993, the Internet has become friendly to consumers, and soon people began taking delight in downloading images, music and videos from it. For digital media, the Internet is the most excellent distribution system, because it is cheap, does not need warehouses to restore materials, and can transmit information in real time. Therefore, digital media are easily copied, restored, distributed and published via the Internet or CD-ROM, which leads to security problems and copyright protection problems during digital information exchange. How to implement valid copyright protection and information security in the network environment has already caused a lot of concern from the international academic community, the business community and relevant government departments, and how to prevent digital products, such as digital publications, audio clips, video clips, cartoons and images, from tort, piracy and random tampering has become a pressing and hot subject all over the world.

Detailed descriptions of the actual distribution mechanism for digital products are very complex, including original authors, editors, multimedia integrators, resellers and official governments. This book presents a simple distribution model as shown in Fig. 1.6. The supplier is a general designation of the copyright owner, editors and retailers, and they try to distribute the digital product $\boldsymbol{x}$ via the network. The consumers, which also can be called customers (clients), hope to receive the digital product $\boldsymbol{x}$ via the network. The pirates are unauthorized suppliers, such as the pirate A, who redistributes the product $\boldsymbol{x}$ without the legal copyright owner's permission, and the pirate B, who intentionally destroys the original product and redistributes the unauthentic edition $\hat{\boldsymbol{x}}$, so it is hard for consumers to avoid receiving the pirate edition $\boldsymbol{x}$ or $\hat{\boldsymbol{x}}$ indirectly. There are three common illegal forms of behavior as follows: (1) Illegal visit, i.e., to copy or pirate digital products without the permission of copyright owners. (2) Intentional tampering, i.e., the pirates maliciously change digital products or insert characteristics and then redistribute them, resulting in the loss of the original copyright information. (3) Copyright destruction, i.e., the pirates, resells digital products without the permission of the copyright owner after receiving them.

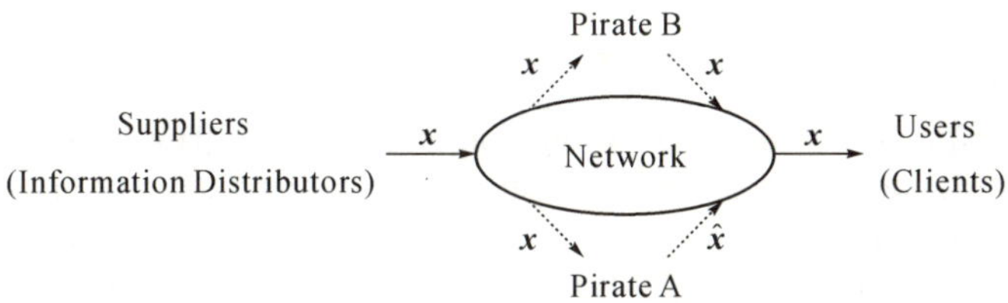

Fig. 1.6. The basic model of digital product distribution over the Internet

To resolve information security and copyright protection problems, the first thing that comes to copyright owners' minds is to use encryption and digital signature techniques. The encryption technique based on private keys and public keys can be used to control data accesses by changing the plaintext information into secret information, which others cannot understand. The encrypted products can be accessed, but only those people who have the right secret keys can decode them. Besides, setting passwords can also make the data unreadable during the transmission process and thereby valid protection can be provided for the data on the way from the sender to the receiver. The digital signature uses the string composed of "0" and "1" instead of the signature or seal, and exerts the same legal effects. The digital signature technique has already been used to testify the reliability of short digital messages, forming the digital signature standard (DSS). It signs each piece of information with private keys, and public detection algorithms are used to testify whether the information content accords with the corresponding signature or not. However, these kinds of digital signatures are neither convenient nor realistic when used in digital images, videos and audios, since plenty of signatures are required to be added to the original data. In addition, with the fast development of computer hardware and software techniques and the gradual growth of decoding techniques with the distributed calculation capability based on the network, the security of these traditional systems has already been compromised. It is no longer a uniquely feasible way to enhance the reliability of security systems by only increasing the length of the secret keys. And if only the people who are authorized to hold secret keys can get the encrypted information, there is no way to make more people obtain their required information via public systems. At the same time, once the information is decoded illegally, there is no direct evidence to prove the information has been illegally copied and resent. Furthermore, for some people, encryption is a challenging task, because people can hardly prevent an encrypted file from being cut during the decoding process. Therefore, it is necessary to seek a more valid method to ensure secure transmission and protect the digital products' copyright.

1.5.2 Concepts of Digital Watermarks

When referring to watermarks, people probably think of the watermarks in bills. Holding a 20-dollar bill, if you observe the side with the portrait of the President Andrew Jackson under lights, you will see a watermark appearing in it. This watermark is directly embedded into the bill during manufacture, so it is hard to fabricate. It also prevents a usual forgery method, i.e., washing off the ink on the 20-dollar bill and then printing "100-dollar" on the same paper. Usually, the bill watermark should have two characteristics. First, watermarks are invisible under normal circumstances, and only appear visible under special observation conditions (here this means putting bills under lights). Second, the watermark information should correlate with carrier objects (here this means watermarks are used to identify bills authenticity).

Besides bills, watermarks can be used in other physical objects, even in electric signals. Fabrics, cloth brands and product packs are all concrete instances, in which watermarks can be embedded with special dyes and inks. The electronic medium, such as music, photos and videos, are some common signal types which can be embedded with watermarks. This book is only concerned with watermarking techniques for electronic signals, and uses the following glossaries to describe these kinds of signals.

Work (or product): a specific song, a video clip, a picture or a copy of one of them. The original work without watermarks is called the "carrier work".

Content: a set of all possible works. For example, music is one kind of "content", and a specific song is one work.

Media: the medium for reproducing, transmitting and recording "content".

Digital watermarking is a kind of information hiding technique [20], and its basic idea is to embed secret information into digital products, such as digital images, audios and videos, in order to protect their copyrights, testify their authenticity, track piracy behavior or supply products' additional information. The secret information can be copyright symbols, users' serial numbers or other relevant information. Usually they need to be embedded into digital products after proper transforms, and usually the transformed information is called a digital watermark. Various watermark signals are referred to in much literature. Usually they can be defined as the following signal $\boldsymbol{w}$:

$$w = \{w_i \mid w_i \in \boldsymbol{O},\ i = 0,\ 1,\ 2,\ \ldots,\ N-1\}, \tag{1.3}$$

where N is the length of the watermark sequence, and $\boldsymbol{O}$ represents the value range. Actually, watermarks can be not only 1D sequences, but also 2D sequences, even multi-dimensional sequences, which are usually decided by the carrier object's dimension. For instance, audio, images and video correspond to 1D, 2D and 3D sequences respectively. For convenience, this book usually uses Eq. (1.3) to represent watermark signals, and for multi-dimensional sequences it is equivalent to expanding them into 1D sequences in a certain order. The range of watermark signals can be in binary forms, such as $\boldsymbol{O} = \{0, 1\}$, $\boldsymbol{O} = \{-1, 1\}$ and $\boldsymbol{O} = \{-r, r\}$, or some other forms, such as white Gaussian noises (with the mean 0 and the variance 1, $N(0, 1)$).

1.5.3 Basic Framework of Digital Watermarking Systems

Roughly speaking, a digital watermarking system contains two main parts, the embedder and the detector. The embedder has at least two inputs, the original information which will be properly transformed into the watermark signal, and the carrier product which will be embedded with watermarks. The output of the embedder is the watermarked product, which will be transmitted or recorded. The input of the detector may be the watermarked work or another random work that has never been embedded with watermarks. Most detectors try their best to estimate whether there are watermarks in the work or not. If the answer is yes, the output will be the watermark signal previously embedded in the carrier product. Fig. 1.7 presents the particular sketch map of the basic framework of digital watermarking systems. It can be defined as a set with nine elements (***M***, ***X***, ***W***, ***K***, *G*, *Em*, *At*, *D*, *Ex*) and they are defined below separately:

(1) ***M*** stands for the set of all possible original information ***m***.

(2) ***X*** is the set of digital products (or works) ***x***, i.e., the content.

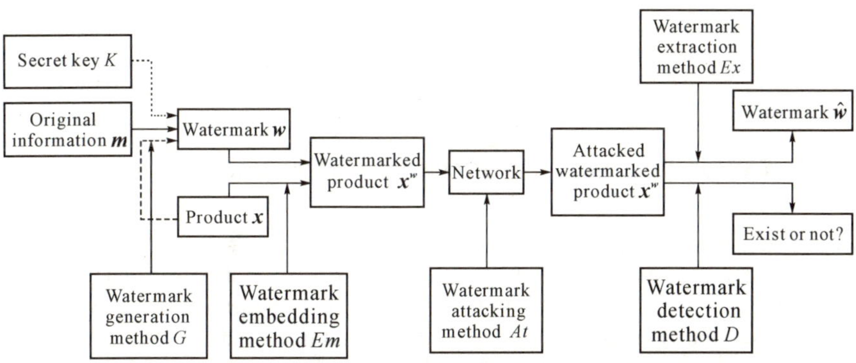

Fig. 1.7. The basic framework of digital watermarking systems

(3) $\boldsymbol{W}$ is the set of all possible watermark signals $\boldsymbol{w}$.

(4) $\boldsymbol{K}$ is the set of watermarking secret keys K.

(5) G is the generation algorithm making use of the original information $\boldsymbol{m}$, the secret key K and the original digital product $\boldsymbol{x}$ together, i.e.,

$$G: \boldsymbol{M} \times \boldsymbol{X} \times \boldsymbol{K} \rightarrow \boldsymbol{W}, \quad \boldsymbol{w} = G(\boldsymbol{m}, \boldsymbol{x}, K). \tag{1.4}$$

It should be pointed out that the original digital product does not necessarily participate in generating watermarks, so we use dashed lines in Fig. 1.7.

(6) Em is the embedding algorithm, which embeds the watermark $\boldsymbol{w}$ into the digital product $\boldsymbol{x}$, i.e.,

$$Em: \boldsymbol{X} \times \boldsymbol{W} \rightarrow \boldsymbol{X}, \quad \boldsymbol{x}^{w} = Em(\boldsymbol{x}, \boldsymbol{w}), \tag{1.5}$$

here $\boldsymbol{x}$ presents the original product and $\boldsymbol{x}^{w}$ presents the watermarked product. To enhance the security, sometimes secret keys are included in the embedding algorithms.

(7) At is the attacking algorithm performed on the watermarked product $\boldsymbol{x}^{w}$, i.e.,

$$At: \boldsymbol{X} \times \boldsymbol{K} \rightarrow \boldsymbol{X}, \quad \hat{x} = At(x^{w}, K'), \tag{1.6}$$

here K' is the secret key fabricated by attackers, and $\hat{\boldsymbol{x}}$ is the attacked watermarked product.

(8) D is the detection algorithm, i.e.,

$$D: \boldsymbol{X} \times \boldsymbol{K} \rightarrow \{0,1\}, \quad D(\hat{\boldsymbol{x}}, K) = \begin{cases} 1, & \text{if } \boldsymbol{w} \text{ exists in } \hat{\boldsymbol{x}}\,(H_1); \\ 0, & \text{if } \boldsymbol{w} \text{ does not exist in } \hat{\boldsymbol{x}}\,(H_0), \end{cases} \tag{1.7}$$

here, H_1 and H_0 stand for binary hypotheses, which indicate the watermark exists or not.

(9) Ex is the extraction algorithm, i.e.,

$$Ex: \boldsymbol{X} \times \boldsymbol{K} \rightarrow \boldsymbol{W}, \quad \hat{\boldsymbol{w}} = Ex(\hat{\boldsymbol{x}}, K). \tag{1.8}$$

1.5.4 Communication-Based Digital Watermarking Models

Essentially speaking, the digital watermarking process is a kind of communication, i.e., delivering a message between the watermark embedder and receiver. Naturally, people try to describe the whole watermarking process with traditional basic communication models. Usually there are three kinds of models and the difference among them is how to introduce the carrier products into traditional communication models. In the first basic model, the carrier work is totally considered as noise. In the second model, the carrier work is still considered as noise but the noise is input into the channel encoder as additional information. In the third model, the carrier work is not considered as noise but the second information. This information and the original information are transmitted in a multiplex manner. Here we only show the first kind of model.

Figs. 1.8 and 1.9 present two basic digital watermarking system models. Fig. 1.8 adopts the non-blind detector and Fig. 1.9 adopts the blind detector. In these two kinds of models, the watermark embedder is considered as a channel. The input information is transmitted via the channel, and the carrier work is a part of it. To depict this conveniently, here the watermark generation algorithm is called the watermark encoder, and it is combined into the watermark embedder. No matter whether adopting the non-blind detector or the blind detector, the first step in the embedding process is mapping the information $\boldsymbol{m}$ to an embedding pattern $\boldsymbol{w}_a$ with the same format and dimension as the original product $\boldsymbol{x}$, which is actually a watermark generation process. For instance, if we embed watermarks into images in the spatial domain, the watermark encoder, i.e., the watermark generator, will generate a 2D image pattern with the same size as the original image. However, when we embed watermarks into audio clips in the time domain, the watermark encoder will generate a 1D pattern with the same length as the original audio clip. This kind of mapping usually needs the aid of the watermarking secret key K. The embedding pattern is calculated with several steps: (1) Predefining one or several reference patterns (represented by $\boldsymbol{w}_r$, e.g., a pseudorandom or chaotic sequence), which depend on some secret key K. (2) These reference patterns are combined together to form a pattern to encode the information $\boldsymbol{m}$, which is usually called the information pattern $\boldsymbol{w}$. In this book, it is called the watermark $\boldsymbol{w}$ to be embedded, which is the output of the watermark generation algorithm. (3) Then this information pattern is scaled proportionally or modified to generate the embedding pattern $\boldsymbol{w}_a$ (In this book this process falls under the first step of the embedding process). The watermark encoders in Figs. 1.8 and 1.9 both do not take carrier works into account, and we call them non-adaptive generators. The watermarked work $\boldsymbol{x}^w$ is gained by embedding the pattern $\boldsymbol{w}_a$ into the work $\boldsymbol{x}$, and it will undergo some kind of processes, whose effect is equal to adding the noise $\boldsymbol{n}$ to the work. Here the processes may be unintentional attacks such as compression, decompression, analog/digital conversion and signal enhancement, or malicious attack behaviors such as wiping off watermarks.

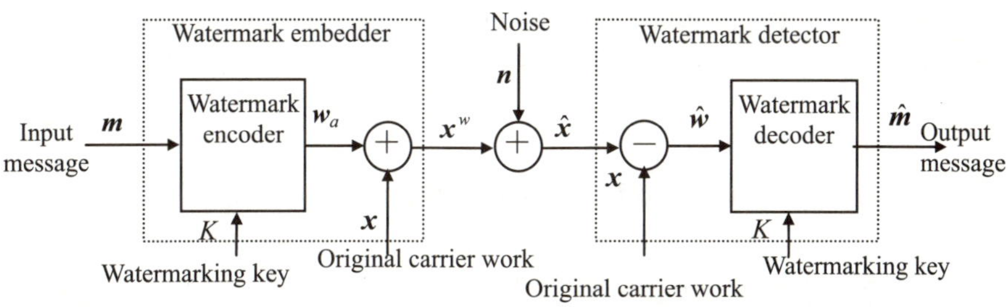

Fig. 1.8. Non-blind watermarking system described by a communication model

There is no essential difference between the watermark detector and the watermark decoder in Fig. 1.9. If using the non-blind detector in Fig. 1.8, the detection process consists of two steps: (1) The carrier work $\boldsymbol{x}$ is subtracted from the receiving work $\hat{\boldsymbol{x}}$ to obtain the watermark pattern $\hat{\boldsymbol{w}}$. (2) The watermark decoder decodes based on the watermarking key. Since adding the carrier work in the embedder is counteracted by the subtraction in the detector, the difference between $\boldsymbol{w}_a$ and $\hat{\boldsymbol{w}}$ is actually aroused by noise. So the influence of the carrier work can be overlooked, which means the watermark encoder, noise adding and the watermark decoder all together compose a system similar to the basic communication model. In some more advanced non-blind detection systems, it is not necessary to have the overall original carrier work; however, a function of $\boldsymbol{x}$, usually a data simplification function, is used to compensate the "noise" effect caused by adding the carrier work in the embedder. In the blind detector of Fig. 1.8, because it is not necessary for the original carrier work to participate in the detection process, it does not need to subtract the original carrier before decoding. In this case, the original carrier work and the combination of attacks can be considered as a single noise. The received watermarked work $\hat{\boldsymbol{x}}$ can be considered as a work edition, in which the embedding pattern $\boldsymbol{w}_a$ has been destroyed and the whole watermark detector can be considered as the channel decoder.

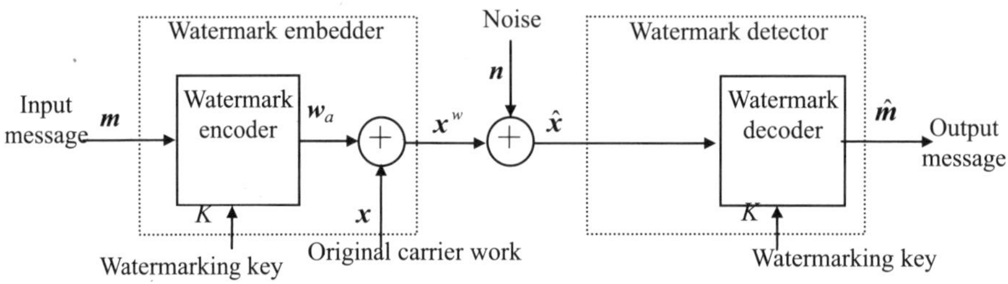

Fig. 1.9. Blind watermarking system described by a communication model

In applications of transaction tracking and copyright protection, people hope the probability that the detected information is the same as the embedded information is maximal, which coincides with the traditional communication system's goal. However, it should be noted that in the application of authentication, because the aim is not delivering information but checking out whether the watermarked work is modified or not and how it is modified, the models shown in Figs. 1.8 and 1.9 are unsuitable for representing authentication systems.

1.5.5 Classification of Digital Watermarking Techniques

Digital watermarks are signals embedded in digital media such as images, audio clips or video clips. These signals enable people to construct products' ownership, identify purchasers and provide some extra information about products. According to the visibility in the carrier work, watermarks can be divided into two categories, visible and invisible watermarks. This book mainly discusses invisible watermarks. Therefore, if there is no special announcement, watermarks in the following discussions refer to invisible watermarks. According to whether the watermark generation process depends on the original carrier work or not, it can be divided into non-adaptive watermarks (independent of the original cover media) and adaptive watermarks. Watermarks dependent on the original cover media can be generated not only randomly or by algorithms, but can also be given in advance, while adaptive watermarks are generated considering the characteristic of the original cover media. According to the watermarked product's ability against attacks, watermarks can be divided into fragile watermarks, semi-fragile watermarks and robust watermarks. Fragile watermarks are very sensitive to any transforms or processing. Semi-fragile watermarks are robust against some special image processing operations while not robust to other operations. Robust watermarks are robust to various popular image processing operations. According to whether the original image is required in the watermark detection process or not, watermarks can be divided into non-blind-detection watermarks (private watermarks) and blind-detection watermarks (public watermarks). Private watermark detection requires the original image, while public watermarks do not. According to different application purposes, watermarks can be divided into copyright protection watermarks, content authentication watermarks, transaction tracking watermarks, copy control watermarks, annotation watermarks, covert communications watermarks, etc.

Accordingly, watermarking algorithms also can be classified into two categories, visible watermarking algorithms and invisible watermarking algorithms. This book mainly discusses invisible watermarking algorithms, which can be mainly classified into three categories, time/spatial-domain-based, transform-domain-based and compression-domain-based schemes. Time/spatial domain watermarking uses various methods to directly modify cover media's time/spatial samples (e.g., pixels' LSB). The robustness of this kind of algorithm is not strong, and the capacity is not very large; otherwise watermarks will become visible. Transform domain watermarking embeds watermarks after various transforms of the original cover media, e.g., DCT transform, DFT transform, wavelet transform, etc. Compression domain watermarking refers to embedding a watermark in the JPEG domain, MPEG domain, VQ compression domain or fractal compression domain. This kind of algorithm is robust against the associated compression attack. Some researchers use public key cryptosystems in watermarking systems where the detection key and the embedding key are different. These kinds of watermarking systems are called public key watermarking systems, or are otherwise called private key watermarking systems. According to whether the original cover media can be losslessly recovered or not, watermarking systems can be classified into two categories, reversible watermarking systems and irreversible watermarking systems. According to different types of original cover media, watermarking processing can be classified into audio watermarking, image watermarking, video watermarking, 3D model or 3D image watermarking, document watermarking, database watermarking, integrated circuit watermarking, software watermarking (The watermark is embedded in program codes or .exe files), etc. According to whether adaptive techniques (including embedding parameter and position adaptivity in watermark generation and embedding) are used in watermarking algorithms or not, digital watermarking systems can be classified into two categories, adaptive digital watermarking systems and non-adaptive digital watermarking systems. In addition, some researchers have also proposed concepts such as the non-linear digital watermarking system (based on chaos, fractals, neural networks or genetic algorithms), the second generation digital watermarking system (based on invariant feature points), multipurpose watermarking systems (embedding multipurpose watermarks at the same time), etc.

1.5.6 Applications of Digital Watermarking Techniques

The application fields of watermarking techniques are very wide. There are mainly the following seven categories: broadcast monitoring, owner identification, ownership verification, transaction tracking, content authentication, copy control and device control. Each application is concretely introduced below. Problem characteristics are analyzed and the reasons for applying watermarking techniques to solve these problems are given.

(1) Broadcast monitoring. The advertiser hopes that his advertisements can be aired completely in the airtime that is bought from the broadcaster, while the broadcaster hopes that he can obtain advertisement dollars from the advertiser. To realize broadcast monitoring, we can hire some people to directly survey and monitor the aired content. But not only does this method cost a lot but also it is easy to make mistakes. We can also use the dynamic monitoring system to put recognition information outside the area of the broadcast signal, e.g., vertical blanking interval (VBI); however there are some compatibility problems to be solved. The watermarking technique can encode recognition information, and it is a good method to replace the dynamic monitoring technique. It uses the characteristic of embedding itself in content and requires no special fragments of the broadcast signal. Thus it is completely compatible with the installed analog or digital broadcast device.

(2) Owner identification. There are some limitations in using the text copyright announcement for product owner recognition. First, during the copying process, this announcement is very easily removed, sometimes accidentally. For example, when a professor copies several pages of a book, the copyright announcement on the topic pages is probably not copied by negligence. Another problem is that it may occupy some parts of the image space, destroying the original image, and it is easy to be cropped. As a watermark is not only invisible, but also cannot be separated from the watermarked product, the watermark is therefore more beneficial than a text announcement in owner identification. If the product user has a watermark detector, he can recognize the watermarked product's owner. Even if the watermarked product is altered by the method that can remove the text

copyright announcement, the watermark can still be detected.

(3) Ownership verification. Besides identification of the copyright owner, applying watermarking techniques for copyright verification is also a particular concern. A conventional text announcement is extremely easy to tamper with and counterfeit, and thus it cannot be used to solve this problem. A solution for this problem is to construct a central information database for digital product registration, but people may not register their products because of the high cost. To save the registration fee, people may use watermarks to protect copyright. And to achieve a certain level of security, the granting of detectors may need to be restricted. If the attacker has no detector, it is quite difficult to remove watermarks. However, even if the watermark cannot be removed, the attacker may also use his own watermarking system. Thus people may feel there is also an attacker's watermark in the same digital product. Therefore, it is not necessary to directly verify the copyright with the embedded watermark. On the contrary, the fact that an image is obtained from another image must be proved. This kind of system can indirectly prove that this disputed image may be owned by the owner instead of the attacker because the copyright owner has the original image. This verification manner is similar to the case where the copyright owner can take out the negative while the attacker can only counterfeit the negative of the disputed image. It is impossible for the attacker to counterfeit the negative of the original image to pass the examination.

(4) Transaction tracking. The watermark can be used to record one or several trades for a certain product copy. For example, the watermark can record each receiver who has been legally sold and sent a product copy. The product owner or producer can embed different watermarks in different copies. If the product is misused (e.g., disclosed to the press or illegally promulgated), the owner can find the people who are responsible for it.

(5) Content authentication. Nowadays, it becomes much easier to tamper with digital products in an inconspicuous manner. Research into the message authentication problem is relatively mature in cryptography. Digital signature is the most popular encryption scheme. It is essentially an encrypted message digest. If we compare the signature of a suspicious message with the original signature and find that they do not match, then we can conclude that the message must have been changed. All of these signatures are source data, and must be transmitted together with the product to be verified. Once the signature is lost, this product cannot be authenticated. It may be a good solution to embed the signature in products with watermarking techniques. This kind of embedded signature is called an authentication mark. If a very small change can make the authentication mark become invalidated, we call this kind of mark a "fragile watermark".

(6) Copy control. Most of the above mentioned watermarking techniques take effect only after the illegal behavior has happened. For example, in the broadcast monitoring system, only when the broadcaster does not broadcast the paid advertisement can we regard the broadcaster dishonest, while in the transaction tracking system, only when the opponent has distributed the illegal copy can we identify the opponent. It is obvious that we had better design the system to prevent the behavior of illegal copying. In copy control, people aim to prevent the

protected content from being illegally copied. The primary defense of illegal copying is encryption. After encrypting the product with a special key, the product simply cannot be used by those without this key. Then this key can be provided to legal users in a secure manner such that the key is difficult to copy or redistribute. However, people usually hope that the media data can be viewed, but cannot be copied by others. At this time, people can embed watermarks in content and play it with the content. If each recording device is installed with a watermark detector, the device can forbid copying when it detects the watermark "copy forbidden".

(7) Device control. In fact, copy control belongs to a larger application category called device control. Device control refers to the phenomenon where a device can react when the watermark is detected. For example, the "media bridge" system of Digimarc can embed the watermark in printed images such as magazines, advertisements, parcels and bills. If this image is captured by a digital camera again, the "media bridge" software and recognition unit in the computer will open a link to related websites.

1.5.7 Characteristics of Watermarking Systems

Ten important characteristics that watermarking systems should possess will be introduced below, according to different applications. The relative importance of each characteristic is determined by application requirements and watermark functions. Even the explanation of each watermark characteristic changes as the application situation changes. First, we discuss several characteristics related to watermark embedding, i.e., effectiveness, fidelity and payload. Then, several characteristics related to watermark detection are discussed, i.e., blind and informed detection, false positive behavior and robustness. Another two properties, security and secret keys, are closely related, for the usage of keys is always an indiscernible part of the security evaluation of watermarking schemes. Next, watermark modification and multiple watermarking are discussed and, finally, the cost of watermark embedding and detection is introduced.

(1) Embedding effectiveness. A product is defined as a watermarked product if a positive result is obtained when it is inputted into the watermark detector. Based on this definition, the effectiveness of a watermarking system refers to the probability that the detector outputs positive results. In other words, effectiveness refers to the probability of obtaining positive results after embedding. In some cases, effectiveness of a watermarking system can be determined by analysis, and also can be determined by the practical results of embedding watermarks in a large scale test image set. As long as the number of images in this set is large enough and their distribution is similar to that of the application situation, the percentage of positive results can be approximately regarded as the probability of effectiveness.

(2) Fidelity. Generally speaking, the fidelity of a watermarking system refers to the perceptual similarity between the original product and its watermarked version. But before the watermarked product is viewed by people, if there is some

quality distortion during transmission, another fidelity definition should be used. In the case that both the watermarked and original products can be obtained by consumers, it can be defined as the perceptual similarity between these two products. When we use the NTSC broadcast standard to transmit watermarked videos or use an AM broadcast to transmit watermarked audios, the difference between the degraded original production due to the channel distortion and its watermarked version is almost unnoticeable because of the relatively bad broadcast quality. But for HDTV/DVD videos and audios, signal quality is very high, and then high fidelity watermarked products are required.

For example, to evaluate the effect of embedded watermarks on the original 3D model, besides qualitative assessments based on perceptual systems, we can also adopt the following quantitative evaluation methods.

(i) Mean squared error (MSE):

$$MSE = \frac{1}{N}\sum_{i=1}^{N}\left\|\boldsymbol{v}_i - \boldsymbol{v}'_i\right\|^2; \tag{1.9}$$

(ii) Peak signal-to-noise ratio (PSNR):

$$PSNR = 10 \cdot \log_{10} \frac{\max\limits_{1 \le i \le N}(\left\|\boldsymbol{v}_i\right\|^2)}{MSE}; \tag{1.10}$$

(iii) Signal-to-noise ratio (SNR):

$$SNR = 10 \cdot \log_{10} \frac{\sum_{i=1}^{N}\left\|\boldsymbol{v}_i\right\|^2}{\sum_{i=1}^{N}\left\|\boldsymbol{v}'_i - \boldsymbol{v}_i\right\|^2}, \tag{1.11}$$

where N is the number of vertices, $\boldsymbol{v}_i$ and $\boldsymbol{v}'_i$ denote the i-th vertex of the original model $\boldsymbol{M}$ and the i-th vertex of the watermarked model $\boldsymbol{M}'$, respectively.

(3) Data capacity. Data capacity refers to the number of bits embedded in unit time or a product. For an image, data capacity refers to the number of bits embedded in this image. For audios, it refers to the number of bits embedded in one second of transmission. For videos, it refers to either the number of bits embedded in each frame, or that embedded in one second. A watermark encoded with N bits is called an N-bit watermark. Such a system can be used to embed 2^N different messages. Many situations require the detector to execute two-layer functions. The first one is to determine whether the watermark exists or not. If it exists, then continue to determine which one of the 2^N messages it is. This kind of detector has 2^N+1 possible output values, i.e., 2^N messages together with the case of "no watermark".

(4) Blind detection and informed detection. The detector that requires the original copy as an input is called an informed detector. This kind of detector also refers to the detector requiring only a small part of the original product information instead of the whole product. The detector that does not require the original product is called a blind detector. To use the blind or informed detector in watermarking systems determines whether it is suitable for some concrete applications. Non-blind detectors can only be used in those situations where the original product can be obtained.

(5) False positive probability. False positive refers to the case where watermarks can be detected in the product without watermarks. There are two definitions for this probability, and their difference lies in that the random variable is a watermark or a product. In the first definition, the false positive probability refers to the probability that the detector finds the watermark, given a product and several randomly selected watermarks. In the second definition, the false positive probability refers to the probability that the detector finds the watermark, given a watermark and several randomly selected products. In most applications, people are more interested in the second definition. But in a few applications, the first definition is also important. For example, in transaction tracking, false pirate accusation often appears when detecting a random watermark in the given product.

(6) Robustness. Robustness refers to the ability for the watermark to be detected if the watermarked product suffers some common signal processing operations, such as spatial filtering, lossy compression, printing and copying, geometry deformation (rotation, translation, scaling and others). In some cases, robustness is useless and even may be avoided. For example, another important research branch of watermarking, fragile watermarking, has an opposite characteristic of robustness. For example, the watermark for content authentication should be fragile, namely any signal processing operation will destroy the watermark. In another kind of extreme application, the watermark must be robust against any distortion that will not destroy the watermarked product.

The three commonly-used evaluation criteria for robustness are given as follows:

(i) Normalized correlation (NC). This criterion is used to quantitatively evaluate the similarity between the extracted watermark and the original watermark, especially for binary watermarks. When the watermarked media is distorted, the robust watermarking algorithm tries to make the NC value maximal, while the fragile watermarking algorithm tries to make the NC value minimal. The definition of NC is as follows:

$$NC(w,\hat{w}) = \frac{\sum_{i=1}^{N_w} w(i)\hat{w}(i)}{\sqrt{\sum_{i=1}^{N_w} w^2(i)}\sqrt{\sum_{i=1}^{N_w} \hat{w}^2(i)}}; \tag{1.12}$$

(ii) Normalized hamming distance (NHD). This criterion is used to quantitatively evaluate the difference between the extracted watermark and the original watermark, only for binary watermarks. The definition of NHD is as follows:

$$\rho = \frac{1}{N_w}\sum_{i=1}^{N_w} w(i) \oplus \hat{w}(i) ; \tag{1.13}$$

(iii) Peak signal-to-noise ratio (PSNR). This criterion is used to quantitatively evaluate the difference between the extracted gray-level watermark and the original gray-level watermark. Its definition is as follows:

$$PSNR = 10 \cdot \log_{10} \frac{w_{\max}^2}{\frac{1}{M \times N}\sum_{\forall (m,n)}[w(m,n) - \hat{w}(m,n)]^2}, \tag{1.14}$$

where N_w is the length of the watermark sequence, $w(i)$ and $\hat{w}(i)$ are the i-th value of the original watermark sequence and the i-th value of the extracted watermark respectively. $w(m,n)$ and $\hat{w}(m,n)$ are the original watermark image and the extracted watermark image respectively. $w_{\max}^2$ denotes the maximal watermark pixel value, and $M \times N$ is the size of the watermark image.

(7) Security. Security indicates the ability of watermarks to resist malicious attacks. The malicious attack refers to any behavior that destroys the function of watermarks. Attacks can be summarized into three categories: unauthorized removing, unauthorized embedding and unauthorized detection. Unauthorized removing and unauthorized embedding may change the watermarked products, and thus they are regarded as active attacks, while unauthorized detection does not change the watermarked products, and thus it is regarded as a passive attack. Unauthorized removing refers to making the watermark in products unable to be detected. Unauthorized embedding also means forgery, namely embedding illegal watermark information in products. Unauthorized detection can be divided into three levels. The most serious level is that the opponent detects and deciphers the embedded message. The second level is that the opponent detects watermarks and recognizes each mark, but he cannot decipher the meaning of these marks. The attack which is not serious is that the opponent can determine the existence of watermarks, but cannot decipher the message or recognize the embedded positions.

(8) Ciphers and watermarking keys. In modern cryptography systems, security depends only on keys instead of algorithms. People hope watermarking systems also have the same standard. In ideal cases, if the key is unknown, it is impossible to detect whether the product contains a watermark or not, even if the watermarking algorithm is known. Even if a part of the keys is known by the opponent, it is impossible to successfully remove the watermark on the precondition that the quality of the watermarked product is well maintained. Since the security of keys used in embedding and extraction is different from that provided in cryptography, two keys are usually used in watermarking systems.

One is used in encoding and the other is used in embedding. To distinguish these two keys, they are called the generation key and the embedding key, respectively.

(9) Content alteration and multiple watermarking. When a watermark is embedded in a product, the watermark transmitter may concern the watermark alteration problem. In some applications, the watermark should not be modified easily, but in some other situations, watermark alteration is necessary. In copy control, broadcast content will be marked with "copy once", and after being recorded, it will be labeled with "copy forbidden". Embedding multiple watermarks in a product is suitable for transaction tracking. Before being obtained by the final user, content is often transmitted by several middlemen. Copy mark first includes the watermark of the copyright owner. After that, the product may be distributed to some music websites. And each product copy may be embedded with a unique watermark to label each distributor's information. Finally, each website may embed the unique watermark to label the associated purchaser.

(10) Cost. It is very complex to economically consider the deploying of watermark embedders and detectors. It depends on the business mode involved. From the technical viewpoint, two main problems are the speed of watermark embedding and detection and the required number of embedders and detectors. Other problems may be whether the embedder and detector are implemented by hardware, software, or by a plug-in unit.

1.6 Overview of Multimedia Retrieval Techniques

Multimedia retrieval techniques include audio, images and video retrieval.

1.6.1 Concepts of Information Retrieval

Information retrieval (IR) [21] is the science of searching for documents, for information within documents and for metadata about documents, as well as that of searching relational databases and the World Wide Web. There is overlap in the usage of the terms data retrieval, document retrieval, information retrieval and text retrieval, but each also has its own body of literature, theory, praxis and technologies. IR is interdisciplinary, based on computer science, mathematics, library science, information science, information architecture, cognitive psychology, linguistics, statistics and physics. Automated information retrieval systems are used to reduce what has been called "information overload". Many universities and public libraries use IR systems to provide access to books, journals and other documents. Web search engines are the most visible IR applications.

The idea of using computers to search for relevant pieces of information was popularized in an article by Vannevar Bush in 1945 [21]. The first implementations of information retrieval systems were introduced in the 1950s and 1960s. By 1990 several different techniques had been shown to perform well on small text corpora (several thousand documents). In 1992 the US Department of Defense, along with the National Institute of Standards and Technology (NIST), co-sponsored the Text Retrieval Conference (TREC) as part of the TIPSTER text program. The aim of this was to look into the information retrieval community by supplying the infrastructure that was needed for evaluation of text retrieval methodologies on a very large text collection. This catalyzed the research into methods that scale to huge corpora. The introduction of web search engines has boosted the need for very large scale retrieval systems even further. The use of digital methods for storing and retrieving information has led to the phenomenon of digital obsolescence, where a digital resource ceases to be readable because the physical media (The reader is required to read the media), the hardware, or the software that runs on it, is no longer available. The information is initially easier to retrieve than if it were on paper, but is then effectively lost.

An information retrieval process begins when a user enters a query into the system. Queries are formal statements of information needs, for example search strings in web search engines. In information retrieval a query does not uniquely identify a single object in the collection. Instead, several objects may match the query, perhaps with different degrees of relevancy. An object is an entity which keeps or stores information in a database. User queries are matched to objects stored in the database. Depending on the application of the data, objects may be, for example, text documents, images or videos. Often the documents themselves are not kept or stored directly in the IR system, but are instead represented in the system by document surrogates. Most IR systems compute a numeric score on how well each object in the database matches the query, and rank the objects according to this value. The top ranking objects are then shown to the user. The process may then be iterated if the user wishes to refine the query.

According to the objects of IR, the techniques used in IR can be classified into three categories: literature retrieval, data retrieval and document retrieval. The main difference between these types of information retrieval systems lies in the following: Data retrieval and document retrieval are required to retrieve the information itself in the literature, while literature retrieval is only required to retrieve the literature including the input information. According to the search means, information retrieval systems can be classified into three categories: manual retrieval systems, mechanical retrieval systems and computer-based retrieval systems. At present, the rapidly developing computer-based retrieval is "network information retrieval", which stands for the behavior of web users to search required information over the Internet with specific network-based searching tools or simple browsing manners. Information retrieval methods can be also classified into direct retrieval and indirect retrieval methods. Currently, the research hotspots in the domain of IR lie in the following three areas.

(1) Knowledge retrieval or intelligent retrieval. Knowledge retrieval (KR) [22] is a field of study which seeks to return information in a structured form, consistent with human cognitive processes as opposed to simple lists of data items. It draws on a range of fields including epistemology (theory of knowledge), cognitive psychology, cognitive neuroscience, logic and inference, machine

learning and knowledge discovery, linguistics, information technology, etc. In the field of retrieval systems, the established approaches include data retrieval systems (DRS), such as database management systems, which are well suitable for the storage and retrieval of structured data, and information retrieval systems (IRS), such as web search engines, which are very effective in finding the relevant documents or web pages that contain the information required by a user. These approaches both require a user to read and often analyze long lists of datasets or documents in order to extract the meaning implicit in them. The goal of knowledge retrieval systems is to reduce the burden of those processes by improved search and representation. This improvement is seen as needed to handle the increasing volumes of data available on the World Wide Web and elsewhere.

KR focuses on the knowledge level. We need to examine how to extract, represent and use the knowledge in data and information. Knowledge retrieval systems provide knowledge for users in a structured way. They are different from data retrieval systems and information retrieval systems in inference models, retrieval methods, result organization, etc. The cores of data retrieval and information retrieval are retrieval subsystems. Data retrieval gets results through Boolean match. Information retrieval uses partial match and best match. KR is also based on partial match and best match. Considering the inference perspective, data retrieval uses deductive inference, and information retrieval uses inductive inference. Considering the limitations from the assumptions of different logics, traditional logic systems cannot make efficient reasoning in a reasonable time. Associative reasoning, analogical reasoning and the idea of unifying reasoning and search may be effective methods of reasoning on the web scale. From the retrieval model perspective, KR systems focus on semantics and better organization of information. Data retrieval and information retrieval organize the data and documents by indexing, while KR organizes information by indicating connections between elements in those documents.

(2) Knowledge mining. Over the past several years, the field of data mining has been rapidly expanding and attracting many new researchers and users. The underlying reason for such a rapid growth is a great need for systems that can automatically derive useful knowledge from the vast volumes of computer data being accumulated worldwide. The field of data mining offers a promise for addressing this need. The major trust of research has been to develop a repertoire of tools for discovering both strong and useful patterns in large databases. The function performed by such tools can be succinctly characterized as a mapping from DATA to PATTERNS. An underlying assumption is that the patterns are created solely from the data, and thus are expressed in terms of attributes and relations appearing in the data. Determining such patterns can be a problem of significant computational complexity, but of a relatively low conceptual complexity, and many efficient algorithms have been developed for this purpose. This approach to the problem of deriving useful knowledge from databases has, however, some fundamental limitations, and new research should address several important tasks. The first task is to integrate a knowledge base within a data mining system, and to develop methods for applying this knowledge during data mining. The second one is to use advanced knowledge representations and be able

to generate many different types of knowledge from a given data source. To address the research direction that aims at achieving all the above-mentioned tasks, we use the term knowledge mining. Knowledge mining [23] can be characterized as concerned with developing and integrating a wide range of data analysis methods that are able to derive directly or incrementally new knowledge from large (or small) volumes of data using relevant prior knowledge. The process of deriving new knowledge has to be guided by criteria inputted to the system defining the type of knowledge a particular user is interested in. Algorithms for generating new knowledge must be not only efficient but also oriented toward producing knowledge satisfying the comprehensibility postulate. This means it must be easy to be understood and interpreted by the users. Knowledge mining can be simply characterized by the mapping from DATA + PRIOR_ KNOWLEDGE + GOAL to NEW_KNOWLEDGE, where GOAL is an encoding of the knowledge needs of the user(s), and NEW_KNOWLEDGE is knowledge satisfying the GOAL. Such knowledge can be in the form of decision rules, association rules, decision trees, conceptual or similarity-based clusters, equations, Bayesian nets, statistical summaries, visualizations, natural language summaries, or other knowledge representations.

(3) Heterogeneous information retrieval. The terms, "parallel", "distributed", "heterogeneity", etc., were really popular in 1990s' computer science research projects and papers. Nowadays those technologies, developed during those years, are actually used and improved. Papers explicitly on those technologies do not appear as frequently as before, but those topics are still present. Ranging from the simple network of a workstation to the more modern and complex grid systems, the adoption of distributed systems instead of massively parallel supercomputers has been preferred due to their reduced cost of ownership. These kinds of systems pose many challenges in terms of information access, storage and retrieval. Usually, in fact, instead of having collections stored at a single site, they are collected, and sometimes managed, at different sites (possibly owned by different institutions). Particular interest is usually expressed in architectures and specifications for information retrieval in the context of heterogeneous distributed computing systems. Under these circumstances, the information retrieval system should be more and more highly open and integrated. The system should be able to search for and integrate the information from different sources and/or with different structures. For example, it should support files with different formats, such as TEXT, HTML, XML, RTF, MS Office, PDF, PS2/PS, MARC and ISO2709, and it should support the retrieval using multiple languages and the uniform processing of structured, semi-structured and non-structured data. It is also required to be seamlessly integrated with the retrieval on relational databases.

1.6.2 Summary of Content-Based Multimedia Retrieval

The growth in the Internet and multimedia technologies brings a huge sea of multimedia information, resulting in very huge multimedia databases, and thus we can hardly describe and search for the multimedia information only by keywords. Therefore, we need an effective retrieval scheme for multimedia. How to help people find the required multimedia information fast and accurately is the key problem to be solved for multimedia information systems.

From the birth of information retrieval in the 1950s to the emergence of multimedia information retrieval in the 1990s, the information retrieval research area has undergone great changes and development, and three stages are traditional text-based information retrieval, current content-based multimedia retrieval and future web-based multimedia retrieval.

Content-based retrieval is a new kind of retrieval technology, which retrieves objects and semantics in multimedia. This technique involves extracting color and texture information in images or scenes and clips in videos, and then performing similarity matching based on these features. Content-based retrieval systems can perform retrieval based on not only discrete media represented by text information but also continuous media represented by images and audio. Content-based multimedia retrieval is a booming research field, and it is at the stage of research and survey. At present, there exist the problems of low processing speed, high false positive and false negative rates, no evaluation criteria for retrieval results and lack of query support for multimedia. On the other hand, with the increase in multimedia content and the improvement in storage technologies, the need for content-based multimedia retrieval techniques will be more and more urgent. Fig. 1.10 describes the academic concerns for content-based multimedia retrieval from the mid-1990's to the 21st century. We can see that researchers are paying more and more attention to this field.

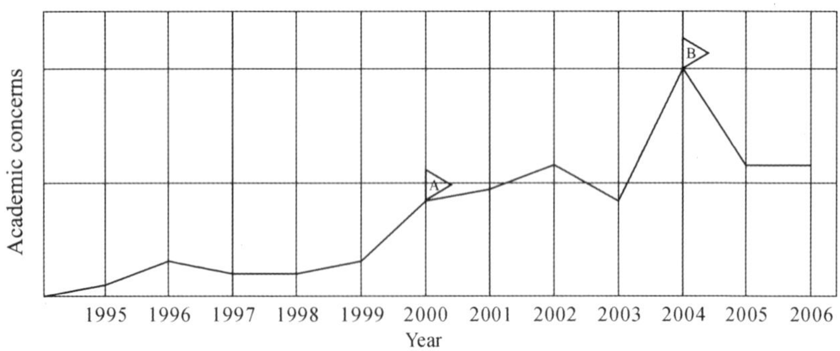

Fig. 1.10. The academic concerns for multimedia information retrieval

According to which kind of media is concerned, content-based multimedia retrieval techniques can be classified into content-based image retrieval, content-based video retrieval, content-based audio retrieval, content-based 3D model retrieval, etc. The following subsections focus on the first three kinds of media, while the fourth one will be discussed in detail in Chapter 4.

1.6.3 Content-Based Image Retrieval

Content-based image retrieval (CBIR) [24] is the application of computer vision to the image retrieval problem, meaning the problem of searching for digital images in large databases. "Content-based" means that the search will analyze the actual contents of the image. The term "content" in this context might refer to colors, shapes, textures, or any other information that can be derived from the image itself. Without the ability to examine image content, searches must rely on metadata such as captions or keywords, which may be laborious or expensive to produce. The term CBIR seems to have originated in 1992, when it was used by Kato to describe experiments into automatic retrieval of images from a database, based on the colors and shapes present. Since then, the term has been used to describe the process of retrieving desired images from a large collection on the basis of syntactical image features. The techniques, tools and algorithms that are used in CBIR originate from fields such as statistics, pattern recognition, signal processing and computer vision.

There is a growing interest in CBIR because of the limitations inherent in metadata-based systems, as well as the large range of possible uses for efficient image retrieval. Textual information about images can be easily searched using existing technologies, but requires people to personally describe every image in the databases. This is impractical for very large databases, or for images that are generated automatically, e.g. from surveillance cameras. It is also possible to miss images that use different synonyms in their descriptions. Systems based on categorizing images in semantic classes like "cat" as a subclass of "animal" can avoid this problem but still face the same scaling issues. Potential uses of CBIR include art collections, photographic archives, retail catalogs, medical diagnosis, crime prevention, military information, intellectual property, architectural and engineering design, geographical information and remote sensing systems. Different implementations of CBIR make use of different types of user queries as follows.

(1) Query by example. Query by example is a query technique that involves providing the CBIR system with an example image that it will then base its search upon. The underlying search algorithms may vary depending on the application, but result images should all share common elements with the provided example. Options for providing example images for the system include: 1) A pre-existing image may be supplied by the user or chosen from a random set. 2) The user draws a rough approximation of the image they are looking for, for example with blobs of color or general shapes. This query technique removes the difficulties that can arise when trying to describe images with words.

(2) Semantic retrieval. The ideal CBIR system from a user perspective would involve what is referred to as semantic retrieval, where the user makes a request like "find pictures of dogs" or even "find pictures of Abraham Lincoln". This type of open-ended task is very difficult for computers to perform, for pictures of Chihuahuas and Great Danes look very different, and Lincoln may not always be facing the camera or in the same pose. Current CBIR systems therefore generally

make use of lower-level features like texture, colors and shapes, although some systems take advantage of very common higher-level features like faces. Not every CBIR system is generic. Some systems are designed for a specific domain, e.g. shape-matching can be used for finding parts inside a CAD-CAM database.

(3) Other query methods. Other query methods include browsing for example images, navigating customized/hierarchical categories, querying by image regions (rather than the entire image), querying by multiple example images, querying by visual sketches, querying by direct specification of image features, and multimodal queries (e.g. combining touch, voice, etc.).

CBIR systems can also make use of relevance feedback, where the user progressively refines the search results by marking images in the results as "relevant", "not relevant", or "neutral" to the search query, then repeating the search with the new information. The following are some commonly-used features for CBIR.

(1) Color. Retrieving images based on color similarity is achieved by computing a color histogram for each image that identifies the proportion of pixels within an image holding specific values. Current research is attempting to segment color proportion by region and by spatial relationships among several color regions. Examining images based on the colors they contain is one of the most widely-used techniques because it does not depend on image sizes or orientations. Color searches will usually involve comparing color histograms, though this is not the only technique in practice.

(2) Texture. Texture measures look for visual patterns in images and how they are spatially defined. Textures are represented by texels which are then placed into a number of sets, depending on how many textures are detected in the image. These sets not only define the texture, but also where the texture is located in the image. Texture is a difficult concept to represent. The identification of specific textures in an image is achieved primarily by modeling texture as a 2D gray level variation. The relative brightness of pairs of pixels is computed such that the degree of contrast, regularity, coarseness and directionality may be estimated. However, the problem is in identifying patterns of co-pixel variation and associating them with particular classes of textures such as "silky" or "rough".

(3) Shape. Shape does not refer to the shape of an image but to the shape of a particular region that is being sought out. Shapes will often be determined by first applying segmentation or edge detection to an image. Other methods use shape filters to identify given shapes of an image. In some cases accurate shape detection will require human intervention because methods like segmentation are very difficult to completely automate.

CBIR belongs to the image analysis research area. Image analysis is a typical domain for which a high degree of abstraction from low-level methods is required, and where the semantic gap immediately affects the user. If image content is to be identified to understand the meaning of an image, the only available independent information is the low-level pixel data. Textual annotations always depend on the knowledge, capability of expression and specific language of the annotator and therefore are unreliable. To recognize the displayed scenes from the raw data of an image the algorithms for selection and manipulation of pixels must be combined and parameterized in an adequate manner and finally linked with the natural description. Even the simple linguistic representation of shape or color, such as round or yellow, requires entirely different mathematical formalization methods, which are neither intuitive nor unique and sound. The above description involves the concept of semantic gap. The semantic gap characterizes the difference between two descriptions of an object by different linguistic representations, for instance, languages or symbols. In computer science, the concept is relevant whenever ordinary human activities, observations and tasks are transferred into a computational representation. More precisely, the gap means the difference between ambiguous formulation of contextual knowledge in a powerful language (e.g. natural language) and its sound, reproducible and computational representation in a formal language (e.g. programming language). The semantics of an object depends on the context it is regarded within. For practical applications, this means any formal representation of real world tasks requires the translation of the contextual expert knowledge of an application (high-level) into the elementary and reproducible operations of a computing machine (low-level). Since natural language allows the expression of tasks which are impossible to compute in a formal language, there is no way to automate this translation in a general way. Moreover, the examination of languages within the Chomsky hierarchy indicates that there is no formal and consequently automated way of translating from one language into another above a certain level of expressional power.

The following are some famous CBIR systems.

(1) QBIC. The earliest CBIR system is the QBIC (query by image content) system, which was developed by IBM Almaden. The QBIC lets you make queries of large image databases based on visual image content, i.e., properties such as color percentages, color layout, and textures occurring in the images. Such queries utilize the visual properties of images, so you can match colors, textures and their positions without describing them in words. Content-based queries are often combined with text and keyword predicates to get powerful retrieval methods for image and multimedia databases.

(2) PhotoBook. PhotoBook is a Facebook photo browser for Mac developed by the MIT Media Lab. It makes it easy and fun to manage, share and view your friends' Facebook photos in one intuitive interface. The key features are: 1) Viewing photos of friends or albums on a single page; 2) Quickly viewing photos with tags and other information all in the same window; 3) Watching slideshows with amazing transitions; 4) Importing photos or entire albums into iPhoto with one click; 5) Filtering through photos or albums instantly with as-you-type search.

(3) VisualSEEK. VisualSEEK is a fully automated content-based image query system developed by Columbia University. VisualSEEk is distinct from other content-based image query systems in that the user may query for images using both the visual properties of regions and their spatial layout. Furthermore, the image analysis for region extraction is fully automated. VisualSEEk uses a novel system for region extraction and representation based upon color sets. Through a process of color set back-projection, the system automatically extracts salient color regions from images.

(4) Other CBIR systems. Some other famous CBIR systems are the MARS

system developed by the University of Illinois at Urbana-Champaign, the Digital Library Project of the University of California, Berkeley, the Retrieval Ware system developed by the Excalibur Technology Corporation and the Virage system developed by the Virage Logic Corporation.

1.6.4 Content-Based Video Retrieval

With technology advances in multimedia, digital TV and information highways, a large amount of video data is now publicly available. However, without an appropriate search technique, all these data are almost unusable. Users are not satisfied with the video retrieval systems that provide analogue VCR (video cassette recording) functionality. They want to query the content instead of raw video data. For example, a user will ask for a specific part of the video, which contains some semantic information. Content-based search and retrieval of these data becomes a challenging and important problem. Therefore, the need for tools that can manipulate the video content in the same way as traditional databases managing numeric and textual data is significant.

1.6.4.1 Basic Concepts and Frameworks

A typical content-based video retrieval (CBVR) [25] is shown in Fig. 1.11. First, we should analyze the video structure and segment the video into shots, and then we select keyframes in each shot, which is the basis and key problem of a highly efficient CBVR system. Second, we extract the motion features from each shot and the visual features from the keyframes in this shot, and store these two kinds of features as a retrieval mechanism in the video database. Finally, we return the retrieval results to users based on their queries according to the similarities between features. If the user is not satisfied with the search results, the system can optimize the retrieval results according to the users' feedback.

1.6.4.2 Video Structure and Related Algorithms

To perform content-based search on video databases, we should first construct a video structure for retrieval. Video data can be divided, from coarse to fine, into four levels: videos, scenes, shots and frames. Frames, shots, scenes, and sequences form a hierarchy of units fundamental to many tasks in the creation of moving-image works. In film, a shot is a continuous strip of motion picture film, composed of a series of frames, which runs for an uninterrupted period of time. Shots are generally filmed with a single camera and can be of any duration. There are several film transitions usually used in film editing to juxtapose adjacent shots. In the context of shot transition detection they are usually grouped into two types:

(1) Abrupt transitions. This is a sudden transition from one shot to another; i.e., one frame belongs to the first shot, and the next frame belongs to the second shot. They are also known as hard cuts or simple cuts. (2) Gradual transitions. In this kind of transition the two shots are combined using chromatic, spatial or spatial-chromatic effects which gradually replace one shot by another. These are also often known as soft transitions and can be of various types, e.g., wipes, dissolves, fades, and so on.

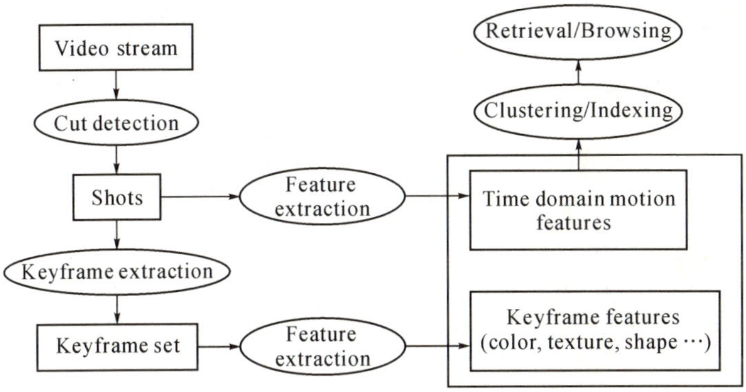

Fig. 1.11. Diagram of the content-based video retrieval system

The entire process of constructing the video structure can be divided into the following three steps: extracting the video shots from the camera, selecting the key frames from the shots and constructing the scenes or groups from the video stream.

(1) Extracting the video shots from the camera (i.e., shot detection). A shot is the basic unit of video data. The first task in video processing or content-based video retrieval is to automatically segment the video into shots and use them as fundamental indexing units. This process is called shot boundary detection. In shot detection, the abrupt transition detection is the keystone, and the related algorithms and ideas can be used in other steps; therefore it is a focus of attention. The main schemes for abrupt transition detection are as follows: 1) color-feature-based methods, such as template matching (sum of absolute differences) and histogram-difference-based schemes; 2) edge-based methods; 3) optical-flow detection-based methods; 4) compressed-domain-based methods; 5) the double-threshold-based method; 6) the sliding window detection method; 7) the dual-window method.

(2) Selecting the keyframes from the shots. A keyframe is a frame that represents the content of a shot or scene. This content must be as representative as possible. In the large amount of video data, we first reduce each video to a set of representative key frames (Though we enrich our representations with shot-level motion-based descriptors as well). In practice, often the first frame or center frame of a shot is chosen, which causes information loss in the case of long shots containing considerable zooming and panning. This is why unsupervised approaches have been suggested that provide multiple key frames per shot. Since

for online videos the structure varies strongly, we use a two-step approach that delivers multiple key frames per shot in an efficient way by following shot boundary detection based on a "divide and conquer" strategy, for which reliable standard techniques exist, which is used to divide keyframe extraction into shot-level sub-problems that are solved separately. Keyframe selection methods can be divided into the following categories: 1) Methods based on the shots. A video clip is first segmented into several shots, and then the first (or last) frame in each shot is viewed as the keyframe. 2) Content-based analysis. This method is based on the change in color, texture and other visual information of each frame to extract the keyframe. When the information changes significantly, the current frame is viewed as a keyframe. 3) Motion-analysis-based methods. 4) Clustering-based methods.

(3) Constructing the scenes or groups from the video stream. First we calculate the similarity between the shots (in fact, the key frames), and then select the appropriate clustering algorithm for analysis. According to the chronological order and the similarity between key frames, we can divide the video stream into scenes, or we can perform the grouping operation only according to the similarity between key frames.

1.6.4.3 Feature Extraction

Various high-level semantic features, concepts such as indoor/outdoor, people and speech, occur frequently in video databases. To date, techniques for video retrieval are mostly extended directly or indirectly from image retrieval techniques. Examples include first selecting key frames from shots and then extracting image features such as color and texture features from those key frames for indexing and retrieval. The success from such an extension, however, is doubtful since the spatio-temporal relationship among video frames is not fully exploited. Motion features that have been used for retrieval include the motion trajectories and motion trails of objects, principle components of MPEG motion vectors and temporal texture. Motion trajectories and trails are used to describe the spatio-temporal relationship of moving objects across time. The relationship can be indexed as 2D or 3D strings to support spatio-temporal search. Principal components are utilized to summarize the motion information in a sequence as several major modes of motion. Temporal textures are employed to model more complex dynamic motion such as the motion of a river, swimming and crowds. An important issue needing to be addressed is the decomposition of camera and object motion prior to feature extraction. Ideally, to fully explore the spatio-temporal relationship in videos, both camera and object motion need to be fully exploited in order to index the foreground and background information separately. Motion segmentation is required, especially when the targets of retrieval are objects of interest. In such applications, camera motion is normally canceled by global motion compensation and foreground objects are segmented by inter-frame subtraction. However, such a task always turns out to be difficult, and most importantly, poor segmentation will always lead to poor retrieval results. Although the motion decomposition is a preferable step prior to the feature extraction of most videos, it may not be necessary for certain videos. If we imagine a camera as a narrative eye, the movement of the eye tells us not only what is to be seen but also the different ways of observing events. Typical examples include sport events that are captured by cameras, which are mounted at fixed locations in a stand. These camera motions are mostly regular and driven by the pace of games and the type of events that are taking place. For these videos, camera motion is always an essential cue for retrieval. Furthermore, fixed motion patterns can always be observed when camera motions are coupled with the object motion of a particular event.

1.6.4.4 Video Retrieval and Browsing

After the keyframe extraction process and the feature extraction operation on keyframes, we need to index video clips based on their characteristics. Through the index, you can use the keyframe-based features or the motion features of the shots, or a combination of both for the video search and browsing. Content-based retrieval is a kind of approximate match, a cycle of stepwise refinement processes, including initial query description, similarity matching, the return of results, the adjustment of features, human-computer interaction, retrieval feedback, and so on, until the results satisfy the customers. The richness and complexity of video content, as well as the subjective evaluation of video content, make it difficult to evaluate the retrieval performance with a uniform standard. This is also a research direction of CBVR. Currently, there are two commonly used criteria, recall and precision, which are defined as:

$$recall = \frac{correct}{correct + missed}, \tag{1.15}$$

$$precision = \frac{correct}{correct + falsepositive}, \tag{1.16}$$

where "*correct*" means the number of correctly detected video clips/shots, "*missed*" is the number of missed video clips/shots, "*falsepositive*" means the number of falsely detected video clips/shots. The following are some typical techniques related to the video retrieval process.

(1) Keyframe-based retrieval. After the keyframes are extracted from the video, the search turns to the process of searching similar keyframes in the database to the query keyframes. The commonly-used query methods are object-feature-description-based queries and visual-sample-based queries. During the retrieval process, users can designate the specific set of features. If a keyframe is returned, users can browse the video clip that is represented by this keyframe. The browsing process can follow the retrieval process to serve as the context connection among retrieved keyframes. Browsing can also be used to initialize a query, so that during the browsing process users can select an image to search for all keyframes that are similar to it.

(2) Shot-motion-based retrieval. To retrieve the shots based on the motion features of shots and main objects is a further requirement of video query. We can use the representations of camera operations to retrieve shots, and use the motion features (directions and scopes) to retrieve moved objects. In the query, we can also combine motion features and keyframe features to retrieve the shots with similar dynamic features but different static features compared to the query.

(3) Video-browsing. For videos, browsing and retrieval with a definite goal are equally important. Browsing requires that the video be described at the semantic level. Some scholars have put forward a concept called scene transition graph (STG), where a node in the directed graph denotes a scene, while the edge stands for the transition in time. Through the simplification of the STG model, we can remove some unimportant shots, resulting in the compact representation of the video. Because it is very difficult to obtain semantic information purely from the images, some scholars have suggested a combination of video images, voice and text information.

(4) Relevance feedback. Several relevance feedback (RF) algorithms have been proposed over the last few years. The idea behind most RF-models is that the distance between image/video shots labeled as relevant and other similar image/video shots in the database should be minimal. The key factor here is that the human visual system does not follow any mathematic metric when looking for similarity in visual content and that the distances used in image/video retrieval systems are well-defined metrics in a feature space.

1.6.5 Content-Based Audio Retrieval

Much previous audio analysis and processing of research was related to speech signal processing, e.g., speech recognition. It is easy for machines to automatically identify isolated words, as used in dictation and telephone applications, while it is relatively hard for machines to perform continuous speech recognition. But recently some breakthrough has been made in this area, and at the same time research into speaker identification has also been carried out. All these advances will provide audio information retrieval systems that are of great help.

1.6.5.1 Some Concepts of Digital Audio

Audio is the important media in multimedia. The frequency range of audio that we can hear is from 60 Hz to 20 kHz, and the speech frequency range is from 300 Hz to 4 kHz, while music and other natural sounds are within the full range of audio frequency. The audio that we can hear is first recorded or regenerated by analog recording equipment, and then digitized into digital audio. During digitalization, the sampling rate must be larger than twice the signal bandwidth in order to correctly restore the signal. Each sample can be represented with 8 or 16 bits.

Audio can be classified into three categories: (1) Waveform sound. We perform the digitization operation on the analog sound to obtain the digital audio signals. It can represent the voice, music, natural and synthetic sounds. (2) Speech. It possesses morphemes such as words and grammars, and it is a kind of highly abstract media for concept communication. Speech can be converted to text through recognition, and text is the script form of speech. (3) Music. It possesses elements such as rhythm, melody or harmony, and it is a kind of sound composed of the human voice and/or sounds from musical instruments.

Overall, the audio content can be divided into three levels: the lowest level of physical samples, the middle level of acoustic characteristics and the most senior level of semantics. From lower levels to higher levels, the content becomes more and more abstract. In the level of physical samples, the audio content is represented in the form of streaming media, and users can retrieve or call the audio data according to the time scale, e.g., the common audio playback API. The middle level is the level of acoustic characteristics. Acoustic characteristics are extracted from audio data automatically. Some auditory features representing users' perception of audio can be used directly for retrieval, and some features can be used for speech recognition or detection, supporting the representation for higher level content. In addition, the space-time structure of audio can also be used. The semantic level is the highest level, i.e., the concept level of representing audio content and objects. Specifically, at this level, the audio content is the result of recognition, detection and identification, or the description of music rhythms, as well as the description of audio objects and concepts. The latter two levels are the most concerned with content-based audio retrieval. In these two levels, the user can submit a concept query or perform the query by auditory perception.

1.6.5.2 Overview of Content-Based Audio Retrieval

Conventional information retrieval research is based mainly on the text, for example, the Yahoo! and AltaVista search engines that we have become very familiar with. The classic IR problem is to use the query text composed of a set of keywords to locate the text documents we need. If a document contains many query items, then it is considered as "more relevant" than any other document that contains fewer query items. Thus, the returned documents can be sorted according to their "relevant" degrees and displayed to users for further search. Although this general process of IR is designed for text, apparently it can be also applied to audio or other multimedia information retrieval. If we view the digital audio as a non-transparent bitstream, although we can give the attributes such as names, file formats and sampling rates, none of them can be identified by words or comparable entities. Therefore, we cannot search the audio content as we can do in text retrieval systems.

As mentioned earlier, CBIR systems should extract color, texture, shape and other features, while CBVR systems should extract the keyframe features. Similarly, content-based audio retrieval (CBAR) [26] should extract the auditory features from audio data. Audio features can be classified into the perceptual auditory features and non-perceptual auditory features (physical characteristics).

The perceptual auditory features include volume, tone and intensity. With respect to speech recognition, IBM's Via Voice has become more and more mature, and the VMR system of the University of Cambridge and Carnegie Mellon University's Informedia are both very good audio processing systems. With respect to content-based audio information retrieval, Muscle Fish of the United States has introduced a prototype of a more comprehensive system for audio retrieval and classification with a high accuracy.

With respect to the query interface, users can adopt the following query types: (1) Query by example. Users choose audio examples to express their queries, searching all sounds similar to the characteristics of query audio, for example, to search for all sounds similar to the roar of aircraft. (2) Simile. A number of acoustic/perceptual features are selected to describe the query, such as loudness, tone and volume. This scheme is similar to the visual query in CBIR or CBVR. (3) Onomatopoeia. We can describe our queries by uttering the sound similar to the sounds we would like to search for. For example, we can search for the bees' hum or electrical noise by uttering buzzes. (4) Subjective features. That means the sound is described by individuals. This method requires training the system to understand the meaning of these terms. For example, the user may search "happy" sounds in the database. (5) Browsing. This is an important means of information discovery, especially for such time-base audio media. Besides the browsing based on pre-classification, it is more important to browse based on the audio structure.

According to the classification of audio media, we know that speech, music and other sound possess significantly different characteristics, so current CBAR approaches can be divided into three categories: retrieval of "speech" audio, retrieval of "non-speech non-music" audio and retrieval of "music" audio. In other words, the first one is mainly based on automatic speech recognition technologies, and the latter two are based on more general audio analysis to suit a wider range of audio media, such as music and sound effects, also including digital speech signals of course. Thus, CBAR can be divided into the following three areas, sound retrieval, speech retrieval and music retrieval.

1.6.5.3 Sound Retrieval

As the use of sounds for computer interfaces, electronic equipment and multimedia contents has increased, the role of sound design tools has become more and more important. In sound retrieval, picking one sound out from huge data is troublesome for users because of the difficulty of simultaneously listening to plural sounds. Consequently, an efficient retrieval method is required for sound databases. Few search engines allow users to search for the Internet with sounds as query inputs. However, users could benefit from the ability to have direct access to these media, which contain rich information but cannot be precisely described in words. It is both challenging and desirable to be able to retrieve sound files relevant to users' interests by searching the Internet. Unlike the traditional way of using keywords as input to search for web pages with relevant texts, query example can be used as input to search for similar sound files. Content-based technology has been applied to automatically retrieve sounds similar to the query-example. Features from time, frequency and coefficients domains are firstly extracted from each sound file. Next, Euclidean distances between the vectors of query and sample audios are measured. An ascending distance list is given as retrieval results.

Feature extraction is the first step towards content-based retrieval. We can extract features from time, frequency and coefficient domains and combine them to form a feature vector for each audio file in the database. Traditional sound retrieval methods have used acoustic features, for example, pitch, harmonicity, loudness, brightness, and spectral peaks, audio databases indexed by using neural nets, etc. These methods have adopted automatic indexing approaches, and have obtained some satisfying results. However, whether the retrieval method is convenient for users has not been verified. By developing the most effective and easy retrieval for users, anyone, even novice users, will be able to intuitively and effectively retrieve the sound regardless of the retrieval situation (whether the user has a concrete idea for the sound or not). After feature extraction, we normalize the feature values across the whole database. Normalization can ensure that contributions of all audio feature elements are adequately represented. The magnitudes of the feature element values are more uniform after normalization and this will prevent a particular feature from dominating the whole feature vector.

When a user inputs a query audio file and requests finding relevant files to the query, both the query and each document in the database are represented as feature vectors. A measure of the similarity between the two vectors is computed, and then a list of files based on the similarity is fed back to the user for listening and browsing. The user may also refine the query to get more audio material relevant to his or her interest by relevant feedback. Users may input at least one type of keyword for retrieval. The system uses each keyword to calculate retrieval points that are dependent on the similarity between the input keyword and the labeled keyword. Retrieval points are calculated for each sound, and then the sounds are preferentially exhibited according to total points.

(1) Retrieval by onomatopoeia. Onomatopoeia is frequently used to specify a sound, mostly as an adverb in Japanese. There is a great variety of onomatopoeias, and one sound can be expressed by different onomatopoeias. Thus, a simple keyword-matching method is insufficient to cope with these variations of onomatopoeia. Onomatopoeia can be treated as a combination of syllables. First, the system retrieves the labeled keywords with the input keyword itself, then by varied keywords composed by cutting one syllable from an input keyword. Retrieval points (0–10 points) are given for each sound, depending on the similarity between the input keyword and the labeled keyword. Here we require a technique for matching two character string values by comparing their phonic sounds, which will be useful for evaluating similarities to English onomatopoeia.

(2) Retrieval by source. The system retrieves the labeled keywords with the input keyword by simple keyword matching. When the input keyword is retrieved in the label, 10 points are given, if no 0 point is given for each sound data.

(3) Retrieval by adjective. This scheme uses adjectives for sound retrieval, and the similarities of these adjectives are analyzed by cluster analysis. A user may

select the keyword from adjectives on retrieval. The adjective values, which are determined for the retrieval keyword, are set to a retrieval point for each sound. This means more retrieval points are given for a sound that is more generally associated with the input adjective.

1.6.5.4 Speech Retrieval

Speech search [27] is concerned with the retrieval of spoken content from collections of speech or multimedia data. The key challenges raised by speech search are indexing via an appropriate process of speech recognition and efficiently accessing specific content elements within spoken data. The specific limitations of speech recognition in terms of vocabulary and word accuracy mean that effective speech search often does not reduce to an application of information retrieval to speech recognition transcripts. Although text information retrieval techniques are clearly helpful, speech retrieval involves confronting issues less apt to arise in the text domain, such as high levels of noise in the indexed data and lack of a clearly defined unit of retrieval. A speech retrieval system accepts vague queries and it performs best-match searches to find speech recordings that are likely to be relevant to the queries. Efficient best-match searches require that the speech recordings be indexed in a previous step. People focus on effective automatic indexing methods that are based on automatic speech recognition. Automatic indexing of speech recordings is a difficult task for several reasons. One main reason is the limited size of vocabularies of speech recognition systems, which are at least one order of magnitude smaller than the indexing vocabularies of text retrieval systems. Another main problem is the deterioration of the retrieval effectiveness due to speech recognition errors that invariably occur when speech recordings are converted into sequences of language units (e.g. words or phonemes).

1.6.5.5 Music Retrieval

The advancement of media computing technology has made the production, storage, transmission and playback of audio-visual information progressively easier. It is very convenient today to purchase and download music from music shopping websites. It can therefore be safely predicted that the size of music databases will rapidly be growing very large. However, without effective and efficient methods of accessing music databases, people could easily get swamped by the huge amount of music information available. The important and traditionally effective way for accessing the music is by the text labels attached to the music data, such as the name of singers or composers, title of the song or music album. But sometimes the text labels might not be characteristic of the piece or may not be remembered by users, and there is a need for accessing the music based on its intrinsic musical content, such as its melody, which is usually more characteristic as well as intuitive than the text labels.

Humming a tune is by far the most straightforward and natural way for normal users to make a melody query. Thus music query-by-humming has attracted much research interest recently. It is a challenging problem since the humming query inevitably contains tremendous variation and inaccuracy. And when the hummed tune corresponds to some arbitrary part in the middle of a melody and is rendered at an unknown speed, the problem becomes even tougher. This is because exhaustive search of location and humming speeds is computationally prohibitive for a feasible music retrieval system. The efficiency of retrieval becomes a key issue when the database is very large. Based on the types of features used for melody representation and matching methods, the past works on query-by-humming can be broadly classified into three categories [28]: the string-matching approach, the beat alignment approach and time-series-matching approach. In the string matching approach, a hummed query is translated into a series of musical notes. The note differences between adjacent notes are then represented by letters or symbols according to the directions and/or the quantity of the differences. The hummed query is thus represented by a string. In the database, the notes of the MIDI music are also translated into strings in the same manner. The retrieval is done by approximate string matching. String edit distance is used for similarity measure. There are many limitations to this approach. It requires precise identification of each note's onset, offset and note values. Any inaccuracies of note articulation in the humming can lead to a large number of wrong notes detected and can result in a poor retrieval accuracy. In the beat alignment approach for query-by-humming, the user expresses the hummed query according to a metronome, by which the hummed tune can be aligned with the notes of the MIDI music clips in the database. Since the timing/speed of humming is controlled, the errors in humming can only come from the pitch/note values and alignment is not affected. By computing the statistical information of the notes in a fixed number of beats, a histogram based feature vector is constructed and used to match the feature vectors for the MIDI music clip database. However, humming with a metronome is a rather restrictive condition for normal use. Many people usually are not very discriminating when it comes to their awareness of the beat of a melody. Different meters (e.g. duple, triple, quadruple meters) of the music can also contribute to the difficulties. In the pitch time-series-matching approaches, a melody is represented by a time series of pitch values. Time-warping distance is used for a similarity metric between the time series. However, current methods have an efficiency problem, especially for matching anywhere in the middle of melodies.

1.7 Overview of Multimedia Perceptual Hashing Techniques

This section briefly introduces multimedia perceptual hashing techniques that can be used in the fields of copyright protection, content authentication and content-based retrieval. In this section, the basic concept of hashing functions is first introduced. Secondly, definitions and properties of perceptual hashing functions are given. Thirdly, the basic framework and state-of-the-art of perceptual hashing techniques are briefly discussed. Finally, some typical applications of perceptual hashing functions are illustrated.

1.7.1 Basic Concept of Hashing Functions

A hashing function is any well-defined procedure or mathematical function which converts a large, possibly variable-sized amount of data into a small datum, usually a single integer that may serve as an index into an array. The values returned by a hash function are called hash values, hash codes, hash sums, or simply hashes. Hash functions are mostly used to speed up table lookup or data comparison tasks, such as finding items in a database, detecting duplicated or similar records in a large file and finding similar stretches in DNA sequences.

A hashing function may map two or more keys to the same hash value. In many applications, it is desirable to minimize the occurrence of such collisions, which means that the hash function must map the keys to the hash values as evenly as possible. Depending on the application, other properties may be required as well. Although the idea was conceived in the 1950s, the design of good hash functions is still a topic of active research.

Hashing functions are related to (and often confused with) checksums, check digits, fingerprints, randomization functions, error correcting codes and cryptographic hash functions. Although these concepts overlap to some extent, each has its own uses and requirements and is designed and optimized differently. The HashKeeper database maintained by the National Drug Intelligence Center, for instance, is more aptly described as a catalog of file fingerprints than of hash values.

Hashing functions are primarily used in hash tables, to quickly locate a data record (for example, a dictionary definition) given its search key (the headword). Specifically, the hash function is used to map the search key to the hash. The index gives the place where the corresponding record should be stored. Hash tables, in turn, are used to implement associative arrays and dynamic sets. Hash functions are also used to build caches for large datasets stored in slow media. A cache is generally simpler than a hashed search table, since any collision can be resolved by discarding or writing back the older of the two collided items. Hash functions are an essential ingredient of the Bloom filter, a compact data structure that provides an enclosing approximation to a set of keys.

1.7.2 Concepts and Properties of Perceptual Hashing Functions

From the above description, we can see that hashing functions can be used to extract the digital digest of the original data irreversibly, and they are one-way and fragile to guarantee the uniqueness and unmodifiability of the original data. Various hashing functions have been successfully used in information retrieval and management, data authentication, and so on. However, with the increasing popularization of multimedia service, traditional hashing functions have no longer satisfied the demand for multimedia information management and protection. The reasons lie in two aspects: (1) The perceptual redundancy of multimedia requires a specific abstraction technique. Traditional hash functions only possess the function of data compression, and they cannot eliminate the redundancy in multimedia perceptual content. Therefore, we need to perform the perceptual abstraction on multimedia information according to human perceptual characteristics, obtaining the concise summary while at the same time retaining the content. (2) The many-to-one mapping properties between digital presentation and multimedia content require that the content digest possess perceptual robustness. We should research the multimedia authentication methods that are fragile to tampering operations but robust to the content-preserved operations. Therefore, according to the distinct properties of multimedia that are different from that of general computer data, we should study the one-way multimedia digest methods and techniques that possess perceptual robustness and the capability of data compression. Thus, perceptual hashing [29] has gradually become a hotspot in the field of multimedia signal processing and multimedia security.

The distinct characteristics of multimedia information that are different from general computer data are determined by the human psychological process of cognizing multimedia. According to the theory of cognitive psychology, this process includes the following stages: sensory input, perceptual content, extraction and cognitive recognition. The theory of perception threshold points out that only when the stimuli brought about by objective things exceed the perceptual threshold can we perceive the objective things and, before that, objective things are just a kind of "data". The kind of elements whose differences are less than the perception threshold is mapped to an element in another collection. The perceptual content of multimedia information is the basic feeling of humans for objective things, and it is also the basis for carrying out high-level mental activities and responding to stimuli. In addition, information processing in the cognitive stage mainly depends on subjective analysis, which has exceeded the current research range of information technology.

The perceptual hash function is an information processing theory based on cognitive psychology, and it is a one-way mapping from a multimedia data set to a multimedia perceptual digest set. The perceptual hash function maps the multimedia data possessing the same perceptual content into one unique segment of digital digest, satisfying the security requirements. We denote the perceptual hashing function by *PH* as shown in Eq.(1.17):

$$PH: M \to H. \tag{1.17}$$

The generated digital digest is called a perceptual hash value. M is a multimedia data set, and H is the set of perceptual hash values.

Assume $a, b, c \in M$, $h_a, h_b, h_c \in H$, $h_a = PH(a)$, $h_b = PH(b)$, $h_c = PH(c)$. $d(h_a, h_b)$ denotes the distance between a and b in the H space, while $d_p(a, b)$ denotes the perceptual distance between a and b in the M space, i.e., perceptual difference. The content-preserved operation of multimedia is defined as $O_{\text{cp}}(\cdot)$. When the perceptual distance between elements is larger than the perceptual threshold T, then the perceptual content is considered to be different between these two elements. $P(A)$ denotes the probability that the event A happens, τ is the decision threshold to judge whether an event happens or not. The perceptual function PH should satisfy the following basic properties.

(1) Collision resistance/discrimination

$$A = \{(a,b) \mid d_p(a,b) > T \,\&\, d(h_a,h_b) < \tau, \forall a,b \in M\} \Rightarrow P(A) \approx 0. \tag{1.18}$$

That means two pieces of multimedia work with different perceptual content should not be mapped to the same perceptual hash value.

(2) Robustness

Assume $a' = O_{\text{cp}}(a) \in M, h_{a'} = PH(a')$, then

$$B = \{(a,a') \mid d_p(a,a') < T \,\&\, d(h_a,h_{a'}) < \tau, \forall a,a' \in M\} \Rightarrow P(B) \approx 1. \tag{1.19}$$

That means two pieces of multimedia work should be mapped into the same hash value if they possess the same content or one is the content-preserved version of another.

(3) One way

Given h_a and $PH(\cdot)$, it is very hard to reversely compute the value a based on $PH(a) = h_a$, or the valid information of a cannot be obtained.

(4) Randomicity

The entropy of perceptual hash values should be equal to the length of the data, meaning the ideal perceptual hash value should be completely random.

(5) Transitivity

$$d(h_a,h_b) < \tau \,\&\, d(h_b,h_c) < \tau \Rightarrow \begin{cases} d(h_a,h_c) < \tau, & \text{if } d_p(a,c) < T; \\ d(h_a,h_c) > \tau, & \text{if } d_p(a,c) > T. \end{cases} \tag{1.20}$$

That means under the perception threshold constraints, perceptual hash functions possess transitivity, otherwise not.

(6) Compactness

Besides the above basic properties, the capacity of perceptual data should be as small as possible.

In addition, easy implementation is also an important evaluation index. Only simple and fast perceptual hash functions can meet the application requirements of massive multimedia data analysis.

1.7.3 The State-of-the-Art of Perceptual Hashing Functions

The overall framework of the perceptual hashing function is shown in Fig. 1.12. Multimedia input cannot only be audios, images, videos, but also biometric templates and 3D models that are stored as the digital sequences in the computer. Perceptual feature extraction is based on the human perceptual model, obtaining the perceptual invariant features resisting content-preserved operations. The preprocessing operations such as framing and filtering can improve the accuracy of feature selection. A variety of signal processing methods in line with the human perception model can remove the perceptual redundancy and select the most perceptually significant characteristic parameters. Furthermore, in order to facilitate hardware implementation and reduce storage requirements, characteristics of these parameters need to be quantized and encoded, i.e., to undergo some post-processing operations. Accurate perceptual feature extraction is the prerequisite for the perceptual hash value to possess a good perceptual robustness. The aim of hash construction is to perform a further dimensionality reduction on the perceptual characteristics, outputting the final result—perceptual hash values. During the design process of hash construction, we should ensure several security requirements such as anti-collision, one-way and randomness. According to different levels of security needs, we may choose not to use perceptual hash keys and to achieve key-dependency at various stages.

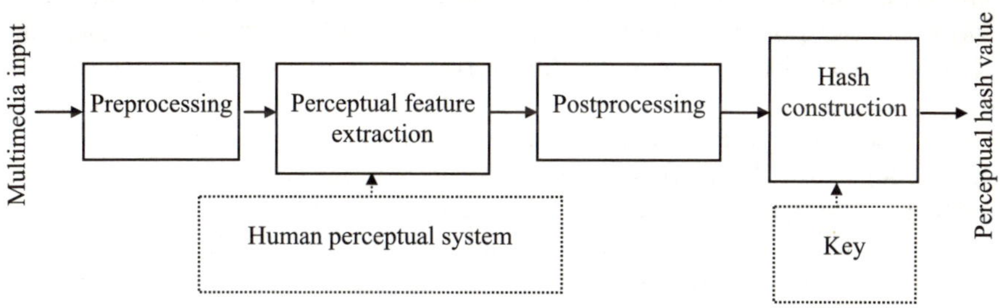

Fig. 1.12. The overall framework of the perceptual hashing function

At present, there are two similar concepts with respect to perceptual hashes. In order to avoid confusion, we make a brief statement on their differences and contacts as follows: (1) Robust hashing. Robust hashing is very close to perceptual hashing in concept, and they both require robust multimedia mapping. However, for robust hashing, the mapping establishment is based on the choice of invariant variables, while for perceptual hashing the invariance is based on multimedia perceptual features in line with the human perceptual model, realizing more accurately multimedia content analysis and protection. (2) Digital fingerprinting. At present, the definition and use of digital fingerprinting is somewhat confusing. There are mainly two types: one is the digital watermarking technique for copyright protection, the other is the media abstraction technique for media content identification. The perceptual hash is similar to a digital fingerprint since it is also a digital digest of multimedia, but it requires more security than the digital fingerprint technology.

The research into perceptual hash functions is still in its infancy. The research content mainly focuses on the one-way mapping from the dataset to the perception data. With in-depth study, it is bound to investigate the perception set in order to achieve deep content protection. At present, a lot of research results in the perceptual hashing area have been published for all kinds of multimedia. Among them, a large number of research results in audio fingerprinting have laid a solid foundation for research into audio perceptual hashing. The perceptual hashing technique for images has been a research hotspot in recent years, and a large number of research results have been published. The research into video perceptual hashing functions is gradually advancing. The state-of-the-art of perceptual hashing research work for these three kinds of multimedia can be given as follows.

(1) Extensive research on audio hashing functions started at the beginning of this century. The PHILIPS Research Institute, Delft University and the NYU-Poly, USA, have achieved significant research results. In China, the research into perceptual audio hashing is still in its infancy. And papers on speech perceptual hashing technology are seldom published. Based on audio signal processing techniques and psychoacoustic models, the audio perceptual feature extraction methods are relatively mature. Mel-frequency cepstrum coefficients and spectral smoothness can be used to evaluate well the quality of pitches and noises of each sub-band. A more common feature is the energy in each critical sub-band. Haitsma and Kalker [30] used 33 sub-band energy values in non-overlapping logarithmic scales to obtain the ultimate digital fingerprint, which is composed of the signs of differential results between adjacent sub-bands (both in the time and frequency axes). The compressed-domain perceptual hashing functions for MPEG audio often adopt MDCT coefficients to calculate the perceptual hash value. This method is prominently robust to MP3 encoding conversion. Performing the post-processing operations such as quantization can further improve the robustness and reduce the amount of data, and discretization is used to enhance the randomness of hash values so as to reduce the probability of their collision.

(2) Image perceptual hashing functions have become research hot spots in the field of perceptual hashing recently. Due to plenty of research results in the field of digital image processing, there are various perceptually-invariant feature extraction methods for images, such as histogram-based, edge-information-based and DCT-coefficient-interrelationship-based methods. Unlike audio perceptual hashing functions, image perceptual hashing functions mainly focus on the image authentication problem. Therefore, the security problem in hashing is also an important research part of image perceptual hashing functions. Currently, there are

mainly two methods for improving the security of image hashing. One is to encrypt the extracted features to assure the security of hashing. However, the encryption mechanism will greatly reduce the robustness of hashing. The other is to perform randomly mapping on the features, for example, to perform random block selection or low-pass projection on features.

(3) How to extract video perceptual features is still the most crucial and most challenging research content in the field of video perceptual hashing. Currently, unlike the spectrum-domain or other transform-domain features extracted from images and audios, many algorithms extract spatial features from video signals. The main aim is to reduce the computational complexity. During the preprocessing process, the video signal is segmented into shots, each shot being composed of frames with similar content. The image perceptual hashing function is adopted to extract the perceptual hash value from keyframes in each shot, and then the final hash value is obtained for the whole video sequence. This kind of method inherits good properties from image perceptual hashing functions. We can select the keyframes with a key, and thus the perceptual hash value is key-dependent. However, the above methods segment the video sequence into isolated images such that the interrelation between frames is neglected, and thus it is hard to completely and accurately describe the video perceptual content. Therefore, the exploitation of spatial-temporal features is the research direction in the field of video perceptual feature extraction. In general, the low-level statistics of the luminance component are viewed as the perceptual features of video, and of course the chromatic components can also be used to extract the perceptual features. However, based on the characteristics of the human visual system, human eyes are more sensitive to the luminance component than to chromatic components, and the luminance component reflects the main feature of videos.

1.7.4 Applications of Perceptual Hashing Functions

The main application fields of perceptual hashing functions include pattern recognition, multimedia retrieval and multimedia authentication.

1.7.4.1 Pattern Recognition

Perceptual hash functions are independent of the subjective evaluation of humans, and thus they can be used for automatic multimedia analysis. In addition, perceptual robustness makes perceptual hash functions applicable to multimedia content identification. For a multimedia recognition system, the most important thing is to provide users with accurate and reliable identification results. Therefore, for the perceptual hashing function applied in the recognition mode, its perceptual anti-collision and robustness are the two most important performance indices. Good compression performance and easy implementation are two preconditions for the widespread use of perceptual hashing functions. Fig. 1.13 shows the identification diagram of a typical audio recognition system.

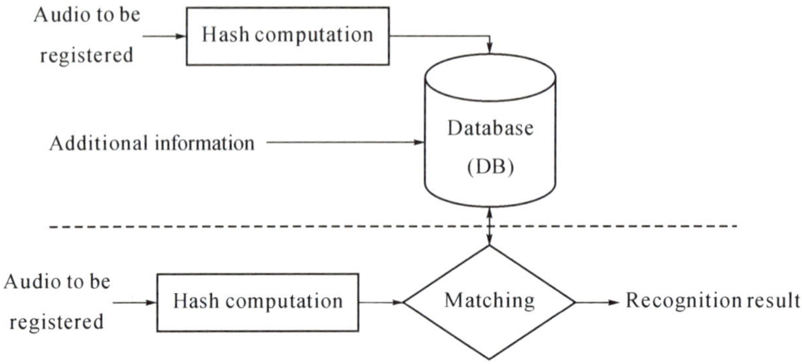

Fig. 1.13. The diagram of audio recognition based on perceptual hashing functions

1.7.4.2 Multimedia Retrieval

Compression capacity and perceptual robustness enable perceptual hashing functions to provide an accurate and efficient technical support for content-based multimedia retrieval. The accuracy requirement for the retrieval application is lower than that for the recognition application, but the efficiency requirement is relatively high. Therefore, the compression capacity is the research focus when perceptual hashing functions are applied to the retrieval field, while the robustness and discrimination are in the next place. Fig. 1.14 shows the diagram of an image retrieval system based on perceptual hashing functions.

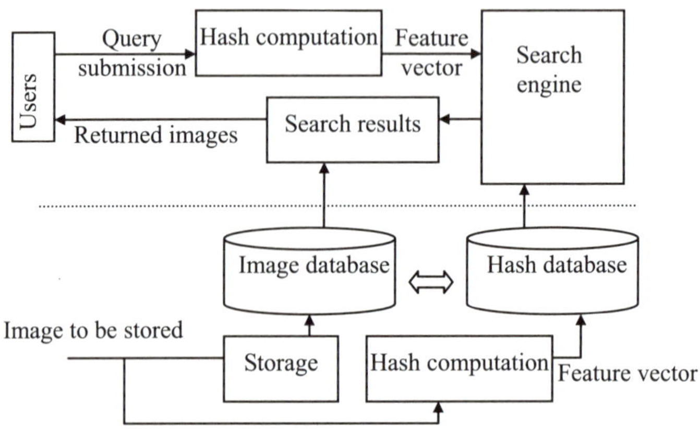

Fig. 1.14. The diagram of image retrieval based on perceptual hashing functions

1.7.4.3 Multimedia Authentication

With the rapid development of multimedia and network communication technologies, the content authentication for multimedia works becomes increasingly important. In order to ensure the security of the authentication process, the security indices such as anti-analysis and anti-counterfeit are the two most important performance indices. In other words, in the authentication application mode, the perceptual hash values must have a highly one-way performance and very good anti-collision. In addition, perceptual hash values should also have the ability of tamper detection. Without the original multimedia, the system should be able to not only judge if the multimedia to be authenticated has suffered alteration, but also point out the location and extent of tampering, by comparing perceptual hash values. Fig. 1.15 shows the block diagram of image authentication based on perceptual hashing functions.

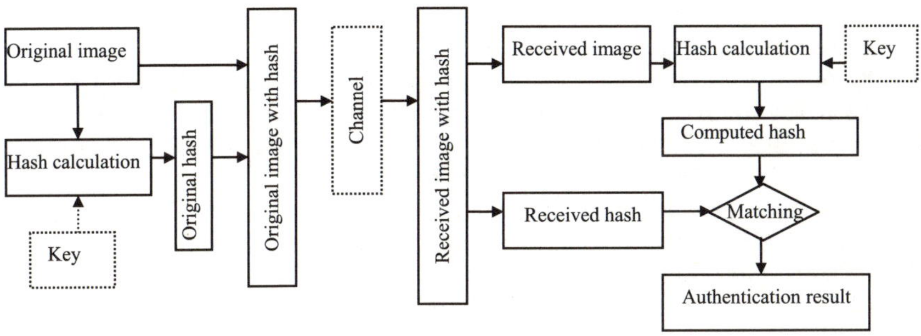

Fig. 1.15. Image authentication based on perceptual hashing functions

The above three aspects are the basic application modes of perceptual hashing functions. In addition, the perceptual hashing technique can also be used in other aspects of multimedia service, including quality assessment of compressed audio, information hiding, 3D image protection and biometric feature template protection, and so on.

1.8 Main Content of This Book

This book mainly focuses on three technical issues: (1) storage and transmission; (2) watermarking and reversible data hiding; (3) retrieval issues for 3D models. Succeeding chapters are organized as follows: From the point of view of lowering the burden of storage and transmission and improving the transmission efficiency, Chapter 2 discusses 3D model compression technology. From the perspective of the application to retrieval, Chapter 3 introduces a variety of 3D model feature extraction techniques, and Chapter 4 is devoted to content-based 3D model retrieval technology. From the perspective of the application of copyright protection and content authentication, Chapter 5 and Chapter 6 discuss 3D digital watermarking techniques, including robust, fragile and reversible watermarking techniques.

References

[1] Z. N. Li and M. S. Drew. Fundamentals of Multimedia. Prentice-Hall, 2004.
[2] J. Williams and J. D. Clark. The information explosion: fact or myth? IEEE Transactions on Engineering Management, 1992, 39(1):79-84.
[3] M. Stamp. Information Security: Principles and Practice. Wiley, 2005.
[4] E. J. Chikofsky and J. H. Cross II. Reverse engineering and design recovery: A taxonomy. IEEE Software, 1990, 7(1):13-17.
[5] M. Attene, S. Katz, M. Mortara, et al. Mesh segmentation: a comparative study. In: Proceedings of Shape Modeling International (SMI'06), 2006, pp. 14-25.
[6] M. Pollefeys. 3D modeling of real-world objects, scenes and events from videos. Paper presented at The 3DTV Conference: The True Vision - Capture, Transmission and Display of 3D Video, 2008, pp. 5-6.
[7] A. Thakur, A. G. Banerjee and S. K. Gupta. A survey of CAD model simplification techniques for physics-based simulation applications. Computer-Aided Design, 2009, 41(2):65-80.
[8] X. Sun, P. L. Rosina, R. R. Martina, et al. Random walks for feature-preserving mesh denoising. Computer Aided Geometric Design, 2008, 25(7):437-456.
[9] A. Kaufman, D. Cohen, R. Yagel, et al. Volume graphics sidebar: fundamentals of voxelization. IEEE Computer, 1993, 26(7):51-64.
[10] P. Heckbert. Fundamentals of Texture Mapping and Image Warping. Master's Thesis, UCB/CSD 89/516, CS Division, U.C. Berkeley, 1989.
[11] J. Peters and U. Reif. The simplest subdivision scheme for smoothing polyhedra. ACM Transactions on Graphics, 1997, 16(4):420-431.
[12] H. Hoppe. Progressive meshes. In: Proceedings of SIGGRAPH'96, 1996, pp. 99-108.
[13] D. Schmalstieg. The Remote Rendering Pipeline. Ph.D Dissertation, Technical University of Vienna, 1997.
[14] T. Funkhouser, P. Min and M. Kazhdan. A search engine for 3D models. ACM Transactions on Graphics, 2003, 22(1):83-105.
[15] N. Nikolaidis and I. Pitas. Still image and video fingerprinting. Paper presented at The Seventh International Conference on Advances in Pattern Recognition (ICAPR'09), 2009, pp. 3-8.
[16] B. van Ginneken, A. F. Frangi, J. J. Staal, et al. Active shape model segmentation with optimal features. IEEE Transactions on Medical Imaging, 2002, 21(8):924-933.
[17] A. Gersho. Advances in speech and audio compression. Proceedings of the IEEE, 1994, 82(6):900-918.
[18] R. J. Clarke. Image and video compression: a survey. Journal of Imaging Systems and Technology, 1999, 10(1):20-32.
[19] G. Voyatzis and I. Pitas. The use of watermarks in the protection of digital multimedia products. Proceedings of the IEEE, 1999, 87(7):1197-1207.
[20] F. A. P. Petitcolas, R. J. Anderson and M. G. Kuhn. Information hiding—a survey. Proceedings of IEEE, 1999, 87(7):1062-1078.
[21] A. Singhal. Modern information retrieval: a brief overview. Bulletin of the IEEE Computer Society Technical Committee on Data Engineering, 2001, 24 (4):35-43.
[22] P. Martin and P. W. Eklund. Knowledge retrieval and the World Wide Web. IEEE

Intelligent Systems, 2000, 15(3):18-25.

[23] R. S. Michalski. Knowledge Mining: a proposed new direction. Paper presented at The 6th Sanken Symposium on Data Mining and Semantic Web, Osaka University, Japan, March 10-11, 2003.

[24] A. W. M. Smeulders, M. Worring, S. Santini, et al. Content based image retrieval at the end of the early years. IEEE Transactions on Pattern Analysis and Machine Intelligence, 2000, 22(12):1349-1380.

[25] M. Petkovic and W. Jonker. Content-Based Video Retrieval: A Database Perspective. Kluwer Academic Publishers, 2003.

[26] P. Wan and L. Lu. Content-based audio retrieval: a comparative study of various features and similarity measures. In: Proceedings of SPIE, Vol. 6015, 2005.

[27] X. Zhuang, J. T. Huang and M. Hasegawa-Johnson. Speech retrieval in unknown languages: a pilot study. Paper presented at NAACL HLT Cross-Lingual Information Access Workshop (CLIAWS), 2009.

[28] Y. Zhu and M. S. Kankanhalli. Melody alignment and similarity metric for content-based music retrieval. In: Proceedings of SPIE–IS&T Electronic Imaging, 2003, Vol. 5021, pp. 112-121.

[29] A. Swaminathan, Y. Mao and M. Wu. Robust and secure image hashing. IEEE Transactions on Information Forensics and Security, 2006, 1(2):211-218.

[30] J. Haitsma and T. Kalker. A highly robust audio fingerprinting system. In: Proceedings of the 3rd International Conference on Music Information Retrieval (ISMIR), 2002, pp. 107-115.

2

3D Mesh Compression

3D meshes have been widely used in graphics and simulation applications for representing 3D objects. They generally require a huge amount of data for storage and/or transmission in the raw data format. Since most applications demand compact storage, fast transmission and efficient processing of 3D meshes, many algorithms have been proposed in the literature to compress 3D meshes efficiently since the early 1990s [1]. Because most of the 3D models in use are polygonal meshes, most of the published papers focus on coding that type of data, which is composed of two main components: connectivity data and geometry data. This chapter discusses 3D mesh compression technologies that have been developed over the last decade, with the main focus on triangle mesh compression technologies.

2.1 Introduction

We first introduce the background, basic concepts and algorithm classification of 3D mesh compression techniques.

2.1.1 Background

Graphics data are more and more widely adopted in various applications, including video games, engineering design, architectural walkthrough, virtual reality, e-commerce and scientific visualization. The emerging demand for visualizing and simulating 3D geometric data in networked environments has aroused research interests in representations of such data. Among various representation tools, triangle meshes provide an effective way to represent 3D models. Typically, connectivity, geometry and property data are together used to represent a 3D polygonal mesh. Connectivity data describe the adjacency relationship between vertices, geometry data specify vertex locations and property data specify several attributes such as normal vectors, material reflectance and texture coordinates. Geometry and property data are often attached to vertices in many cases, where they are often called vertex data, and most 3D triangle mesh compression algorithms handle geometry and property data in a similar way. Therefore, we focus on the compression of connectivity and geometry data in this chapter.

As the number and the complexity of existing 3D meshes increase explosively, higher resource demands are placed on the storage space, computing power and network bandwidth. Among these resources, the network bandwidth is the most severe bottleneck in network-based graphics that demands real-time interactivity. Thus, it is essential to compress graphics data efficiently. This research area has received a lot of attention since the early 1990s, and there has been a significant amount of progress in this direction over the last decade [2].

Due to the significance of 3D mesh compression, it has been incorporated into several international standards. VRML [3] has established a standard for transmitting 3D models over the Internet. Originally, a 3D mesh was represented in ASCII format without any compression in VRML. To implement efficient transmission, Taubin et al. developed a compressed binary format for VRML [4] based on the topological surgery algorithm [5], which can easily achieve a compression ratio of 50 over the VRML ASCII format. MPEG-4 [6], which is an ISO/IEC multimedia standard developed by the Moving Picture Experts Group for digital TV, interactive graphics and interactive multimedia applications, also includes the 3D mesh coding (3DMC) algorithm to encode graphics data. The 3DMC algorithm is also based on the topological surgery algorithm, which is basically a single-rate coder for manifold triangle meshes. Furthermore, MPEG-4 3DMC incorporates progressive 3D mesh compression, non-manifold 3D mesh encoding, error resiliency and quality scalability as optional modes. In this book, we intend to review various 3D mesh compression technologies with the main focus on triangle mesh compression.

With respect to 3D mesh compression, there have been several survey papers. Taubin and Rossignac [5] briefly summarized prior schemes on vertex data compression and connectivity data compression for triangle meshes. Taubin [8] gave a survey on various geometry and progressive compression schemes, but the focus was on two schemes in the MPEG-4 standard. Shikhare [9] classified and described mesh compression schemes, but progressive schemes were not discussed in enough depth. Gotsman et al. [10] gave an overview on mesh simplification, connectivity compression and geometry compression techniques, but the review on connectivity coding algorithms focused mostly on single-rate region-growing schemes. Recently, Alliez and Gotsman [1] surveyed techniques for both single-rate and progressive compression of 3D meshes, but the review focused only on static (single-rate) compression. Compared with previous survey papers, this chapter attempts to achieve the following three goals: (1) To be comprehensive. This chapter covers both single-rate and progressive mesh compression schemes. (2) To be in-depth. This chapter attempts to make a more detailed classification and explanation of different algorithms. For example, techniques based on vector quantization (VQ) are discussed in a whole section. (3)

To use performance analysis and comparisons. Compression efficiency is compared between different methods to assist engineers in the selection of schemes based on application requirements.

2.1.2 Basic Concepts and Definitions

Several definitions and concepts required to understand 3D mesh compression algorithms are presented as follows.

2.1.2.1 Surface-Based Models

Definition 2.1 (Homeomorphic) *We say that two objects A and B are homeomorphic, if A can be stretched or bent without tearing B.*

The surface-based characterization of solids looks at the boundary of a solid object and composes it into a collection of faces, which are glued together such that they form a complete and closed skin around the object. A surface can be viewed as a 2D subset of R^3. Each surface point is surrounded by a "2D region" of surface points. The "2-manifold" definition gives a more abstract notion to a surface.

Definition 2.2 (2-Manifold) *A 2-manifold is a topological space, where every point has a neighborhood topologically equivalent to an open disk of* R^2.

In fact, here "topologically equivalent" means "homeomorphic". Thus, a 3D mesh is called a manifold if its every point has a neighborhood homeomorphic to an open disk or a half disk. In a manifold, the boundary consists of the points that have no neighborhoods homeomorphic to an open disk but have neighborhoods homeomorphic to a half disk. In 3D mesh compression, a manifold with boundary is often pre-converted into a manifold without boundary by adding a dummy vertex to each boundary loop and then connecting the dummy vertex to every vertex on the boundary loop. A manifold surface mesh is shown in Fig. 2.1(a). In computer graphics, it is also quite common to handle surfaces with boundaries, e.g., the lamp shade shown in Fig. 2.1(b). Thus one also allows points with a neighborhood topologically equivalent to a half disk and calls these surfaces

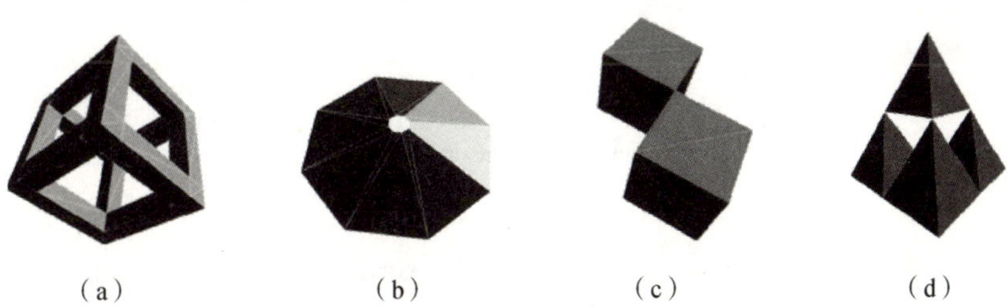

Fig. 2.1. Manifold and non-manifold meshes
(a) Manifold mesh; (b) Manifold with border; (c) Non-manifold because of edge with more than two incident faces; (d) Non-manifold because of vertices with more than one connected face loop

manifold with boundary. However, there are also quite common surface models that are not manifold, e.g., the other two examples in Fig. 2.1. In Fig. 2.1(c), the two cubes touch at a common edge, which contains points with a neighborhood not equivalent to a disk or a half disk. And in Fig. 2.1(d), the tetrahedra touch at points with a non-manifold neighborhood.

2.1.2.2 Connectivity

In order to analyze and represent complex surfaces, we subdivide the surfaces into polygonal patches enclosed by edges and vertices. Fig. 2.2(a) shows the subdivision of the torus surface into four patches p_1, p_2, p_3, p_4. Each patch can be embedded into the Euclidean plane resulting in four planar polygons as shown in Fig. 2.2(b). The embedding allows the mapping of the Euclidean topology to the interior of each patch on the surface. The collection of polygons can represent the same topology as the surface if the edges and vertices of adjacent patches are identified. In Fig. 2.2(b), identified edges and vertices are labeled with the same specifier. The topology of the points on two identified edges is defined as follows. The points on the edges are parameterized over the interval [0, 1], where zero corresponds to the vertex with a smaller index and one to the vertex with a larger index. The points on the identified edges with the same parameter value are identified and the neighborhood of the unified point is composed of the unions of half-disks with the same diameter in both adjacent patches. In this way, the identified edges are treated as one edge. The topology around vertices is defined similarly. Here the neighborhood is composed of disks put together from several pies with the same radius of all incident patches.

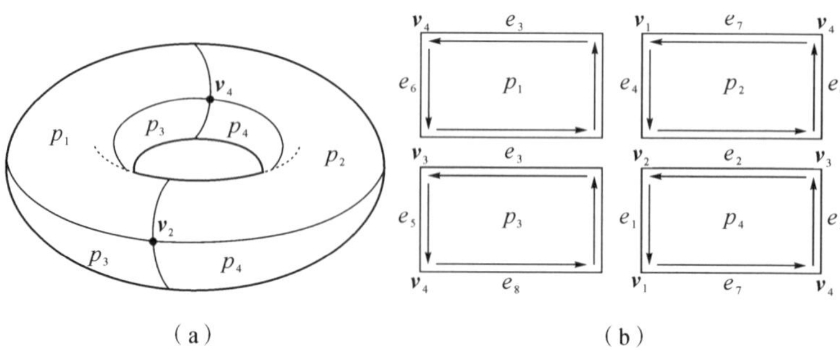

Fig. 2.2. Polygonal patches enclosed by edges and vertices
(a) Torus subdivided into four patches; (b) Planar embedding of patches with identified edges and vertices

We are now in the position to split the surface into two constitutes: the connectivity and the geometry. The connectivity $\boldsymbol{C}$ defines the polygons, edges and vertices and their incidence relation. The geometry $\boldsymbol{G}$ on the other hand defines the mappings from the polygons, edges and vertices to patches, possibly bent edges and vertices in the 3D Euclidean space. The pair $\boldsymbol{M} = (\boldsymbol{C}, \boldsymbol{G})$ defines a polygonal mesh and allows the representation of solids via their surface. First we discuss the connectivity, which defines the incidence among polygons, edges and vertices and which is independent of the geometric realization.

Definition 2.3 (Polygonal Connectivity) *The polygonal connectivity is a quadruple (V, E, F, I) of the set of vertices V, the set of edges E, the set of faces F and the incidence relation I, such that: 1) each edge is incident to its two end vertices; 2) each face is incident to an ordered closed loop of edges* $(e_1, e_2, \ldots, e_n)$ *with* $e_i \in E$, *such that* e_1 *is incident to* $\boldsymbol{v}_1$ *and* $\boldsymbol{v}_2$, ..., e_i *is incident to* $\boldsymbol{v}_i$ *and* $\boldsymbol{v}_{i+1}$, $\forall i = 2, \ldots, n-1$, *and* e_n *is incident to* $\boldsymbol{v}_n$ *and* $\boldsymbol{v}_1$; *3) in the notation of the previous item, the face is also incident to the vertices* $\boldsymbol{v}_1, \ldots, \boldsymbol{v}_n$; *4) the incidence relation is reflexive.*

The collection of all vertices, all edges and all faces are called the mesh elements. We next define the relation "adjacent", which is defined on pairs of mesh elements of the same type.

Definition 2.4 (Adjacent) *Two faces are adjacent, if there exists an edge incident to both of them. Two edges are adjacent, if there exists a vertex incident to both. Two vertices are adjacent, if there exists an edge incident to both.*

Up to now we defined only terms for very local properties among the mesh elements. Now we move on to global properties.

Definition 2.5 (Edge-connected) *A polygonal connectivity is edge-connected, if each two faces are connected by a path of faces such that two successive faces in the path are adjacent.*

Definition 2.6 (Valence, Degree and Ring) *The valence of a vertex is the number of edges incident to it, and the degree of a face is the number of edges incident to it. The ring of a vertex is the ordered list of all its incident faces.*

Fig. 2.3 gives an example to show the valence of a vertex and the degree of a face.

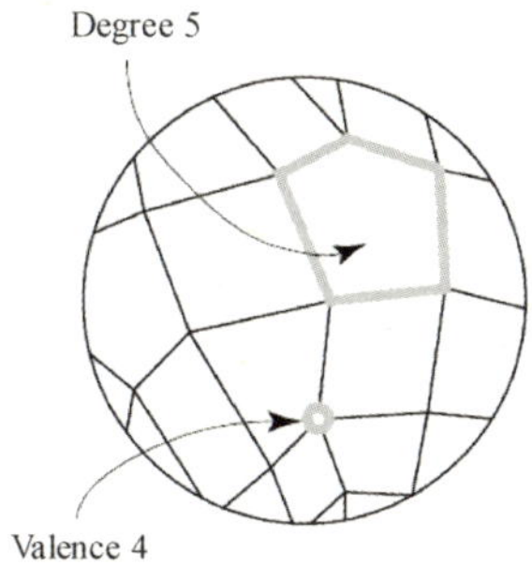

Fig. 2.3. Close-up of a polygon mesh: the valence of a vertex is the number of edges incident to this vertex, while the degree of a face is the number of edges enclosing it

As the connectivity is used to define the topology of the mesh and the represented surface, one can define the following criterion for the surface to be manifold.

Definition 2.7 (Potentially Manifold) *A polygonal connectivity is potentially*

manifold, if 1) each edge is incident to exactly two faces; 2) the non-empty set of faces around each vertex forms a closed cycle.

Definition 2.8 (Potentially Manifold with Border) *A polygonal connectivity is potentially manifold with border, if 1) each edge is incident to one or two faces; 2) the non-empty set of faces around each vertex forms an open or closed cycle.*

A surface defined by a mesh is manifold, if the connectivity is potentially manifold and no patch has a self-intersection and the intersection of two different patches is either empty or equal to the identified edges and vertices. All the non-manifold meshes in Fig. 2.1 are not potentially manifold.

Definition 2.9 (Genus of a Manifold) *The genus of a connected orientable manifold without boundary is defined as the number of handles.*

As we know, there is no handle in a sphere, one handle in a torus, and two handles in an eight-shaped surface as shown in Fig. 2.4. Thus, their genera are 0, 1 and 2, respectively. For a connected orientable manifold without boundary, Euler's formula is given by

$$N_v - N_e + N_f = 2 - 2G\,, \tag{2.1}$$

where G is the genus of the manifold, and the total number of vertices, edges and faces of a mesh are denoted as N_v, N_e, and N_f respectively.

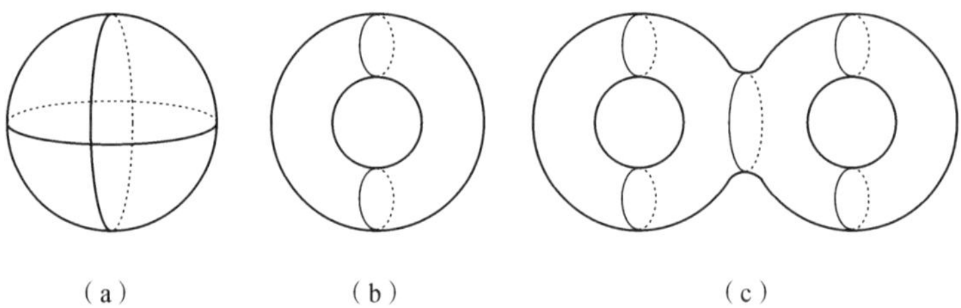

Fig. 2.4. Examples to show the genus of a manifold. (a) Sphere; (b) Torus; (c) Eight-shaped mesh

Suppose that a triangular manifold mesh consists of a sufficiently large number of edges and triangles, and that the ratio of the number of boundary edges to the number of non-boundary edges is negligible. Then, considering that an edge is shared by two triangles in general, we can estimate the number of edges by

$$N_e \cong 3N_f / 2\,. \tag{2.2}$$

Substituting Eq.(2.2) into Eq.(2.1), we have $N_v \cong N_f / 2 + 2 - 2G$. Since $N_f/2$ is much larger than $2–2G$, we have

$$N_v \cong N_f / 2\,. \tag{2.3}$$

That is to say, a typical triangle mesh has twice as many triangles as vertices.

According to Eqs.(2.2) and (2.3), we furthermore have an approximate relationship

$$N_e \cong 3N_v. \tag{2.4}$$

As defined above, the valence of a vertex is the number of edges incident on that vertex. It can be shown that the sum of valences is twice the number of edges [11]. Thus, we have

$$\sum \text{valence} = 2N_e \cong 6N_v. \tag{2.5}$$

Therefore, in a typical triangle mesh, the average vertex valence is 6.

In order to determine whether a potentially manifold mesh can be embedded without self-intersections in the 3D Euclidean space, the orientability plays the crucial role. The orientation of each face has been defined with the connectivity in the order of the edges and vertices. From the face orientation, each incident edge inherits an orientation as illustrated in Fig. 2.2(b). In fact, the orientation of a polygon can be specified by the ordering of its bounding vertices.

Definition 2.10 (Compatible) *The orientations of two adjacent polygons are called compatible if they impose opposite directions on their common edges.*

With the inherit orientation of the edges, the orientability of a mesh can be defined.

Definition 2.11 (Orientable) *A polygonal connectivity is orientable if the face orientations can be chosen in a way that for each two adjacent faces the common incident edges inherit different orientations from the different faces. That is, a 3D mesh is said to be orientable if there is an arrangement of polygon orientations such that each pair of adjacent polygons are compatible.*

The orientation of a face in a polygonal mesh can be used to define the outside of a mesh or to calculate the surface normal. It is also important during the navigation through the mesh, which is essential for most connectivity compression techniques. The problem with non-orientable meshes is that we cannot choose the orientation of the faces consistently. Thus surface normals cannot be computed consistently and no inside or outside relation makes sense. Furthermore, it complicates the navigation in the mesh, as we must know during the traversal between two adjacent faces, whether the orientation of the face changes. Meshes in Figs. 2.5(a) and 2.5(c) are orientable with the compatible orientations marked by arrows. In contrast, Fig. 2.5(b) is not orientable, for three polygons share the same edge (v_1, v_2). Note that, after we make polygons B and C compatible, it is impossible to find an orientation of polygon A such that A is compatible with both B and C. A manifold mesh is orientable if and only if there is a choice of orientations that makes all pairs of adjacent triangles compatible.

So far we have restricted the definition of a mesh to the 2D case. We also want to describe volumetric meshes and in particular tetrahedral meshes. The vertices are zero dimensional mesh elements, the edges one dimensional and the faces two dimensional. The embedding of a 3D mesh element is a subset of the Euclidean space with non zero volume. For this we define the topological polyhedron as follows.

Definition 2.12 (Topological Polyhedron) *A topological polyhedron is a potentially manifold and edge-connected polygonal connectivity.*

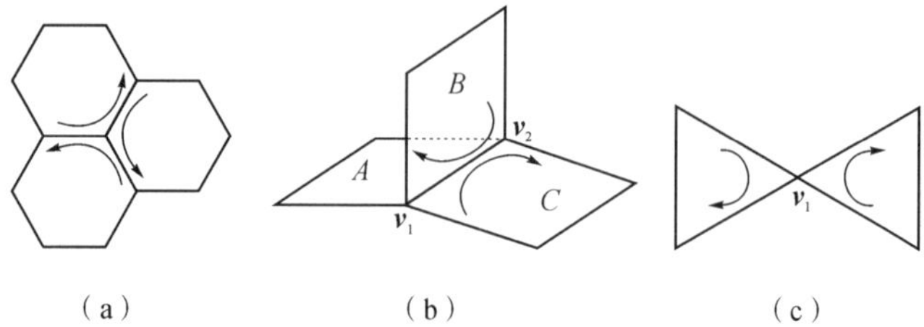

Fig. 2.5. Examples of orientable and non-orientable meshes. (a) Orientable manifold mesh; (b) Non-orientable non-manifold mesh; (c) Orientable non-manifold mesh

Based on the definition of a topological polyhedron, we can define the polyhedral connectivity as a quintuple (V, E, F, P, I) of vertices, edges, faces and polyhedra. Each polyhedron is incident to a set of oriented faces that form a topological polyhedron. The local and global relations of adjacent, face-connected, manifold and manifold with border are direct generalizations of the corresponding attributes in a polygonal connectivity. We do not want to define all these terms in detail, but want to mention that the roll of the face orientation is taken by the outside relation of the topological polyhedron. Note that in a pure polyhedral connectivity the border is always a closed polygonal connectivity and therefore the number of faces incident on an edge is always larger than two. Polyhedral meshes that are embedded self-intersection free in the 3D Euclidean space are always orientable as polygonal meshes in the plane.

2.1.2.3 Geometry

It is now time to add some geometry to the connectivity. We want to describe this procedure only for the typical case of polygonal and polyhedral geometry in the Euclidean space. Similarly, meshes with curved edges and surfaces could be defined.

Definition 2.13 (Euclidean Polygonal/Polyhedral Geometry) *The Euclidean geometry* $\boldsymbol{G}$ *of a polygonal/polyhedral mesh* $\boldsymbol{M} = (\boldsymbol{C}, \boldsymbol{G})$ *is a mapping from the mesh elements in* $\boldsymbol{C}$ *to* R^3 *with the following properties: 1) a vertex is mapped to a point in* R^3*; 2) an edge is mapped to the line segment connecting the points of its incident vertices; 3) a face is mapped to the inside of the polygon formed by the line segments of the incident edges; 4) a topological polyhedron is mapped to the sub-volume of* R^3 *enclosed by its incident faces.*

Here arises a problem that also often arises in practice. In R^3, the edges of a face often do not lie in the same plane. Therefore, the geometric representation of a face is not defined properly and also a sound 2D parameterization of the polygon is not easily defined. In practice, this is often ignored and the polygon is split into triangles for which a unique plane is given in the Euclidean space. Often further attributes like physical properties of the described surface/volume, the surface color, the surface normal or a parameterization of the surface are necessary. In practice, we often simplify the problem to the simplest types of mesh elements, the simplices. The k-dimensional simplex (or for short k-simplex) is formed by the convex hull of k+1 points in the Euclidean space. A 0-simplex is just a point, a 1-simplex is a line segment, a 2-simplex is a triangle and the 3-simplex forms a tetrahedron. For simplices, the linear and quadratic interpolations of vertex and edge attributes are simply defined via the barycentric coordinates.

In some applications, the handling of mixed dimensional meshes is necessary. As the handling of mixed dimensional polygonal/polyhedral meshes becomes very complicated, one often gives up polygons and polyhedra and restricts oneself to simplicial complexes, which allow for singleton vertices and edges and non-manifold mesh elements. A simplicial complex is defined as follows.

Definition 2.14 (Simplicial Complex) *A k dimensional simplicial complex is a (k+1)-tuple (S_0, ..., S_k), where S_i contains all i-simplices of the complex. The simplices fulfill the condition that the intersection of two i-simplices is either empty or equal to a simplex of lower dimension.*

As a simplex and therefore a simplicial complex is only a geometric description, we have to define the connectivity of a simplicial complex, which is easily done by specifying the incidence relation among the simplices of different dimensions. An i-simplex is incident to a j-simplex with $i < j$ if the i-simplex forms a sub-simplex of the j-simplex.

2.1.2.4 Triangle Meshes

A triangle mesh is defined by a set of vertices and by its triangle-vertex incidence graph. The vertex description comprises geometry (3 coordinates per vertex) and optionally photometry (surface normals, vertex colors, or texture coordinates), which will not be discussed here. Incidence, sometimes referred to as topology, defines each triangle by the 3 integer indices that identify its vertices. We define $|X|$ as the number of elements in the set X, and T denotes a set of topologically closed triangles, T_i, for the integer i in $[1, |T|]$. $\{T_i\}$ is the closed point set of T_i. $\{T\}$ is the union of these point sets for all triangles in T. V is the set of the vertices that bound the triangles of T. For simplicity, and without loss of generality, we assume that the vertices of V may be uniquely identified by integer labels between 1 and $|V|$. The connectivity may be represented by a triangle-vertex incidence table, which associates each triangle with three integer labels that reference its bounding vertices.

Definition 2.15 (Interior and Exterior Edges) *Edges that bound two triangles are called interior edges. Edges that bound exactly one triangle are called exterior edges.*

The union of interior and exterior edges is denoted as $b\{T\}$ and called the boundary of $\{T\}$. The connected components of $b\{T\}$ are one-manifold polygonal curves, called loops. Vertices of T that do not bind any exterior edge are called interior vertices. The set of all interior vertices is denoted as V_I. The other vertices are called exterior vertices and their set is denoted as V_E.

2.1.2.5 Simple Meshes

Definition 2.16 (Simple Mesh) *A simple mesh is a triangle mesh that forms a connected, orientable, manifold surface that is homeomorphic to a sphere or to a half-sphere. Such meshes have no handle and either have no boundary or have a boundary that is a connected, manifold, closed curve, i.e., a simple loop.*

For simple meshes, the Euler equation yields

$$N_t - N_e + N_v = 1, \tag{2.6}$$

where N_t=$|T|$ is the number of triangles, N_v =$|V_I|$ + $|V_E|$, and N_e is the total number of the external and internal edges. Since there are $|V_E|$ external edges and $(3|T|-|V_E|)/2$ internal edges, we have $N_e = (3|T|+|V_E|)/2$. Thus, based on Eq.(2.6), we can easily have

$$|T| = 2|V_I| + |V_E| - 2. \tag{2.7}$$

When $|V_E| << |V_I|$, there are approximately twice as many triangles as vertices.

2.1.2.6 Compression Performance

When reporting the compression performance, some papers employ the measure of bits per triangle (bpt) while others use bits per vertex (bpv). For consistency, we adopt the bpv measure exclusively, and convert the bpt metric to the bpv metric by assuming that a mesh has twice as many triangles as vertices.

2.1.3 Algorithm Classification

Recently, 3D model compression has been an important branch of multimedia data compression. In fact, there are primarily three different approaches for reducing the size of a mesh: compression, simplification and remeshing. In the compression approach, the goal is to find an encoding bitstream for a mesh that is as short as possible. Compression is especially useful not only for the efficient encoding of databases with a lot of small models, but also as an encoding tool for simplification and remeshing approaches, which typically end up with a small mesh that also has to be encoded efficiently. Large and regular models often contain more information than necessary or maybe even redundant information. Then it cannot be justified anymore that the connectivity of the mesh should be preserved and mesh simplification should be utilized. The most commonly adopted idea in mesh simplification is to simplify the mesh through a sequence of local operations that eliminate a small number of adjacent mesh elements. An also very interesting idea is remeshing, where a second very regular mesh is generated that approximates the original mesh. The regularity of the approximation allows the storing of the new mesh much more efficiently.

Because most 3D models in use are polygonal meshes, this chapter mainly focuses on compression techniques for 3D polygon meshes. Typically, connectivity, geometry and property data are together used to represent a 3D polygonal mesh. Connectivity data describe the adjacency relationship between vertices, geometry data specify vertex locations and property data specify several attributes such as normal vectors, material reflectance and texture coordinates. Thus, according to which part of 3D polygon mesh data are concerned, 3D model compression methods can be classified into three categories, i.e., connectivity compression, geometry data compression and geometry property compression. Currently, the research emphasis of 3D mesh compression is on geometry data compression. This chapter ascribes geometry data compression and geometry property compression to a larger category, i.e., geometry compression. A typical mesh compression algorithm encodes connectivity data and geometry data separately. Of course, connectivity compression and geometry compression may be both used in a specific compression scheme. Most early work focused on the connectivity coding. Then, the coding order of geometry data is determined by the underlying connectivity coding. However, since geometry data demand more bits than topology data, some methods have been proposed recently for efficient compression of geometry data without reference to topology data. According to whether the reconstructed data can be used to completely restore the original 3D geometry data or not, geometry compression techniques can be classified into lossless geometry compression and lossy geometry compression. Lossless compression can completely restore the original geometry information from the compressed data, while in the case of lossy compression there are some differences between the decoded geometry information and the original geometry information. In lossy compression, the loss is introduced by quantization. According to whether the compression scheme requires altering the connectivity or not, geometry compression techniques can be classified into non- reconstruction-based compression and reconstruction-based compression. Non-reconstruction-based compression schemes directly perform the compression operation on the original model, while reconstruction-based compression methods first perform mesh reconstruction on the original model and then perform compression on the reconstructed mesh. Obviously, most reconstruction-based compression methods are lossy. According to which domain is adopted to perform the compression operation, we can classify the 3D mesh compression methods into two categories, i.e., spatial-domain based and transform-domain-based methods.

Slow networks require data compression to reduce the latency and progressive representations to transform 3D objects into streams manageable by the networks. Depending on whether the model is decoded during, or only after, the transmission, we classify mesh compression methods into single-rate (single-resolution or static) compression schemes and progressive compression techniques. Single-resolution compression schemes for 3D meshes usually create a single bitstream, which can be split into two parts: the connectivity bitstream (which describes the mesh connectivity graph) and the geometry bitstream (the vertices' coordinates). Progressive transmission of meshes involves splitting both the bitstreams into several components. The connectivity bitstream usually contains a base mesh which is further refined by reading the successive bitstreams. The geometry bitstream is also decomposed into a base geometry and several geometrical refinements. In the case of single-rate lossless coding, the goal is to remove the redundancy present in the original description of the data. In the case of progressive compression, the problem is more challenging, aiming for the best trade-off between data size and approximation accuracy (the so-called rate-distortion tradeoff). Single-rate lossy coding may also be achieved by modifying the data set, making it more amenable to coding, without losing too much information. Early research on 3D mesh compression focused on single-rate compression techniques to save the bandwidth between the CPU and the graphics card. In a single-rate 3D mesh compression algorithm, all connectivity and geometry data are compressed and decompressed as a whole. The graphics card cannot render the original mesh until the entire bitstream has been wholly received. Later, with the popularity of the Internet, progressive compression and transmission has been intensively researched. When progressively compressed and transmitted, a 3D mesh can be reconstructed continuously from coarse to fine levels of detail (LODs) by the decoder while the bitstream is being received. Moreover, progressive compression can enhance the interaction capability, since the transmission can be stopped whenever a user finds out that the mesh being downloaded is not what he/she wants or the resolution is already good enough for his/her purposes.

From the point of view of development trends, the research focus of 3D mesh compression techniques is being gradually changed from former topology-driven compression techniques to current geometry-driven compression techniques. This chapter introduces connectivity compression methods in two categories, i.e., single-rate and progressive compression schemes, while discussing the geometry compression techniques in three categories, i.e., spatial-domain-based, transform-domain-based and vector-quantization (VQ)-based methods. Here, VQ can be performed in the spatial domain or transform domains, and several studies have been done by the authors of this book. Thus we separately introduce VQ-based geometry compression in Section 2.6.

2.2 Single-Rate Connectivity Compression

Single resolution mesh compression methods are important for encoding large data bases of small objects, base meshes of progressive representations or for fast transmission of meshes over the Internet. We can classify the single resolution techniques into two classes: (1) techniques aiming at coding the original mesh without making any assumption about its complexity, regularity or uniformity; (2) techniques which remesh the model before compression. The original mesh is considered as just one instance of the shape geometry.

Single-rate or static connectivity compression methods perform the single-rate compression only on the connectivity data, without considering the geometry data. Single-rate connectivity compression can be roughly divided into two types: edge-based and vertex-based coders. Here, we classify existing typical single-rate connectivity compression algorithms into six classes: the indexed face set, the triangle strip, the spanning tree, the layered decomposition, the valence-driven approach and the triangle conquest method. They can be described in detail as follows.

2.2.1 Representation of Indexed Face Set

In the VRML ASCII format [3], a triangle mesh is represented with an indexed face set that is composed of a coordinate array and a face array. The coordinate array gives the coordinates of all vertices, and the face array shows each face by indexing its three vertices in the coordinate array. Fig. 2.6 gives a mesh example and its face array.

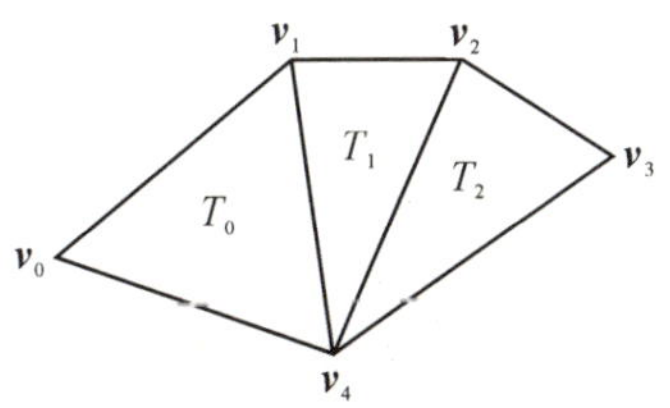

(a)

Index	Face
0	(0,1,4)
1	(1,2,4)
2	(2,3,4)

(b)

Fig. 2.6. The indexed face set representation of a mesh. (a) A mesh example; (b) Its face array

If the number of vertices in a mesh is N_v, then we need $\log_2 N_v$ bits to represent the index of each vertex. Thus, $3\log_2 N_v$ bits are required to represent the connectivity information of a triangular face. Since there are about twice as many triangles as vertices in a typical triangle mesh, the connectivity information costs about $6\log_2 N_v$ bpv in the indexed face set method. This method provides a straightforward way for the representation of triangle meshes. There is actually no compression applied in this method, but we still list it here to provide a basis of comparison for the following compression schemes.

Obviously, in this representation, each vertex may be indexed several times by all its adjacent triangles. Repeated vertex references will definitely degrade the efficiency of connectivity representation. In other words, a good connectivity compression method should reduce the number of repeated vertex references. This observation motivates researchers to develop the following triangle strip scheme.

2.2.2 Triangle-Strip-Based Connectivity Coding

The triangle strip scheme attempts to segment a 3D mesh into long strips of triangles, and then encode them. The main aim of this method is to reduce the amount of data transmitted between the CPU and the graphic card, for triangle strips are well supported by most graphic cards. Although this method requires less storage space and transmission bandwidth than the indexed face set, it is still not very efficient for the compression purpose.

Fig. 2.7(a) shows a triangle strip, where each vertex is combined with the previous two vertices in a vertex sequence to form a new triangle. Fig. 2.7(b) shows a triangle fan, where each vertex after the first two forms a new triangle with the previous vertex and the first vertex. Fig. 2.7(c) shows a generalized triangle strip that is a mixture of triangle strips and triangle fans. Note that, in a generalized triangle strip, a new triangle is introduced by each vertex after the first two in a vertex sequence. However, in an indexed face set, a new triangle is introduced by three vertices. Therefore, the generalized triangle strip provides a more compact representation than the indexed face set, especially when the strip length is long. In a rather long generalized triangle strip, the ratio of the number of triangles to the number of vertices is very close to 1, meaning that a triangle can be represented by almost exactly 1 vertex index.

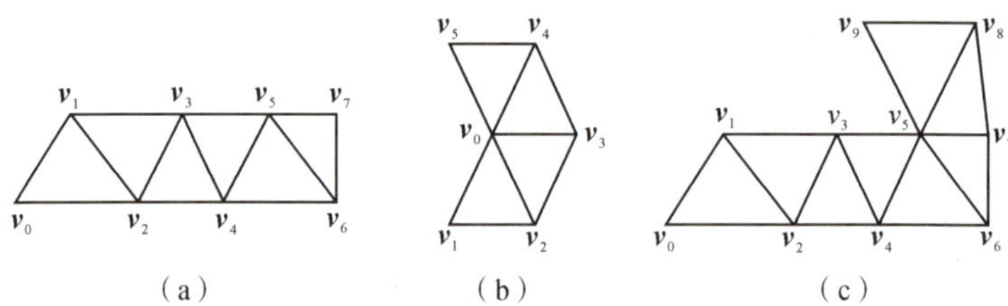

Fig. 2.7. Example of triangle trips. (a) Triangle strip; (b) Triangle fan; (c) Generalized triangle strip

However, since there are about twice as many triangles as vertices in a typical mesh, some vertex indices should be repeated in the generalized triangle strip representation of the mesh, which indicates a waste of storage. To alleviate this problem, several schemes have been developed, where a vertex buffer is utilized to store the indices of recently traversed vertices. Deering [12] first introduced the concept of the generalized triangle mesh. A generalized triangle mesh is formed by combining generalized triangle strips with a vertex buffer. He used a first-in-first-out (FIFO) buffer to store the indices of up to 16 recently-visited vertices. If a vertex is saved in the vertex buffer, it can be represented with the buffer index that requires a lower number of bits than the global vertex index. Assuming that each vertex is reused by the buffer index only once, Taubin and Rossignac [5] showed that the generalized triangle mesh representation requires approximately 11 bpv to encode the connectivity data for large meshes. Deering, however, did not propose a method to decompose a mesh into triangle strips.

Based on Deering's work, Chow [13] proposed a mesh compression scheme optimized for real-time rendering. He proposed a mesh decomposition method as illustrated in Fig. 2.8. First, it finds a set of boundary edges. Then, it finds a fan of triangles around each vertex incident to two consecutive boundary edges. These triangle fans are combined to form the first generalized triangle strip. The triangles in this strip are marked as discovered, and a new set of boundary edges is generated to separate discovered triangles from undiscovered triangles. The next generalized triangle strip is similarly formed from the new set of boundary edges. With the vertex buffer, the vertices in the previous generalized triangle strip can be reused in the next one. This process continues until all triangles in a mesh are traversed.

The triangle strip representation can be applied to a triangle mesh of arbitrary topology. However, it is effective only if the triangle mesh is decomposed into long triangle strips. It is a challenging computational geometry problem to obtain optimal triangle strip decomposition [14]. Several heuristics have been proposed to obtain sub-optimal decompositions at a moderate computational cost [15].

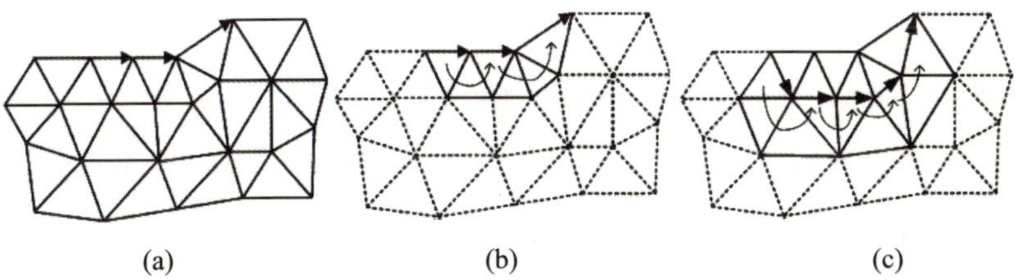

Fig. 2.8. The mesh decomposition method proposed by Chow [13]. (a) A set of boundary edges; (b) Triangle fans for the first strip; (c) Triangle fans for the second strip. Thick arrows show selected boundary edges and thin arrows show the triangle fans associated with each inner boundary vertex (©[1997]IEEE)

2.2.3 Spanning-Tree-Based Connectivity Coding

Turan [16] observed that the connectivity of a planar graph can be encoded with a constant number of bpv using two spanning trees: a vertex spanning tree and a triangle spanning tree. Based on this observation, Taubin and Rossignac [5] presented a topological surgery approach to encode mesh connectivity. The basic idea is to cut a given mesh along a selected set of cut edges to make a planar polygon. The mesh connectivity is then represented by the structures of cut edges and the polygon. In a simple mesh, any vertex spanning tree can be selected as the set of cut edges.

Fig. 2.9 illustrates the encoding process. Fig. 2.9(a) is an octahedron mesh. First, the encoder constructs a vertex spanning tree as shown in Fig. 2.9(b), where each node corresponds to a vertex in the input mesh. Then, it cuts the mesh along the edges of the vertex spanning tree. Fig. 2.9(c) shows the resulting planar polygon and the triangle spanning tree. Each node in the triangle spanning tree corresponds to a triangle in the polygon, and two nodes are connected if and only if the corresponding triangles share an edge.

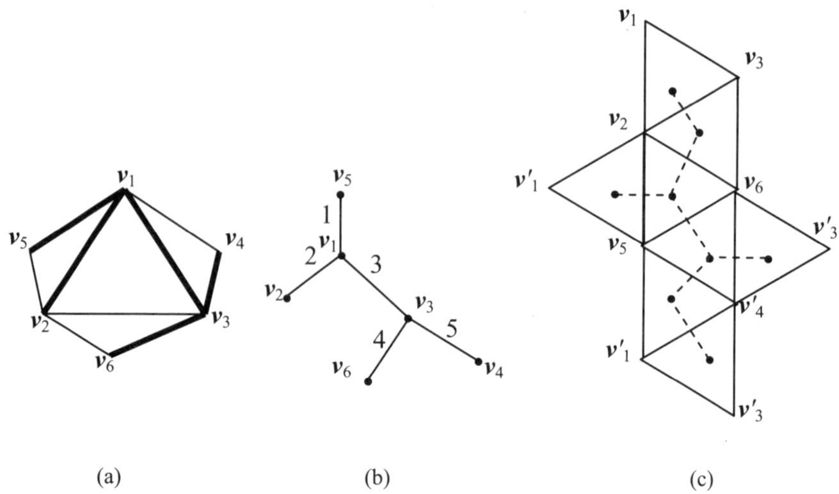

Fig. 2.9. Encoding process of the topological surgery approach [5]. (a) An octahedron mesh; (b) Its vertex spanning tree; (c) The cut and flattened mesh with its triangle spanning tree shown by dashed lines (©1998 Association for Computing Machinery, Inc. Reprinted by permission)

Then, the two spanning trees are run-length encoded. A run is defined as a tree segment between two nodes with degrees not equal to 2. For each run of the vertex spanning tree, the encoder records its length with two additional flags. The first flag is the branching bit indicating whether a run subsequent to the current run starts at the same branching node, and the second flag is the leaf bit indicating whether the current run ends at a leaf node. For example, let us encode the vertex spanning tree in Fig. 2.9(b), where the edges are labeled with their run indices. The first run is represented by (1, 0, 0), since its length is 1, the next run does not start at the same node and it does not end at a leaf node. In this way, the vertex spanning tree in Fig. 2.9(b) is represented by (1,0,0), (1,1,1), (1,0,0), (1,1,1), (1,0,1). Similarly, for each run of the triangle spanning tree, the encoder writes its length and the leaf bit. Note that the triangle spanning tree is always binary so that it does not need the branching bit. Furthermore, the encoder records the marching pattern with one bit per triangle to indicate how to triangulate the planar polygon internally. The decoder can reconstruct the original mesh connectivity from this set of information.

In both vertex and triangle spanning trees, a run is a basic coding unit. Thus, the coding cost is proportional to the number of runs, which in turn depends on how the vertex spanning tree is constructed. Taubin and Rossignac's algorithm builds the vertex spanning tree based on layered decomposition, which is similar to the way we peel an orange along a spiral path, to maximize the length of each run and minimize the number of runs generated.

Taubin and Rossignac also presented several modifications so that their algorithm can encode general manifold meshes: meshes with arbitrary genus, meshes with boundary and non-orientable meshes. However, their algorithm cannot directly deal with non-manifold meshes. As a preprocessing step, the encoder should segment a non-manifold mesh into several manifold components, thereby duplicating non-manifold vertices, edges and faces. Experimentally, Taubin and Rossignac's algorithm requires 2.48–7.0 bpv for mesh connectivity. It was also shown that the time as well as the space complexities of their algorithm is $O(N)$, where N is the maximum value among N_v, N_e and N_f. It demands a large memory buffer due to its global random vertex access at the decompression stage.

2.2.4 Layered-Decomposition-Based Connectivity Coding

Bajaj et al. [17] proposed a connectivity coding method based on a layered structure of vertices. The main idea is to first decompose a triangle mesh into several concentric layers of vertices, and then construct triangle layers within each pair of adjacent vertex layers. The mesh connectivity is represented by the total number of vertex layers, the layout of each vertex layer and the layout of triangles in each triangle layer. Ideally, a vertex layer does not intersect itself and a triangle layer is a generalized triangle strip. In such a case, the connectivity compression is reduced to the coding of the number of vertex layers, the number of vertices in each vertex layer and the generalized triangle strip in each triangle layer. However, in practice, overhead bits are introduced due to the existence of branching points, bubble triangles and triangle fans.

Branching points are produced when a vertex layer intersects itself. In Fig. 2.10(a), the middle layer intersects itself at the branching point indicated by a big dot. Branching points partition a vertex layer into several segments called contours. To encode the layout of a vertex layer, we have to encode the information of both contours and branching points. In addition, as shown in Figs. 2.10(b)–(d), each triangle in a triangle layer can be categorized into three cases: (1) Its vertices are located on two adjacent vertex layers. A generalized triangle strip consists of a sequence of triangles of this kind. (2) All its vertices belong to one contour. It is called a bubble triangle. (3) Its vertices are located on two or three contours in one vertex layer. A cross-contour triangle fan is composed of a sequence of triangles of this kind. Therefore, besides encoding generalized triangle strips between two adjacent vertex layers, this algorithm requires additional bits to encode bubble triangles and cross-contour triangle fans.

Taubin and Rossignac [5] also utilized layered decomposition in the vertex spanning tree construction. However, Bajaj et al.'s algorithm [17] is different from Taubin and Rossignac's scheme [5] in the following three aspects: (1) It does not combine vertex layers into the vertex spanning tree. (2) Its decoder does not need a large memory buffer, since it accesses only a small portion of vertices at each decompression step. (3) It is applicable to any kind of mesh topology, while Taubin and Rossignac's scheme [5] cannot encode non-manifold meshes directly.

The layered decomposition method encodes the connectivity information with about 1.40–6.08 bpv. Moreover, it has the desirable property that each triangle depends on at most two adjacent vertex layers and each vertex is referenced by at most two triangle layers. This property enables the error-resilient transmission of mesh data, for the effects of transmission errors can be localized by encoding different vertex and triangle layers independently. Based on the layered decomposition method, Bajaj et al. [18] also proposed an algorithm to encode large CAD models. This algorithm extends the layered decomposition method to compress quadrilateral and general polygonal models as well as CAD models with smooth non-uniform rational B-splines (NURBS) patches.

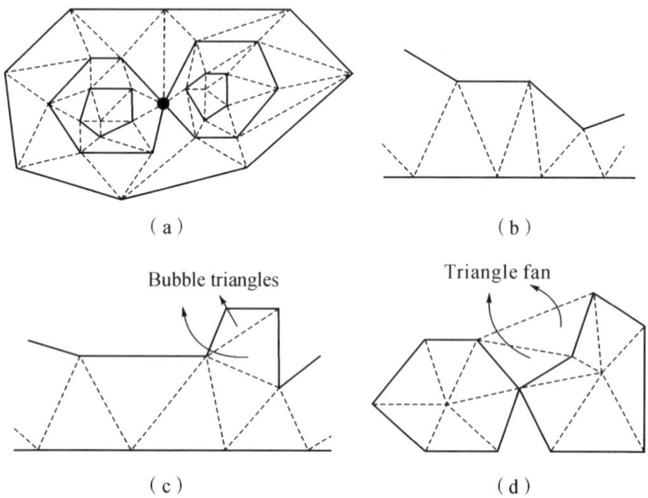

Fig. 2.10. Three cases in the triangle layer, where contours are depicted with solid lines and other edges with dashed lines. (a) The layered vertex structure and the branching point depicted by a black dot; (b) A triangle strip; (c) Bubble triangles; (d) A cross-contour triangle fan

2.2.5 Valence-Driven Connectivity Coding Approach

The main idea of the valence-driven approach is as follows. First, it selects a seed triangle whose three edges form the initial borderline. Then, the borderline partitions the whole mesh into two parts, i.e., the inner part that has been processed and the outer part that is to be processed. Next, the borderline gradually expands outwards until the whole mesh is processed. The output is a stream of vertex valences, from which the original connectivity can be reconstructed.

In [19], Touma and Gotsman presented a pioneering algorithm known as the valence-driven approach. It starts from an arbitrary triangle, and pushes its three vertices into a list called the active list. Then, it pops up a vertex from the active list, traverses all untraversed edges connected to that vertex, and pushes the new vertices into the end of the list. For each processed vertex, it outputs the valence. Sometimes it needs to split the current active list or merge it with another active list. These cases are encoded with special codes. Before encoding, for each boundary loop, a dummy vertex is added and connected to all the vertices in that boundary loop, making the topology closed. Fig. 2.11 shows an example of the encoding process, where the active list is depicted by thick lines, and the focus vertex by the black dot, and the dummy vertex by the gray dot. Table 2.1 lists the output of each step associated with Fig. 2.11.

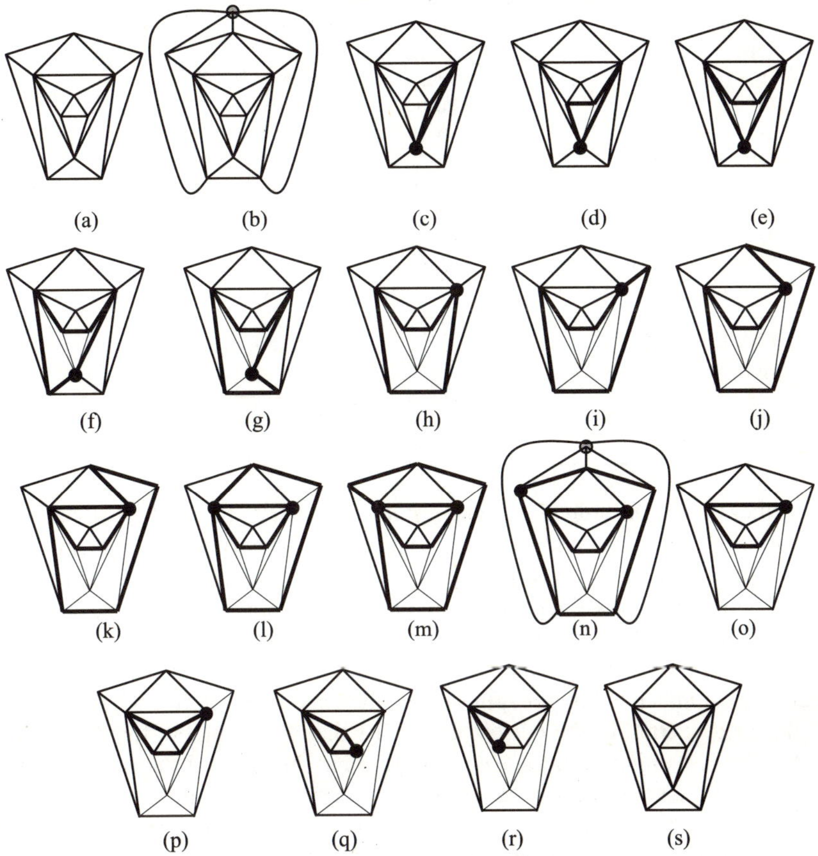

Fig. 2.11. (a)–(s) showing a mesh connectivity encoding example by Touma and Gotsman [19], where the active list is shown with thick lines, the focus vertex with the black dot and the dummy vertex with the gray dot (With courtesy of Touma and Gotsman)

Since vertex valences are compactly distributed around 6 in a typical mesh, arithmetic coding can be utilized to encode the valence information of a vertex effectively [19]. The resulting algorithm costs less than 1.5 bpv on average to encode mesh connectivity. This is the state-of-the-art compression ratio that has not been seriously challenged up to now. However, it is only applicable to orientable manifold meshes.

Table 2.1 The output of each step in Fig. 2.11

Subfigure	Output	Comments
(a)		An input mesh is given
(b)		Add a dummy vertex
(c)	Add 6, add 7, add 4	Output the valences of starting vertices
(d)	Add 4	Expand the active list
(e)	Add 7	Expand the active list
(f)	Add 5	Expand the active list
(g)	Add 5	Expand the active list
(h)		Choose the next focus vertex
(i)	Add 4	Expand the active list
(j)	Add 5	Expand the active list
(k)	Split 5	Split the active list, and push the new active list into stack
(l)		Choose the next focus vertex
(m)	Add 4	Expand the active list
(n)	Add dummy 5	Choose the next focus vertex and conquer the dummy vertex
(o)		Pop the new active list from the stack
(p)	Add 4	Expand the active list
(q)		Choose the next focus vertex
(r)		Choose the next focus vertex
(s)		The whole mesh is conquered

Alliez and Desbrun [20] suggested a method to further improve the performance of Touma and Gotsman's algorithm. They observed that split codes, split offsets and dummy vertices consume a non-trivial portion of coding bits in Touma and Gotsman's algorithm. To reduce the number of split codes, they used a heuristic method that selects the vertex with the minimal number of free edges as the next focus vertex, instead of choosing the next vertex in the active list. To reduce the number of bits for split offsets, they excluded the two adjacent vertices of the focus vertex in the current active list that are ineligible for the split, and sorted the remaining vertices according to their Euclidean distances to the focus vertex. Then, a split offset is represented with an index into this sorted list, which is further added by 6 and encoded in the same way as a normal valence. To reduce the number of dummy vertices, they adopted one common dummy vertex for all boundaries in the input mesh. Furthermore, they encoded the output symbols with the range encoder [21], an effective adaptive arithmetic encoder.

Alliez and Desbrun's algorithm is also applicable only to orientable manifold meshes. It outperforms Touma and Gotsman's algorithm, especially for irregular meshes. Alliez and Desbrun proved that if the number of splits is negligible, the performance of their algorithm is upper-bounded by 3.24 bpv, which is exactly the same as the theoretical bpv value computed by enumerating all possible planar graphs [22].

Recently, Gotsman [23] has shown that the average entropy of the distribution of valences in valence sequences for the class of manifold 3D triangle meshes and the class of manifold 3D polygon meshes is strictly less than the entropy of these classes themselves. This fact indicates that some of the bits per vertex in the valence-based connectivity code must be due to the split operations (or some other essential piece of information). In other words, the number of split operations in the code is linear in the size of the mesh, albeit with a very small constant. This means that the empirical observation that the number of split operations is negligible is incorrect, and is probably due to the experiments being performed on a small subset of relatively "well-behaved" mesh connectivities. At present, there is no way of bounding this number, meaning that even if the coding algorithms minimize the number of split operations, there is no way for us to eliminate the possibility that the size of the code may actually exceed the Tutte entropy (due to these split operations). The question of the optimality of valence-based coding of 3D meshes will remain open until more concrete information on the expected number of split operations incurred during the mesh conquest is available. We do believe, nonetheless, that even if the valence-based coding is not optimal, it is probably not far from this.

2.2.6 Triangle-Conquest-Based Connectivity Coding

Similar to the valence-driven approach, the triangle conquest approach starts from the initial borderline, which partitions the whole mesh into conquered and unconquered parts, and then inserts triangle by triangle into the conquered parts. The main difference is that the triangle conquest scheme outputs the building operations of new triangles, while the valence-driven approach outputs the valences of new vertices. Gumhold and Straßer [24] first presented a triangle conquest approach, called the cut-border machine. At each step, this scheme inserts a new triangle into the conquered part, closed by the cut-border, with one of the five building operations: "new vertex", "forward", "backward", "split" and "close". The sequence of building operations is encoded with Huffman codes. This method is applicable to manifold meshes that are either orientable or non-orientable. Experimentally, its compression cost lies within 3.22–8.94 bpv, mostly around 4 bpv. The most important advantage of this scheme is that the decompression speed is very fast and the decompression method is easy to implement with hardware. Furthermore, compression and decompression operations can be performed in parallel. These properties make this method very attractive in real-time coding applications. In [25], Gumhold further improved the compression performance by using an adaptive arithmetic coder to optimize the border encoding. The experimental compression ratio is within the range of 0.3–2.7 bpv, and on average 1.9 bpv.

Rossignac [26] proposed another triangle conquest approach called the edgebreaker algorithm. It is nearly equivalent to the cut-border machine, except that it does not encode the offset data associated with the split operation. The triangle traversal is controlled by edge loops as shown in Fig. 2.12(a). Each edge loop bounds a conquered region and contains a gate edge. At each step, this approach focuses on one edge loop and its gate edge is called the active gate, while the other edge loops are stored in a stack and will be processed later. Initially, for each connected component, one edge loop is defined. If the component has no physical boundary, two half edges corresponding to one edge are set as the edge loop. For example, in Fig. 2.12(b), the mesh has no boundary and the initial edge loop is formed by g and $g{\cdot}o$, where $g{\cdot}o$ is the opposite half edge of g. In Fig. 2.12(c), the initial edge loop is the mesh boundary.

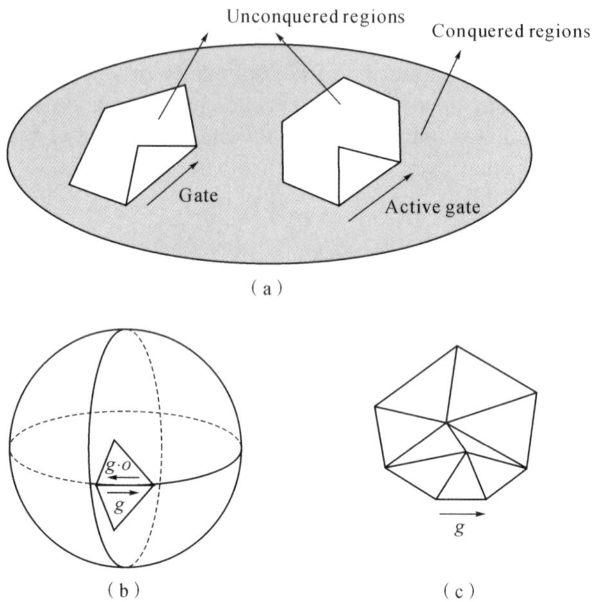

Fig. 2.12. Illustration of the Edgebreaker algorithm, where thick lines depict edge loops, and g denotes the gate. (a) Edge loops; (b) Gates and initial edge loops for a mesh without boundary; (c) Gates and initial edge loops for a mesh with boundary

At each step, this scheme conquers a triangle incident on the active gate, updates the current loop, and moves the active gate to the next edge in the updated loop. For each conquered triangle, this algorithm outputs an op-code. Assume that the triangle to be removed is enclosed by the active gate g and the vertex $\boldsymbol{v}$, there are five kinds of possible op-codes as shown in Fig. 2.13(a): (1) C (loop extension), if $\boldsymbol{v}$ is not on the edge loop; (2) L (left), if $\boldsymbol{v}$ immediately precedes g in the edge loop; (3) R (right), if $\boldsymbol{v}$ immediately follows g; (4) E (end), if $\boldsymbol{v}$ precedes and follows g; (5) S (split), otherwise. Essentially, the compression process is a depth-first traversal of the dual graph of the mesh. When the split case is encountered, the current loop is split into two, and one of them is pushed into the stack while the other is further traced. Fig. 2.13(b) shows an example of the encoding process, where the arrows and the numbers give the order of the triangle conquest. The triangles are filled with different patterns to represent different op-codes, which are produced when they are conquered. In this case, the encoder outputs the series of op-codes as *CCRSRLLRSEERLRE*.

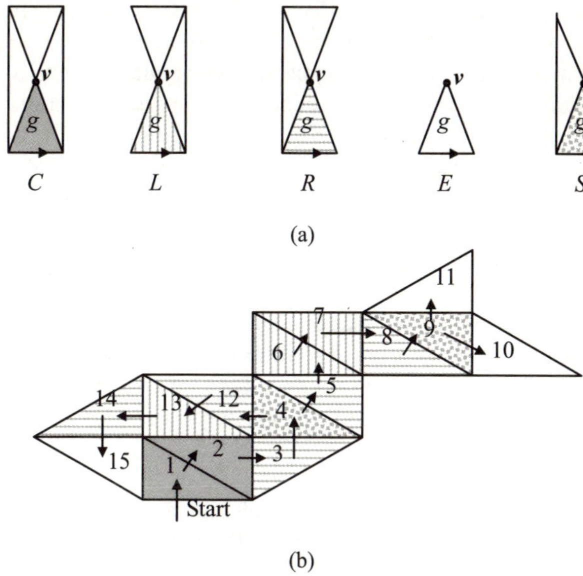

Fig. 2.13. Five op-codes used in the Edgebreaker algorithm. (a) Five op-codes *C*, *L*, *R*, *E*, and *S*, where the gate *g* is marked with an arrow; (b) An example of the encoding process in the Edgebreaker algorithm, where the arrows and the numbers show the traversal order and different filling patterns are used to represent different op-codes

The Edgebreaker method can encode the topology data of orientable manifold meshes with multiple boundary loops or with arbitrary genus, and guarantee a worst-case coding cost of 4 bpv for simple meshes. However, it is unsuitable for streaming applications, since it requires a two-pass process for decompression, and the decompression time is $O(N_v^2)$. Another disadvantage is that, even for regular meshes, it requires about the same bitrate as that for non-regular meshes.

King and Rossignac [27] modified the Edgebreaker method to guarantee a worst-case coding cost of 3.67 bpv for simple meshes, and Gumhold [28] further improved this upper bound to 3.522 bpv. The decoding efficiency of the Edgebreaker method was also improved to exhibit linear time and space complexities in [27, 29, 30]. Furthermore, Szymczak et al. [31] optimized the Edgebreaker method for meshes with high regularity by exploiting dependencies of output symbols. It guarantees a worst-case performance of 1.622 bpv for sufficiently large meshes with high regularity.

As mentioned earlier, we can reduce the amount of data transmission between the CPU and the graphic card by decomposing a mesh into long triangle strips, but finding a good decomposition is often computationally intensive. Thus, it is often desirable to generate long strips from a given mesh only once and distribute the stripification information together with the mesh. Based on this observation, Isenburg [32] presented an approach to encode the mesh connectivity together with its stripification information. It is basically a modification of the Edgebreaker method, but its traversal order is guided by strips obtained by the STRIPE algorithm [15]. When a new triangle is included, its relation to the underlying triangle strip is encoded with a label. The label sequences are then entropy encoded. The experimental compression performance ranges from 3.0 to 5.0 bpv.

Recently, Jong et al. proposed an edge-based single-resolution compression scheme [33] to encode and decode 3D models straightforwardly via single pass traversal in a sequential order. Most algorithms use the split operation to separate the 3D model into two components; however, the displacement is recorded or an extra operator is required for identifying the branch. This study suggested using the *J* operator to skip to the next edge of the active boundary, and thus it does not require split overhead. With all sorts of conditions of active gates and third vertices, this study adopted five operators, *QCRLJ*, and then used them to encode and decode triangular meshes. This algorithm adopts Rossignac's *CRL* operators [26] as shown in Fig. 2.13(a), and two new operators are proposed, *Q* and *J*, as illustrated in Fig. 2.14(a). For explanatory purposes, *Q* and *J* operators are described as follows:

(1) *Q*. The third vertex is a new vertex and its consecutive triangle is *R*. These two triangles, which comprise a quadrilateral, are then shifted from the un-compressed area into the compressed area. The active gate is then removed and the other two sides of the quadrilateral that are not on the active boundary are moved to the active boundary, then the right side is allowed to serve as the new active gate. The geometric characteristics demonstrate that the *Q* operator represents two triangles which are coded *CR*. Different from the further context-based encoding for *CR* codes conducted by Rossignac, this approach only requires us to read *Q* at the decompression process, and treats it as two triangles. However, using the context-based coder requires transforming the code to *CR*, and then acknowledges these two triangles.

(2) *J*. The third vertex lies on the active boundary and is not the previous or next vertex of the active gate. This operator does not compress any triangle and the next side of active boundary is allowed to serve as the new active gate. The active gate skips to the next edge of the active boundary. Since the third vertex that corresponds with the active gate comprises one triangle, and this triangle divides the un-compressed area into two, numerous indications for the third vertex

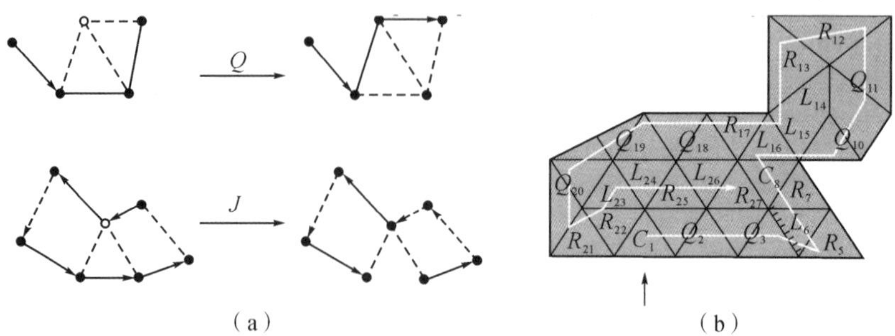

Fig. 2.14. Two new operators and the corresponding compression process adopted in [33]. (a) Operators *Q* and *J*; (b) A compression example (©[2005]IEEE)

are stumped up under this condition. Thus, this triangle is not compressed and is eventually compressed by "*R*" or "*L*".

Fig. 2.14(b) illustrates the compression course of Jong et al.'s algorithm, where the dotted lines represent *J* operators. A total of 27 operators are calculated as *CQQJRLRCJQ QRRLLLRQQQ RRLLRLR* using Jong et al.'s algorithm. Furthermore, the adaptive arithmetic coder is applied in Jong et al.'s algorithm to achieve an improved compression ratio.

2.2.7 Summary

Table 2.2 summarizes the bitrates of various connectivity coding schemes introduced above. The bitrates marked by "*" are the theoretical upper bounds obtained by the worst-case analysis, while the others are experimental bitrates. Among these methods, Touma and Gotsman's algorithm [19] is viewed as the state-of-the-art technique for single-rate 3D mesh compression. With some minor improvements on Touma and Gotsman's algorithm, Alliez and Desbrun's algorithm [20] yields an improved compression ratio. The indexed face set, triangle strip and layered decomposition methods can encode meshes with arbitrary topology. In contrast, other approaches can handle only manifold meshes with additional constraints. For instance, the valence-driven approach [19, 20]

Table 2.2 Comparisons of bitrates for various single-rate connectivity coding algorithms

Category	Algorithm	Bitrate (bpv)	Comment
Indexed face set	VRML ASCII Format [3]	$6\log_2 N_v$	No compression
Triangle strip	Deering [12]	11	
Spanning tree	Taubin and Rossignac [5]	2.48–7.0	
Layered decomposition	Bajaj et al. [17]	1.40–6.08	
Valence-driven approach	Touma and Gotsman [19]	0.2–2.4, 1.5 on average	Especially good for regular meshes
	Alliez and Desbrun [20]	0.024—2.96, 3.24*	
Triangle conquest	Gumhold and Straßer [24]	3.22–8.94, 4 on average	Optimized for real-time applications
	Gumhold [25]	0.3–2.7, 1.9 on average	
	Rossignac [26]	4*	
	King and Rossignac [27]	3.67*	
	Gumhold [28]	3.522*	
	Szymczak et al. [31]	1.622* for sufficiently large meshes with high regularity	Optimized for regular meshes
	Jong et al. [33]	1.19 on average	An adaptive arithmetic coder is used

* Theoretical upper bounds obtained by the worst-case analysis

requires that the manifold be also orientable. Szymczak et al.'s algorithm [31] requires that the manifold have neither boundary nor handles. Note that using these algorithms, a non-manifold mesh can be handled only if it is pre-converted to a manifold mesh by replicating non-manifold vertices, edges and faces as in [34].

2.3 Progressive Connectivity Compression

Progressive compression of 3D meshes is desirable for transmission of complex meshes over networks with limited bandwidth. The main idea is as follows: a coarse mesh is first transmitted and rendered. Then, the refinement data are progressively transmitted to perfect the mesh representation until the received mesh is rendered in its full resolution or the transmission task is canceled by users. The main advantage of progressive compression is that we can have access to intermediate meshes of the object during its transmission over the network, as illustrated in Fig. 2.15. Furthermore, progressive compression allows transmission and rendering of different levels of details (LOD). However, there is a tradeoff between the compression ratio and the number of LODs. In general, a progressive coder is less effective than a single-rate coder in terms of the coding gain, for it cannot make full use of the correlation among mesh data as freely as the single-rate coder. The challenge is then composed of reconstructing a least distorted object at all points in time during transmission (i.e., optimization of rate-distortion tradeoff).

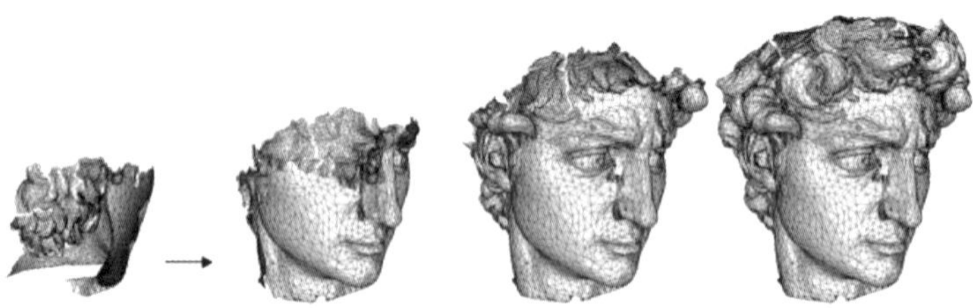

(a)

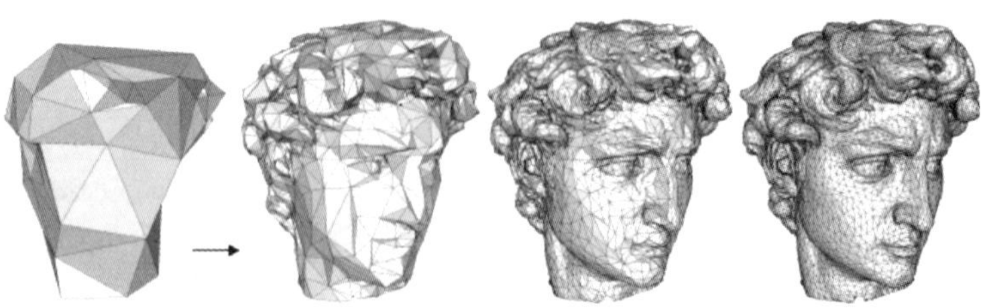

(b)

Fig. 2.15. Intermediate meshes [1]. (a) Based on a single-rate technique; (b) Using a progressive technique (With courtesy of Alliez and Gotsman)

Progressive mesh compression is highly related to the research work on mesh simplification. Typically, to encode a 3D mesh progressively, we gradually simplify it to a base mesh that has a much smaller number of vertices, edges and faces than the original one. During the simplification process, we record each operation. By reversing the series of simplification operations, we can restore the base mesh to the original one. Progressive coders attempt to compress the base mesh and the series of reversed simplification operations. However, progressive coders differ in three aspects, i.e., mesh simplification techniques, geometry coding methods and interaction between connectivity coding and geometry coding.

We call a mesh compression technique "lossless" if the method can restore the original mesh connectivity and geometry data once the transmission is complete, even though intermediate stages are obviously lossy. Most of these techniques proceed by decimating the mesh while recording the minimally redundant information required for reversing this process. The three basic ingredients behind most of progressive mesh compression techniques are: (1) the selection of an atomic mesh decimation operator; (2) the choice of a geometric distance metric to determine the elements to be decimated; (3) the design of an efficient coding scheme for the information required to reverse the decimation process. Intuitively, we have to encode for the decoder both the locations of the refinement and the parameters to perform the refinement itself.

Similar to single-rate compression techniques, in many traditional progressive coding schemes, the compact representation of connectivity data is given a priority and then geometry coding is driven, but restrained at the same time, by connectivity coding. However, three types of new approaches have emerged: the first type is to compress geometry data with little reference to connectivity data, the second type is to drive connectivity coding with geometry coding, and the third type is to even change mesh connectivity in favor of a better compression of geometry data. Therefore, we can classify the progressive coding schemes into two classes, i.e., connectivity-driven compression and geometry-driven compression. In this section, we discuss several typical progressive connectivity-driven compression methods.

2.3.1 Progressive Meshes

Hoppe [35] first introduced the progressive mesh (PM) representation, a new scheme for storing and transmitting arbitrary triangle meshes. This efficient, lossless, continuous-resolution representation addresses several practical problems in graphics: smooth geomorphing of level-of-detail approximations, progressive transmission, mesh compression and selective refinement. This scheme simplifies a given orientable manifold mesh with successive edge collapse operations. As shown in Fig. 2.16, if an edge is collapsed, its two end points are merged into one, and two triangles (or one triangle if the collapsed edge is on the boundary) incident to this edge are removed, and all vertices previously connected to the two end points are re-connected to the merged vertex. The inverse operation of edge collapse (e_col as shown in Fig. 2.16) is vertex split (v_split as shown in Fig. 2.16) that inserts a new vertex into the mesh together with corresponding edges and triangles.

An original mesh $\boldsymbol{M} = \boldsymbol{M}_k$ can be simplified into a coarser mesh $\boldsymbol{M}_0$ by performing k successive edge collapse operations. Each edge collapse operation $ecol_i$ transforms the mesh $\boldsymbol{M}_i$ to $\boldsymbol{M}_{i-1}$, with $i = k, k-1, \ldots, 1$. Since edge collapse operations are invertible, we can represent an arbitrary triangle mesh $\boldsymbol{M}$ with its base mesh $\boldsymbol{M}_0$ together with a sequence of vertex split operations. Each vertex split operation $vsplit_i$ refines the mesh $\boldsymbol{M}_{i-1}$ back to $\boldsymbol{M}_i$, with $i = 1, 2, \ldots, k$. Thus, we can view ($\boldsymbol{M}_0$, $vsplit_1$, …, $vsplit_k$) as the progressive mesh representation of $\boldsymbol{M}$.

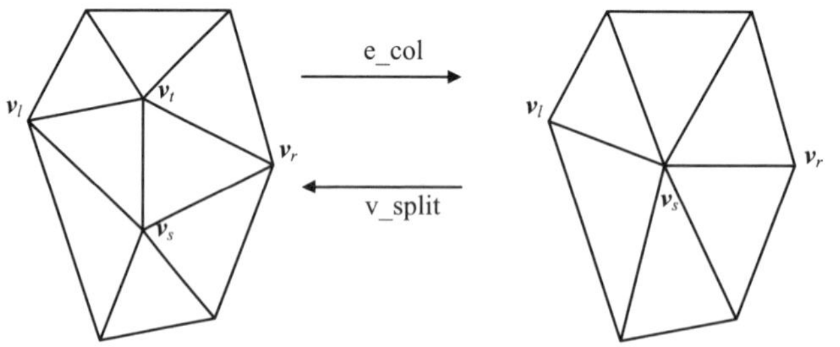

Fig. 2.16. Illustration of the edge collapse and vertex split processes

During the construction of a progressive mesh, it is important to select a proper edge to be collapsed at each step. Similar to Hoppe et al.'s mesh optimization scheme [36], we can adopt an energy function E that takes several aspects into account, i.e., distance accuracy, attribute accuracy, regularization and discontinuity curves. Each edge is put into a priority queue, where the priority value is its estimated energy cost ΔE. Initially, we calculate the priority value for each edge. Then, at each iteration, we collapse the edge with the smallest priority value and then update the priorities of its neighboring edges.

The connectivity of the base mesh $\boldsymbol{M}_0$ can be encoded using any single-rate coder as introduced in the last section. The vertex split in Fig. 2.16 can be specified by the indices of the split vertex $\boldsymbol{v}_s$ and its left and right vertices, $\boldsymbol{v}_l$ and $\boldsymbol{v}_r$. If there are N_{vi} vertices in the intermediate mesh $\boldsymbol{M}_i$, the index of $\boldsymbol{v}_s$ can be encoded with $\log_2 N_{vi}$ bits. Then, the two indices of $\boldsymbol{v}_l$ and $\boldsymbol{v}_r$ can be encoded with $\log_2(\beta(\beta-1))$ bits, where β is the number of vertices connected to $\boldsymbol{v}_s$. Since the average vertex valence is 6 in a typical mesh, the indices of $\boldsymbol{v}_l$ and $\boldsymbol{v}_r$ can be encoded with about 5 ($\approx\log_2(6\times5)$) bits. Thus, we require about ($\log_2 N_{vi}$+5) bits to represent the vertex split operation. Overall, PM requires $O(N_v \log N_v)$ bits to represent the topology of a mesh with N_v vertices. Accompanied with the vertex split operation, positions of $\boldsymbol{v}_t$ and $\boldsymbol{v}_s$ are Huffman-coded after delta prediction.

Although the original PM is innovative in nature, it is not a very efficient compression scheme. To improve its coding efficiency, Hoppe proposed another PM implementation method in [37]. It reorders the vertex split operations to increase the compression ratio at the cost of quality degradation of intermediate meshes. It requires about 10.4 bits to represent each vertex split operation. Furthermore, Hoppe's PM method has been extended or improved by several researchers as discussed below.

2.3.1.1 Progressive Simplicial Complex

Popovic and Hoppe [38] observed that the original PM has two restrictions: (1) It is applicable only to orientable manifold meshes; (2) It does not possess the freedom to change the topological type of a given mesh during the simplification and refinement, which limits its coding efficiency. To alleviate these problems, they presented a method called progressive simplicial complex (PSC). In this scheme, a more general vertex split operation is exploited to encode the changes in both geometry and topology. A PSC representation consists of a single-vertex base model followed by a sequence of generalized vertex split operations. PSC can be used to compress meshes of any topology type.

To construct a PSC representation, a sequence of vertex merging operations are performed to simplify a given mesh model. Each vertex merging operation merges an arbitrary pair of vertices, which are not necessarily connected by an edge, into a single vertex. The inverse operation of vertex merging is the generalized vertex split operation that splits a vertex into two. Suppose that the vertex $\boldsymbol{v}_i$ in the mesh $\boldsymbol{M}_i$ is to be split to generate a new vertex whose index is i+1 in the mesh $\boldsymbol{M}_{i+1}$. Each simplex adjacent to $\boldsymbol{v}_i$ in $\boldsymbol{M}_i$ is the merging result of one of four cases as shown in Fig. 2.17. For a rigorous definition of simplex, readers can refer to [38]. Intuitively, a 0-dimensional simplex is a point, a 1D simplex is an edge and a 2D simplex is a triangle face, and so on. For each simplex adjacent to $\boldsymbol{v}_i$, PSC assigns a code to indicate one of the four cases as given in Fig. 2.17.

Since the generalized vertex split operation is more flexible than the original vertex split operation in PM, PSC may require more bits in connectivity coding than PM. Specifically, PSC requires about ($\log_2 N_{vi}$+8) bits to specify the connectivity change around the split vertex, while PM requires only about ($\log_2 N_{vi}$+5) bits. However, the main advantage of PSC is its capability to handle arbitrary triangular models without any topology constraint. Similar to PM, the geometry data in PSC are also encoded based on delta prediction.

2.3.1.2 Progressive Forest Split

Taubin et al. [39] suggested the progressive forest split (PFS) representation for manifold meshes. Similar to the PM representation [35], a triangle mesh is represented with a low resolution base model and a series of refinement operations in PFS. Instead of the vertex split operation, the PFS scheme exploits the forest split operation as illustrated in Fig. 2.18. The forest split operation cuts a mesh along the edges in the forest and fills in the resulting crevice with triangles. For the sake of simplicity, the forest contains only one tree in Fig. 2.18. In practice, a forest may be composed of many complex trees, and a single forest split operation may double the number of triangles in a mesh. Therefore, PFS can obtain a much higher compression ratio than PM at the cost of reduced granularity.

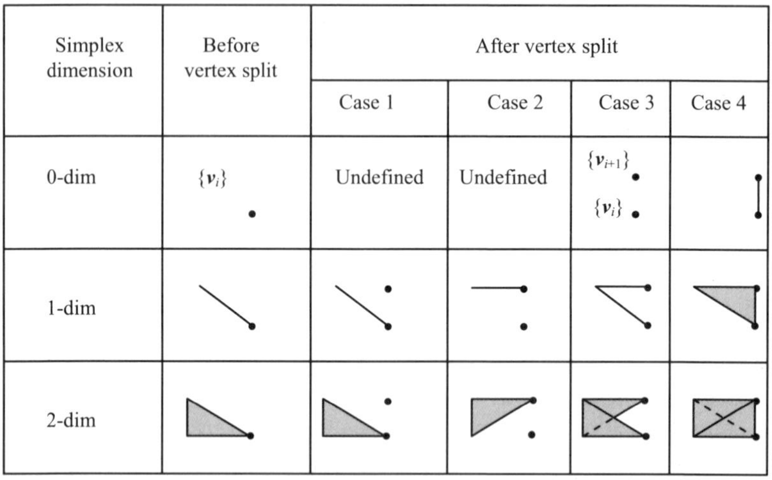

Fig. 2.17. Possible cases after a generalized vertex split for different-dimensional simplices

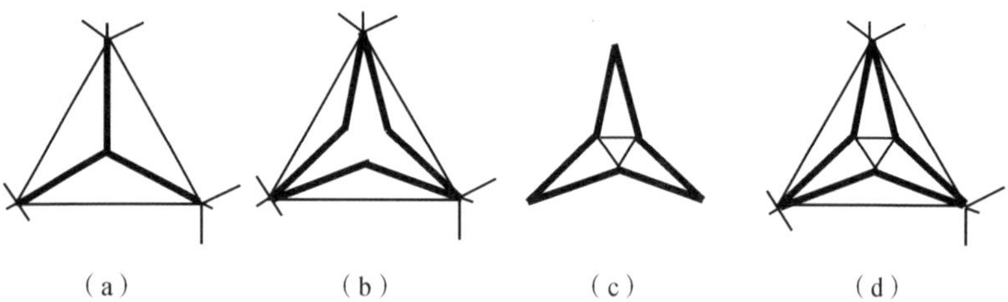

Fig. 2.18. Illustration of a forest split process. (a) The original mesh with a forest marked with thick lines; (b) The cut of the original mesh along the forest edges; (c) Triangulation of the crevice; (d) The cut mesh in (b) filled with the triangulation in (c)

For each forest split operation, the forest structure, the triangulation information of the crevices and the vertex displacements are encoded. To encode the forest structure, one bit is required for each edge indicating whether it belongs to the forest or not. To encode the triangulation of the crevices, the triangle spanning tree and the marching patterns can be adopted as in Taubin and Rossignac's algorithm [5], or a simple constant-length encoding scheme can be employed, which requires exactly 2 bits per new triangle. To encode the vertex displacements, a smoothing algorithm [40] is first applied after connectivity refinement, and then the difference between the original vertex position and the smoothed vertex position is Huffman-coded.

With respect to the coding efficiency, to progressively encode a given mesh with four or five LODs, PFS requires about 7–10 bpv for the connectivity data and

20–40 bpv for the geometry data at the 6-bit quantization resolution. Here, we should point out that the bpv performance is measured with respect to the number of vertices in the original mesh. PFS has been adopted in MPEG-4 3DMC [6] as an optional mode for progressive mesh coding.

2.3.1.3 Compressed Progressive Mesh

Pajarola and Rossignac [41] suggested a modified PM called the compressed progressive mesh (CPM), which is applicable to manifold meshes. Similar to PFS, CPM also improves the compression performance at the expense of reduced granularity. To use fewer bits for connectivity data, CPM groups vertex splits into batches. CPM adopts a sequence of marking bits to specify the vertices to be split in one batch, while PM uses $\log_2 N_{vi}$ bits for each vertex split in the intermediate mesh $\boldsymbol{M}_i$. For geometry coding, an edge ($\boldsymbol{v}_1$, $\boldsymbol{v}_2$) is collapsed to its midpoint $\boldsymbol{v} = (\boldsymbol{v}_1+\boldsymbol{v}_2)/2$. Thus, if the vector $\boldsymbol{d} = \boldsymbol{v}_2-\boldsymbol{v}_1$ is known, the positions of $\boldsymbol{v}_1$ and $\boldsymbol{v}_2$ can be reconstructed from $\boldsymbol{v}$ and $\boldsymbol{d}$. CPM obtains the prediction $\hat{\boldsymbol{d}}$ of $\boldsymbol{d}$ based on the vertices that have a topological distance of 1 or 2 from the vertex $\boldsymbol{v}$ in a similar manner to the butterfly subdivision technique [42, 43]. The prediction error $\boldsymbol{d}-\hat{\boldsymbol{d}}$ is then Huffman-coded.

CPM adopts the Laplacian distribution to approximate the prediction error histogram. For each batch, it computes and transmits the variance of the Laplacian distribution for the decoder to reconstruct the Huffman coding table, thus alleviating the need to transmit the table. CPM can encode all connectivity data with about 7.0 bpv and all geometry data with about 12–15 bpv at 8-bit to 12-bit quantization resolutions. Overall, CPM requires about 22 bpv, that is approximately half the bitrate of PFS [39].

Further, Pajarola and Rossignac [44] optimized CPM for real-time applications. They adopted the so-called half-edge collapse operation to collapse an edge into one of its ending points instead of its midpoint, since the midpoint may not lie on the quantized coordinate grid which makes geometry coding more complex. In addition, to reduce the overhead computational complexity, a new vertex position is estimated by averaging only over the adjacent vertices within the topological distance of 1. Furthermore, a faster Huffman decoder [45] and a series of pre-computed Huffman coding tables are utilized. With the above means of optimization, this algorithm possesses a faster decoding speed than Hoppe's efficient implementation of PM [37].

2.3.2 Patch Coloring

As we know, a triangle mesh can be simplified and hierarchically represented through vertex decimation [46, 47]. Unlike the edge collapse approach, the vertex decimation approach removes a vertex and its adjacent edges, and then re-triangulates the resulting hole. The topology data record the way of re-triangulation after each vertex is decimated, or equivalently, the neighborhood of each new vertex before it is inserted.

Cohen-Or et al. [48] suggested the patch coloring algorithm for progressive mesh compression based on vertex decimation. First, the original mesh is simplified by iteratively decimating a set of vertices. At each iteration, decimated vertices are selected such that they are not adjacent to one another. Each vertex decimation results in a hole, which is then re-triangulated. The set of new triangles filling in this hole is called a patch. By reversing the simplification process, a hierarchical progressive reconstruction process can be obtained. In order to identify the patches in the decoding process, two patch coloring techniques were proposed: 4-coloring and 2-coloring. The 4-coloring scheme colors adjacent patches with distinct colors, requiring 2 bits per triangle. It is applicable to patches of any degree. The 2-coloring scheme further saves topology bits by coloring the whole mesh with only two colors. It enforces the re-triangulation of each patch in a zigzag manner and encodes the two outer triangles with the bit "1", and the other triangles with the bit "0". Therefore, it requires only 1 bit per triangle but applies only to the patches with a degree greater than 4. During the encoding process, at each level of detail, either the 2-coloring or 4-coloring scheme is selected based on the distribution of patch degrees. Then, the coloring bitstream is encoded with the famous Ziv-Lempel coder. For geometry coding, the position of a new vertex is simply predicted by averaging over its direct neighboring vertices. Experimentally, this approach requires about 6 bpv for connectivity data and about 16–22 bpv for geometry data at the 12-bit quantization resolution.

2.3.3 Valence-Driven Conquest

Alliez and Desbrun [49] proposed a progressive mesh coder for manifold 3D meshes. Observing the fact that the entropy of mesh connectivity is dependent on the distribution of vertex valences, they iteratively applied the valence-driven decimating conquest and the cleaning conquest in pair to get multiresolution meshes. The vertex valences are output and entropy encoded during this process.

The decimating conquest is a mesh simplification process based on vertex decimation. It only decimates vertices with valences not larger than 6 to maintain a statistical concentration of valences around 6. In the decimating conquest, a 3D mesh is traversed from patch to patch. A degree-n patch is a set of triangles incident to a common vertex of valence n, and a gate is an oriented boundary edge of a patch, storing the reference to its front vertex. The encoder enters a patch through one of its boundary edges, called the input gate. If the front vertex of the input gate has a valence not larger than 6, the encoder decimates the front vertex, re-triangulates the remaining polygon, and outputs the front vertex valence. Then, it pushes the other boundary edges, called output gates, into a FIFO list, and replaces the current input gate with the next available gate in the FIFO list. This procedure is repeated until the FIFO list becomes empty.

In fact, a breadth-first patch traversal is performed in the decimating conquest. Fig. 2.19(a) illustrates the decimating conquest on a 6-regular mesh. An initial input gate g_1 is chosen, a degree-6 patch is conquered and the output gates, g_2–g_6, are pushed into the FIFO list. Next, g_2 is chosen as the new input gate and another patch is conquered, and so on. Each conquered patch is re-triangulated so that the valences with half of the vertices on the patch boundary become lower. Therefore, the mesh after the decimating conquest has many vertices with valence 3 as shown in Fig. 2.19(b), and the vertex valences are no more concentrated around 6.

To maintain the statistical concentration of valences, a cleaning conquest is applied after each decimating conquest. The cleaning conquest is almost the same as the decimating conquest, except that the output gates are placed on the two edges of each face adjacent to the patch border, instead of on the patch border itself, and that only valence-three vertices are decimated. For example, in Fig. 2.19(b), suppose that an initial input gate g_1 is chosen. Then, its front vertex of valence 3 is decimated, and g_2–g_5 are chosen as the output gates. Fig. 2.19(c) shows the resulting mesh after a pair of decimating and cleaning conquests. We can

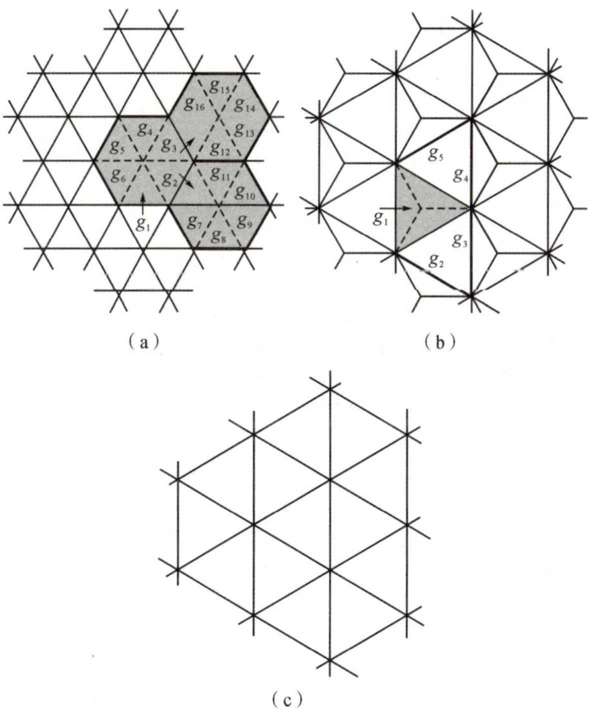

Fig. 2.19. An example to explain valence-driven conquests. (a) The decimating conquest; (b) The cleaning conquest; (c) The resulting mesh after the decimating conquest and the cleaning conquest. The shaded areas represent the conquered patches and the thick lines represent the gates. The gates to be processed are depicted in black, while the gates already processed are in normal color. Each arrow represents the direction of entrance into a patch

see that the resulting mesh is also a 6-regular mesh as the original mesh in Fig. 2.19(a). If an input mesh is irregular, it may not be completely covered by patches in the decimating conquest. In such a case, null patches are generated. For geometry coding, Alliez and Desbrun [49] adopted the barycentric prediction and the approximate Frenet coordinate frame. The normal and the barycenter of a patch approximate the tangent plane of the surface. Then, the position of the inserted vertex is encoded as an offset from the tangent plane.

Experimentally, for connectivity coding, this scheme requires about 2–5 bpv, on average 3.7 bpv, which is about 40% lower than the results reported in [41, 48]. For geometry coding, the performance typically ranges from 10 to 16 bpv with quantization resolutions between 10 and 12 bits. In particular, the geometry coding rate is much less than 10 bpv for meshes with high-connectivity regularity and geometry uniformity. Furthermore, this scheme has a comparable performance with that of the state-of-the-art single-rate coder. This scheme yields a compressed file size only about 1.1 times larger than Touma and Gotsman's algorithm [19], even though it supports full progressiveness.

2.3.4 Embedded Coding

Li and Kuo [50] suggested the concept of embedded coding to encode connectivity and geometry data in an interwoven manner. The geometry data together with the connectivity data are encoded progressively. Thus, when the coded data stream is received and decoded by the receiver, not only new vertices are added to the model, but also the precision of each old vertex position is progressively improved. This coding scheme is applicable to triangle meshes of any topology and it preserves the topology during mesh simplification.

With respect to mesh simplification, Li and Kuo also adopted the vertex decimation method. To record the neighborhood of each new vertex before it is inserted, their algorithm exploits a pattern table. It encodes the index to the pattern table and the indices of one marked triangle and one marked edge to locate the selected pattern within the mesh. For each vertex insertion, the topology data requires about ($\log_2 N_{vi}$+6) bits experimentally, where N_{vi} is the number of vertices in the current mesh $\boldsymbol{M}_i$.

The position of each vertex is predicted from the average position of its adjacent vertices, and the residue is obtained. Then, the encoder multiplexes topology data and geometry residual data into one data bitstream. Suppose that a residue is quantized as $0a_0a_1$… in the binary format. Fig. 2.20 shows the integration process, where each column represents the data associated with a vertex insertion. "*" denotes the topology data, a_0a_1… denotes the residue data for that vertex, and the flags "0" and "1" determine the order of bits in the final bitstream, which is depicted by the zigzag lines in Fig. 2.20. As more bits are received and decoded, more vertices are inserted and the precision of each vertex position is increased. The order of bits, determined by the flags, is selected by the encoder to achieve the rate-distortion tradeoff.

This algorithm requires about 20 bpv to decode a mesh model at an acceptable quality. However, at this bitrate, only one-third of the total number of vertices and triangles are reconstructed, since a significant portion of bits are used to increase the precisions of important vertices rather than to increase the number of reconstructed vertices.

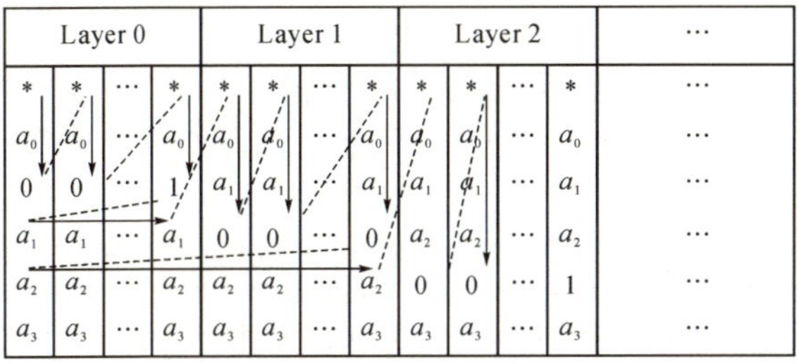

Fig. 2.20. The multiplexing of topology and geometry data, where the zigzag lines illustrate the bit order

2.3.5 *Layered Decomposition*

In [51], Bajaj et al. generalized their single-rate mesh coder [17] based on layered decomposition to a progressive mesh coder that is applicable to arbitrary meshes. An input mesh is decomposed into layers of vertices and triangles. Then the mesh is simplified through three stages: intra-layer simplification, inter-layer simplification and generalized triangle contraction. The former two are topology-preserving, whereas the last one may change the mesh topology.

The intra-layer simplification operation selects vertices to be removed from each contour. After those vertices are removed, re-triangulation is performed in the region between the simplified contour and its adjacent contours. A bit string is encoded to indicate which vertices are removed, and extra bits are encoded to reconstruct the original connectivity between the decimated vertex and its neighbors in the refinement process.

In the inter-layer simplification stage, a contour can be totally removed. Then, the two triangle strips sharing the removed contour are replaced by a single coarse strip [52]. Fig. 2.21 illustrates the process of contour removal and re-triangulation. A dashed line in Fig. 2.21(b), called a constraining chord, is associated with each edge in the contour to be removed, which is illustrated with a thick line. The simplification process is encoded as (0, 6, 2, 3, 1, 3), where the first bit indicates whether the contour is open or closed, the second value denotes the number of vertices in the removed contour, and the remaining values indicate the number of triangles between every two consecutive constraining chords in the coarse strip.

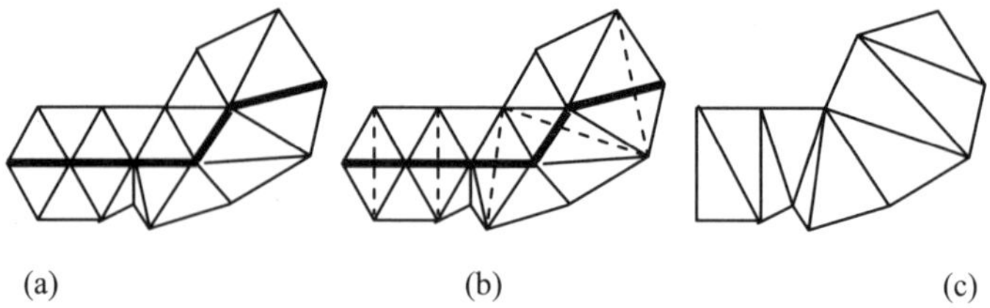

Fig. 2.21. Illustration of the inter-layer simplification process. (a) The fine level; (b) Constraining chords; (c) The coarse strip. Dashed lines depict constraining chords and thick lines depict the contour to be removed

After intra-layer and inter-layer simplification processes, the mesh can be further simplified using the generalized triangle contraction process [53], which contracts a triangle into a single point. To reduce the storage overhead, this point is chosen as the barycenter of the triangle. By allowing generalized triangle contraction, this scheme can simplify even a very complex model into a single triangle or vertex, achieving a guaranteed size of the mesh at the coarsest level.

The connectivity coding cost for the whole mesh is $O(N_v)$ due to the locality of the layering structure, which is much better than PM that requires $O(N_v \log_2 N_v)$ bits. Experimentally, it requires about 10–17 bpv for connectivity coding and 30 bpv for geometry coding at 10-bit or 12-bit quantization resolution. For geometry coding, similar to the single-rate algorithm [17], the second-order prediction is used to exploit the correlation between consecutive correction vectors.

2.3.6 Summary

In Table 2.3, we summarize the bitrates of progressive connectivity coding algorithms, which are extracted from experimental results reported in the original papers. Those explicit bitrates stand for the final bitrates required to decode meshes at the most refined level. The progressive mesh (PM) coder [35] is a pioneering algorithm that has a connectivity cost of $O(N_v \log_2 N_v)$. PFS [39], CPM [41], the patch coloring technique [48] and the layered decomposition algorithm [51] reduce the coding cost to $O(N_v)$. The valance-driven conquest algorithm [49] requires less than 4 bpv on the average for the connectivity coding. "Bitrate C: G (Q)" means the bit rate of connectivity coding in bpv: the bit rate of geometry coding in bpv (quantization resolutions in bits).

Table 2.3 Comparisons of bitrates for typical progressive connectivity coding algorithms

Category	Algorithm	Bitrate C:G (Q)	Comment
Progressive meshes	Hoppe [35]	$O(N_v \log_2 N_v)$:N/A	
	Popovic and Hoppe [38]	$O(N_v \log_2 N_v)$:N/A	
	Taubin et al. [39]	(7–10):(20–40) (6)	
	Pajarola and Rossignac [41]	7(12–15) (8, 10, 12)	
Patch coloring	Cohen-Or et al. [48]	6(16–22) (12)	
Valence-driven conquest	Alliez and Desbrun [49]	3.7(10–16) (10, 12)	
Embedded coding	Li and Kuo [50]	$O(N_v \log_2 N_v)$:N/A	Embedded multiplexing
Layered decomposition	Bajaj et al. [51]	(10–17):30 (10, 12)	

N/A: Not available

2.4 Spatial-Domain Geometry Compression

As described in the above two sections, the state-of-the-art connectivity coding algorithms cost only a few bits per vertex, and their performance has been approaching the optimal case. By comparison, geometry coding techniques received much less attention in the past. However, since geometry data dominate the total mesh data, more attention has been shifted to geometry coding recently. In most traditional mesh compression techniques, geometry coding is driven by the underlying connectivity coding. However, since geometry data require more bits than topology data, many methods have been suggested recently to efficiently compress the geometry data without reference to topology data.

Basically, single-rate mesh compression schemes compress the connectivity data in a lossless manner. In contrast, geometry data are generally compressed in a lossy manner. Although the geometry data are often provided in precise floating point representation for representing vertex positions, some applications may accept the reduction of this precision in order to obtain higher compression ratios. To exploit high correlation between adjacent vertices, most single-rate geometry compression methods are based on the spatial domain and generally follow a three-step procedure: quantization of vertex positions, prediction of quantized positions exploiting the neighboring vertices based on some data smoothness assumptions, and entropy coding of prediction residuals. With regard to progressive geometry coding, some techniques are based on the spatial domain, and others are based on transform domains.

This section focuses on the spatial domain geometry compression techniques for 3D triangle meshes. Among these techniques, scalar quantization, prediction, vector quantization (VQ) are single-rate methods, while *k-d* tree-based and octree-based methods are progressive methods. Note that VQ can not only be performed in the spatial domain but also in transform domains. Secondly, the utilization manner of VQ methods in geometry compression is much more different from that of other spatial-domain-based methods. In addition, the authors of this book have achieved several research results in VQ-based mesh compression. Thus we introduce VQ-based geometry techniques in a separate section.

2.4.1 Scalar Quantization

Geometry data without compression typically specify each coordinate component with a 32-bit floating-point number. However, this precision is beyond human perception with the naked eye and is far more than required for most applications. Thus, quantization can be performed to reduce the data amount without a serious reduction in visual quality. Quantization is a lossy approach for it attempts to encode a large or infinite set of values with a smaller set.

In signal processing, quantization refers to approximating the output by one of a discrete and finite set of values, while replacing the input by a discrete set is called discretization and is done by sampling: the resulting sampled signal is called a discrete signal (discrete time), and need not be quantized (it can have continuous values). To produce a digital signal (discrete time and discrete values), one both samples (discrete time) and quantizes the resulting sample values (discrete values). In digital signal processing, quantization is the process of approximating ("mapping") a continuous range of values (or a very large set of possible discrete values) by a relatively small ("finite") set of ("values which can still take on continuous range") discrete symbols or integer values. For example, this means rounding a real number in the interval [0, 100] to an integer among 0, 1, …, 100. Here, quantization means the latter.

From the point of view of the object to be quantized, quantization techniques can be classified into scalar quantization and vector quantization techniques. According to whether the quantization step is uniform or not, quantization techniques can be classified into uniform and non-uniform quantization techniques [54]. Each cell is of the same length in the uniform scalar quantizer, while cells have different lengths in the non-uniform scalar quantizer. Compared with non-uniform vector quantization, uniform scalar quantization is simple and computationally efficient even though it is not optimal in the rate-distortion performance.

Typical geometry coding algorithms quantize uniformly the vertex positions for each coordinate component separately in the Cartesian space at 8- to 16-bit quantization resolutions. In most scalar-quantization-based geometry compression methods, the same quantization resolution is globally applied. However, in [13], a mesh was first segmented into several regions, and then different resolutions were adaptively applied for different regions according to the local curvature and triangle sizes. Within each region, the vertex coordinates are still uniformly quantized.

2.4.2 Prediction

After the quantization of vertex coordinates, the resulting values are then typically compressed by entropy coding after prediction relying on some data smoothness assumptions. A prediction is a mathematical operation where future values of a discrete-time signal are estimated as a certain function of previous samples. In 3D mesh compression, the prediction step makes full use of the correlation between adjacent vertex coordinates and it is most crucial in reducing the amount of geometry data. A good prediction scheme produces prediction errors with a highly skewed distribution, which are then encoded with entropy coders, such as the Huffman coder or the arithmetic coder.

Different types of prediction schemes for 3D mesh geometry coding have been proposed in the literature, such as delta prediction [12, 13], linear prediction [5], parallelogram prediction [19] and second-order prediction [17]. All these prediction methods can be treated as a special case of the linear prediction scheme with carefully selected coefficients.

2.4.2.1 Delta Prediction

The early work employed simple delta coding or linear prediction along a vertex ordering guided by connectivity coding. Delta coding or delta prediction is based on the fact that adjacent vertices tend to have slightly different coordinates, and the differences (or deltas) between them are usually very small. Deering's work [12] and Chow's work [13] encode the deltas of coordinates instead of the original coordinates with variable length codes according to the distribution of deltas. Deering's scheme adopts the quantization resolutions between 10 and 16 bits per coordinate component and its coding cost is roughly between 36 and 17 bpv. In Chow's geometry coder, bitrates of 13–18 bpv can be achieved at quantization resolutions of 9–12 bits per coordinate component.

2.4.2.2 Linear Prediction

Linear prediction is a mathematical operation where future values of a discrete-time signal are estimated as a linear function of previous samples. In digital signal processing, linear prediction is often called linear predictive coding (LPC) and can thus be viewed as a subset of filter theory. In system analysis (a subfield of mathematics), linear prediction can be viewed as a part of mathematical modeling or optimization.

In Taubin and Rossignac's scheme [5], the position of a vertex is predicted from a linear combination of positions of K uniquely-selected previous vertices along the path from the root to the current vertex in the vertex spanning tree. Concretely, the position $\boldsymbol{v}_n$ of the n-th vertex can be given by

$$\boldsymbol{v}_n = \sum_{i=1}^{K} \lambda_i \cdot \boldsymbol{v}_{n-i} + \varepsilon(\boldsymbol{v}_n), \tag{2.8}$$

where $\lambda_1, \lambda_2, \ldots, \lambda_K$ are carefully selected to minimize the mean square error

$$E\left\{\left\|\varepsilon(\boldsymbol{v}_n)\right\|^2\right\} = E\left\{\left\|\boldsymbol{v}_n - \sum_{i=1}^{K}(\lambda_i \cdot \boldsymbol{v}_{n-i})\right\|^2\right\} \tag{2.9}$$

and transmitted to the decoder as the side information. The bitrate of this method is not directly reported in [5]. However, as estimated by Touma and Gotsman [19], it costs about 13 bpv at the 8-bit quantization resolution. Note that the delta prediction is a special case of linear prediction with $K = 1$ and $\lambda_1=1$.

The approach proposed by Lee et al. [55] consists of quantizing in the angle space after prediction. By applying different levels of precision while quantizing the dihedral or the internal angles between or inside each facet, this method achieves better visual appearance by allocating more precision to the dihedral angles, since they are more related to the geometry and normals.

2.4.2.3 Parallelogram Prediction

Touma and Gotsman [19] used a more sophisticated prediction scheme. To encode a new vertex $\boldsymbol{v}_n$, it considers a triangle with two vertices $\hat{\boldsymbol{v}}_{n-1}$ and $\hat{\boldsymbol{v}}_{n-2}$ on the active list, where triangle ($\hat{\boldsymbol{v}}_{n-1}, \hat{\boldsymbol{v}}_{n-2}, \hat{\boldsymbol{v}}_{n-3}$) is already encoded as shown in Fig. 2.22. The parallelogram prediction assumes that the four vertices $\hat{\boldsymbol{v}}_{n-1}$, $\hat{\boldsymbol{v}}_{n-2}$, $\hat{\boldsymbol{v}}_{n-3}$ and $\boldsymbol{v}_n$ form a parallelogram. Therefore, the new vertex position can be predicted as

$$\tilde{\boldsymbol{v}}_n = \hat{\boldsymbol{v}}_{n-1} + \hat{\boldsymbol{v}}_{n-2} - \hat{\boldsymbol{v}}_{n-3}. \tag{2.10}$$

This method performs well only if the four vertices are exactly or nearly co-planar. To further improve the prediction accuracy, the crease angle between the two triangles ($\hat{\boldsymbol{v}}_{n-1}, \hat{\boldsymbol{v}}_{n-2}, \hat{\boldsymbol{v}}_{n-3}$) and ($\hat{\boldsymbol{v}}_{n-1}, \hat{\boldsymbol{v}}_{n-2}, \hat{\boldsymbol{v}}_{n}$) can also be estimated using the crease angle θ between the two triangles ($\hat{\boldsymbol{v}}_{n-2}, \hat{\boldsymbol{v}}_{n-3}, \hat{\boldsymbol{v}}_{n-4}$) and ($\hat{\boldsymbol{v}}_{n-2}, \hat{\boldsymbol{v}}_{n-4}, \hat{\boldsymbol{v}}_{n-5}$). In Fig. 2.22, $\tilde{\boldsymbol{v}}'_n$ is the predicted position of $\boldsymbol{v}_n$ using the crease angle estimation. This work achieves an average bitrate of 9 bpv at 8-bit quantization resolution. The parallelogram prediction is also a linear prediction in essence, since the predicted vertex position is a linear combination of the three previously visited vertex positions.

Inspired by the above TG parallelogram prediction scheme, Isenburg and Alliez [56] generalized it to polygon mesh geometry compression. They let the polygon information dictate where to apply the parallelogram rule that they use to predict vertex positions. Since polygons tend to be fairly planar and fairly convex, it is beneficial to make predictions within a polygon rather than across polygons.

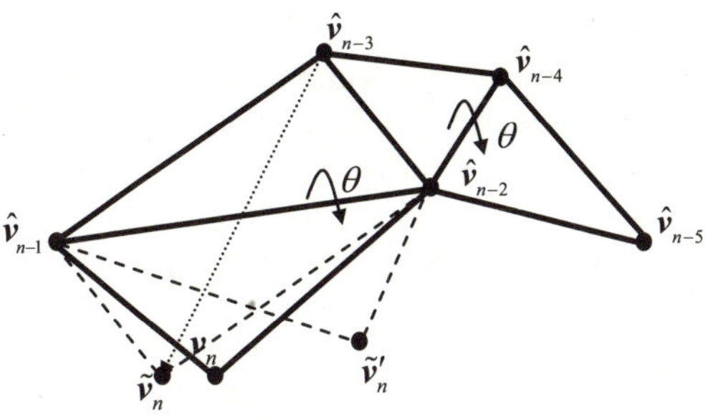

Fig. 2.22. Illustration of the parallelogram prediction scheme

This, for example, avoids poor predictions due to a crease angle between polygons. Up to 90% of the vertices can be predicted in this way. Their strategy improves geometry compression performance by 10%–40%, depending on how polygonal the mesh is and the quality (planarity/convexity) of the polygons.

2.4.2.4 Second-Order Prediction

Linear prediction removes redundancy by identifying similar bit values between coordinates of adjacent vertices. However, it is not an optimal way, especially for models without many sharp features. In [17], a second-order prediction is proposed to encode the vertices along contours, whereas the coordinates of branching points are encoded directly. This is done in two steps. The first step computes and quantizes the differences between adjacent vertex positions. This first step alone is equivalent to delta prediction. The second step calculates the difference between quantized difference codes. It was confirmed experimentally that the second-order prediction provides a better performance than the delta prediction, when incorporated with entropy coding techniques. The geometry coding bitrate is about 11 bpv at the 8-bit quantization resolution and about 14 bpv at the 15-bit quantization resolution. Since the second-order prediction scheme predicts $\boldsymbol{v}_n - \boldsymbol{v}_{n-1}$ from $\boldsymbol{v}_{n-1} - \boldsymbol{v}_{n-2}$, it is still a linear predictor, which is equivalent to predicting $\boldsymbol{v}_n$ from $2\boldsymbol{v}_{n-1} - \boldsymbol{v}_{n-2}$.

2.4.2.5 Other Improved Prediction Methods

Since polygons tend to be fairly planar and convex, it is more appropriate to perform prediction operations within polygons rather than across them. Intuitively, this idea avoids poor predictions resulting from a crease angle between polygons. Despite the effectiveness of the published predictive geometry schemes, they are not optimal because the mesh traversal is still controlled by the connectivity coding scheme. Since the traversal order is independent of the geometry data, and the prediction from one polygon to the next is performed along this order, it cannot be expected to do the best job.

The first approach to improve the prediction is called prediction trees [57], where the geometry drives the traversal instead of the connectivity as before. This is based on the solution of an optimization problem. In some cases, it results in a reduction of up to 50% in the geometry code entropy, particularly in meshes with significant creases and corners, e.g. CAD models. The main drawback of this method is the complexity of the encoder. Due to the need to run an optimization procedure at the encoder, it is up to one order of magnitude slower than, for example, the TG encoder. The decoder, however, is very fast, so for many applications where the encoding is done offline, the encoder speed is not an impediment. Cohen-Or et al. [58] suggested a multi-way prediction technique, where each vertex position was predicted from all its neighboring vertices, as opposed to the one-way parallelogram prediction. In addition, an extreme approach to prediction is the feature discovery approach by Shikhare et al. [59], which removes the redundancy by detecting similar geometric patterns. However, this technique works well only for a certain class of models and involves expensive matching computations.

2.4.3 k-d Tree

Now we turn to introduce progressive geometry coding schemes in this and the next subsections. In most mesh compression techniques, geometry coding is guided by the underlying connectivity coding. Gandoin and Devillers [60] proposed a fundamentally different strategy, where connectivity coding is guided by geometry coding. Their algorithm works in two passes: the first pass encodes geometry data progressively without considering connectivity data. The second pass encodes connectivity changes between two successive LODs. Their algorithm can encode arbitrary simplicial complexes without any topological constraint.

For geometry coding, their algorithm employs a *k-d* tree decomposition based on cell subdivisions [61]. At each iteration, it subdivides a cell into two child cells, and then it encodes the number of vertices in one of the two child cells. If the parent cell contains N_{vp} vertices, the number of vertices in one of the child cells can be encoded using $\log_2(N_{vp}+1)$ bits with the arithmetic coder [62]. This subdivision is recursively applied, until each nonempty cell is small enough to contain only one vertex and enables a sufficiently precise reconstruction of the vertex position. Fig. 2.23 illustrates the geometry coding process based on a 2D example. First, the total number of vertices, 7, is encoded using a fixed number of bits (32 in this example). Then, the entire cell is divided vertically into two cells, and the number of vertices in the left cell, 4, is encoded using $\log_2(7+1)$ bits. Note that the number of vertices in the right cell is not encoded, since it is deducible from the number of vertices in the entire cell and the number of vertices in the left cell. The left and right cells are then horizontally divided, respectively, and the numbers of vertices in the upper cells are encoded, and so on. To improve the coding gain, the number of vertices in a cell can be predicted from the point distribution in its neighborhood.

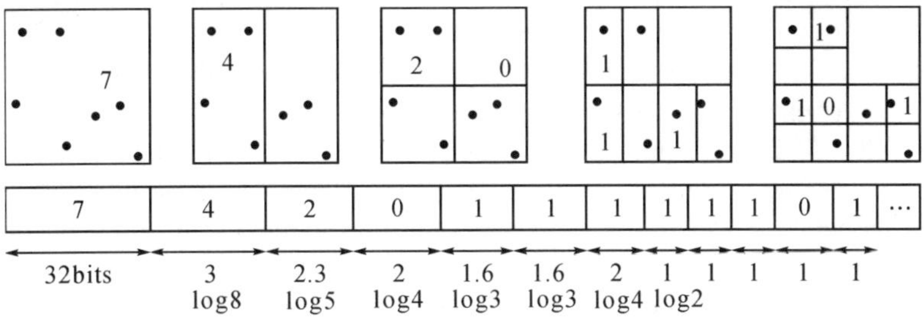

Fig. 2.23. Illustration of *k-d* tree geometry coding in the 2D case

For connectivity coding, their algorithm encodes the topology change after each cell subdivision using one of two operations: vertex split [35] or generalized vertex split [38]. Specifically, after each cell subdivision, the connectivity coder records a symbol, indicating which operation is used, and parameters specific to that operation. Compared to [35, 38], their algorithm has the advantage that split vertices are implicitly determined by the subdivision order given in geometry coding, resulting in a reduction in the topology coding cost. Moreover, to improve the coding gain further, they proposed several rules, which predict the parameters for vertex split operations efficiently using already encoded geometry data.

On average, this scheme requires 3.5 bpv for connectivity coding and 15.7 bpv for geometry coding at the 10-bit or 12-bit quantization resolution, which outperforms progressive mesh coders presented in [44, 49]. This scheme is even comparable to the single-rate mesh coder given in [19], achieving a full progressiveness at a cost of only 5% overhead bitrate. It is also worthwhile to point out that this scheme is especially useful for terrain models and densely sampled objects, where topology data can be losslessly reconstructed from geometry data. Besides its good coding gain, it can be easily extended to compress tetrahedral meshes.

2.4.4 Octree Decomposition

Peng and Kuo [63] proposed a progressive lossless mesh coder based on the octree decomposition, which can encode triangle meshes with arbitrary topology. Given a 3D mesh, an octree structure is first constructed through recursive partitioning of the bounding box. The mesh coder traverses the octree in a top-down fashion and encodes the local changes of geometry and connectivity associated with each octree cell subdivision.

In [63], the geometry coder does not encode the vertex number in each cell, but encodes the information whether each cell is empty or not, which is usually more concise in the top levels of the octree. For connectivity coding, a uniform approach is adopted, which is efficient and easily extendable to arbitrary polygonal meshes.

For each octree cell subdivision, the geometry coder encodes the number, T ($1 \leq T \leq 8$), of non-empty-child cells and the configuration of non-empty-child cells among $K_T = C_8^T$ possible combinations. When the data are encoded straightforwardly, T takes 3 bits and the non-empty-child-cell configuration takes $\log_2 K_T$ bits. To further improve the coding efficiency, T is arithmetic coded using the context of the parent cell's octree level and valence, resulting in a 30%–50% bitrate reduction. Furthermore, all K_T possible configurations are sorted according to their estimated probability values, and the index of the configuration in the sorted array is arithmetic coded. The probability estimation is based on the observation that non-empty-child cells tend to gather around the centroid of the parent-cell's neighbors. This technique leads to a more than 20% improvement.

For the connectivity coding, each octree cell subdivision is simulated by a sequence of *k-d* tree cell subdivisions. Each vertex split corresponds to a *k-d* tree cell subdivision, which generates two non-empty-child cells. Let the vertex to be split be denoted by $\boldsymbol{v}$, the neighboring vertices before the vertex split by $P = \{\boldsymbol{p}_1, \boldsymbol{p}_2, \ldots, \boldsymbol{p}_K\}$ and the two new vertices from the vertex split by $\boldsymbol{v}_1$ and $\boldsymbol{v}_2$. Then, the following information will be encoded: (1) Vertices among P that are connected to both $\boldsymbol{v}_1$ and $\boldsymbol{v}_2$ (called the pivot vertices); (2) Whether each non-pivot vertex in P is connected to $\boldsymbol{v}_1$ or $\boldsymbol{v}_2$; and (3) Whether $\boldsymbol{v}_1$ and $\boldsymbol{v}_2$ are connected in the refined mesh. During the coding process, a triangle regularity metric is used to predict each neighboring vertex's probability of being a pivot vertex, and a spatial distance metric is used to predict the connectivity of non-pivot neighbor vertices to the new vertices. At the decoder side, the facets are constructed from the edge-based connectivity without an extra coding cost. To further improve the R-D performance, the prioritized cell subdivision is applied. Higher priorities are given to cells of a bigger size, a bigger valence and a larger distance from neighbors.

The octree-based mesh coder outperforms the *k-d* tree algorithm [60] in both geometry and connectivity coding efficiency. For geometry coding, it provides about a 10%–20% improvement for typical meshes, but up to 50%–60% improvement for meshes with highly regular geometry data and/or tightly clustered vertices. With respect to connectivity coding, the improvement ranges from 10% to 60%.

2.5 Transform-Based Geometric Compression

Transform coding is a type of data compression for "natural" data like audio signals or photographic images [64]. The transformation is typically lossy, resulting in a lower quality copy of the original input. In transform coding, knowledge of the application is used to choose information to discard, thereby lowering its bandwidth. The remaining information can then be compressed using a variety of methods. When the output is decoded, the result may not be identical to the original input, but is expected to be close enough for the purpose of applications. The discrete cosine transform (DCT) or the discrete Fourier transform (DFT) is often used to represent a sequence of source samples to another sequence of transform coefficients, whose energy is concentrated in relatively few low-frequency coefficients. Thus, great degradation can be obtained if we encode low-frequency coefficients while discarding higher frequency ones. The common JPEG image format is an example of transform coding, one that examines small blocks of the image and "averages out" the color using a discrete cosine transform to form an image with far fewer colors in total. MPEG modifies this across frames in a motion image, further reducing the size compared to a series of JPEGs. MPEG audio compression analyzes the transformed data according to a psychoacoustic model that describes the human ear's sensitivity to parts of the signal, similar to the TV model. In this section, we briefly introduce several typical 3D mesh geometry compression methods based on DFT and wavelet transforms. Some are single-rate compression techniques, and others are progressive schemes.

2.5.1 Single-Rate Spectral Compression of Mesh Geometry

Karni and Gotsman [65] used the spectral theory on meshes [40] to compress geometry data. It is a single-rate geometry compression scheme. Suppose that a mesh consists of N_v vertices. Then the mesh Laplacian matrix $\boldsymbol{L}$ of size $N_v \times N_v$ is derived from the mesh connectivity as follows:

$$L_{ij} = \begin{cases} 1, & \boldsymbol{v}_l = \boldsymbol{v}_j; \\ -1/d_i, & \boldsymbol{v}_i \text{ and } \boldsymbol{v}_j \text{ are adjacent}; \\ 0, & \text{otherwise}, \end{cases} \tag{2.11}$$

where d_i is the valence of vertex $\boldsymbol{v}_i$. The eigenvectors of $\boldsymbol{L}$ form an orthogonal basis of R^{N_v} and the associated eigenvalues represent the frequencies of those basis functions. The encoder projects the x, y, and z coordinate vectors of the mesh onto the basis functions to obtain the geometry spectra, respectively. Then, the encoder quantizes these spectra, truncates high-frequency coefficients, and entropy encodes the quantized coefficients. This approach can naturally support progressiveness by transmitting the coefficients in the increasing order of frequencies.

Experimentally, this approach requires only 1/2–1/3 of the bitrate of Touma and Gotsman's algorithm [19] to achieve a similar visual quality. This approach is especially suitable for smooth meshes, which can be faithfully represented with a fewer number of low-frequency coefficients.

Finding the eigenvectors of an $N_v \times N_v$ matrix requires $O(N_v^3)$ computational complexity. To reduce the complexity, an input mesh can be partitioned into several segments and each segment can be independently encoded. However, the eigenvectors should be computed in the decoder as well. Thus, even though the partitioning is incorporated, the decoding complexity is too high for real-time applications. To alleviate this problem, Karni and Gotsman [66] proposed to use fixed basis functions, which are computed from a 6-regular connectivity. Those basis functions are actually the Fourier basis functions. Therefore, the encoding and decoding processes can be performed with the fast Fourier transform (FFT) efficiently. Before encoding, the connectivity of an input mesh is mapped into a 6-regular connectivity. No geometry information is used during the mapping. Thus, the decoder can perform the same mapping with separately received connectivity data and determine the correct ordering of vertices. The exploitation of fixed basis functions is obviously not optimal, but provides an acceptable performance at much lower complexity.

In addition, Sorkine et al. [67] addressed the issue of reducing the visual effect of quantization errors. Considering the fact that the human visual system is more sensitive to normal distortion than to geometric distortion, they propose to apply quantization not in the coordinate space as usual, but rather in a transformed coordinate space obtained by applying a so-called "k-anchor invertible Laplacian transformation" over the original vertex coordinates. This concentrates the quantization error at the low-frequency end of the spectrum, thus preserving the normal variations over the surface, even after aggressive quantization. To avoid significant low-frequency errors, a set of anchor vertex positions are also selected to "nail down" the geometry at a selected number of vertex locations.

2.5.2 Progressive Compression Based on Wavelet Transform

It is well known from image coding that wavelet representations are very effective in decorrelating the original data, greatly facilitating subsequent entropy coding. In essence, coarser level data provides excellent predictors for finer level data, leaving only generally small prediction residuals for the coding step. For tensor product surfaces, many of these ideas can be applied in a straightforward fashion. However, the arbitrary topology surface case is much more challenging. To begin with, wavelet decompositions of general surfaces were not known until the pioneering work by Lounsbery [68]. These constructions were subsequently applied to progressive approximation of surfaces as well as data on surfaces.

Khodakovsky et al. [69] proposed a progressive geometry compression (PGC) algorithm based on the wavelet transform. It first remeshes an arbitrary manifold mesh $\boldsymbol{M}$ into a semi-regular mesh, where most vertices are of degree 6, using the MAPS algorithm [70]. MAPS generates a semi-regular approximation of $\boldsymbol{M}$ by finding a coarse base mesh and successively subdividing each triangle into four triangles. Fig. 2.24 shows a remeshing example. In this figure, vertices within the region bounded by white curves in Fig. 2.24(a) are projected onto a base triangle.

These projected vertices are depicted by black dots in Fig. 2.24(b). Each vertex projected onto the base triangle contains the information of the original vertex position. By interpolating these original vertex positions, each subdivision point can be mapped approximately to a point (not necessarily a vertex) in the original mesh. Note that the connectivity information of the semi-regular mesh can be efficiently encoded, since it can be reconstructed using only the connectivity of the base mesh and the number of subdivisions. However, this algorithm attempts to preserve only the geometry information. Thus, the original connectivity of $\boldsymbol{M}$ cannot be reconstructed at the decoder.

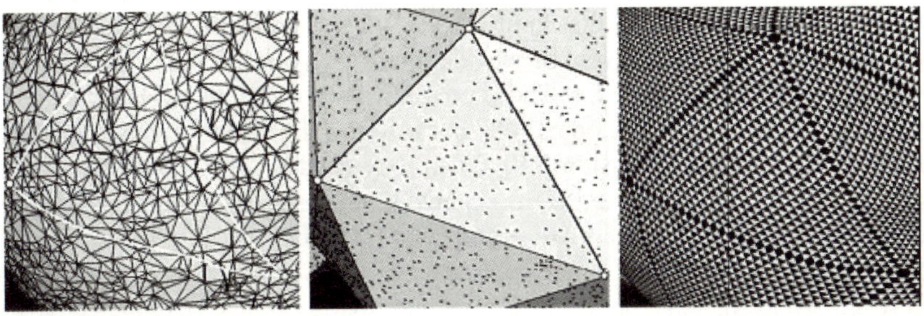

(a) (b) (c)

Fig. 2.24. A remeshing example [2]. (a) An irregular mesh; (b) The corresponding base mesh; (c) The corresponding semi-regular mesh. Triangles are illustrated with a normal flipping pattern to clarify the semi-regular connectivity (With permission of Elsevier)

Based on the Loop algorithm [71], this algorithm then represents the semi-regular mesh geometry with the base mesh geometry and a sequence of wavelet coefficients. These coefficients represent the differences between successive LODs with a concentrated distribution around zero, which is suitable for entropy coding. The wavelet coefficients are encoded using a zerotree approach, introducing progressiveness into the geometry data. More specifically, they modified the SPIHT algorithm [72], which is one of the successful 2D image coders, to compress the Loop wavelet coefficients. Their algorithm provides about 12 dB or four times better image quality than CPM [41], and even a better performance than Touma and Gotsman's single-rate coder [19]. This is mainly due to the fact that they employed semi-regular meshes, enabling the wavelet coding approach.

Khodakovsky and Guskov [73] later proposed another wavelet coder based on the normal mesh representation [74]. In the subdivision, their algorithm restricts the offset vector which should be in the normal direction of the surface. Therefore, whereas 3D coefficients are used in [69], 1D coefficients are used in the normal mesh algorithm. Furthermore, their algorithm employs the uplifted version of butterfly wavelets [42, 43] as the transform. As a result, it achieves about 2–5 dB quality improvement over that in [69].

In addition, Payan and Antonini [75] proposed an efficient low complexity compression scheme for densely sampled irregular 3D meshes. This scheme is based on 3D multiresolution analysis (3D discrete wavelet transform) and includes a model-based bit allocation process across the wavelet sub-bands. Coordinates of 3D wavelet coefficients are processed separately and statistically modeled by a generalized Gaussian distribution. This permits an efficient allocation even at a low bitrate and with a very low complexity. They introduced a predictive geometry coding of LF sub-bands and topology coding is made by using an original edge-based method. The main idea of their approach is the model-based bit allocation adapted to 3D wavelet coefficients and the use of EBCOT coder to efficiently encode the quantized coefficients. The first step of their compression scheme (see Fig. 2.25) is to obtain a semi-regular mesh of the original irregular mesh based on the MAPS technique [70]. Hence, a discrete wavelet transform (DWT) can be applied on the semi-regular mesh to obtain a multi-resolution representation, resolution levels of wavelet coefficients (HF coefficients) and the coarsest level (LF coefficients). These coefficients are tridimensional vectors. In their work, they chose the Loop DWT because this transform gives good visual results in 3D mesh compression [69]. Then they used an optimal nearly uniform scalar quantizer with non-uniform quantization steps described in [76]. The quantized wavelet coefficients are entropy coded using the EBCOT coder [77]. This lossless context based coder, included in JPEG 2000, creates an embedded bitstream. Also it will be used to encode the topology. Compared to the well-known PGC method [69], the compression ratio is improved for similar reconstruction quality.

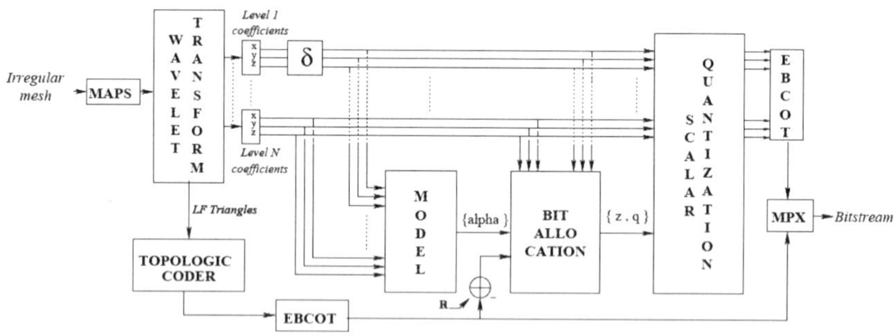

Fig. 2.25. Payan and Antonini's compression scheme [75] (©[2002]IEEE)

Recently, Chen et al. [78] proposed a progressive compression method based on quadrilateral remeshing, wavelet transform and zerotree coding. It is applicable to arbitrary topology with highly detailed triangle meshes. They firstly parameterized the original triangle mesh to a regular quadrilateral approximation. A wavelet transform was then applied to the approximation to remove a large amount of correlation between neighboring vertices. Finally, they used low cost zerotree coding and subdivision based reconstruction to build a sequence of progressive models. Their method can greatly reduce the cost of transportation with acceptable quality loss. By applying a quadrilateral subdivision scheme, they subdivided a mesh into a denser one. Each face was split into four new faces. The simplification process will just act in a reverse way, joining four faces into a new one and eliminating redundant points. Their method for constructing the wavelet transform requires three steps: vertex split, prediction and update. With respect to zerotree coding, they adopted a new approach. In their approach, vertices do not have a tree structure, but the edges and faces do. Each edge and each face is the parent of four edges of the same orientation in the finer mesh. Hence, each edge and face of the coarsest domain mesh forms the root of each zerotree, and it groups all the wavelet coefficients of a fixed wavelet subband from its incident based domain faces. No coefficient is accounted for multiple times or left out by this grouping.

2.5.3 Geometry Image Coding

Surface geometry is often modeled with irregular triangle meshes. The process of remeshing refers to approximating such geometry using a mesh with (semi)-regular connectivity, which has advantages for many graphics applications. However, current techniques for remeshing arbitrary surfaces create only semi-regular meshes. The original mesh is typically decomposed into a set of disk-like charts, onto which the geometry is parameterized and sampled. Unlike this approach, Gu et al. [79] proposed to remesh an arbitrary surface onto a completely regular structure called a geometry image. It captures geometry as a simple 2D array of quantized points. Surface signals like normals and colors are stored in similar 2D arrays using the same implicit surface parameterization, where texture coordinates are absent. Each pixel value in the geometry image represents a 3D position vector (x, y, z). Fig. 2.26 shows the geometry image of the Stanford Bunny. Due to its regular structure, the geometry image representation can facilitate the compression and rendering of 3D data.

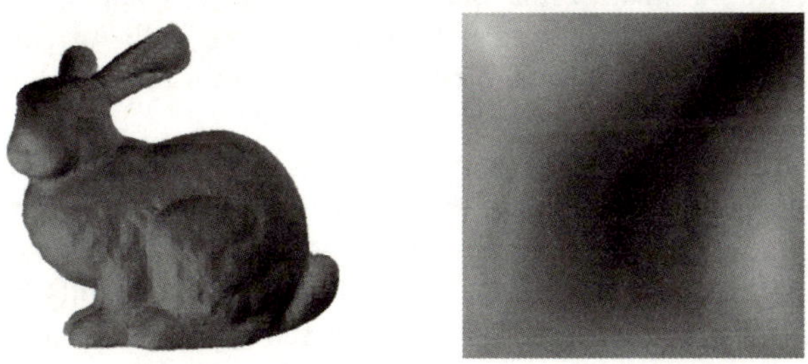

(a) (b)

Fig. 2.26. The geometry image of the Stanford Bunny. (a) The Stanford Bunny; (b) Its geometry image

To generate the geometry image, an input manifold mesh is cut and opened to be homeomorphic to a disk. The cut mesh is then parameterized onto a 2D square, which is in turn regularly sampled. In the cut process, an initial cut is first selected and then iteratively refined. At each iteration, it selects a vertex of the triangle with the biggest geometric stretch and inserts the path, connecting the selected vertex to the previous cut, into the refined cut. After the final cut is determined, the boundary of the square domain is parameterized with special constraints to prevent cracks along the cut, and the interior is parameterized using geometry-stretch parameterization in [80], which attempts to distribute vertex samples evenly over the 3D surface.

Geometry images can be compressed using standard 2D image compression techniques, such as wavelet-based coders. To seamlessly zip the cut in the reconstructed 3D surface, especially when the geometry image is compressed in a lossy manner, it encodes the sideband signal, which records the topological structure of the cut boundary and its alignment with the boundary of the square domain.

The geometry image compression provides about 3 dB worse R-D performance than the wavelet mesh coder [69]. Also, since it maps complex 3D shapes onto a simple square, it may yield large distortions for high-genus meshes and unwanted smoothing of 3D features. References [81] and [82] proposed an approach to parameterize a manifold 3D mesh with genus 0 onto a spherical domain. Compared with the square domain approach [79], this approach leads to a simple cut topology and an easy-to-extend image boundary. It was shown by experiments that the spherical geometry image coder achieves better R-D performance than the square domain approach [79] and the wavelet mesh coder [69], but slightly worse performance than the normal mesh coder [73].

2.5.4 Summary

In Table 2.4, we summarize the bitrates of geometry compression algorithms, which are extracted from experimental results reported in the original papers. For progressive compression, those explicit bitrates stand for the final bitrates required to decode meshes at the most refined level.

For the geometry coding, a bitrate of 15 bpv at a quantization resolution of around 10 bits has been achieved by the *k-d* tree decomposition [60]. These progressive coders [49, 60] have excellent performance in the sense that they support the progressive coding property at a bitrate that is slightly higher than the state-of-the-art single-rate coder [19]. The octree decomposition algorithm [63] further reduces the overall bitrate of [60] by 10%–60%. The spectral coding [65], the wavelet coding [69, 73] and the geometry image coding methods [79, 81, 82] improve the coding gain and provide even better compression performance than the single-rate coder in [19]. It is worthwhile to point out that these coding algorithms are generalizations of successful 2D image coding techniques, e.g., JPEG and JPEG-2000. The *k-d* tree decomposition algorithm [60] can compress arbitrary simplicial complexes. The octree decomposition algorithm [63] can encode triangular meshes with arbitrary topology. All the remaining algorithms can

Table 2.4 Comparisons of bitrates for typical geometry coding algorithms

Category	Algorithm	Bitrate C:G (Q)	Comments
k-d tree decomposition	Gandoin and Devillers [60]	3.5:15.7 (10, 12) for manifold meshes	Capable of encoding triangle soups
Octree decomposition	Peng and Kuo [63]	40%–90% bitrate of [60] for similar quality	
Spectral coding	Karni and Gotsman [65]	30%–50% bitrate of [19] for similar quality	
Wavelet coding	Khodakovsky et al. [69]	12 dB better quality than [41] at the same bitrate	
	Khodakovsky and Guskov [73]	2–5 dB better quality than [69] at the same bitrate	Loss of original connectivity
Geometry image coding	Gu et al. [79]	3 dB worse quality than [69]	
	Praun and Hoppe [81, 82]	Better R-D than [79, 69], slightly worse R-D than [73]	Loss of original connectivity

deal with manifold triangular meshes only. In the wavelet coding methods [69, 73] and the geometry image coding methods [79, 81, 82], the original connectivity is lost due to the remeshing procedure.

2.6 Geometry Compression Based on Vector Quantization

Recently, vector quantization (VQ) has been proposed for geometry compression, which does not follow the conventional "quantization+prediction+entropy coding" approach. The conventional approach pre-quantizes each vertex coordinate using a scalar quantizer and then predictively encodes the quantized coordinates. In contrast, typical VQ approaches first predict vertex positions and then jointly compress the three components of each prediction residual. Thus, it can utilize the correlation between different coordinate components of the residual. Compared with scalar quantization, the main advantages of VQ include a superior rate-distortion performance, more freedom in choosing shapes of quantization cells, and better exploitation of redundancy between vector components. In this section, we first introduce some basic concepts of VQ and then introduce several typical VQ-based geometry compression methods.

2.6.1 Vector Quantization

VQ has become an attractive block-based encoding method for data compression in recent years. It can achieve a high compression ratio. In environments such as image archiving and one-to-many communications, the simplicity of the decoder makes VQ very efficient. In brief, VQ can be defined as a mapping from k-dimensional Euclidean space R^k into a finite subset $C = \{\boldsymbol{c}_i \mid i = 0, 1, \ldots, N-1\}$ that is generally called a codebook, where $\boldsymbol{c}_i$ is a codeword and N is the codebook size. VQ first generates a representative codebook from a number of training vectors using, for example, the well-known iterative clustering algorithm [83] that is often referred to as the generalized Lloyd algorithm (GLA). In VQ, the image to be encoded is first decomposed into vectors and then sequentially encoded vector by vector. In the encoding phase, each k-dimensional input vector $\boldsymbol{x} = (x_1, x_2, \ldots, x_k)$ is compared with the codewords in the codebook $C = \{\boldsymbol{c}_0, \boldsymbol{c}_1, \ldots, \boldsymbol{c}_{N-1}\}$ to find the best matching codeword $\boldsymbol{c}_i = (c_{i1}, c_{i2}, \ldots, c_{ik})$ satisfying the following condition:

$$d(\boldsymbol{x}, \boldsymbol{c}_i) = \min_{0 \le j \le N-1} d(\boldsymbol{x}, \boldsymbol{c}_j) \cdot \tag{2.12}$$

That is, the distance between $\boldsymbol{x}$ and $\boldsymbol{c}_i$ is the smallest. In Eq.(2.12) $d(\boldsymbol{x}, \boldsymbol{c}_j)$ is the distortion of representing the input vector $\boldsymbol{x}$ by the codeword $\boldsymbol{c}_j$, which is often measured by the squared Euclidean distance, i.e.,

$$d(\boldsymbol{x}, \boldsymbol{c}_j) = \sum_{l=1}^{k} (x_l - c_{jl})^2 \cdot \tag{2.13}$$

And then the index i of the best matching codeword assigned to the input vector $\boldsymbol{x}$ is transmitted over the channel to the decoder. The decoder has the same codebook as the encoder. In the decoding phase, for each index i, the decoder merely performs a simple table look-up operation to obtain $\boldsymbol{c}_i$ and then uses $\boldsymbol{c}_i$ to reconstruct the input vector $\boldsymbol{x}$. Compression is achieved by transmitting or storing the index of a codeword rather than the codeword itself. The compression ratio is determined by the codebook size and the dimension of the input vectors, and the overall distortion is dependent on the codebook size and the selection of codewords.

2.6.2 Quantization of 3D Model Space Vectors

In Lee and Ko's work [84], the Cartesian coordinates of a vertex were transformed into a model space vector using the three previous vertex positions. In fact, the model space transformation is a kind of prediction and the model space vector can be regarded as a prediction residual. Then the model space vector was quantized using the generalized Lloyd algorithm [83]. Since they used the original positions of previous vertices in the model space transform, the quantization errors will be accumulated in the decoder. To overcome this encoder-decoder mismatch problem, they periodically inserted correction vectors into the bitstream. Experimentally, this scheme requires about 6.7 bpv on average to achieve the same visual quality as conventional methods at 8-bit quantization resolution. Note that Touma and Gotsman's work requires about 9 bpv at 8-bit resolution [19]. This method is especially efficient for 3D meshes with high-geometry regularity.

2.6.3 PVQ-Based Geometry Compression

In predictive 3D mesh geometry coding, the position of each vertex is predicted from the previously coded neighboring vertices and the resultant prediction error vectors are coded. Predictive VQ yields good compression performance at medium to high coding rates by exploiting the statistical dependencies among the components of the vertex prediction error vector. In addition, the mapping of the prediction error vectors to the channel indices by the VQ encoder is very suitable for parallel hardware implementation and the mapping of these indices to the reconstruction vectors by the VQ decoder requires low computational complexity. Predictive VQ may be preferred to transform based coding in applications where low complexity is desired along with high reconstruction fidelity.

Chou and Meng [85] first proposed a predictive VQ (PVQ) scheme for mesh geometry compression. To ensure a linear time complexity, a simple predictor is adopted to predict a new vertex from the midpoint of two previously traversed vertices. Several VQ techniques, including the open loop VQ, the asymptotic closed loop VQ and the product code pyramid VQ are applied for residual vector quantization. All these VQ techniques yield a better rate-distortion performance than Deering's work [12], which employs the uniform scalar quantizer and delta coding. A beneficial side effect of this PVQ scheme is that linear vertex transformation forms a rendering pipeline and can be greatly accelerated.

In Bayazit et al.'s work [86], the prediction error vectors are represented in a local coordinate system in order to cluster them around a subset of a 2D planar subspace and thereby increase block coding efficiency. Alphabet entropy constrained vector quantization (AECVQ) [87] is preferred to the previously employed minimum distortion vector quantization (MDVQ) for block coding the prediction error vectors with high coding efficiency and low implementation complexity. Estimation and compensation of the bias in the parallelogram prediction rule and partial adaptation of the AECVQ codebook to the encoded vector source by normalization using source statistics are the other salient features of the proposed coding system. Experimental results verify the advantage of the use of the local coordinate system over the global one. The visual error of the proposed coding system is lower than that of the predictive coding method of Touma and Gotsman [19], especially at low rates.

2.6.4 Fast VQ Compression for 3D Mesh Models

As we know, the main disadvantage of VQ is its high complexity during the encoding process. Assume the number of codewords is N and the vector dimension is k, when quantizing an input vector with the full search (FS) method, kN multiplications, $(2k-1)N$ additions and N comparisons are required. To reduce the computational burden of the FS algorithm, researchers have presented many efficient fast codevector search algorithms. Among these algorithms, Hadamard transform partial distortion search (HTPDS) [88] is a typical one. In [88], all the codevectors are first Hadamard transformed and sorted in terms of their first elements. Though this technique is efficient for image data compression, Hadamard transform can only be applied to vector quantization in a 2^n dimensional space. Thus it is not applicable to 3D vector quantization.

To alleviate the above problems, a fast approach to the nearest codevector search for 3D mesh compression using an orthonormal transformed codebook is proposed by Li and Lu [89]. The algorithm uses the coefficients of an input vector along a set of orthonormal bases as the criteria to reject impossible codevectors. Compared to the full search algorithm, a great deal of computational time is saved without extra distortion and additional storage requirement. This method can be illustrated as follows:

Let us consider a set of orthonormal base vectors $V = \{\boldsymbol{v}_1, \boldsymbol{v}_2, \ldots, \boldsymbol{v}_k\}$ for the Euclidean vector space R^k. For any k-dimensional vector $\boldsymbol{x} = (x_1, x_2, \ldots, x_k)$, it can be transformed to another Euclidean space defined by the k orthonormal base vectors, i.e., $\boldsymbol{x} = \sum_{i=1}^{k} X_i \boldsymbol{v}_i$, where $\boldsymbol{X} = (X_1, X_2, \ldots, X_k)$ is the coefficient vector in the transformed space.

Our aim is to find an appropriate set of orthonormal base vectors $V = \{\boldsymbol{v}_1, \boldsymbol{v}_2, \ldots, \boldsymbol{v}_k\}$ so that the coefficient along each base vector is a criterion for rejecting impossible codevectors. Since the possible nearest codevectors for an input vector locate in the hypersphere with centre at $\boldsymbol{x}$ and radius $d_{\min}$ that is the distortion between $\boldsymbol{x}$ and the current best matched codevector, and the hypersphere can be confined by k pairs of parallelogram hyperplanes that are tangential to the hypersphere in the Euclidean space R^k, we can use these parallelogram hyperplanes to form a hypercube which encloses the hypersphere, thus reducing the search space to a great extent. It follows that if we select the k different unit normal vectors of these hyperplanes as V, we can reject impossible codevectors according to each component of $\boldsymbol{X}$.

In Li and Lu's work [89], 3D meshes are vector quantized based on the parallelogram prediction, so each input vector is a 3D residual vector. They set V to be the unit normal vectors of 3 pairs of parallelogram hyperplanes enclosing the sphere on which all the possible nearest codevectors lie, i.e., $\boldsymbol{v}_1 = \{1/\sqrt{3},\ 1/\sqrt{3},\ 1/\sqrt{3}\}$, $\boldsymbol{v}_2 = \{1/\sqrt{6},\ 1/\sqrt{6},\ -2/\sqrt{6}\}$ and $\boldsymbol{v}_3 = \{1/\sqrt{2},\ -1/\sqrt{2},\ 0\}$. So the kick-out conditions for judging possible nearest codevectors are:

$$X_{i,\min} \le Y_{ji} \le X_{i,\max},\ 1 \le i \le 3, \tag{2.14}$$

where $\boldsymbol{Y}_j = (Y_{j1}, Y_{j2}, Y_{j3})$ is the coefficient vector of y_j in the transformed space and

$$X_{i,\min} = X_i - d_{\min}, \tag{2.15}$$

$$X_{i,\max} = X_i + d_{\min}. \tag{2.16}$$

Then, Li and Lu's algorithm can be illustrated as follows.

2.6.4.1 Preprocessing

The first step is to transform each codevector of the codebook into the space with base vectors $V = \{\boldsymbol{v}_1, \boldsymbol{v}_2, \boldsymbol{v}_3\}$ in order that each input vector can be quantized in the transformed space with the transformed codebook. This process involves $3N$ multiplications and $6N$ additions.

Then, the transformed codevectors are sorted in the ascending order of their first elements, i.e., the coefficients along the base vector $\boldsymbol{v}_1$.

2.6.4.2 Online Steps

Step 1: To carry out the codevector search in the transformed space, we first perform the transformation on the input vector $\boldsymbol{x}$ to obtain $\boldsymbol{X}$. This process involves 3 multiplications and 6 additions.

Step 2: A probable nearby codevector $\boldsymbol{Y}_i$ is guessed, based on the minimum first element difference criterion. This is easy to implement with the bisection technique. $d_{\min}$, $X_{i,\min}$ and $X_{i,\max}$ are calculated.

Step 3: For each codevector $\boldsymbol{Y}_j$, we check if Eq.(2.14) is satisfied. If not, then $\boldsymbol{Y}_j$ is rejected, thus discarding those codevectors which are far away from $\boldsymbol{X}$, resulting in a reduced cube search space containing the sphere centered at $\boldsymbol{X}$ with radius $d_{\min}$; else we proceed to the next step.

Step 4: If $\boldsymbol{Y}_j$ is not rejected in the third step, then $d(\boldsymbol{X},\boldsymbol{Y}_j)$ is calculated. If $d(\boldsymbol{X},\boldsymbol{Y}_j) < d_{\min}$, then the current closest codevector to $\boldsymbol{X}$ is taken as $\boldsymbol{Y}_j$ with $d_{\min}$ set to be $d(\boldsymbol{X},\boldsymbol{Y}_j)$, and $X_{i,\min}$ and $X_{i,\max}$ are updated accordingly. The procedure is repeated until we arrive at the best matched codevector $\boldsymbol{Y}_p$ for $\boldsymbol{X}$.

Step 5: Inversely transform $\boldsymbol{Y}_p$ to $\boldsymbol{y}_p$ in the original space. This process needs 3 multiplications and 6 additions.

In the codevector search process, we expect the "so far" $d_{\min}$ to be as small as possible in order to reject $\boldsymbol{x}$ with lighter computation. The projection of $\boldsymbol{x}$ on $\boldsymbol{v}_1$ is proportional to the mean of $\boldsymbol{x}$, so it has a clear physical meaning and is regarded as the best value to represent $\boldsymbol{x}$. In this sense, the initial $d_{\min}$ in Step 2 is minimized, and further rejection of $\boldsymbol{x}$ based on Eq.(2.14) is more likely to occur.

It is obvious that this fast method can be extended to VQ in a Euclidean space of any dimension by finding an orthonormal transform of the original space. The number of the kick-out conditions for nearest codevectors can either be equal or be less than the dimension of the space.

The computational efficiency of the proposed algorithm in compressing 3D mesh geometry data, in comparison to PDS [90], ENNS [91] and EENNS [92] algorithms, was evaluated in [89]. In the fast VQ scheme [89], 20 meshes were randomly selected from the famous Princeton 3D mesh library and 42,507 3D residual vectors were generated from these meshes based on the parallelogram prediction. The residual vectors are then used to generate the codebook, and the sizes of the codebooks are 256, 1,024 and 8,192. Table 2.5 shows the time needed for quantizing the geometry of two 3D mesh models, Stanford Dragon (100,250 vertices and 202,520 triangles) and Stanford Bunny (35,947 vertices and 69,451 triangles). The time is the average of three experiments. The encoding qualities for different codebooks are also shown. The coding quality remains the same for all the algorithms since they are full-search equivalent. No extra memory is demanded for Full Search (FS), PDS and Li and Lu's approach while ENNS and EENNS need N and $2N$ pre-stored float data respectively, where N is the size of the codebook. The platform is Visual C++ 6.0 and PC 2.0 GHz.

The search efficiency in the form of a ratio is evaluated by how many times the Euclidean distance computation is averagely performed compared to the size of codebook, as shown in Table 2.6. The ratio is a relative baseline rather than encoding time to exclude the effect of programming skills, but it ignores the online computation complexity for non-winner rejection. A smaller ratio is better.

Table 2.5 Performance comparison among the algorithms on the time used to quantize the Dragon and Bunny meshes

Mesh	Codebook size	PSNR (dB)	Time (s)				
			FS	PDS	ENNS	EENNS	Li and Lu's approach
Dragon	256	41.00	1.45	0.86	0.25	0.28	0.15
	1,024	48.25	5.34	2.89	0.44	0.41	0.20
	8,192	56.40	43.12	26.13	1.58	0.95	0.55
Bunny	256	41.72	0.49	0.30	0.08	0.09	0.04
	1,024	49.96	1.94	1.02	0.16	0.14	0.07
	8,192	58.47	15.41	10.70	0.50	0.27	0.17

Table 2.6 Ratio of the reduced search space after each check step compared to FS (100%) for Dragon and Bunny meshes

Mesh	Codebook size	Ratio compared to FS			
		PDS	ENNS	EENNS	Li and Lu's approach
Dragon	256	11.90	7.60	3.00	1.52
	1,024	3.67	3.65	1.00	0.43
	8,192	5.43	1.83	0.26	0.08
Bunny	256	11.26	7.20	2.79	1.50
	1,024	3.59	3.19	0.84	0.40
	8,192	5.31	1.47	0.19	0.07

Evident in Table 2.5 and Table 2.6, Li and Lu's approach [89] is a computation efficient algorithm in terms of both encoding time and the effect of search space reduction, compared to state-of-art fast search algorithms that can be extended to mesh VQ.

2.6.5 VQ Scheme Based on Dynamically Restricted Codebook

When vertex positions are VQ compressed based on full search in a stationary codebook, the encoding performance will be fixed. So if we desire a higher compression rate, a lower level of codebook is needed. It is not convenient to transmit a unique codebook with the compressed mesh bit stream or pre-store codebooks of many different sizes in all terminals over the Internet. However, it is possible to use a parameter which controls the encoding quality to get any desired compression rate in a range with only one codebook and a better rate-distortion performance (R-D) can be expected. To address this issue, Lu and Li [93] presented a novel vertex encoding algorithm using the dynamically restricted codebook based vector quantization (DRCVQ).

2.6.5.1 Basic DRCVQ Idea

In DRCVQ, a parameter is used to control the encoding quality to get the desired compression rate in a range with only one codebook, instead of using different levels of codebooks to get a different compression rate. During the encoding process, the indexes of the preceding encoded residual vectors which have high correlation with the current input vector are pre-stored in an FIFO so both the codevector searching range and bit rate are averagely reduced. The proposed scheme also incorporates a very effective Laplacian smooth operator. A unique feature of this scheme is that there is an adjustable parameter in the proposed scheme, so the user can get a desired rate-distortion performance conveniently, without encoding the vertex data with a codebook of another quality level. In addition, it permits compatibility with most of the existing algorithms for geometry data compression. Combined with other schemes, the rate-distortion performance may be further improved.

The DRCVQ approach uses a fixed-length first-in-first-out (FIFO) buffer to store the previously encoded codevector indexes. The sequence of vertices encountered during a mesh traversal defines which vector is to be coded and the correlation between codevectors of the processed input vectors is also employed. When the encoding procedure begins, the approach sets FIFO to be null, and then appends the index of the current encoded vertex to the buffer if it is not found in the buffer.

Using a fixed-length FIFO, the codevector search range of an input vector can be reduced so the bit rate is reduced, as illustrated as follows. First we define the stationary codebook C_0 which has N_0 codevectors and its restricted part C_1. The restricted codebook C_1 contains the N_1 most likely codevector indexes when the stationary codebook C_0 is applied to the source. Here, the restricted codebook C_1 is dynamic for each encoded vertex and is regenerated by buffering a series of codevector indices since the statistics of the ongoing sequence of vectors may undergo a sudden and substantial change. As each of the input vectors is encoded using codebook C_0, there are in total N_0 possible codevector indexes for each input vector. If the input vectors are highly correlated, then we are lucky to specify an input vector by one of the codevector index in C_1, and $\log_2 N_1$ bits are sufficient to represent the input vector instead of $\log_2 N_0$ bits. Since N_1 is normally much smaller than N_0, bpv can be greatly reduced.

2.6.5.2 Vector Quantizer Design

The first issue in designing a VQ scheme for compressing any kind of source is how to map the source data into a vector sequence as the input of the vector quantizer. For 2D signals such as images, the vector sequence is commonly formed from blocks of neighboring pixels. The blocks can be directly used as the input vector for the quantizer. In the case of triangle meshes, neighboring vertices are also likely to be correlated. However, blocking multiple vertices is not as straightforward as the case for images. The coordinate vector of a vertex cannot be directly regarded as an input vector to the quantizer because if multiple vertices are mapped into the same vertex, the distortion of the mesh will be unacceptable and the connectivity of the mesh will also disappear.

Since the principle of the vector quantizer design method remains the same in both ordinary VQ and DRCVQ, we only discuss ordinary VQ here. In order to exploit the correlation between vertices, it is necessary to use a vector quantizer with memory. Thus, Lu and Li [93] employed predictive vector quantization. The index identifying this residual vector in PVQ was then stored or transmitted to the decoder.

There are two components in a PVQ system: prediction and residual vector quantization. We first discuss the design of the predictor. The goal of the predictor is to minimize the variance of the residuals, as well as maintaining low computation complexity, allowing them to be coded more efficiently by the vector quantizer.

Lu and Li [93] used the principle of the "parallelogram" prediction illustrated in Fig. 2.22. The three vertices of the initial triangle in the traversal order are uniformly scalar quantized at 10 bits per coordinate and then Huffman encoded. Any other vertex can be predicted by its neighboring triangles, enabling exploitation of the tendency for neighboring triangles to be roughly coplanar and similar in size. This is particularly true for high-resolution, scanned models, which have little variation in the triangle size. As shown in Fig. 2.22 and Eq.(2.10), the prediction error between v_n and $\tilde{v}_n$ may be accumulated to the subsequent encoded vertices. When the number of vertices in a mesh is large enough, the accumulated error may be unacceptable. To permit reconstruction of the vertices by the decoder, the prediction must only be based on previous reconstructed vertices. Thus, the encoder also needs to replace the processed vertex to be its quantized vertex for predicting subsequent vertices. The residual vectors are then used to generate the codebook.

In fact, there are many variations of VQ that could be employed for quantizing the residuals. Lu and Li [93] focused on the conventional unconstrained VQ. The disadvantages of this unconstrained VQ generation scheme mainly include the time required to train the codebook and the time consumption for transmitting a codebook with the mesh. In Lu and Li's scheme, 20 meshes were randomly selected from the famous Princeton 3D mesh library and 42,507 training vectors were generated from these meshes for training the approximate universal codebook off-line, and its size ranges from 64 to 8,192. In this way, we expect the codebook to be suitable for nearly all triangle meshes for VQ compression and it can be pre-stored in terminals over the network. Thus the compressed bit stream can be transmitted alone with convenience.

2.6.5.3 Adjustable Parameter

In order to achieve the desired compression ratio, Lu and Li assumed that some applications can tolerate a little degradation of PSNR to reduce the bpv. They set a threshold T as the parameter to control the PSNR degradation. Note that T is the parameter for additional distortion control because the compression is always lossy due to the restriction to N_0 codevectors in the global codebook. When the Euclidcan distancc of thc input vcctor and its closest codevector specified by the index stored in C_1 is not more than the desired T, we assign the index in C_1 to the input vector as its encoded index and its corresponding codevector is easily found. This method has the advantage of adjusting T by the user to get a satisfactory R-D performance, rather than changing the codebook to another size as in conventional VQ compression methods. In Lu and Li's scheme, 1 bit side information is needed for identifying whether a codevector index is for C_0 or C_1.

The correlation of consecutive subsets of residual vectors in the connectivity traversal order that the algorithm is taking advantage of is shown in a graphical way in Fig. 2.27. Stars represent an example of typical 16 consecutive residual vectors generated from the Caltech Feline mesh model compression, and their bounding sphere radius is 0.02, while the dots indicate part of the codevectors of the universal codebook consisting of 8,192 codevectors whose bounding sphere radius is 2.00. It is evident that consecutive residual vectors concentrate in a small region relative to the whole codevectors. Thus it may happen that multiple residual vectors of the 16 consecutive vectors are mapped to the same codevector and, if we increase T for further distortion tolerance, any residual vectors in the sphere with radius T and centered at that codevector will be mapped to it, resulting in more likelihood of the local search in the FIFO and thus bit rate reduction.

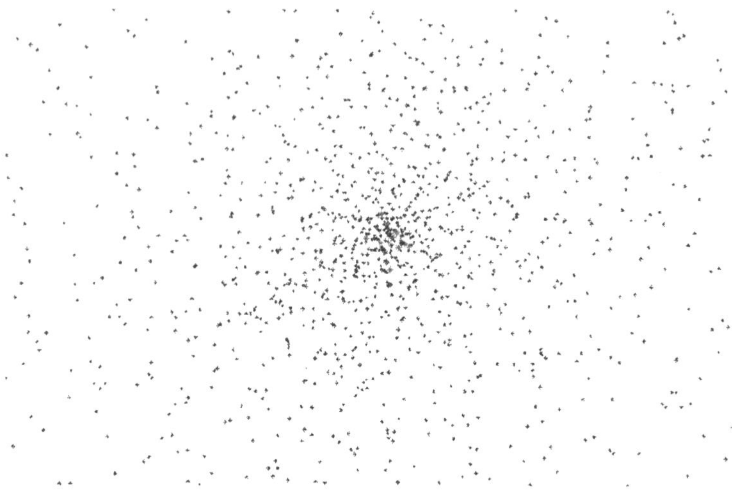

Fig. 2.27. Zoom-in of an example of consecutive residual vectors (in stars) and codevectors (in dots)

2.6.5.4 Other Considerations

The most computationally intensive part of the DRCVQ algorithm is the distortion calculation between an input vector and each codevector in the stationary codebook C_0 for finding the closest codevector for the input vector. The distance computation in R^3 Euclidean space needs $3N_0$ multiplications, $5N_0$ additions and N_0 comparisons to encode each input vector in the full search VQ. Lu and Li [93] adopted the mean-distance-ordered partial codebook search (MPS) [94] as an efficient fast codevector search algorithm which uses the mean of the input vector to reduce the computational burden of the full search algorithm without sacrificing performance. In [94], the codevectors are sorted according to their component means, and the search for the codevector having the minimum Euclidean distance to a given input vector starts with the one having the minimum mean distance to it. The search is then made to terminate as soon as possible since the mean distance out of a range should correspond to a larger Euclidean distance.

The mesh distortion metric is also an important issue. Let $d(\boldsymbol{x}, Y)$ be the Euclidean distance from a point $\boldsymbol{x}$ on X to its closest point on Y, then the distance from X to Y is defined as follows:

$$d(X,Y) = \sqrt{1/A(X) \cdot \int_{x \in X} d(\mathbf{x},Y)^2 \mathrm{d}x}\,, \tag{2.17}$$

where $A(X)$ is the area of X. Since this distance is not symmetric, the distortion between X and Y is given as:

$$d = \max\{d(X,Y), d(Y,X)\}. \tag{2.18}$$

This distance is called symmetric face-to-face Hausdorff distance. All the distortion errors reported in Lu and Li's work are in terms of the percentage of the mesh bounding box.

In order to further reduce the bit rate without affecting the mesh quality, Lu and Li used entropy coding to encode the residual vector indexes before they are transmitted through the channel. Lu and Li simply divided the indexes bit sequence into groups of 8 bits, and then encoded them using arithmetic coding.

The "parallelogram" prediction rule assumes that neighboring vertices are coplanar. However, since a universal codebook contains codevectors uniformly in all directions, when a vertex is reconstructed from its prediction vector and its quantized residual vector with a universal codebook, it deviates from the original plane. So vector quantization introduces high frequencies to the original mesh. In order to improve the visual quality of the decoded meshes, a Laplacian low frequency pass filter is adopted which is derived from the mesh connectivity that has already been received and decoded before residual vectors are decoded. The Mesh Laplacian operator is defined in Eq.(2.11), and then the filtered vertex is defined as:

$$\boldsymbol{v}_i' = \sum_j |L_{ij}| \cdot \boldsymbol{v}_j / 2, \tag{2.19}$$

where $\boldsymbol{v}_i'$ is the filtered version of $\boldsymbol{v}_i$. This filter can be operated iteratively. Based on the assumption that similar mesh models should have similar surface area, the criterion for terminating the Laplacian filter is set to be:

$$|area(\boldsymbol{M}^{(i)}) - area(\boldsymbol{M})| / area(\boldsymbol{M}) < \delta, \tag{2.20}$$

where $\boldsymbol{M}^{(i)}$ is the i-th version of filtered original $\boldsymbol{M}$, area ($\boldsymbol{M}$) is a 32-bit float value which can be transmitted along with the compressed mesh bit stream, and δ is set to be 0.03.

Since the above geometry compression scheme does not alter any connectivity of the original mesh and the vertex coding order only depends on the connectivity encoder, the connectivity encoding algorithm can be freely chosen in Lu and Li's work. Alliez's valence-driven connectivity encoder is adopted as an effective method which reaches the optimal upper bound (3.24 bpv) for the bit rate per vertex for large, arbitrary meshes. In addition, Lu and Li also proposed a similar method based on dynamic extended codebook based vector quantization (DECVQ) in [95]. Readers can refer to it for detailed information.

2.6.5.5 Simulation Results

Rate-distortion performances of “Wavemesh” [96] and the conventional VQ are compared with Lu and Li’s work. In the conventional VQ method, all the prediction error vectors based on the parallelogram prediction are quantized with the stationary codebook C_0 using full search method. Wavemesh is combined with Wavelet Geometrical Criterion (WGC) if it improves the result. As expected, the proposed dynamically restricted scheme produces a better bpv-PSNR curve, outperforming the conventional VQ method, as shown in Fig. 2.28. For fair comparison, DRCVQ here is not combined with entropy coding or Laplacian smoothener. The size of the additional codebook C_1 is set to be 16. The PSNR measure is defined as $20\log_{10}peak/d$, where *peak* is the mesh bounding box diagonal and d is the root mean square error. The rate is represented as bits per vertex in terms of mesh geometry. When the distortion threshold T for Lu and Li’s scheme is set to be 0, the bpv of DRCVQ is higher than the conventional method because of the 1 bit side information stored, indicating whether or not an input vector is encoded using C_1, the restricted codebook. However, with the increasing of the threshold T, bpv decreases relatively more with only a little bit of PSNR degradation. When the bit rate is 10 bpv, Lu and Li’s method performs as much as about 6dB better than the conventional VQ for Stanford Bunny, Caltech Feline and Fandisk models, because the high resolution results in a high correlation among vertices along the traversal order and thus input vectors are more likely to be encoded in codebook C_1. However, when DRCVQ is applied to the heavily simplified version of Stanford Bunny, only about 2.5 dB is gained at 10 bpv. This is mainly because residual vectors generated from models with low definition vary much and locate in a large range so DRCVQ does not work very well. From Fig. 2.28, it is evident that by using DRCVQ we can use the codebook of 8,192 codevectors alone to encode triangle meshes instead of using the conventional method with stationary codebooks of sizes from 64 to 4,096.

Fig. 2.29 shows 3 different curves on the Fandisk, Venus head and Venus body models for Wavemesh (optionally with WGC), DRCVQ without entropy coding or filtering and DRCVQ with entropy coding and filtering. The bit rate consists of mesh connectivity and geometry, and is represented by bits per vertex. For Fandisk and Venus head models, DRCVQ performs much better than Wavemesh, though the proposed method is always lossy while Wavemesh can achieve lossless coding. All the bpv values given by DRCVQ in the experiments are more than about 7 bpv, because of about 1.5 bpv for connectivity coding and at least about 5.0 bpv for geometry coding (the length of FIFO is fixed to be 16 and 1 extra bit). As expected, mesh compression methods in the spectral domain perform better for mesh models with high definition and uniformity while vector quantizers introduce high frequency noises and are slightly worse for this type of model. In the Venus body experiment, the rate-distortion curve of DRCVQ cannot outperform Wavemesh.

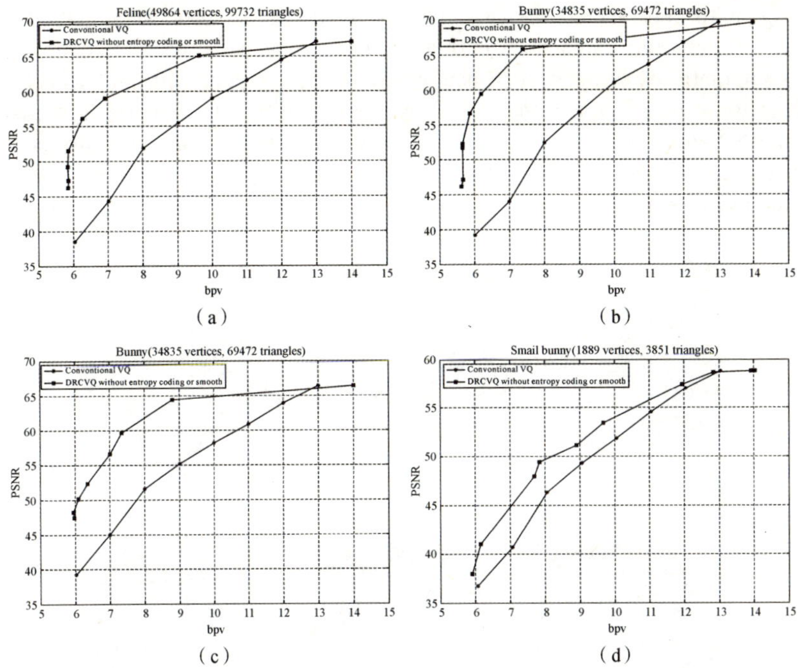

Fig. 2.28. DRCVQ compared with conventional VQ. (a) Caltech Feline; (b) Stanford Bunny; (c) Fandisk; (d) Stanford simplified Bunny

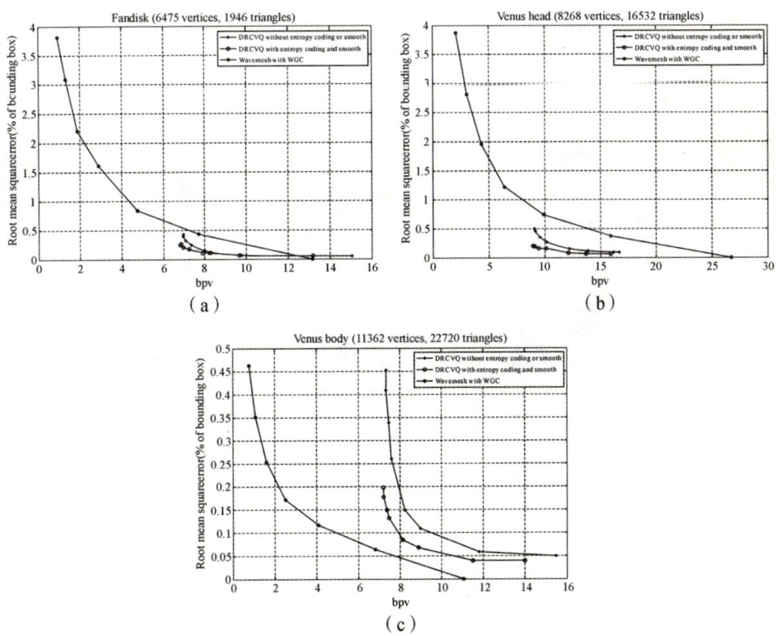

Fig. 2.29. Comparisons with Wavemesh. (a) Fandisk; (b) Venus head; (c) Venus body

Fig. 2.30 shows reconstructed meshes by using the proposed method with entropy coding and Laplacian filtering. Lu and Li's scheme has the advantage of low computational complexity. Since they have incorporated MPS in DRCVQ, the codevector search time is rather low. With T increasing from 0 to 1E−3 relative to the mesh bounding box diagonal, the geometry compression time ranges from 0.15 to 0.05 s for Bunny and 0.20 to 0.07 s for Feline. The platform is Visual C++ 6.0 and PC 2.0 GHz.

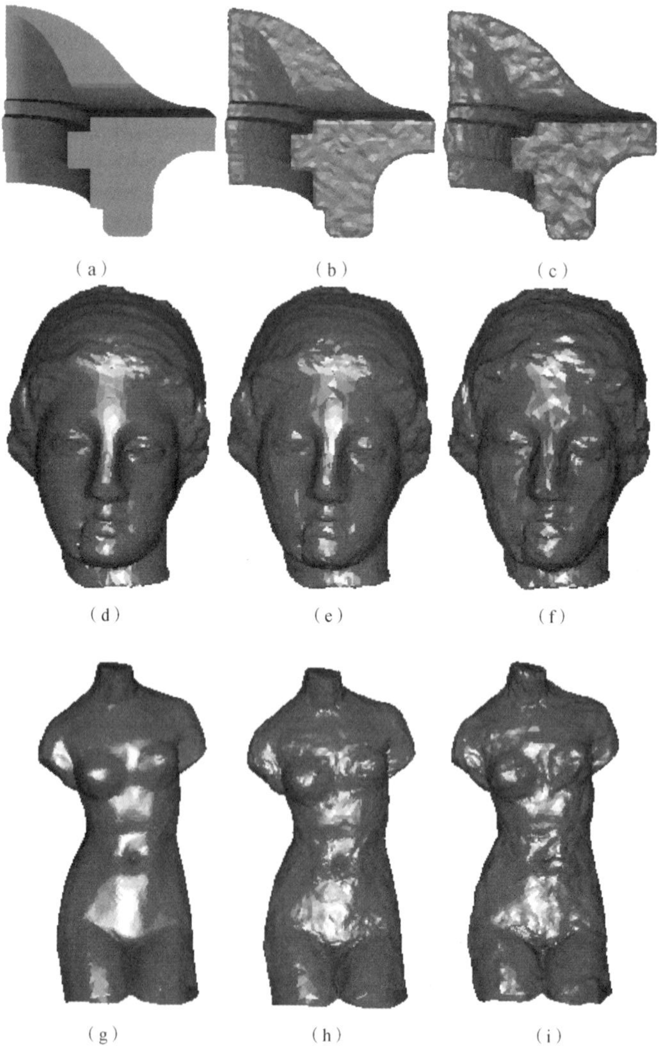

Fig. 2.30. Reconstructed meshes of typical models using DRCVQ with entropy coding and Laplacian smooth. (a) Original Fandisk; (b) 7.22 bpv, 59.24 dB; (c) 5.94 bpv, 53.79 dB; (d) Original Venus head; (e) 11.00 bpv, 62.85 dB; (f) 6.76 bpv, 55.86 dB; (g) Original Venus body; (h) 7.39 bpv, 63.43 dB; (i) 5.86 bpv, 56.54 dB

2.7 Summary

This chapter performed a relatively detailed survey of current 3D mesh compression techniques by classifying major algorithms, describing the main ideas behind each category, and comparing their strength and weakness. First, the background, basic concepts and algorithm classification of 3D mesh compression techniques were briefly introduced. Then, the connectivity compression methods were introduced in two sections, i.e., single-rate and progressive compression schemes. Next, the geometry compression techniques were discussed in three sections, i.e., spatial-domain based, transform-domain based and vector quantization-based (VQ-based) methods.

For single-rate connectivity coding, the best schemes are those based on the valence-driven approach. For progressive connectivity compression, the valence-driven conquest approach is still among the best ones. For spatial-domain geometry compression, *k-d* tree, octree and VQ based methods are the state-of-the-art methods. For transform based geometry compression, Khodakovsky and Gusko's wavelet coding method is the best one.

In early mesh coding schemes, geometry coding was tightly coupled with, and restrained by, connectivity coding. However, this dependence has been weakened or even reversed. Geometry data tend to consume a dominant portion of the storage space, and their correlation can be exploited more effectively without the restraint of connectivity. In addition, remesh-based progressive mesh coders completely discard the irregular connectivity of an input mesh and resample the surface with a regular pattern. Due to regular resampling, connectivity coding requires almost no information while geometry data can be efficiently compressed.

Research on single-rate coding seems to be mature except for further improvement of geometry coding. Progressive coding has been thought to be inferior to single-rate coding in terms of the coding gain. However, high-performance progressive codecs have emerged these days and they often outperform some of the state-of-the-art single-rate codecs. In other words, a progressive mesh representation seems to be a natural choice, which demands no extra burden in the coding process. There is still room to improve progressive coding to provide better R-D performance at a lower computational cost.

Future mesh coding schemes will be inspired by new 3D representations such as the normal mesh representation and the point cloud-based geometry representation. Another promising research area may be animated-mesh coding that was overlooked in the past but has been getting more attention recently.

References

[1] P. Alliez and C. Gotsman. Recent advances in compression of 3D meshes. In: Proceedings of the Symposium on Multiresolution in Geometric Modeling,

2003.
[2] J. L. Peng, C. S. Kim and C. C. Jay Kuo. Technologies for 3D mesh compression: A survey. Journal of Visual Communication and Image Representation, 2005, 16(6):688-733.
[3] ISO/IEC 14772-1. The Virtual Reality Modeling Language VRML. 1997.
[4] G. Taubin, W. Horn, F. Lazaru, et al. Geometry coding and VRML. Proceedings of the IEEE, 1998, 96(6):1228-1243.
[5] G. Taubin and J. Rossignac. Geometric compression through topological surgery. ACM Trans. Graph., 1998, 17(2):84-115.
[6] ISO/IEC 14496-2. Coding of Audio-Visual Objects: Visual. 2001.
[7] O. Devillers and P. Gandoin. Geometric compression for interactive transmission. In: Proceedings of the IEEE Conference on Visualization, 2000, pp. 319-326.
[8] G. Taubin. 3D geometry compression and progressive transmission. EUROGRAPHICS—State of the Art Report, 1999.
[9] D. Shikhare. State of the art in geometry compression. Technical Report, National Centre for Software Technology, India, 2000.
[10] C. Gotsman, S. Gumhold and L. Kobbelt. Simplification and compression of 3D meshes. Tutorials on Multiresolution in Geometric Modelling, 2002.
[11] J. Gross and J. Yellen. Graph Theory and Its Applications. CRC Press, 1998.
[12] M. Deering. Geometry compression. ACM SIGGRAPH, 1995, pp. 13-20.
[13] M. Chow. Optimized geometry compression for real-time rendering. IEEE Visualization, 1997, pp. 347-354.
[14] E. M. Arkin, M. Held, J. S. B. Mitchell, et al. Hamiltonian triangulations for fast rendering. Visual Computation, 1996, 12(9):429-444.
[15] F. Evans, S. S. Skiena and A. Varshney. Optimizing triangle strips for fast rendering. IEEE Visualization, 1996, pp. 319-326.
[16] G. Turan. On the succinct representations of graphs. Discr. Appl. Math, 1984, 8:289-294.
[17] C. L. Bajaj, V. Pascucci and G. Zhuang. Single resolution compression of arbitrary triangular meshes with properties. Comput. Geom. Theor. Appl., 1999, 14:167-186.
[18] C. Bajaj, V. Pascucci and G. Zhuang. Compression and coding of large CAD models. Technical Report, University of Texas, 1998.
[19] C. Touma and C. Gotsman. Triangle mesh compression. In: Proceedings of Graphics Interface, 1998, pp. 26-34.
[20] P. Alliez and M. Desbrun. Valence-driven connectivity encoding for 3D meshes. EUROGRAPHICS, 2001, pp. 480-489.
[21] M. Schindler. A fast renormalization for arithmetic coding. In: Proceedings of IEEE Data Compression Conference, 1998, p. 572.
[22] W. Tutte. A census of planar triangulations. Can. J. Math., 1962, 14:21-38.
[23] C. Gotsman. On the optimality of valence-based connectivity coding. Computer Graphics Forum, 2003, 22(1):99-102.
[24] S. Gumhold and W. Straßer. Real time compression of triangle mesh connectivity. ACM SIGGRAPH, 1998, pp. 133-140.
[25] S. Gumhold. Improved cut-border machine for triangle mesh compression. Paper presented at The Erlangen Workshop'99 on Vision, Modeling and Visualization, 1999.
[26] J. Rossignac. Edgebreaker: connectivity compression for triangle meshes. IEEE

Trans. Vis. Comput. Graph., 1999, 5(1):47-61.
[27] D. King and J. Rossignac. Guaranteed 3.67v bit encoding of planar triangle graphs. Paper presented at The 11th Canadian Conference on Computational Geometry, 1999, pp. 146-149.
[28] S. Gumhold. New bounds on the encoding of planar triangulations. Technical Report WSI-2000-1, Wilhelm-Schickard-Institut für Informatik, University of Tübingen, Germany, 2000.
[29] J. Rossignac and A. Szymczak. Wrap and zip decompression of the connectivity of triangle meshes compressed with edgebreaker. Comput. Geom., 1999, 14(1-3):119-135.
[30] M. Isenburg and J. Snoeyink. Spirale reversi: reverse decoding of the Edgebreaker encoding. Paper presented at The 12th Canadian Conference on Computational Geometry, 2000, pp. 247-256.
[31] A. Szymczak, D. King and J. Rossignac. An Edgebreaker-based efficient compression scheme for regular meshes. In: Proceedings of 12th Canadian Conference on Computational Geometry, 2000, pp. 257-264.
[32] M. Isenburg. Triangle strip compression. In: Proceedings of the Graphics Interface, 2000, pp. 197-204.
[33] B. S. Jong, W. H. Yang, J. L. Tseng, et al. An efficient connectivity compression for triangular meshes. In: Proceedings of the Fourth Annual ACIS International Conference on Computer and Information Science (ICIS'05), 2005.
[34] A. Guéziec, G. Taubin, F. Lazarus, et al. Converting sets of polygons to manifold surfaces by cutting and stitching. IEEE Visualization, 1998, pp. 383-390.
[35] H. Hoppe. Progressive meshes. ACM SIGGRAPH, 1996, pp. 99-108.
[36] H. Hoppe, T. DeRose, T. Duchamp, et al. Mesh optimization. ACM SIGGRAPH, 1993, pp. 19-25.
[37] H. Hoppe. Efficient implementation of progressive meshes. Comput. Graph, 1998, 22(1):27-36.
[38] J. Popovic and H. Hoppe. Progressive simplicial complexes. ACM SIGGRAPH, 1997, pp. 217-224.
[39] G. Taubin, A. Gueziec, W. Horn, et al. Progressive forest split compression. ACM SIGGRAPH, 1998, pp. 123-132.
[40] G. Taubin. A signal processing approach to fair surface design. ACM SIGGRAPH, 1995, pp. 351-358.
[41] R. Pajarola and J. Rossignac. Compressed progressive meshes. IEEE Trans. Vis. Comput. Graph., 2000, 6(1):79-93.
[42] N. Dyn, D. Levin and J. A. Gregory. A butterfly subdivision scheme for surface interpolation with tension control. ACM Trans. Graph., 1990, 9(2):160-169.
[43] D. Zorin, P. Schröder and W. Sweldens. Interpolating subdivision for meshes with arbitrary topology. ACM SIGGRAPH, 1996, pp. 189-192.
[44] R. Pajarola and J. Rossignac. Squeeze: fast and progressive decompression of triangle meshes. In: Proceedings of Computer Graphics International Conference, 2000, pp. 173-182.
[45] R. Pajarola. Fast Huffman code processing. Technical Report UCI-ICS-99-43, Information and Computer Science, UCI, 1999.
[46] W. J. Schroeder, J. A. Zarge and W. E. Lorensen. Decimation of triangle meshes. ACM SIGGRAPH, 1992, pp. 65-70.
[47] M. Soucy and D. Laurendeau. Multiresolution surface modeling based on

hierarchical triangulation. Comput. Vis. Image Understand., 1996, 63(1):1-14.
[48] D. Cohen-Or, D. Levin and O. Remez. Progressive compression of arbitrary triangular meshes. IEEE Visualization, 1999, pp. 67-72.
[49] P. Alliez and M. Desbrun. Progressive encoding for lossless transmission of triangle meshes. ACM SIGGRAPH, 2001, pp. 198-205.
[50] J. Li and C. C. J. Kuo. Progressive coding of 3-D graphic models. In: Proc. of the IEEE, 1998, 86(6):1052-1063.
[51] C. Bajaj, V. Pascucci and G. Zhuang. Progressive compression and transmission of arbitrary triangular meshes. IEEE Visualization, 1999, pp. 307-316.
[52] C. L. Bajaj, E. J. Coyle and K. N. Lin. Arbitrary topology shape reconstruction from planar cross sections. Graph. Models Image Proc., 1996, 58(6):524-543.
[53] T. S. Gieng, B. Hamann, K. I. Joy, et al. Constructing hierarchies for triangle meshes. IEEE Trans. Vis. Comput. Graph., 1998, 4(2):145-161.
[54] A. Gersho and R. M. Gray. Vector Quantization and Signal Compression. Kluwer Academic Publishers, 1992.
[55] H. Lee, P. Alliez and M. Desbrun. Angle-analyzer: a triangle-quad mesh codec. In: Eurographics Conference Proceedings, 2002, pp. 383-392.
[56] M. Isenburg and P. Alliez. Compressing polygon mesh geometry with parallelogram prediction. In: IEEE Visualization Conference Proceedings, 2002, pp. 141-146.
[57] B. Kronrod and C. Gotsman. Optimized compression of triangle mesh geometry using prediction trees. In: Proceedings of 1st International Symposium on 3D Data Processing, Visualization and Transmission, 2002, pp. 602-608.
[58] R. Cohen, D. Cohen-Or and T. Ironi. Multi-way geometry encoding. Technical Report, 2002.
[59] D. Shikhare, S. Bhakar and S. P. Mudur. Compression of large 3D engineering models using automatic discovery of repeating geometric features. In: Proceedings of 6th International Fall Workshop on Vision, Modeling and Visualization, 2001.
[60] P. M. Gandoin and O. Devillers. Progressive lossless compression of arbitrary simplicial complexes. ACM Trans. Graph., 2002, 21(3):372-379.
[61] O. Devillers and P. Gandoin. Geometric compression for interactive transmission. IEEE Visualization, 2000, pp. 319-326.
[62] I. H. Witten, R. M. Neal and J. G. Cleary. Arithmetic coding for data compression. Commun. ACM, 1987, 30(6):520-540.
[63] J. Peng and C. C. J. Kuo. Geometry-guided progressive lossless 3D mesh coding with octree (OT) decomposition. ACM Trans. Graph., 2005, 24(3):609-616.
[64] N. S. Jayant and P. Noll. Digital Coding of Waveforms—Principles and Applications to Speech and Video. Prentice Hall, 1984.
[65] Z. Karni and C. Gotsman. Spectral compression of mesh geometry. ACM SIGGRAPH, 2000, pp. 279-286.
[66] Z. Karni and C. Gotsman. 3D mesh compression using fixed spectral bases. In: Proceedings of the Graphics Interface, 2001, pp. 1-8.
[67] O. Sorkine, D. Cohen-Or and S. Toldeo. High-pass quantization for mesh encoding. In: Proceedings of Eurographics Symposium on Geometry Processing, 2003.
[68] M. Lounsbery, T. D. Derose and J. Warren. Multiresolution analysis for surfaces of arbitrary topological type. ACM Transactions on Graphics, 1997, 16(1):34-73.

[69] A. Khodakovsky, P. Schröder and W. Sweldens. Progressive geometry compression. ACM SIGGRAPH, 2000, pp. 271-278.
[70] A. W. F. Lee, W. Sweldens, P. Schröder, et al. MAPS: multiresolution adaptive parametrization of surfaces. ACM SIGGRAPH, 1998, pp. 95-104.
[71] C. Loop. Smooth subdivision surfaces based on triangles. Master's Thesis, Department of Mathematics, University of Utah, 1987.
[72] A. Said and W. A. Pearlman. A new, fast and efficient image codec based on set partitioning in hierarchical trees. IEEE Trans. Circuits Syst. Video Technol., 1996, 6(3):243-250.
[73] A. Khodakovsky and I. Guskov. Normal mesh compression. Geometric Modeling for Scientific Visualization, Springer-Verlag, 2002.
[74] I. Guskov, K. Vidimce, W. Sweldens, et al. Normal meshes. ACM SIGGRAPH, 2000, pp. 95-102.
[75] F. Payan and M. Antonini. Multiresolution 3D mesh compression. Proceedings of IEEE International Conference in Image Processing, 2002, pp. 245-248.
[76] C. Parisot, M. Antonini and M. Barlaud. Optimal nearly uniform scalar quantizer design for wavelet coding. In: Proc. of SPIE VCIP Conference, 2002.
[77] C. Parisot, M. Antonini and M. Barlaud. Model-based bit allocation for JPEG 2000. In: Proc. of EUSIPCO, 2002.
[78] R. Chen, X. Luo and H. Xu. Geometric compression of a quadrilateral mesh. Computers and Mathematics with Applications, 2008, 56:1597-1603.
[79] X. Gu, S. J. Gortler and H. Hoppe. Geometry images. ACM SIGGRAPH, 2002, pp. 355-361.
[80] P. Sander, S. Gortler, J. Snyder, et al. Signal-specialized parametrization. Technical Report MSR-TR-2002-27, Microsoft Research, 2002.
[81] E. Praun and H. Hoppe. Spherical parametrization and remeshing. ACM Trans. Graph., 2003, 22(3):340-349.
[82] H. Hoppe and E. Praun. Shape compression using spherical geometry images. In: N. Dodgson, M. Floater, M. Sabin (Eds.), Advances in Multiresolution for Geometric Modelling, Springer-Verlag, 2005, pp. 27-46.
[83] Y. Linde, A. Buzo and R. M. Gray. An algorithm for vector quantizer design. IEEE Trans. Commun., 1980, 28(1):84-95.
[84] E. S. Lee and H. S. Ko. Vertex data compression for triangular meshes. In: Proceedings of the 8th Pacific Conference on Computer Graphics and Applications, 2000, pp. 225-234.
[85] P. H. Chou and T. H. Meng. Vertex data compression through vector quantization. IEEE Trans. Vis. Comput. Graph., 2002, 8(4):373-382.
[86] U. Bayazit, O. Orcay, U. Konurand, et al. Predictive vector quantization of 3-D mesh geometry by representation of vertices in local coordinate systems. Journal of Visual Communication & Image Representation, 2007, 18(4):341-353.
[87] R. P. Rao and W. A. Pearlman. Alphabet- and entropy-constrained vector quantization of image pyramids. Opt. Eng., 1991, 30:865-872.
[88] Z. M. Lu, J. S. Pan and S. H. Sun. Efficient codevector search algorithm based on Hadamard transform. Electronics Letters, 2000, 36(16):1364-1365.
[89] Z. Li and Z. M. Lu. Fast codevector search scheme for 3D mesh model vector quantization. IET Electronics Letters, 2008, 44(2):104-105.
[90] C. D. Bei and R. M. Gray. An improvement of the minimum distortion encoding algorithm for vector quantization. IEEE Trans. Commun., 1985,

33(10):1132-1133.
[91] L. Guan and M. Kamel. Equal-average hyperplane partitioning method for vector quantization of image data. Pattern Recognition Letters, 1992, 13(10):693-699.
[92] H. Lee and L. H. Chen. Fast closest codevector search algorithms for vector quantization. Signal Processing, 1995, 43:323-331.
[93] Z. M. Lu and Z. Li. Dynamically restricted codebook based vector quantization scheme for mesh geometry compression. Signal Image and Video Processing, 2008, 2(3):251-260.
[94] S. W. Ra and J. K. Kim. Fast mean-distance-ordered partial codebook search algorithm for image vector quantization. IEEE. Transactions on Circuits and Systems-II, 1993, 40(9):576-579.
[95] Z. Li, Z. M. Lu and L. Sun. Dynamic extended codebook based vector quantization scheme for mesh geometry compression. Paper presented at The IEEE Third International Conference on Intelligent Information Hiding and Multimedia Signal Processing (IIHMSP2007), 2007, Vol. 1, pp. 178-181.
[96] S. Valette and R. Prost. Wavelet-based progressive compression scheme for triangle meshes: Wavemesh. IEEE Transactions on Visualizations and Computer Graphics, 2004, 10(2):123-129.

3

3D Model Feature Extraction

Features are important parts of geometric models. They come in different varieties [1]: sharp edges, smoothed edges, ridges or valleys, prongs, bridges and others, as shown in Fig. 3.1. The crucial role of features for a correct appearance and an accurate representation of a geometric model have led to increasing activity in research on feature extraction. Feature extraction from 3D models is an essential and beforehand task for subsequent analysis, retrieval, recognition, classification and tracking processes. This chapter focuses on the techniques of feature extraction from 3D models.

3.1 Introduction

First, the background, basic concepts and algorithm classification related to 3D model feature extraction are introduced.

3.1.1 Background

As surface acquisition methods such as LADAR or range scanners are becoming more and more popular, there is an increasing interest in the use of 3D geometric data in various computer vision applications, such as computer graphics, computer-aided design, medical imaging, molecular analysis, the cultural heritage in virtual environments, the movie industry, military target detection and industrial quality control. However, the processing of 3D datasets, such as range images, is a demanding job due to not only the huge amount of surface data but also the noise and non-uniform sampling introduced by the sensors or the reconstruction process. It is therefore desirable to have a more compact intermediate representation (i.e. features) of 3D objects or images that can be used efficiently in computer vision tasks [2] such as content-based retrieval, 3D scene registration or object recognition.

Fig. 3.1. Example of automatic feature classification: ridges (orange), valleys (blue), and prongs (pink) [1] (©[2007]IEEE)

3.1.1.1 Content-Based 3D Model Retrieval

The development of modeling tools, such as 3D scanners and 3D graphics hardware, has enabled access to 3D materials of high quality both over the Internet and in domain-specific databases. 3D models now play an important role in many applications, such as mechanical manufacture, games, biochemistry, art and virtual reality. Efficient organization and access to these databases demand effective tools for indexing, categorization, classification and representation of 3D objects. All these database activities hinge on the development of 3D object similarity measures [3]. How to find the desired models quickly and accurately from 3D model databases and how to classify the 3D models have become practical problems. So, the development of the technology for content-based retrieval of 3D models has become an important issue. More and more researchers have been involved in the research about the retrieval of 3D models. As opposed to the conventional text-based search algorithms, the content-based search requires deep understanding of the specific data representation. Researchers in many well-known institutions and universities all over the world are dedicating themselves to this research field, which has led to the development of experimental search engines for 3D shapes, such as the 3D model search engine at Princeton University, and the 3D model retrieval system at the National Taiwan University. A typical method for model similarity search and retrieval of 3D models usually consists of three steps [4]: (1) The feature extraction of the model; (2) The computation of distance among the features of the models; (3) The retrieval of the models based on the computed distance values, where the feature extraction of the model is the critical step. Because 3D models are usually defined as the collection of vertex and polygon, a similarity measure between two 3D models cannot be done directly upon such representations. Indeed, content-based search algorithms share the need to define an effective feature space representing the data. Because most 3D models are used in data visualization, the 3D object file only consists of geometry data, connectivity data and appearance data, and there are few descriptions of high-level semantic features for automatic matching. How to describe 3D models appropriately (i.e., feature extraction) is the issue to be urgently solved, and it has been hard to obtain a satisfying solution up to now.

Building correct feature correspondence for 3D models is more difficult and time-consuming [5]. 3D models possess more complex and excessive poses than 2D media, with different translations, rotations, scales and reflections. This gives 3D models many more arbitrary and unpredictable positions, orientations and measurements and makes 3D models difficult to parameterize and search. The new adopted features in content-based 3D model retrieval include 2D shape projections, 3D shapes, 3D appearances and even high-level semantics, which are required not only to be extracted, represented and indexed easily and efficiently, but also for effectively distinguishing similar models from dissimilar models, invariant to typical affine transformations.

3.1.1.2 3D Scene Registration

Scan registration [6] can be defined as finding the translation and rotation of a projected scan contour that produces maximum overlap with a reference scan or a previous model. Scan matching is a highly non-linear problem, with no analytical solution, which requires an initial estimation to be solved iteratively. In addition, some applications of registration with 3D laser range-finders, like mobile robotics, impose time constraints on this problem, in spite of the large amount of raw data to be processed.

Registration of 3D scenes from laser range data is more complex than matching 2D views: (1) The amount of raw data is substantially bigger; (2) The number of degrees of freedom increases twofold. Moreover, registration of 3D scenes is different from modeling single objects in several aspects: (1) The scene can have more occlusions and more invalid ranges; (2) The scene may contain points from unconnected regions; (3) All scan directions in the scene may contain relevant information.

There are two general approaches for 3D scan registration: feature matching and point matching. The goal of feature matching is to find correspondences between singular points, edges or surfaces from range images. The segmentation process used to extract and select image primitives determines computation time and maximum accuracy. On the other hand, point matching techniques try to directly establish correspondences between spatial points from two views. Exact point correspondence from different scans is impossible due to a number of facts: spurious ranges, random noise, mixed pixels, occluded areas and discrete angular resolution. This is why point matching is usually regarded as an optimization problem, where the maximum expected precision is intrinsically limited by the working environment and by the rangefinder performance.

3.1.1.3 Object Recognition

Feature extraction is also an essential step in 3D single object recognition, involving recognizing and determining the pose of user-chosen 3D objects in a photograph or range scan. Typically, an example of the object to be recognized is presented to a vision system in a controlled environment and then, for an arbitrary input such as a video stream, the system locates the previously presented object. This can be done either off-line, or in real-time. The algorithms for solving this problem are specialized for locating a single pre-identified object, and can be contrasted with algorithms which operate on general classes of objects, such as face recognition systems or 3D generic object recognition. Due to the low cost and ease of acquiring photographs, a significant amount of research has been devoted to 3D object recognition in photographs. The method of recognizing a 3D object depends on the properties of an object. For simplicity, many existing algorithms have focused on recognizing rigid objects consisting of a single part, that is objects whose spatial transformation is an Euclidean motion. Two general approaches have been taken to the problem: Pattern recognition approaches use low-level image appearance information to locate an object, while feature-based geometric approaches construct a model for the object to be recognized and match the model against the photograph. Pattern recognition approaches use appearance information gathered from pre-captured or pre-computed projections of an object to match the object in the potentially cluttered scene. However, they do not take the 3D geometric constraints of the object into consideration during matching, and typically also do not handle occlusion as well as feature-based approaches. Feature-based approaches work well for objects which have distinctive features. Thus far, objects which have good edge features or blob features have been successfully recognized with the Harris affine region detector and SIFT. Due to lack of the appropriate feature detectors, objects without textured, smooth surfaces cannot currently be handled by this approach. Feature-based object recognizers generally work by pre-capturing a number of fixed views of the object to be recognized, extracting features from these views and then, in the recognition process, matching these features to the scene and enforcing geometric constraints.

3.1.2 Basic Concepts and Definitions

We introduce some basic concepts and definitions, such as features, feature extraction, 3D shape descriptor, and requirements for 3D feature extraction.

3.1.2.1 Features

In pattern recognition, features are the individual measurable heuristic properties of the phenomena being observed. In 3D models, feature is something that can be used to identify the objective. We can further narrow it to be something that can be easily understood and processed by computers, meaning the feature of regular geometric shape. Choosing discriminating and independent features is essential to any pattern recognition algorithm being successful in classification. Features are usually numeric, but structural features such as strings and graphs are used in syntactic pattern recognition. While different areas of pattern recognition obviously have different features, once the features are decided, they are classified by a much smaller set of algorithms. These include nearest neighbor classification in multiple dimensions, neural networks or statistical techniques such as Bayesian approaches.

In character recognition, features may include horizontal and vertical profiles, the number of internal holes, stroke detection and many others. In speech recognition, features for recognizing phonemes can include noise ratios, length of sounds, relative power, filter matches and many others. In spam detection algorithms, features may include whether certain email headers are present or absent, whether they are well formed, what language the email appears to be, the grammatical correctness of the text, Markovian frequency analysis and many others. In all these cases and many others, extracting features that are measurable by a computer is an art and, with the exception of some neural networking and genetic techniques that automatically intuit "features", hand selection of good features forms the basis of almost all classification algorithms.

3.1.2.2 Feature Extraction

In pattern recognition and multimedia processing, feature extraction is a special form of dimensionality reduction. When the input data to an algorithm is too large to be processed and it is suspected to be notoriously redundant (much data, but not much information) then the input data will be transformed into a reduced representation set of features (also named feature vector). Transforming the input data into the set of features is called feature extraction. If the features extracted are carefully chosen, it is expected that the feature set will extract the relevant information from the input data in order to perform the desired task using this reduced representation instead of the full size input.

Feature extraction involves simplifying the amount of resources required to describe a large set of data accurately. When performing an analysis of complex data, one of the major problem stems from the number of variables involved. An analysis with a large number of variables generally requires a large amount of memory and computation power or a classification algorithm which overfits the training sample and generalizes poorly to new samples. Feature extraction is a general term for methods of constructing combinations of the variables to get around these problems, while still describing the data with sufficient accuracy.

The best result is achieved when an expert constructs a set of application-dependent features. Nevertheless, if no such expert knowledge is available, general dimensionality reduction techniques may help. These include principal components analysis, semi-definite embedding, multifactor dimensionality reduction, nonlinear dimensionality reduction, isomap, kernel PCA, latent semantic analysis, partial least squares and independent component analysis.

3.1.2.3 3D Shape Descriptor

As we know, shape is easy for humans to perceive directly. Many feature extraction methods are based on the shape of 3D models, which often use the surface geometric features to describe models. The shape of the model is fundamental and the lowest level feature. So there are many methods that extract features through the models' surface shape attribute. Distance or geodesic distance on the surface, area of pieces, volume and normal direction are all the shape characteristics.

Representations used for shape matching are often referred to as 3D shape descriptors and they usually differ substantially from those intended for 3D object rendering and visualization. Shape descriptors aim at encoding geometrical and topological properties of an object in a discriminative and compact manner. The diversity of shape descriptors range from 3D moments to shape distributions, from spherical harmonics to ray-based sampling and from point clouds to voxelized volume transforms.

3.1.2.4 Requirements for 3D Feature Extraction

The shape of a 3D object is described by the feature vector that serves as a search key in the database. If an unsuitable feature extraction method had been used, the whole retrieval system would not be usable. Therefore, the following text is dedicated to properties that an ideal feature extraction method should have [7]:

(1) Independence of 3D object representations. At first we have to realize that 3D objects can be saved in many representations such as polyhedral meshes, volumetric data, parametric or implicit equations. The method for feature extraction should accept this fact and it should be independent of data representations.

(2) Invariance under transformations. The computed descriptor values have to be invariant under an application dependent set of transformations. Usually, these are the similarity transformations, but some applications like retrieval of articulated objects may additionally demand invariance under certain deformations. Perhaps it is the most important requirement, because the 3D objects are usually saved in various poses and scales.

(3) Insensitiveness to noise. The 3D object can be obtained either from a 3D graphics program or from a 3D input device. The second way is more susceptible to some errors. Thus, the feature extraction method should also be insensitive to noise.

(4) Descriptive power. The similarity measure based on the descriptor should deliver a similarity ordering that is close to the application driven notion of resemblance. The features between different models should be distinguishable.

(5) Conciseness and ease of indexing. The database can contain thousands of objects and the agility of the system would also be one of the main requirements. The descriptor should be compact in order to minimize the storage requirements and accelerate the search by reducing the dimensionality of the problem. Very

importantly, it should provide some means of indexing and thereby structuring the database in order to further accelerate the search process.

The feature extraction method that would have all the above mentioned requirements probably does not exist. For all that, some methods that try to find a compromise among ideal properties exist.

3.1.3 Classification of 3D Feature Extraction Algorithms

According to different aspects of the content they represent, features of 3D models can be roughly categorized into two main types [5]: (1) shape features, namely geometry and topology features and (2) appearance features, which represent some important cognitive characteristics such as material colors, reflection coefficients and textures mapping.

According to different feature representation data formats, Akgül et al. [3] pointed out that there are two paradigms for 3D object database operations and design of similarity measures, namely the feature vector approach and the non-feature vector approach. The feature vector paradigm aims at obtaining numerical values of certain shape descriptors and measuring the distances between these vectors. On the other hand, a typical example of the non-feature-based approach is to describe the object as a graph and then use graph similarity metrics. From the same point of view, Akgül et al. [3] pointed out that there are two main paradigms of 3D shape description, namely graph-based and vector-based. Graph-based representations, on one hand, are more elaborate and complex, harder to obtain, but represent shape properties in a more faithful and intuitive manner. Shock graphs [8], multiresolution Reeb graphs [9] and skeletal graphs [10] are methods that fall in this category. However, they do not generalize easily and hence they are not very convenient to use in unsupervised learning, for example for searching for natural shape classes in a database. Vector-based representations, on the other hand, are more easily computed. Although they are not necessarily conducive to plausible topological visualizations, they can be naturally employed in both supervised and unsupervised classification tasks. Typical vector-based representations are extended Gaussian images [11], cord and angle histograms [12], 3D shape histograms [13], spherical harmonics [14] and shape distributions [15].

It is necessary to search 3D models invariantly with respect to translation, rotation, scaling and reflection. Therefore, in many cases, more additional alignment-normalization (pose registration) processes may be required to align 3D objects to their canonical coordinate frame, or more intricate mappings or transformations for extracting invariant feature representations of a 3D model before a similarity match. From this point of view, we can classify 3D features into two categories: rotation-variant feature (RVF) and rotation-invariant features (RIF).

According to different types of 3D models, 3D feature extraction schemes can be also classified into mesh-based feature extraction and point-based feature extraction [16]. Many techniques have investigated the identification of feature edges on polygonal models. However, for point-based models, the underlying assumption of connectivity and normals associated with the vertices of the mesh is not available. In order to extract feature lines from point clouds using these techniques, a connectivity construction method (surface reconstruction) must be applied in a preprocessing step. The construction of connectivity is non-trivial, computationally expensive and, moreover, the success of feature extraction relies on the ability of the polygonal meshing procedure to accurately build the sharp edges. For point-based feature extraction methods, extracting features from point-based models is not straightforward in the absence of connectivity and normal information. Pauly et al. [17] used covariance analysis of the distance-driven local neighborhoods to flag potential feature points. By varying the radius of the neighborhoods, they developed a multi-resolution scheme capable of processing noisy input data. Gumhold et al. [18] constructed a Riemann graph over local neighborhoods and use covariance analysis to compute weights that flag points as potential creases, boundaries, or corners. Both techniques [17, 18] connect the flagged points using a minimum spanning tree and fit curves to approximate sharp edges. Demarsin et al. [19] computed point normals using principal component analysis and segment the points into groups based on the normal variation in local neighborhoods. A minimum spanning tree is constructed between the boundary points of the assorted clusters, which was used to build the final feature curves. These techniques are capable of extracting features on point clouds by connecting existing points. However, their accuracy depends on the sampling quality of the input model.

In this chapter, according to the technique, we classify the 3D feature extraction schemes into six categories: statistical-data-based, global-geometrical analysis-based, signal-analysis-based, topology-based, visual-image-based and appearance-based feature extraction algorithms. Note that we introduce statistical-data-based methods in three sections, where the authors of this book propose two statistical-based methods, i.e., rotation-based and vector-quantization based. To describe our own methods more clearly, we introduce our methods in separate sections. From Section 3.2 to Section 3.9, we will discuss these types of techniques respectively.

3.2 Statistical Feature Extraction

At present, the parameterization of 3D models is a very complicated issue. Furthermore, since 3D surfaces may possess arbitrary topology, some widely used methods (e.g., Fourier-transform-based methods) in image processing are not directly applicable to 3D models. Thus, it is hard for us to acquire 3D model features with explicit meaning of geometry or shapes. From the point of view of statistics, researchers show preference for the statistical feature with high distinguishability. Currently, the research work in this field mainly adopts the following statistical features: the geometric relationship between vertices (distances, angles, normal directions), curvature distribution of vertices, moments with various orders of vertices and feature coefficients of various transforms, and so on.

Statistical-data-based feature extraction approaches sample points on the surface of 3D models and extracts characteristics from the sample points. These characteristics are typically organized in the form of histograms or distributions representing frequency of occurrence. The most extensively used statistical property is the "moments", such as Hu's image moments [20]. There are also many other kinds of statistical property features expressed in the form of different discrete histograms of geometrical statistics [21]. The shape representation is simplified as a probability distribution problem by using histograms and avoids the model normalization process.

Compared with other methods, most statistical feature extraction methods are not only fast and easy to implement, but also have some desired properties, such as robustness and invariance. In many cases, they are also robust against noise, or the small cracks and holes that exist in a 3D model. Unfortunately, as an inherent drawback of a histogram representation, they provide only limited discrimination between objects: they neither preserve nor construct spatial information. Thus, they are often not discriminating enough to make small differences between dissimilar 3D shapes, and usually fail to distinguish different shapes having the same histogram. In this section, we mainly introduce several typical moment-based and histogram-based feature descriptors for 3D models, including one method proposed by the authors of this book.

3.2.1 3D Moments of Surface

Assume that an object is given in VRML, i.e., it is a 3D object represented by a set of vertices and a set of polygonal faces embedded in 3D. The features Elad et al. [22] chose to represent the objects are the moments computed for object surfaces, assuming that the 3D model is a hollow model bounded by its surfaces. 3D moments of surfaces can be calculated as follows:

$$m_{pqr} = \int_{\partial \boldsymbol{M}} x^p y^q z^r \mathrm{d}x\mathrm{d}y\mathrm{d}z, \tag{3.1}$$

where $\boldsymbol{M}$ is the 3D model, $\partial\boldsymbol{M}$ is the surface of $\boldsymbol{M}$, and m_{pqr} is the (p, q, r)-th 3D moment. For a 3D model, the set of moments m_{pqr} is unique so that it constitutes a full and complete description of $\boldsymbol{M}$, and a partial object description can also be obtained by using some subset of these moments [23].

3.2.1.1 Sampling to Approximate the Moments

The crux of Elad et al.'s algorithm lies in the computation of a subset of the (p, q, r)-th moments of each object, which are used as the feature set. Thus, it is necessary to perform a pre-processing stage where the features are calculated for each database object. A practical way to evaluate the integral defining moments is to compute this analytically for each facet of the object, and then sum over all the facets. They use an alternative approach, yielding an approximation of the moments. The algorithm draws a sequence of points (x, y, z) distributed uniformly over the object's surface. The number of points drawn from each of the object's facets is proportional to its relative surface area. If we denote the list of points for a given object by $\{x_i, y_i, z_i\}$, $i = 1, 2, \ldots, N$, then the (p, q, r)-th moment is approximated by

$$\hat{m}_{pqr} = \frac{1}{N}\sum_{i=1}^{N} x_i^{\ p} y_i^{\ q} z_i^{\ r} . \tag{3.2}$$

3.2.1.2 Normalizing the Objects

The similarity measure should be invariant to spatial position, scale and rotation of the different objects. One is therefore required to normalize the feature vectors of all objects. The first moments m_{100}, m_{010} and m_{001} represent the object's center of mass. Thus, the normalization starts by estimating the first moments for each object represented as a set of surface sample points, and subtracting them from each of these points

$$\forall i = 1, 2, \ldots, N, \quad [\boldsymbol{x}_i, \boldsymbol{y}_i, \boldsymbol{z}_i]^{\mathrm{T}} \leftarrow [\boldsymbol{x}_i - \hat{m}_{100}, \ \boldsymbol{y}_i - \hat{m}_{010}, \ \boldsymbol{z}_i - \hat{m}_{001}]^{\mathrm{T}} . \tag{3.3}$$

This amounts to positioning all objects so that their center of mass is at coordinates (0,0,0), thus removing any dependence on translation, or spatial position. This also sets each of $\hat{m}_{100}$, $\hat{m}_{010}$ and $\hat{m}_{001}$ to 0 for all objects, and thus renders them useless for further computations.

The second moments $-m_{200}$, m_{020}, m_{002}, m_{110}, m_{011} and m_{101} represent the object's rotation and scale in the following manner. The second moments, calculated for the object re-centered at (0, 0, 0), can be ordered into a matrix

$$\boldsymbol{Z} = \begin{bmatrix} m_{200} & m_{110} & m_{101} \\ m_{110} & m_{020} & m_{011} \\ m_{101} & m_{011} & m_{002} \end{bmatrix} . \tag{3.4}$$

Singular value decomposition (SVD) is then performed on this matrix, obtaining the result as follows:

$$\boldsymbol{U}\boldsymbol{\Delta}\boldsymbol{\Delta}^{\mathrm{T}} = SVD(\boldsymbol{Z}) , \tag{3.5}$$

where the unitary matrix $\boldsymbol{U}$ represents the rotation and the diagonal matrix $\boldsymbol{\Delta}$ represents the scale in each axis, ordered in decreasing size.

The normalization continues with a second stage approximating the second moments for each object, by computing them from the updated surface point data sets, using Eq.(3.2) into $\hat{\boldsymbol{Z}}$. After performing the SVD decomposition of the second moment matrix $\hat{\boldsymbol{Z}}$, we multiply each point by $\boldsymbol{U}$ to rotate the object back to a canonic position. We also divide each point by $\Delta(1,1)$ to rescale the object so that its largest scale is 1. To summarize, each point is replaced by

$$[\boldsymbol{x}_i, \boldsymbol{y}_i, \boldsymbol{z}_i]^{\mathrm{T}} \leftarrow \frac{1}{\Delta(1,1)} \cdot \boldsymbol{U} \cdot [\boldsymbol{x}_i, \boldsymbol{y}_i, \boldsymbol{z}_i]^{\mathrm{T}}. \tag{3.6}$$

Finally, the algorithm should also determine each object's orientation, relative to each axis. To do this, we count the number of points on each side of the center of the body. In order to normalize such that all the objects have the same orientation, we flip each object so that it is "heavier" on the positive side. In counting the number of points and flipping according to it, we are actually forcing the median center to be on a predetermined side relative to the center of mass. After applying all the normalization stages to each object, the moments are computed once more, up to the pre-specified order. Obviously, the normalization process fixes $\hat{m}_{100}$, $\hat{m}_{010}$, $\hat{m}_{001}$ and $\hat{m}_{200}$ to 0, 0, 0 and 1, respectively, for each and every object. These are therefore no longer useful as object features.

3.2.2 3D Zernike Moments

The main drawback of the method in Subsection 3.2.1 is that a unit-scale coordinate frame of 3D models has to be acquired prior to the feature computation process. To address this issue, some new statistical feature extraction approaches without pose registration have been proposed. Shape feature based on 3D Zernike moments [24] is an example. Novotni et al. [25] demonstrated that 3D Zernike moments are computed as a projection from the function defining the 3D object onto a set of orthonormal functions within a unit sphere, which have simple representation but good retrieval performance. They further presented 3D Zernike invariants as the 3D shape descriptor. The steps needed to compute the 3D Zernike moments and descriptors can be expressed as follows:

(1) Normalization. Compute the center of gravity of the object, transform it to the origin, and scale the object so that it will be mapped into the unit ball.

(2) Geometrical moment computation. Compute all geometrical moments

$$m_{pqr} = \int_{|x^2+y^2+z^2|\le 1} f(x,y,z) x^p y^q z^r \mathrm{d}x\mathrm{d}y\mathrm{d}z \tag{3.7}$$

for each combination of indices, such that $p, q, r \ge 0$ and $p + q + r \le N$. Note that the computation of the geometrical moments is of central importance with respect to the overall computational efficiency and numerical accuracy. A typical approach to computing the geometrical moments of an object represented by a 3D voxel grid is as follows: 1) Fix a coordinate system with its origin at a corner of the grid and axes aligned with the grid axes. Subsequently, sample all monomials of order up to N at the grid point positions. 2) Compute the geometrical moments according to Eq.(3.7) but integrating over the whole voxel grid. 3) Transform the geometrical moments according to the normalization transformation of the object. This can be easily accomplished, since scaling can be achieved by scaling the moments, and the moments of the translated object can be represented in terms of a linear combination of original moments of not greater order. The first two steps introduce numerical problems. First, the sampling at grid points implies that we treat the monomial as a function having a constant value within a voxel, which is determined by the value of the monomial, e.g., in the center of the voxel. For rapidly changing functions, like the monomials of high order, this results in inaccuracy. Second, for a 643-grid for instance, the precision of the double precision floating point number is exceeded already at the order of 9. According to experience, moments up to the order of 20 are required to provide a good descriptor. The first issue can be treated by computing the geometrical moments in terms of monomials integrated over the voxels. Since for high orders the 3D Zernike descriptors seem to discard the values of voxels close to the origin, the object is normalized prior to computation of moments, thus obtaining considerably better numerical accuracy and providing a cure to the second problem. For the detailed procedure, readers can refer to [25].

(3) 3D Zernike moment computation. The 3D Zernike invariants can be extracted on the basis of those computed geometrical moments. Zernike moments can be written in a compact form as a linear combination of monomials of order up to n as follows:

$$\Omega_{nl}^{m} = \frac{3}{4\pi} \sum_{p+q+r\le n} \overline{\chi}_{nlm}^{pqr} \cdot m_{pqr}, \tag{3.8}$$

where $\overline{\chi}_{nlm}^{pqr}$ is the intermediate monomial that can be found in [25] for more details. Note that the summation has to be conducted only for the nonzero coefficients $\overline{\chi}_{nlm}^{pqr}$. Also note that for $m \le 0$, Ω_{nl}^{m} may be computed using the symmetry relation $\Omega_{nl}^{-m} = (-1)^m \overline{\Omega_{nl}^{m}}$.

(4) 3D Zernike descriptor generation. Compute the rotationally invariant 3D Zernike descriptors as norms of vectors $\boldsymbol{\Omega}_{nl}$ as follows:

$$F_{nl} = \left\| \boldsymbol{\Omega}_{nl} \right\|, \tag{3.9}$$

here, $\boldsymbol{\Omega}_{nl}$ is a $(2l+1)$-dimensional vector consisting of $2l+1$ moments $\boldsymbol{\Omega}_{nl}$, $\boldsymbol{\Omega}_{nl}^{l-1}$, $\boldsymbol{\Omega}_{nl}^{l-2}$, …, $\boldsymbol{\Omega}_{nl}^{-l}$.

The 3D Zernike invariants were reported [25] to gain robustness against both topological and geometrical deformations.

3.2.3 3D Shape Histograms

The definition of an appropriate distance function is crucial for the effectiveness of any nearest neighbor classifier. A common approach for similarity models is based on the paradigm of feature vectors. A feature transform maps a complex object onto a feature vector in a multidimensional space.

3.2.3.1 3D Shape Histogram

The similarity of two objects is then defined as the vicinity of their feature vectors in the feature space. Ankerst et al. [26] introduced 3D shape histograms as intuitive feature vectors. In general, histograms are based on a partitioning of the space in which the objects reside, i.e., a complete and disjoint decomposition into cells which correspond to the bins of the histograms. The space may be geometric (2D, 3D), thematic (e.g., physical or chemical properties), or temporal (modeling the behavior of objects). They suggested three techniques for decomposing the space: a shell model, a sector model and a spiderweb model as the combination of the former two, as shown in Fig. 3.2. In the preprocessing step, a 3D solid is moved to the origin. Thus the models are aligned to the center of mass of the solid.

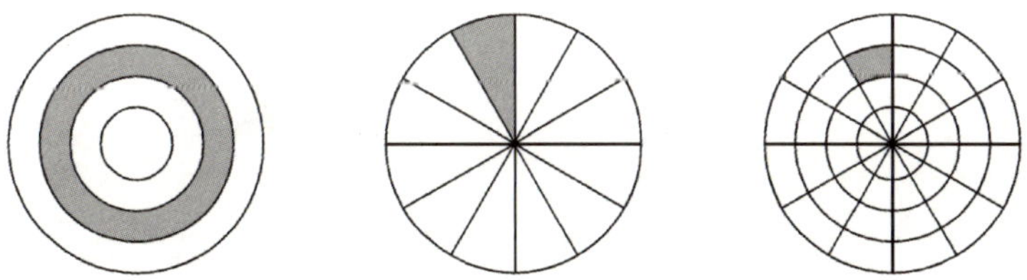

(a) (b) (c)

Fig. 3.2. Shells and sectors as basic space decompositions for shape histograms. (a) 4 shell bins; (b) 12 sector bins; (c) 48 combined bins. In each of the 2D examples, a single bin is marked

(1) Shell model

The 3D model is decomposed into concentric shells around the center point. This representation is particularly independent of a rotation of the objects, i.e., any rotation of an object around the center point of the model results in the same histogram. The radii of the shells are determined from the extensions of the objects in the database. The outermost shell is left unbound in order to cover objects that exceed the size of the largest known object.

(2) Sector model

The 3D model is decomposed into sectors that emerge from the center point of the model. This approach is closely related to the 2D section coding method.

However, the definition and computation of 3D sector histograms is more sophisticated, and they define the sectors as follows: To distribute the desired number of points uniformly on the surface of a sphere. For this purpose, we use the vertices of regular polyhedrons and their recursive refinements. Once the points are distributed, the Voronoi diagram of the points immediately defines an appropriate decomposition of the space. Since the points are regularly distributed on the sphere, the Voronoi cells meet at the center point of the model. For the computation of sector-based shape histograms, we need not materialize the complex Voronoi diagram but simply apply a nearest neighbor search in the 3D model since the typical number of sectors is not very large.

(3) Combined model

The combined model represents more detailed information than pure shell models and pure sector models. A simple combination of two fine-grained 3D decompositions results in a high dimensionality. However, since the resolution of the space decomposition is a parameter in any case, the number of dimensions may easily be adapted to the particular application.

In Fig. 3.3, Ankerst et al. [26] illustrated various shape histograms for the example protein, 1SER-B, which is depicted on the left of the figure. In the middle, the various space decompositions are indicated schematically and, on the right, the corresponding shape histograms are depicted. The top histogram is purely based on shell bins, and the bottom histogram is defined by 122 sector bins. The histograms in the middle follow the combined model, and they are defined by 20 shell bins and 6 sector bins, and by 6 shell bins and 20 sector bins, respectively. In this example, all the different histograms have approximately the same dimension

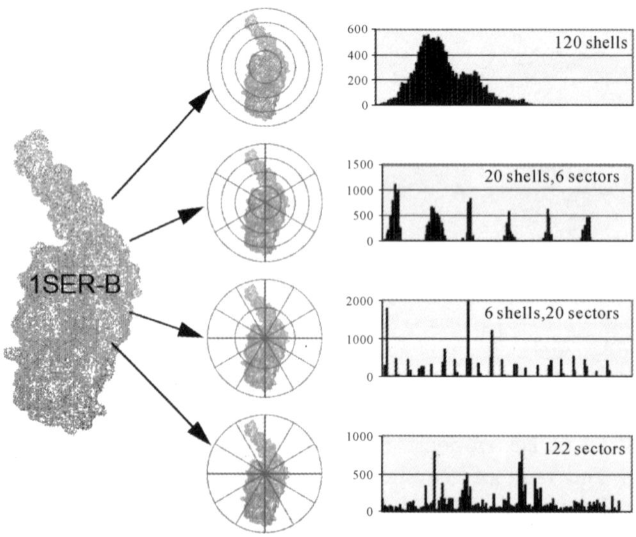

Fig. 3.3. Several 3D shape histograms of the example protein 1SER-B. From top to bottom, the number of shells decreases and the number of sectors increases [13] (With kind permission of Springer Science+Business Media)

of around 120. Note that the histograms are not built from volume elements but from uniformly distributed surface points taken from the molecular surfaces.

3.2.3.2 Crease Angle Histogram

Besl [27] constructed 3D histograms on the crease angles for all edges in a 3D triangular mesh to match 3D shapes. Fig. 3.4 shows the crease angle histograms (CAHs) and hidden line drawings for eight simple shapes: a block, a cylinder, a sphere, a block with channel, a "soap-shape" superquadric, two blocks glued together, a "double horn" superquadric, and a "jack-shaped" superquadric. Working from the bottom up, we see the block CAH consists of two simple peaks: one peak at 90 degrees for the 12 edges and one peak at zero for the adjacent triangles within a face. The cylinder's creases will have angles that are zero or small and positive as well as a peak at 90 degrees. The three ideal peaks, one for flatness, one for convex curvature, and one for 90 angles, are the signature for the cylinder. An ideal cone's histogram will look very, very similar except the peak at 90 degrees should be half the size.

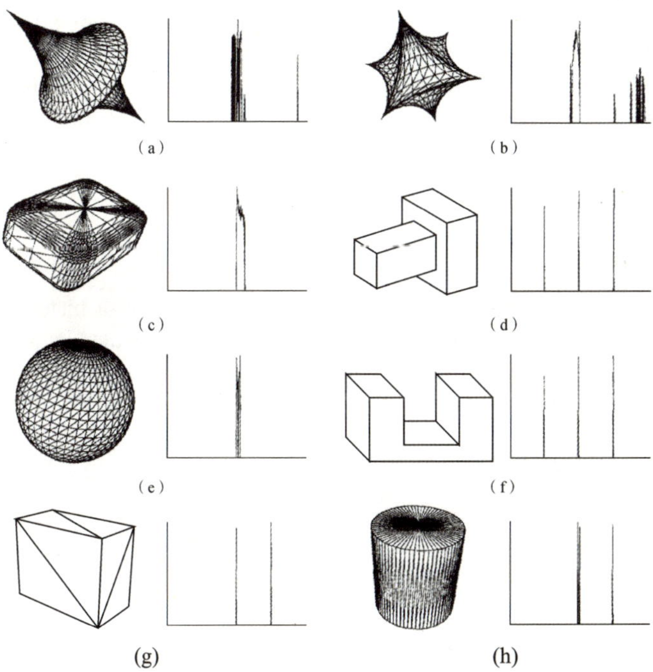

Fig. 3.4. Crease angle histograms for simple shapes. (a) Double-horn superquadric; (b) Jack-shaped superquadric; Soap superquadric (c); (d) Two blocks glued; (e) Sphere; (f) Block with channel; (g) Block; (h) Cylinder [27] (With kind permission of Springer Science+Business Media)

3.2.3.3 Distance Histogram

For rigid 3D shapes, Novotni et al. [28] introduced the so-called "distance histograms" as a basic representation. Their fundamental idea is that if two objects were similar, only a small part of the volume of one of the objects would be outside the boundary of the other one, and the average distance from the boundary would also be small. They first computed the offset hulls of each object based on a 3D distance field, and then constructed the distance histograms for each object to indicate how much of the volume of one object is inside the offset hull of the other.

3.2.3.4 Multiresolution Shape Descriptor

The introduction of geometrical properties into the histogram makes multiresolution shape representation possible. Ohbuchi et al. [29] proposed a multiresolution shape descriptor, represented in the form of an ordered set of histograms. They first defined a multiresolution representation (MRR) feature, specified as a set of 3D α-shapes [30], which was defined by using a group of α-values spaced at power of two intervals. α-shapes are a generalization of the convex hull of a point set, which shrinks by gradually developing cavities until it is identical to the convex hull when $\alpha = \infty$ [30]. Next, a 2D histogram was generated for each MRR so that an ordered set of histograms could be produced as the shape descriptor.

3.2.3.5 Other Histograms

Paquet et al. [31] presented histogram features, including color histogram, normal vector histogram and material histogram to represent 3D shapes. Paquet et al. also pointed out that a histogram can represent the 3D data distributions, based on voxels, and is transformation invariant. In the MPEG-7 standard, there is also a shape histogram descriptor for 3D mesh model known as the 3D shape spectrum descriptor (3-DSSD) [32].

3.2.4 Point Density

Suzuki et al. [33] presented another kind of 3D model feature representation method, called point density. We introduce its basic idea, equivalent classes and algorithm description.

3.2.4.1 Basic Idea

Suzuki et al. [33] suggested that several steps are required to create rotation invariant feature descriptors: (1) Information associated with shape features has to be extracted from data files; (2) The extracted information is converted to feature vectors as indices of the database; (3) Feature vectors are grouped into equivalence classes, so that these vectors can be converted into rotation invariant feature vectors. In their paper, only 3D model shapes are of concern, thus only information related to vertices is used.

When a 3D graphical object is displayed, a set of points is used to represent the shape. This set of points is connected by lines to form a wireframe. This wireframe shows a set of polygons. Once polygons have been created, the rendering algorithm can shade the individual polygons to produce a solid object. Suzuki et al. [33] used the density of the point clouds as feature vectors. Each 3D model is placed into the unit cube, and then the unit cube is divided into coarse grids. The number of points is counted in each grid cell to compute the density of the point clouds. In their paper, only the density of the point clouds is used. However, other features can also be used, such as normal vectors of polygon faces.

Since the distributions of the point clouds depend on how the 3D model is generated, they normalized point positions by using polygon triangulation programs. The density of the point clouds gives us rough shape descriptors of the 3D models which include curvature, height, width and positions. These feature descriptors are not rotation invariant, because orientations of 3D models are defined by those who designed the 3D models. Orientations may be normalized by rules. Suitable rules to set 3D model orientations depend on the purpose of the applications.

3.2.4.2 Equivalent Classes

To explain the concept of equivalent classes, Fig. 3.5 illustrates the rotations that are parallel to one of the coordinate axes in the order of 90 degrees. Each cell can be moved to a new position by rotation. When rotations are repeated, eventually each cell can return to its original position. In this moving cell process, some unique paths are generated. For example, the coordinates of the 8 cells which lie along the edge of the grid are as follows: (−1, −1, −1), (−1, −1, +1), (−1, +1, −1), (−1, +1, +1), (+1, −1, −1), (+1, −1, +1), (+1, +1, −1), (+1, +1, +1). When we apply the rotation to the cell which has one of the above coordinates, the calculated new coordinate is also one of the above. This means that these 8 cells have no path to any other cells. For instance, the cell which lies at the origin can keep its own position even if rotations are applied, so it has an independent path.

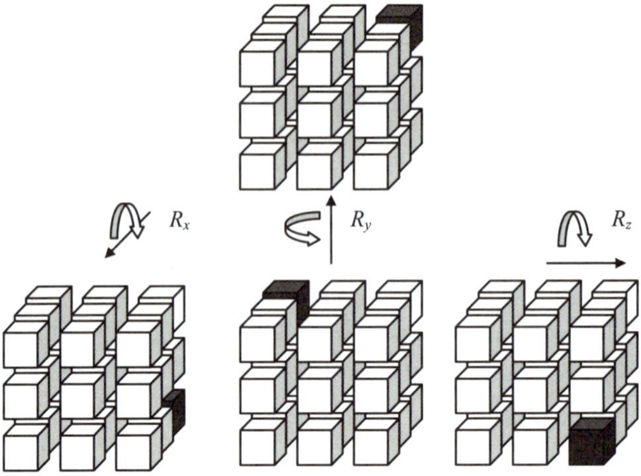

Fig. 3.5. Illustration of rotations parallel to coordinate axes

Each cell can be classified by the unique paths. Rotation operations are needed to find the unique path. The rotation matrices with respect to X, Y and Z axes are:

$$\boldsymbol{R}_x = \begin{pmatrix} 1 & 0 & 0 & 0 \\ 0 & +\cos\theta & +\sin\theta & 0 \\ 0 & -\sin\theta & +\cos\theta & 0 \\ 0 & 0 & 0 & 1 \end{pmatrix}, \tag{3.10}$$

$$\boldsymbol{R}_y = \begin{pmatrix} +\cos\theta & 0 & -\sin\theta & 0 \\ 0 & 1 & 0 & 0 \\ +\sin\theta & 0 & +\cos\theta & 0 \\ 0 & 0 & 0 & 1 \end{pmatrix}, \tag{3.11}$$

$$\boldsymbol{R}_z = \begin{pmatrix} +\cos\theta & +\sin\theta & 0 & 0 \\ -\sin\theta & +\cos\theta & 0 & 0 \\ 0 & 0 & 1 & 0 \\ 0 & 0 & 0 & 1 \end{pmatrix}. \tag{3.12}$$

Cells have equivalent relations if they belong to the same paths. The cell sets that have equivalent relations are called equivalence classes. Fig. 3.6 shows the equivalence classes of the 3×3×3 grid. Each cell is classified into one of four equivalence classes. The 3×3×3 grid contains 27 cells. Since we define each class of cells as having an identical relation, the summation of cells in the same class can be calculated. Each cell contains the density of the point clouds. The $P_n(x, y, z)$ contains values for the density of the point clouds for the cell located at coordinates (x, y, z), where n is the index of each cell as shown in Fig. 3.6. In the case of the 3×3×3 grid, we can define the following four functions to calculate the rotation invariant feature vectors in the order of 90 degrees. Twenty seven vectors are reduced to 4 vectors by these equations. Since these 4 vectors are recalculated to be rotation invariant vectors, some of the fine details of the feature descriptors are lost.

$$\begin{aligned} f_1 = P_0(-1,\ -1,\ -1) + P_2(-1,\ -1,\ 1) + P_6(-1,\ 1,\ -1) + P_8(-1,\ 1,\ 1) \\ + P_{18}(1,\ -1,\ -1) + P_{20}(1,\ -1,\ 1) + P_{24}(1,\ 1,\ -1) + P_{26}(1,\ 1,\ 1), \end{aligned} \tag{3.13}$$

$$\begin{aligned} f_2 = P_1(-1,\ -1,\ 0) + P_3(-1,\ 0,\ -1) + P_5(-1,\ 0,\ 1) + P_7(-1,\ 1,\ 0) \\ + P_9(0,\ -1,\ -1) + P_{11}(0,\ -1,\ 1) + P_{15}(0,\ 1,\ -1) + P_{17}(0,\ 1,\ 1) \\ + P_{19}(1,\ -1,\ 0) + P_{21}(1,\ 0,\ -1) + P_{23}(1,\ 0,\ 1) + P_{25}(1,\ 1,\ 0), \end{aligned} \tag{3.14}$$

$$\begin{aligned} f_3 = P_4(-1,\ 0,\ 0) + P_{10}(0,\ -1,\ 0) + P_{12}(0,\ 0,\ -1) + P_{14}(0,\ 0,\ 1) \\ + P_{16}(0,\ 1,\ 0) + P_{22}(1,\ 0,\ 0), \end{aligned} \tag{3.15}$$

$$f_4 = P_{13}(0,\ 0,\ 0). \tag{3.16}$$

The number of equivalence classes Q_{num} in an $N \times N \times N$ grid can be calculated by the following equation [33]:

$$Q_{num} = \begin{cases} \sum_{j=0}^{n} F_j + \sum_{j=0}^{n-2} F_j, & n > 3; \\ \sum_{j=0}^{n} F_j, & n \le 3, \end{cases} \tag{3.17}$$

with

$$F_j = \sum_{k=0}^{j} (j-k). \tag{3.18}$$

Here, an $N \times N \times N$ grid has $N = 2n + 1$ relations. Thus, if the grid size is larger than 7×7×7, the first part of Eq.(3.17) is used, otherwise the second part is used. We can easily see that the number of cells increases rapidly for the higher resolutions of the $N \times N \times N$ grid compared to the number of equivalent classes. Comparisons of the huge number of vectors cause inefficient retrieval, and it requires more memory to store the vectors. Statistical approaches such as principal component analysis (PCA), multidimensional scaling and multiple regression analysis can be used to reduce the size of the vectors for similarity retrieval. However, these approaches need a sufficient number of data samples and processes to determine which vectors can be eliminated.

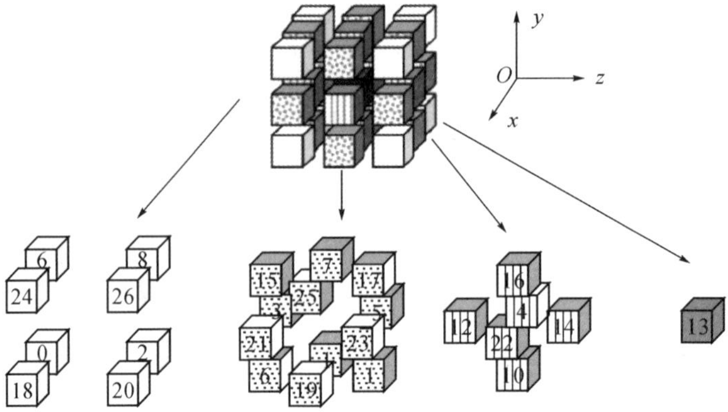

Fig. 3.6. Four equivalence classes for the 3×3×3 grid

3.2.4.3 Algorithm Description

In fact, the basic idea of this method is similar to 3D shape histograms. They both calculate the point distribution, but their implementation methods are different. The detailed procedure of Suzuki et al.'s method [33] can be expressed as follows:

Step 1: Transform the 3D model into the normalized coordinate system by the PCA method.

Step 2: Partition the cube into $N \times N \times N$ cells.

Step 3: Classify each cell into the equivalent class it belongs to.

Step 4: Compute the number of vertices in each class, and divide it by the total number of vertices in the 3D model, composing a feature vector for the 3D model.

Experimentally, it has been shown that the computational complexity of the point density approach is low, and in the retrieval application, based on this feature, we can obtain good retrieval performance in terms of precision and recall.

3.2.5 Shape Distribution Functions

Osada et al. [34] described and analyzed a method for computing 3D shape signatures and dissimilarity measures for arbitrary objects described by possibly degenerate 3D polygonal models. The key idea is to represent the signature of an object as a shape distribution sampled from a shape function measuring global geometric properties of the object. The primary motivation for this approach is that the shape matching problem is reduced to the comparison of two probability distributions, which is a relatively simple problem when compared to the more difficult problems encountered by traditional shape matching methods, such as pose registration, parameterization, feature correspondence and model fitting. The challenges of this approach are to select discriminating shape functions, to develop efficient methods for sampling them, and to robustly compute the dissimilarity of probability distributions.

3.2.5.1 Selecting a Shape Function

The first and most interesting issue is to select a function whose distribution provides a good signature for the shape of a 3D polygonal model. Ideally, the distribution should be invariant under similarity transformations, and it should be insensitive to noise, cracks, tessellation and insertion/removal of small polygons.

In general, any function could be sampled to form a shape distribution, including ones that incorporate domain-specific knowledge, visibility information (e.g., the distance between random but mutually visible points), and/or surface attributes (e.g., color, texture coordinates, normals and curvature). However, for the sake of clarity, Osada et al. focused on a small set of shape functions based on geometric measurements (e.g., angles, distances, areas, and volumes). Specifically, in their initial investigation, they have experimented with the following shape functions (see Fig. 3.7):

(1) A3: Measures the angle between three random points on the surface of a 3D model.

(2) D1: Measures the distance between a fixed point and one random point on the surface. We use the centroid of the boundary of the model as the fixed point.

(3) D2: Measures the distance between two random points on the surface.

(4) D3: Measures the square root of the area of the triangle between three random points on the surface.

(5) D4: Measures the cube root of the volume of the tetrahedron between four random points on the surface.

These five shape functions were chosen mostly for their simplicity and invariance. In particular, they are quick to compute, easy to understand, and produce distributions that are invariant to rigid motions (translations and rotations). They are invariant to tessellation of the 3D polygonal model, since points are selected randomly from the surface. They are insensitive to small perturbations due to noise, cracks, and insertion/removal of polygons, since sampling is area weighted. In addition, the A3 shape function is invariant to scale, while the others have to be normalized to enable comparisons. Finally, the D2, D3, and D4 shape functions provide a nice comparison of 1D, 2D, and 3D geometric measurements.

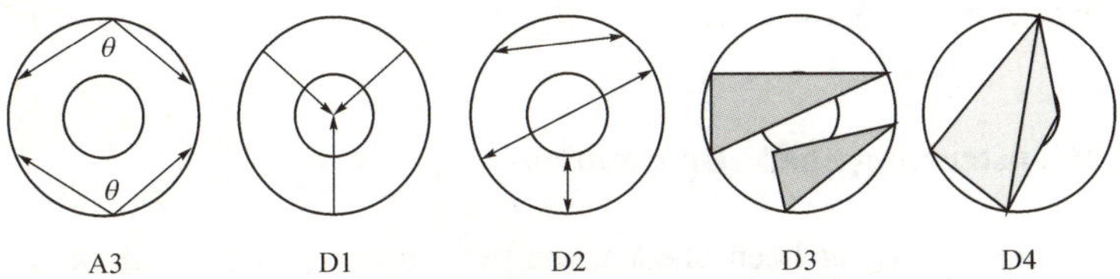

Fig. 3.7. Five simple shape functions based on angles (A3), lengths (D1, D2), areas (D3) and volumes (D4)

In spite of their simplicity, Osada et al. found these general-purpose shape functions to be fairly distinguishing as signatures for 3D shape, as significant changes to the rigid structures in the 3D model affect the geometric relationships between points on their surfaces. For instance, It can be noticed that distributions the D2 shape function are shown for a few canonical shapes in Figs. 3.8(a)–(f). Each distribution is distinctive. And continuous changes to the 3D model affect of the D2 distributions. For instance, Fig. 3.8(g) shows the distance distributions for ellipsoids of different semi-axis lengths overlaid on the same plot. The leftmost curve represents the D2 distribution for a line segment-ellipsoid (0, 0, 1); the rightmost curve represents the D2 distribution for a sphere-ellipsoid (1, 1, 1); and the remaining curves show the D2 distribution for ellipsoids in between-ellipsoid (r, r, 1) with $0 < r < 1$. Note how the change from sphere to line segment is continuous. Similarly, Figs. 3.8(h)–(i) show the D2 distributions of two unit spheres as they move 0, 1, 2, 3, and 4 units apart. In each distribution, the first hump resembles the linear distribution of a sphere, while the second hump is the cross-term of distances between the two spheres. As the spheres move further apart, the D2 distribution changes continuously.

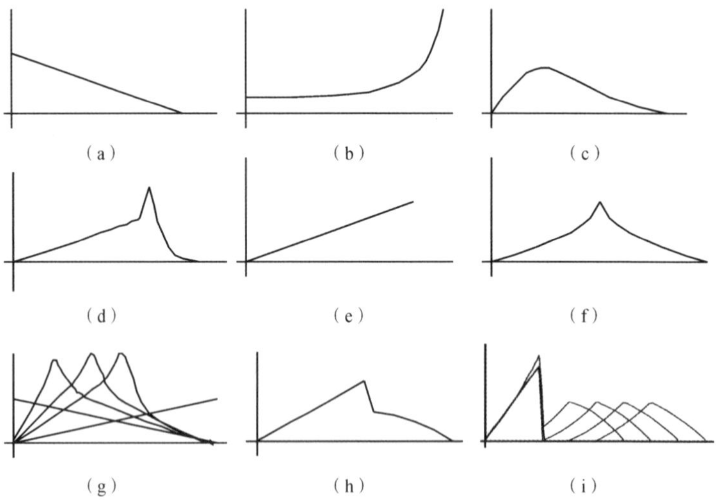

Fig. 3.8. Example D2 shape distributions. In each plot, the horizontal axis represents distance, and the vertical axis represents the probability of that distance between two points on the surface. (a) Line segment; (b) Circle (perimeter only); (c) Triangle; (d) Cube; (e) Sphere; (f) Cylinder (without caps); (g) Ellipsoids of different radii; (h) Two adjacent unit spheres; (i) Two unit spheres separated by 1, 2, 3, and 4 units

3.2.5.2 Constructing Shape Distributions

A shape function having been chosen, the next issue is to compute and store a representation of its distribution. Analytic calculation of the distribution is feasible only for certain combinations of shape functions and models (e.g., the D2 function for a sphere or line). Thus, in general, Osada et al. employed stochastic methods. Specifically, Osada et al. evaluated N samples from the shape distribution and construct a histogram by counting how many samples fall into each of B fixed sized bins. From the histogram, Osada et al. reconstructed a piecewise linear function with V ($\leq B$) equally spaced vertices, which forms the representation for the shape distribution. Osada et al. computed the shape distribution once for each model and stored it as a sequence of V integers.

One issue we must be concerned with is the sampling density. On one hand, the more samples we take, the more accurately and precisely we can reconstruct the shape distribution. On the other hand, the time to sample a shape distribution is linearly proportional to the number of samples, so there is an accuracy/time tradeoff in the choice of N. Similarly, a larger number of vertices yield higher resolution distributions, while increasing the storage and comparison costs of the shape signature. In Osada et al.'s experiments, they have chosen to err on the side of robustness, taking a large number of samples for each histogram bin. Empirically, they have found that using $N = 1{,}024^2$ samples, $B = 1{,}024$ bins, and $V = 64$ vertices yields shape distributions with low enough variance and high enough resolution to be useful for our initial experiments. Adaptive sampling methods could be used in future work to make robust construction of shape distributions more efficient.

A second issue is sample generation. Although it would be simplest to sample vertices of the 3D model directly, the resulting shape distributions would be biased and sensitive to changes in tessellation. Instead, Osada et al.'s shape functions are sampled from random points on the surface of a 3D model. The method for generating unbiased random points with respect to the surface area of a polygonal model proceeds as follows. First, Osada et al. iterated through all polygons, splitting them into triangles as necessary. Then, for each triangle, Osada et al. computed its area and store it in an array along with the cumulative area of triangles visited so far. Next, Osada et al. selected a triangle with probability proportional to its area by generating a random number between 0 and the total cumulative area and performed a binary search on the array of cumulative areas. For each selected triangle with vertices ($\boldsymbol{A}$, $\boldsymbol{B}$, $\boldsymbol{C}$), Osada et al. constructed a point on its surface by generating two random numbers, r_1 and r_2, between 0 and 1, and evaluate the following equation:

$$\boldsymbol{P} = (1-\sqrt{r_1})\boldsymbol{A} + \sqrt{r_1}(1-r_2)\boldsymbol{B} + \sqrt{r_1}\,r_2\boldsymbol{C}\,. \tag{3.19}$$

Intuitively, $\sqrt{r_1}$ sets the percentage from vertex A to the opposing edge, while r_2 represents the percentage along that edge (see Fig. 3.9). Taking the square-root of r_1 gives a uniform random point with respect to surface area.

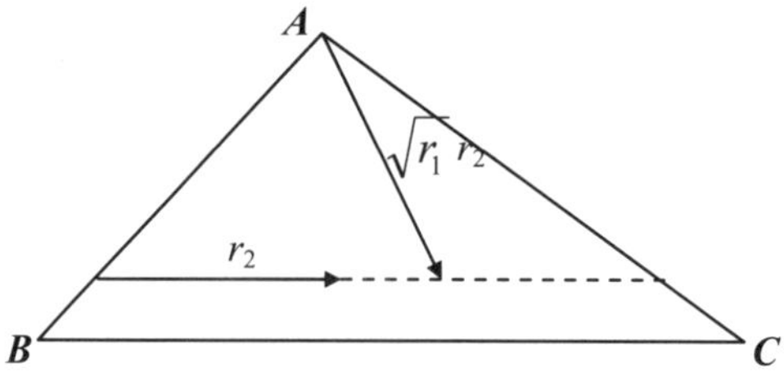

Fig. 3.9. Sampling a random point in a triangle

Osada et al.'s experimental results demonstrated that shape distributions can be fairly effective at discriminating between groups of 3D models. Overall, they achieved 66% accuracy in their classification experiments with a diverse database of degenerate 3D models assigned to functional groups. The D2 shape distribution was more effective than moments during their classification tests. Unfortunately, it is difficult to evaluate the quality of this result as compared to other methods, as it depends largely on the details of the test database. However, they believe that their method is demonstrated to be useful for the discrimination of 3D shapes, at least for pre-classification prior to more exact similarity comparisons with more expensive methods.

3.2.5.3 Improved Methods

Osada et al. have shown that D2 is the best feature among their five features. It represents the distribution of distances between two random points. This feature is invariant to tessellation of 3D polygonal models, since points are randomly selected from the object's surface. However, it is sensitive to small deformation due to noise, cracks, or insertion/removal of polygons, since sampling is area weighted. To finely represent the complex components of a 3D object, a 3D model often requires many polygons. The random sampling of a 3D model would be dominated by those complex components. Thus, a novel feature, called grid D2, is proposed by Shih et al. [35] to improve the performance of the traditional D2. First, the 3D model is decomposed by a voxel grid. A voxel is regarded as valid if there is a polygonal surface located within it, and invalid otherwise. Then the distribution of distances between two valid voxels instead of two points on the surface is calculated. Therefore, the area weighted defect in the sampling process will be greatly reduced since each valid voxel is weighted equally irrespective of how many points are located within this voxel. The main steps for computing the grid D2 are described as follows:

(1) First, a 3D model is segmented into a $2R\times2R\times2R$ voxel grid. To be invariant to translation and scaling, the object's mass centre is moved to the location (R, R, R) and the average distance from valid voxels to the mass centre is scaled to be $R/2$. R is set as 32, which provides adequate resolution for discriminating objects while filtering out those high-frequency polygonal surfaces

in the complex components of a 3D object.

(2) Two valid voxels are randomly selected and their distance is measured. A total of U distances are evaluated from the set of valid voxels. A histogram containing 256 bins is constructed: $H = \{B_1, B_2, ..., B_{256}\}$, where B_i denotes the number of distances within the range of the i-th bin. To normalize the distribution, the grid D2 (GD2) is defined as:

$$GD2 = \left\{ \frac{B_1}{U}, \frac{B_2}{U}, \frac{B_3}{U}, \dots, \frac{B_{256}}{U} \right\}, \tag{3.20}$$

where U is set as 64^3. From Fig. 3.10 we can see that the D2 distributions are clearly different while GD2 distributions are similar for these two similar airplanes. Experimental results show that Shih et al.'s method is superior to others, and the new shape descriptor is both discriminating and robust.

In addition, Song et al. [36] also adopted a histogram representation, based on shape functions to match 3D shapes by generating histograms using the discrete Gaussian curvature and discrete mean curvature of every vertex of a 3D triangle mesh.

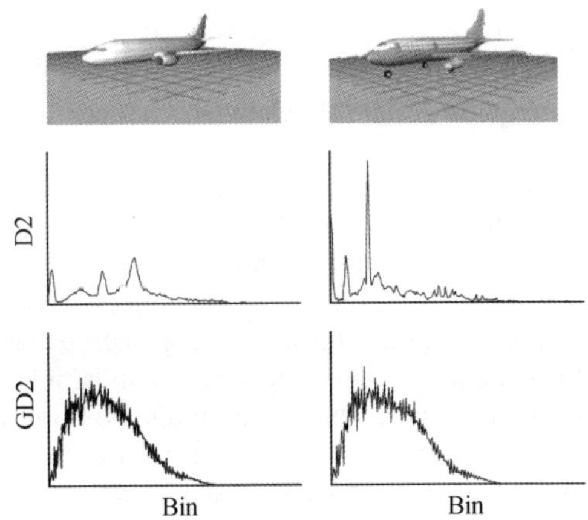

Fig. 3.10. D2 and GD2 distributions for two similar airplane objects [35] (©[2005]IEEE)

3.2.6 Extended Gaussian Image

In [37], Horn defined the extended Gaussian image (EGI), discussed its properties, and gave examples. Methods for determining the extended Gaussian images of polyhedra, solids of revolution and smoothly curved objects in general were shown. The orientation histogram, a discrete approximation of the extended Gaussian image, was described along with a variety of ways of tessellating the sphere. The detailed concepts and properties of EGI can be described as follows.

3.2.6.1 Definitions of Extended Gaussian Image for Convex Polyhedra

Minkowski showed in 1897 that a convex polyhedron is fully specified by the area and orientation of its faces. Surface normal vector information for any object can be mapped onto a unit sphere, called the Gaussian sphere. We can represent area and orientation of the faces conveniently by point masses on this sphere. A weight is assigned to each point on the Gaussian sphere equal to the area of the surface having the given normal. Weights are represented by vectors parallel to the surface normals, with length equal to the weight. Imagine moving the unit surface normal of each face so that its tail is at the center of a unit sphere. The head of the unit normal then lies on the surface of the unit sphere. Each point on the Gaussian sphere corresponds to a particular surface orientation. The extended Gaussian image of the polyhedron is obtained by placing a mass at each point equal to the surface area of the corresponding face.

It seems at first as if some information is lost in this mapping, since the position of the surface normals is discarded. Viewed from another angle, no note is made of the shape of the faces or their adjacency relationships. It can nevertheless be shown that the extended Gaussian image uniquely defines a convex polyhedron. Iterative algorithms can be used for recovering a convex polyhedron from its extended Gaussian image.

3.2.6.2 Gaussian Image for Smoothly Curved Surfaces

One can associate a point on the Gaussian sphere with a given point on a surface by finding the point on the sphere which has the same surface normal. Thus it is possible to map information associated with points on the surface onto points on the Gaussian sphere. In the case of a convex object with positive Gaussian curvature everywhere, no two points have the same surface normal. The mapping from the object to the Gaussian sphere in this case is invertible: Corresponding to each point on the Gaussian sphere, there is a unique point on the surface. If the convex surface has patches with zero Gaussian curvature, curves or even areas on it may correspond to a single point on the Gaussian sphere.

One useful property of the Gaussian image is that it rotates with the object. Consider two parallel surface normals, one on the object and the other on the Gaussian sphere. The two normals will remain parallel if the object and the Gaussian sphere are rotated in the same fashion. A rotation of the object thus corresponds to an equal rotation of the Gaussian sphere.

3.2.6.3 Gaussian Curvature for Smoothly Curved Surfaces

Consider a small patch δO on the object. Each point in this patch corresponds to a particular point on the Gaussian sphere. The patch δO on the object maps into a patch, δS say, on the Gaussian sphere. On one hand, if the surface is strongly curved, the normals of points in the patch will point into a wide fan of directions. The corresponding points on the Gaussian sphere will be spread out. On the other hand, if the surface is planar, the surface normals are parallel and map into a single point.

These considerations suggest a suitable definition of curvature. The Gaussian curvature is defined to be equal to the limit of the ratio of the two areas as they tend to zero. That is,

$$K = \lim_{\delta O \to 0} \frac{\delta S}{\delta O} = \frac{\mathrm{d}S}{\mathrm{d}O}. \tag{3.21}$$

From this differential relationship we can obtain two useful integrals. Consider first integrating K over a finite patch O on the object:

$$\iint_O K \mathrm{d}O = \iint_S \mathrm{d}S = A_S, \tag{3.22}$$

where A_S is the area of the corresponding patch on the Gaussian sphere. The expression on the left is called the integral curvature. This relationship allows one to deal with surfaces which have discontinuities in surface normal.

Now consider instead integrating $1/K$ over a patch S on the Gaussian sphere

$$\iint_S (1/K)\mathrm{d}S = \iint_O \mathrm{d}O = A_O, \tag{3.23}$$

where A_O is the area of the corresponding patch on the object. This relationship suggests the use of the inverse of the Gaussian curvature in the definition of the extended Gaussian image of a smoothly curved object, as we shall see. It also shows, by the way, that the integral of $1/K$ over the whole Gaussian sphere equals the total area of the object.

3.2.6.4 Extended Gaussian Image Definition for Smoothly Curved Surfaces

We can define a mapping which associates the inverse of the Gaussian curvature at a point on the surface of the object with the corresponding point on the Gaussian sphere. Let u and v be parameters used to identify points on the original surface. Similarly, let ξ and η be parameters used to identify points on the Gaussian sphere. These could be longitude and latitude, for example. Then we define the extended Gaussian image as

$$G(\xi, \eta) = \frac{1}{K(u,v)}, \tag{3.24}$$

where (ξ, η) is the point on the Gaussian sphere which has the same normal as the point (u, v) on the original surface. It can be shown that this mapping is unique for convex objects. That is, there is only one convex object corresponding to a particular extended Gaussian image. The proof is unfortunately non-constructive and no direct method for recovering the object is known.

3.2.6.5 Properties of the Extended Gaussian Image for Convex Polyhedra

The extended Gaussian image is not affected by translation of the object. Rotation of the object induces an equal rotation of the extended Gaussian image, since the unit surface normals rotate with the object.

Mass distributions, which lie entirely within one hemisphere, are zero in the complementary hemisphere and do not correspond to closed objects. We can demonstrate that the center of mass of an extended Gaussian image has to lie at the origin. This is clearly impossible if the whole hemisphere is empty. Also, a mass distribution which is nonzero only on a great circle of the sphere corresponds to the limit of a sequence of cylindrical objects of increasing length and decreasing diameter. Here, such pathological cases are excluded and our attention is confined to closed, bounded objects.

Some properties of the extended Gaussian image are important. First, the total mass of the extended Gaussian image is obviously just equal to the total surface area of the polyhedron. If the polyhedron is closed, it will have the same projected area when viewed from any pair of opposite directions. This allows us to compute the location of the center of mass of the extended Gaussian image.

An equivalent representation, called a spike model, is a collection of vectors each of which is parallel to one of the surface normals and of length equal to the area of the corresponding face. The result regarding the center of mass is equivalent to the statement that these vectors must form a closed chain when placed end to end.

3.3 Rotation-Based Shape Descriptor

Recently, the authors of this book [38] presented a new shape descriptor based on rotation. The proposed method is designed for 3D mesh models. Our approach is to represent 3D shape as a 1D histogram. The motivation originates from a question such as this: As a 3D model rotates in the spatial domain, why is the human vision system, from the fixed viewing angle, sensitive to the fact that the shape after rotation differs from the initial shape, as shown in Fig. 3.11? If points are sampled uniformly on the model surface, we notice that the orientation of the normal vector of points is changed after rotation. As Fig. 3.12 shows, regardless of the position of point $\boldsymbol{p}$, we translate its normal vector $\boldsymbol{n}$ so that its origin coincides with the origin of the coordinate system, and the end of the unit normal lies on a

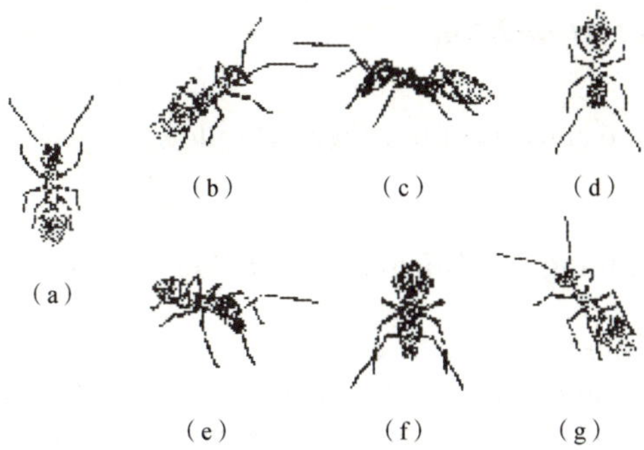

Fig. 3.11. Shape of a 3D model viewing from the same angle after various rotations. (a) The shape of the original model; (b)–(g) Shapes after various random rotations

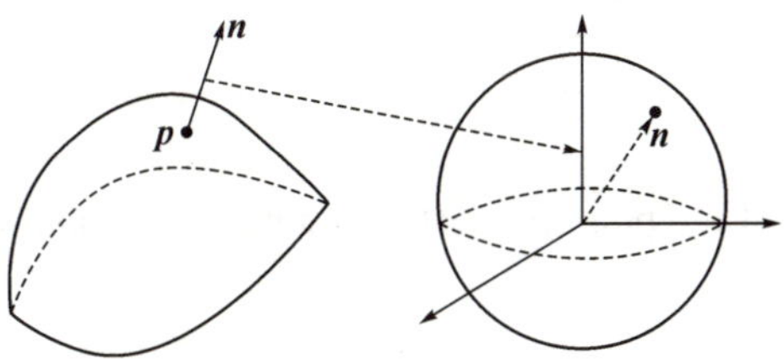

Fig. 3.12. Gaussian mapping

unit sphere. As mentioned in Subsection 3.2.6, this process is called Gaussian mapping, and the sphere is called a Gaussian sphere.

Let us assume that considerable points are sampled on the surface of a model. Repeating Gaussian mapping, we attain a sphere distributed with normal vectors of sample points. Thus shape feature extraction can be transformed into analyzing normal distributions on the sphere. Once randomly rotating a model K times, we attain K different shapes and corresponding spheres with different normal distributions. To describe the shape with a histogram, our approach statistically analyzes the normal distribution on K spheres.

The intrinsic properties of our proposed descriptor are as follows: (1) Generality. The description scope of the method is for all classes of shapes. It can be applied to extract shape features of popular models, such as meshes, solid models and other geometric representations. (2) Invariance to rotation, translation and scaling. In order to capture features, a model is usually placed into a canonical coordinate frame. This is called pose estimation or normalization. Nowadays normalization is an important task in preprocessing a 3D model. However, it is still a difficult problem. The proposed descriptor does not need to normalize the 3D model to speed up shape extraction. The proposed descriptor is invariant to transformations such as rotation, translation and scaling. The reason for this is that we only consider the orientation of normal, instead of the position of sample points. (3) Robustness. Random sampling ensures the descriptor is insensitive to noises. In other words, as a statistical method, the descriptor lays emphasis on the global shape feature.

3.3.1 Proposed Algorithm

The proposed method consists of four steps as follows.

3.3.1.1 Point Sampling and Normal Vector Computation

For a triangulated mesh model, N random points are sampled uniformly on the surface. Suppose s_i and k denote the area of the triangle i and the number of triangles, respectively. Then we can compute n_i, namely the number of sample points on the triangle i as follows:

$$n_i = \frac{Ns_i}{\sum_{i=1}^{k} s_i}. \tag{3.25}$$

The normal vector of the point $\boldsymbol{p}$ is estimated by the normal of $\triangle ABC$, where $\boldsymbol{p}$ lies, as follows:

$$n_p = n_{\Delta ABC}. \tag{3.26}$$

Hereto a mesh model is translated into a point set with orientations. Notice that the proposed method does not need to accurately determine positions of random points, but only needs to attain the orientation of normals. Different from this, positions of sample points must be obtained in Osada's D2 [34] and Ohbuchi's improvement [39]. Consequently computational complexity of our descriptor is lower than that in [34] and [39].

3.3.1.2 Rotation of the Model

We randomly rotate models, controlled by α, β, γ, namely rotation angles with respect to x-, y-, z-axes, respectively.

$$\boldsymbol{R} = \begin{pmatrix} \cos\beta\cos\gamma & -\cos\beta\sin\gamma & \sin\beta \\ \sin\alpha\sin\beta\cos\gamma + \cos\alpha\sin\gamma & -\sin\alpha\sin\beta\sin\gamma + \cos\alpha\cos\gamma & -\sin\alpha\cos\beta \\ -\cos\alpha\sin\beta\cos\gamma + \sin\alpha\sin\gamma & \cos\alpha\sin\beta\sin\gamma + \sin\alpha\cos\gamma & \cos\alpha\cos\beta \end{pmatrix}. \tag{3.27}$$

As shown in Eq.(3.27), $\boldsymbol{R}$ is the general 3D rotation matrix. When a 3D point $\boldsymbol{p}$ is rotated by $\boldsymbol{R}$, $\boldsymbol{p}$ is transformed into $\boldsymbol{p}'$ as follows:

$$\boldsymbol{p}' = \boldsymbol{R}\boldsymbol{p} \tag{3.28}$$

Actually, we rotate a model in order to find the shape difference after rotation. This can be translated into analyzing normal distributions on the unit sphere. Let us assume we rotate a model T times with T groups of rotation angles; α, β, γ are randomly selected in the range of $[0, 2\pi]$. When rotating a model, the normal distribution of points is changed accordingly.

As shown in Fig. 3.13, the triangle ABC and point $\boldsymbol{p}$ are rotated to $A'B'C'$ and $\boldsymbol{p}'$, respectively. Then $\boldsymbol{n}_p$ and $\boldsymbol{n}'_p$ have the relationship as follows:

$$\boldsymbol{n}'_p = \boldsymbol{R}\boldsymbol{n}_p . \tag{3.29}$$

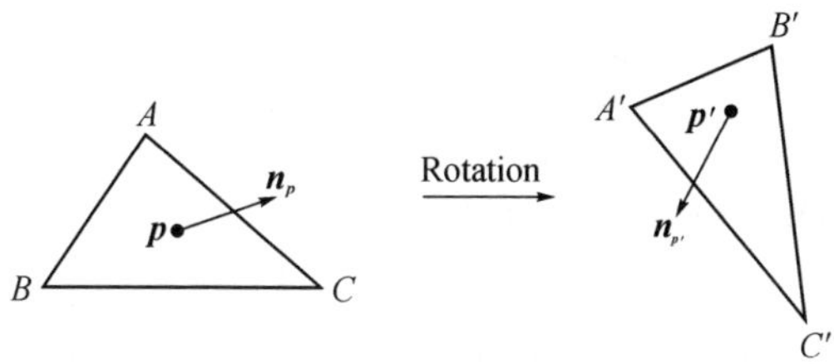

Fig. 3.13. Rotation of a triangle on the surface

3.3.1.3 Calculation of Normal Distributions

As a model is rotated T times, we obtain T Gaussian spheres, each being distributed by N normal vectors. To analyze the distributions, we segment the surface of a Gaussian sphere into L sections. As an example, the spherical surface is segmented into 8 sections by x-y, y-z, and x-z planes, as shown in Fig. 3.14(a). We count the normal on each section in turn. To determine which section a normal belongs to, we only need to capture signs of each component of a normal, as shown in Fig. 3.15(a). Thus we obtain T groups of 8-dimensional vectors, as shown in Eqs.(3.30) and (3.31). The element v_i is the number of the normal distributed in the i-th section.

$$V = (v_1, v_2, v_3, v_4, v_5, v_6, v_7, v_8), \tag{3.30}$$

$$N = \sum_{i=1}^{8} v_i . \tag{3.31}$$

Based on these 8 sections, the spherical surface also can be further segmented into 24 sections. As shown in Fig. 3.14(b), one eighth of the surface is divided into three subsections by finding the maximum absolute value of three components of the normal.

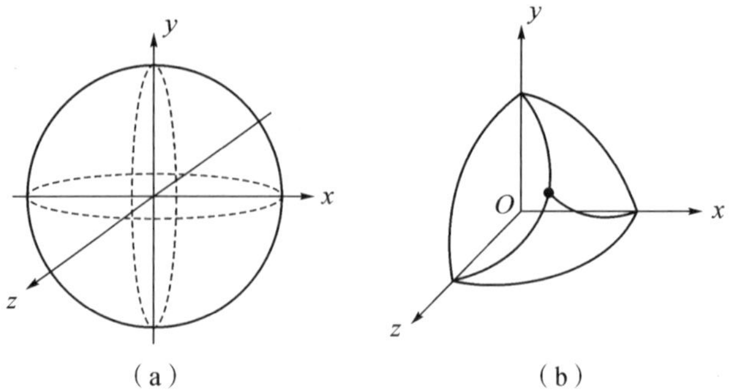

Fig. 3.14. Segmentation of Gaussian sphere. (a) 8 sections; (b) 24 sections

3.3.1.4 Construction of Histograms

To construct a 1D histogram, we compute the Euclidean distance L_2 between two vectors $\boldsymbol{V}_x$ and $\boldsymbol{V}_y$, as shown in Eq.(3.32). Thus, we obtain $T(T-1)/2$ distances for T groups of vectors, and a histogram is then constructed:

$$L_2(\boldsymbol{V}_x, \boldsymbol{V}_y) = \left(\sum_{i=1}^{k} \left| \boldsymbol{V}_x(i) - \boldsymbol{V}_y(i) \right|^2 \right)^{\frac{1}{2}}. \tag{3.32}$$

x	y	z	Section
−	−	−	1
−	−	+	2
−	+	−	3
−	+	+	4
+	−	−	5
+	−	+	6
+	+	−	7
+	+	+	8

(a)

−0.2673	−0.5345	−0.8018
−0.1231	−0.4924	−0.8616
−0.2338	−0.9584	−0.1636
…		
…		
…		

(b)

Fig. 3.15. Calculation of normal distribution. (a) Signs and corresponding section; (b) Example normals

3.3.2 Experimental Results

In the experiment, we test the descriptor with a set of 18 parameter combinations; N = {32,768, 65,536, 131,072}, T = {1,000, 2,000, 3,000}, L = {8, 24}. Empirically, considering lower computational complexity, we find that N = 65,536, T = 2,000 and L = 24 yields a histogram with good discrimination ability. Experimental models are randomly selected from the database of the Princeton Shape Benchmark (PSB), a publicly available 3D model database with 1,814 mesh models. We classify the experimental models into 10 classes, each class containing 2–9 models. All histograms are normalized under the same mode with 256 bins. From Fig. 3.16 we can find that: Models in the same class have similar histograms, while models in the different classes have dissimilar histograms. Experimental results show that its discriminating ability is good enough to classify different models. Therefore, the descriptor can be applied to specific applications such as 3D model retrieval, 3D object classification, 3D object recognition, etc.

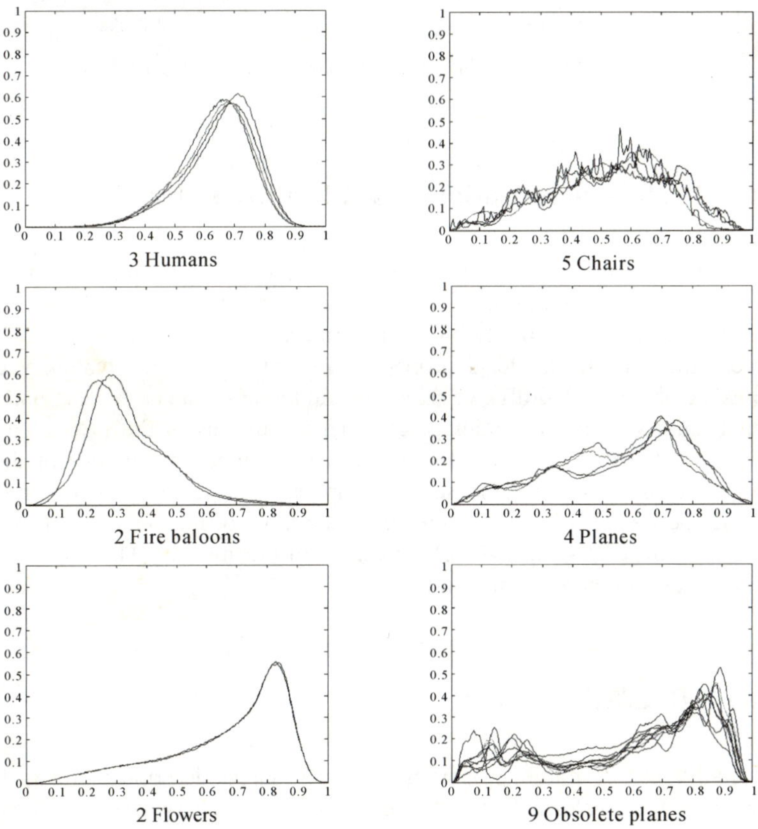

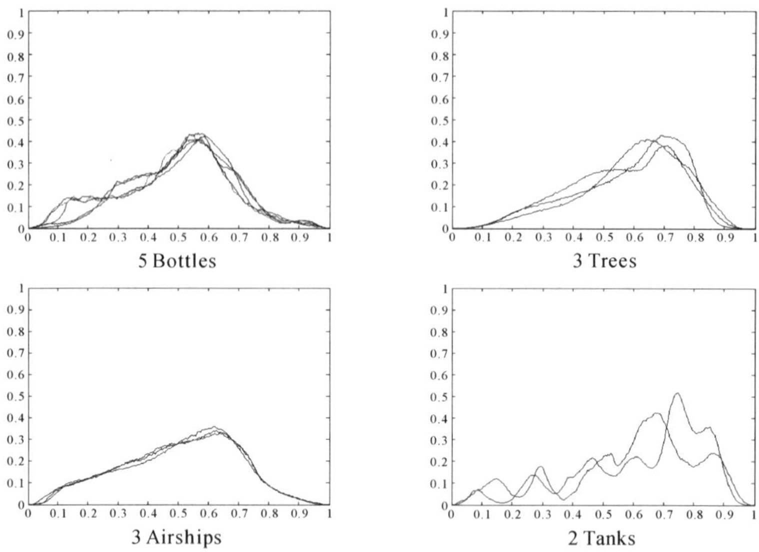

Fig. 3.16. Shape histograms for models grouped into 10 classes

3.4 Vector-Quantization-Based Feature Extraction

The authors of this book proposed a novel feature for 3D mesh models, i.e., a vector quantization index histogram [40]. The main idea is as follows: Firstly, points are sampled uniformly on mesh surface. Secondly, to a point five features representing global and local properties are extracted. Thus feature vectors of points are obtained. Thirdly, we select several models from each class, and employ their feature vectors as a training set. After training using the LBG algorithm, a public codebook is constructed. Next, codeword index histograms of the query model and those in the database are computed. The last step is to compute the distance between histograms of the query and those of the models in the database. Experimental results show the effectiveness of our method. The following is the detailed description of our method.

3.4.1 Detailed Procedure

Generally, the desirable properties of a 3D shape descriptor are as follows: invariance to transformation, robustness to noise, conciseness for storage, less computational complexity, shape discrimination, etc. In this subsection, we give a novel 3D shape description method with the above properties. The detailed steps can be described below.

3.4.1.1 Sample Points Uniformly on Surface

A 3D mesh consists of vertices coordinates and their connectivity information. Since different models may contain a different number of vertices, we randomly sample points on the model surface to guarantee all models including the query model and those in the database have the same number of points. We use Osada's method [34] to generate sample points on the model surface. For each selected triangle $T_{(A,\ B,\ C)}$ with vertices (A, B, C), we sample a point on its surface by generating two random numbers, r_1 and r_2 and using Eq.(3.33):

$$\boldsymbol{p} = (1-\sqrt{r_1})\boldsymbol{A} + \sqrt{r_1}(1-\sqrt{r_2})\boldsymbol{B} + \sqrt{r_1 r_2}\boldsymbol{C}, \tag{3.33}$$

where the random numbers r_1 and r_2 are uniformly distributed between 0 and 1. Clearly, the number of sample points on a triangle is proportional to its area. This step aims to guarantee that the number of sample points of all models is exactly the same. Suppose n denotes it.

3.4.1.2 Computation of Subfeatures

This step is to compute subfeature vectors of sample points. After sampling, we perform principle component analysis (PCA) on the model first. Using the point mass on the surface, the covariance matrix $\boldsymbol{CV}$ can be computed as

$$\boldsymbol{CV} = \frac{1}{n}\sum_{i=1}^{n}(\boldsymbol{p}_i - \boldsymbol{m}) * (\boldsymbol{p}_i - \boldsymbol{m})^{\mathrm{T}}, \tag{3.34}$$

where $\boldsymbol{p}_i$ is a sample point, and $\boldsymbol{m}$ is the center of mass. The center of mass is computed as follows:

$$\boldsymbol{m} = \frac{1}{S}\sum_{i=1}^{k} s_i \boldsymbol{g}_i, \tag{3.35}$$

where s_i and $\boldsymbol{g}_i$ is the area and gravity of triangle T_i. Three eigenvectors of the covariance matrix $\boldsymbol{CV}$ are the principal axes of inertia of the model. The first, the second and the third significant principle axes correspond to the associated magnitude of the eigenvalues in decreasing order.

Next, five sub-features are extracted for each point. Suppose a cord $\boldsymbol{c}_i$ is defined to be a vector that goes from the center of mass $\boldsymbol{m}$ to the sample point $\boldsymbol{p}_i$.

D_1: the Euclidean distance between $\boldsymbol{p}_i$ and $\boldsymbol{m}$, i.e. the length of $\boldsymbol{c}_i$.

α: the angle between $\boldsymbol{c}_i$ and the first most significant principle axis.

β: the angle between $\boldsymbol{c}_i$ and the second most significant principle axis.

γ: the angle between $\boldsymbol{c}_i$ and the third most significant principle axis.

θ: the angle between $\boldsymbol{c}_i$ and the normal vector of $\boldsymbol{p}_i$.

VI: visual importance of the point $\boldsymbol{p}_i$.

Here the normal vector of a point is estimated as the normal of the triangle it lies on. Clearly, D_1, α, β, γ and θ describe the relationship between the local points and the global properties, while VI denotes the local characteristics.

Suppose ϕ is the inclination of two vectors $\boldsymbol{OM}$, $\boldsymbol{ON}$. The cosine of this inclination is computed as

$$\cos\phi = \frac{\boldsymbol{OM}\cdot\boldsymbol{ON}}{|\boldsymbol{OM}||\boldsymbol{ON}|}. \tag{3.36}$$

Thus the $\cos\alpha$, $\cos\beta$, $\cos\gamma$ and $\cos\theta$ can be computed like this.

We associate a vertex $\boldsymbol{v}$ with a value that represents its visual importance [13], defined by:

$$VI_v = 1 - \left\| \frac{\sum_i \Delta_i \boldsymbol{n}_i}{\sum_i \Delta_i} \right\|, \tag{3.37}$$

where $\boldsymbol{n}_i$ is the unit normal of one of neighboring triangles of vertex $\boldsymbol{v}$ and Δ_i is the area of the neighboring triangle. VI of $\boldsymbol{p}_i$ is estimated as the mean of visual importance of three vertices of the triangle it lies on. Thus the final VI of $\boldsymbol{p}_i$ can be calculated as follows:

$$VI_{p_i} = \frac{1}{3}(VI_A + VI_B + VI_C). \tag{3.38}$$

It is obvious that VI is in the range of [0, 1], which can indicate the local curvature around $\boldsymbol{p}_i$. When VI is equal to 0, the vertex $\boldsymbol{v}$ is on a flat plane. The increase of VI is coupled with the increase of curvature.

After calculating the above five sub-features, we can construct a feature vector for each point as follows:

$$\boldsymbol{f}_i = [D_1, |\cos\alpha|, |\cos\beta|, |\cos\gamma|, |\cos\theta|, VI], \tag{3.39}$$

where $1 \le i \le N$ and the sub-feature D_1 of a specific model has been normalized. Thus, N feature vectors for each model are obtained, in which five components are real values in the range of [0, 1]. For each model, we can obtain its feature matrix as

$$\boldsymbol{F} = [\boldsymbol{f}_1, \boldsymbol{f}_2, \ldots, \boldsymbol{f}_N]^{\mathrm{T}}. \tag{3.40}$$

Obviously, for any model, the size of $\boldsymbol{F}$ is $N \times 5$.

3.4.1.3 Codebook Generation

Suppose there are K categories of models in the database. We randomly selected L models from each class to construct a training set. The feature matrices of these models are regarded as entries of the LBG algorithm [41]. In other words, totally $N \cdot L$ sub-feature vectors as input vectors are trained. After training, a public codebook is constructed.

3.4.1.4 Index Histogram Construction

For all of the models in the database, we construct their codeword index histograms offline, while that of the query model is obtained online, all based on the public codebook. As the sample points in all histograms are equal to N, there is no normalization operation required before comparison. Suppose all index histograms contain B bins.

3.4.1.5 Feature Comparison

This step is to measure the similarity between the histogram of the query and those of the models in the database. We employ the Euclidean distance as the similarity metric. Suppose $Q = \{q_1, q_2, \ldots, q_B\}$ denotes the index histogram of the query, $H = \{h_1, h_2, \ldots, h_B\}$ is the histogram of a model from the database, we have

$$D = \sum_{i=1}^{B} (q_i - h_i)^2 . \tag{3.41}$$

After computing the distances, retrieval results can be returned, which are ranked in the descending order of the distances between the query and models in the database.

3.4.2 Experimental Results

In the experiment, the test database contains 95 models, which are classified into 10 categories. The names of the categories are: bottles (5 models), cars (8), dogs (6), human bodies (24), planes (8), tanks (5), televisions (7), fire balloons (19), helicopters (5) and chess (8). From each class, we randomly select one model and thus our training set has ten models. For each model, we sample 30,000 points on its surface, thus there are 300,000 sub-feature vectors as training vectors. The codebook contains 500 codewords. Each index histogram also consists of 500 bins.

Some samples of 3D model retrieval results are shown in Fig. 3.17, from which we can see our method is effective.

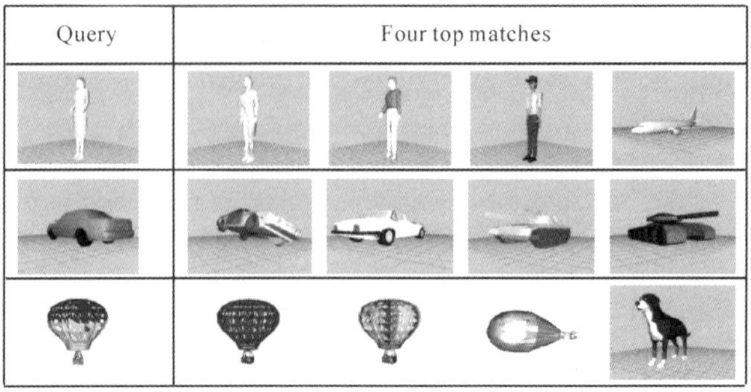

Fig. 3.17. 3D query models and the four top matches listed from left to right

In the experiments, we find that the retrieval performance is closely related to the number of sample points. On the one hand, sampling more points can improve the retrieval precision. The reason is that our method is based on statistics. In addition, adopting more sub-features of sample points can also result in higher precision. On the other hand, these improvements are at the cost of larger computational complexity. Therefore, it is necessary to achieve a good tradeoff between precision and computational complexity according to different requirements.

3.5 Global Geometry Feature Extraction

The global geometry of a 3D model is analyzed by directly sampling the vertex set, the polygon mesh set, or the voxel set in the spatial domain. Aspect ratio, binary 3D voxel bitmap, and 3D angles of vertices or edges may be considered as the most simple and straightforward features [42], although their discriminative powers are limited. These types of analyses generally use PCA-like methods to align the model into a canonical coordinate frame at first, and then define the shape representation on this normalized orientation.

The common characteristic of these methods is that they are almost all derived directly from the elementary unit of a 3D model, that is the vertex, polygon, or voxel, and a 3D model is viewed and handled as a vertex set, a polygon mesh set or a voxel set. Their advantages lie in their easy and direct derivation from 3D data structures, together with their relatively good representation power. However, the computation processes are usually too time-consuming and sensitive for small features. Also, the storage requirements are too high due to the difficulties in building a concise and efficient indexing mechanism for them in large model databases.

3.5.1 Ray-Based Geometrical Feature Representation

Vranić et al. [43] proposed a ray-based geometrical feature representation. They sampled a 3D model in its canonical coordinate frame as a set of regular spaced direction vectors and set rays along each direction vector from the coordinate origin, which intersected with the triangle mesh of a polyhedron surrounding the 3D model. For each direction, the maximum distance from the intersected triangle mesh to the coordinate origin was computed and all the distance samples composed a feature vector. The detailed process can be expressed as follows.

3.5.1.1 Preprocessing with the Modified PCA Technology

Vranić et al. incorporated a modification of principal component analysis (PCA) in the geometrical feature extraction module. This transformation changes the coordinate system axes to new ones which coincide with the directions of the three largest spreads of the point (i.e. vertex) distribution. A 3D object representing a triangle mesh consists of geometry, topology and attributes. Geometry is determined by the vertex coordinates, information about how vertices are connected in order to form triangles is called topology and attributes are color, texture, etc. In their system, attributes are still not under consideration because the stress is on representing spatial relations within a 3D model, i.e., geometry and topology.

The aim of principal component analysis applied to the 3D model is to make the resulting shape feature vector independent of translation and rotation as much as possible. The PCA will be based on the collection of vertex vectors. To account for the differing sizes of the corresponding triangles, Vranić et al. introduced weighting factors proportional to the corresponding surface area.

3.5.1.2 Feature Extraction

Suppose we have a given set of L directional vectors $\{\boldsymbol{u}_1, \boldsymbol{u}_2, \ldots, \boldsymbol{u}_L\}$, as shown in Fig. 3.18. Then the triangle mesh is intersected with the ray emanating from the origin of the PCA coordinate system and traveling in the direction $\boldsymbol{u}_i$ ($i \in \{1, \ldots, L\}$). The distance to the farthest intersection is taken as the i-th component of the feature vector which is scaled to the Euclidean unit length to ensure scale invariance. In Vranić et al.'s experiment, L is set to be 20. The vertices of a dodecahedron, with the center in the coordinate origin, are taken as directions. This feature is invariant with respect to rotation and translation because of the fact that initial coordinate axes are transformed. The scaling invariance is accomplished by normalizing the feature vector.

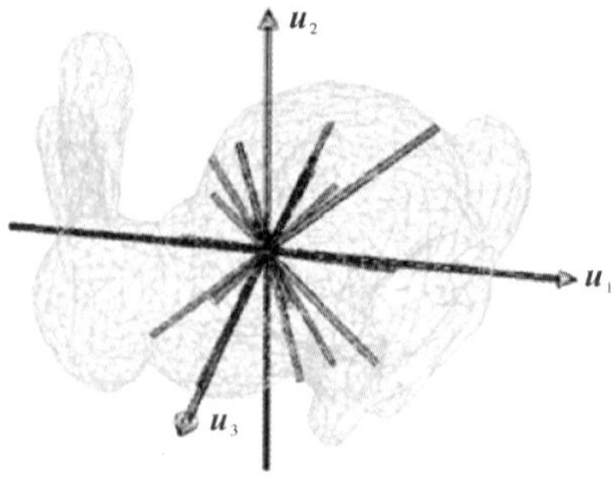

Fig. 3.18. Illustration of ray-based shape descriptor [53] (With permission of Comenius University Press)

3.5.1.3 Feature Description

After extraction of features, the next step is their formal description. As we know, the MPEG-7 standard provides a rich set of standardized mechanisms and means aimed at describing multimedia content. The MPEG-7 terminology has been adopted and the mutual relation between a descriptor and a feature is explained in the following definition: A descriptor is a representation of a feature. A descriptor is used to define the syntax and the semantics of the feature representation [44].

Therefore, the descriptor of the above feature vector is determined with 20 non-negative real numbers, where the i-th component is the object extension in the direction of the i-th vertex of the mentioned dodecahedron, which is defined (the vertex coordinates and the numbering) internally. This defines the semantics of the descriptor. The syntax is defined by description schemes (DS) for real vectors.

MPEG-7 is not a restrictive system for audio-visual content description. It is a flexible and extensible scope for describing multimedia data with a developed set of methods and tools. As mentioned in MPEG-7, the 3D Model DS should support "the hierarchical representation of different descriptors in order that queries may be processed more efficiently at successive levels (where N level descriptors complement (N−1) level descriptors)". Hence, different features at different levels of detail should be considered. Vranić et al. were encouraged by the reflector of the MPEG-7 DS group to implement their own DS for 3D models.

This DS should comply with MPEG-7 specification [44].

3.5.1.4 Other Methods

Using a similar idea, Yu et al. [45] extracted the 3D global geometry as a distance map and surface penetration map features. These two spatial feature maps describe the geometry and topology of the surface patches on the object, while preserving the spatial information of the patches in the maps. The feature maps capture the amount of effort required to morph a 3D object into a canonical sphere, without performing explicit 3D morphing. Given a 3D object, it is first scaled and embedded in a sphere of unit radius such that the center of the sphere coincides with the object's centroid. Then, a ray is shot from the center of the sphere through each point of the object to the sphere's surface, as shown in Fig. 3.19. The distance traveled by the ray from an object point to the sphere's surface is recorded in the distance map (DM). Fourier transforms of the feature maps are used for object comparison so as to achieve invariant retrieval under arbitrary rotation, reflection, and non-uniform scaling of the objects. Experimental results show that their method of retrieving 3D models is very accurate, achieving a precision of above 0.86, even at a recall rate of 1.0.

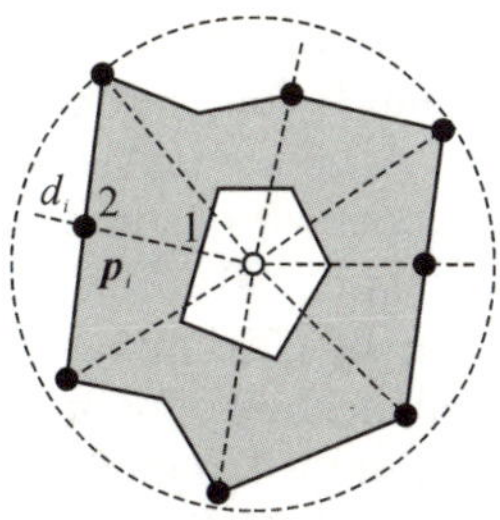

Fig. 3.19. Computing feature maps. Rays (dashed lines) are shot from the center (white dot) of a bounding sphere (dashed circle) through the object points (black dots) to the sphere's surface. The distance d_i traveled by the ray from a point $\boldsymbol{p}_i$ to the sphere's surface and the number of object surfaces (solid lines; 2, in this case) penetrated by the ray since it leaves the sphere's center are recorded in the feature maps [45] (©[2003]IEEE)

3.5.2 Weighted Point Sets

Tangelder et al. proposed a method using weighted point sets as the shape descriptor for a 3D polygon mesh [46]. They assumed that a 3D shape is represented by a polyhedral mesh. They do not require the polyhedral mesh to be closed. Therefore, their method can also handle polyhedral models that may contain gaps. They also enveloped the object in a 3D voxel grid and represented the shape as a weighted point set by selecting one representative point for each non-empty grid cell. They then selected the vertex with the highest Gaussian curvature or the area-weighted mean of all the vertices in a grid cell, to represent the model's geometry features.

Many methods mentioned in previous sections do not take the overall relative spatial location into account, but throw away some of this information, in order to deal with data of lower complexity, e.g. 2D views or 1D histograms. What is new in Tangelder et al.'s method is that they use the overall relative spatial position by representing the 3D shape as a weighted point set, without taking the connectivity relations into account. The weighted point sets, which can be viewed as 3D probability distributions, are compared using a new transportation distance that is a variant of the Earth Mover's Distance [47]. In contrast, histogram-based approaches can be viewed as methods comparing 1D probability distributions. Unlike the Earth Mover's Distance, the transportation distance in Tangelder et al.'s approach satisfies the triangle inequality, and thus their method can be used in indexing schemes that employ this property. Their experiments demonstrate that the retrieval performance of their method compares favorably with some other shape matching methods.

To compare two objects independently of orientation, position and scaling, Tangelder et al. first applied principal components analysis to bring the objects into a standard pose defined by the principal axes of inertia. Also, in the preprocessing step, they enclose each object by a 3D grid and generate for each object a signature representing a weighted point set, which contains for each non-empty grid cell a salient point. Below they compare three methods to obtain in each grid cell a salient point. All three methods use only the vertices and the facets adjacent to the vertices to obtain a salient point. Therefore, they can handle models that contain gaps. Note that models containing polygons that are wrongly oriented are only handled correctly by the third method.

(1) Gaussian-curvature-based method. For a smooth surface, the Gaussian curvature at a point is the product of the minimal and maximal principal curvature at that point. The vertex in the cell with the highest Gaussian curvature can be chosen as the salient point.

(2) Normal-variation-based method. Another approach to obtain a measure related to the curvature is the normal variation method. In this approach we estimate the curvature in a grid cell by the normal variation in the grid cell. We choose the area-weighted mean of the vertices in the grid cell as a salient point.

(3) Midpoint-based method. The two methods described above may fail if the 3D models contain wrongly oriented polygons. This is the case for models that are represented by "polygonal soups", i.e. unorganized and degenerate sets of polygons. To handle such degenerate models, we can adopt a simple approach called midpoint method that is similar to Rossignac's polygon simplification algorithm [48]. The midpoint method obtains a signature S by adding for each grid cell the centre of mass of all vertices in the cell with unit weight to the signature S.

Finally, they compute the similarity between two shapes by comparing their signatures using a shape similarity measure that is a new variation of the Earth Mover's Distance. The experimental results given by Tangelder et al. are very promising, but their main shortcoming is the long time it took to compute the descriptors.

3.5.3 Other Methods

Heczko et al. [49] implemented an octree-structure-based method to represent the shape features of 3D volumetric models by fulfilling a multi-resolution subdivision of the 3D model space. For each grid cell, they took the sum of mesh sizes bounded by the grid cell as the feature components, which formed a feature descriptor of $2r \times 2r \times 2r$ dimensions, where r is the resolution of octree representation.

As for 3D industrial solid models, Cicirello et al. [50] and McWherter et al. [51] both compared 3D shapes by extracting the geometrical and engineering features of 3D models in spatial domains.

In order to improve the overall performance, the "divide-and-conquer" strategy can be adopted in the feature extraction process. In some cases, the low efficiency is mainly caused because some of the feature representations cannot be computed directly from the 3D meshes, which are required to be transformed into a 3D voxel space first. This process is time-consuming and requires a large amount of storage space. To address this issue, Zhang et al. [52] proposed a global geometrical analysis algorithm using the "divide-and-conquer" strategy without volumetric transformation. They first computed the features for each elementary surface (a triangle or a tetrahedron) of a 3D mesh model, and then summed them up to form the global feature vector.

3.6 Signal-Analysis-Based Feature Extraction

Feature extraction methods based on signal analysis analyze 3D models from the point of view of the frequency domain. However, because the 3D model is not a regularly sampled signal, the preprocessing process before feature extraction is generally complicated. In this section, we would like to introduce three typical shape descriptors based on transform domains.

3.6.1 Fourier Descriptor

We introduce discrete Fourier transform, Vranić and Soupe's Scheme and other schemes.

3.6.1.1 Discrete Fourier Transform

In mathematics, the discrete Fourier transform (DFT) is a specific kind of Fourier transform, used in Fourier analysis. It transforms one function into another, which is called the frequency domain representation, or simply the DFT, of the original function (which is often a function in the time domain). But the DFT requires an input function that is discrete and whose non-zero values have a limited (finite) duration. Such inputs are often created by sampling a continuous function, like a person's voice. And unlike the discrete-time Fourier transform (DTFT), it only evaluates enough frequency components to reconstruct the finite segment that was analyzed. Its inverse transform cannot reproduce the entire time domain, unless the input happens to be periodic (forever). Therefore, it is often said that the DFT is a transform for Fourier analysis of finite-domain discrete-time functions. The sinusoidal basis functions of the decomposition have the same properties. Since the input function is a finite sequence of real or complex numbers, the DFT is ideal for processing information stored in computers. In particular, the DFT is widely employed in signal processing and related fields to analyze the frequencies contained in a sampled signal, to solve partial differential equations and to perform other operations such as convolutions. The DFT can be computed efficiently in practice using a fast Fourier transform (FFT) algorithm.

The sequence of N complex numbers x_0, ..., x_{N-1} is transformed into the sequence of N complex numbers X_0, ..., X_{N-1} by the DFT according to the formula:

$$X_k = \sum_{n=0}^{N-1} x_n \mathrm{e}^{-\frac{2\pi j}{N}kn}, \quad k = 0, \ ..., \ N-1, \tag{3.42}$$

where $\mathrm{e}^{-\frac{2\pi j}{N}}$ is a primitive N-th root of unity. The inverse discrete Fourier transform (IDFT) is given by

$$x_n = \frac{1}{N}\sum_{k=0}^{N-1} X_k \mathrm{e}^{\frac{2\pi j}{N}kn}, \quad n = 0, \ ..., \ N-1. \tag{3.43}$$

3.6.1.2 Vranić and Soupe's Scheme

In 3D model analysis, the fourier descriptor decomposes the 3D model into frequency components and extracts features from DFT coefficients. Vranić and Soupe [53] applied 3D-DFT to extract features. The steps include pose normalization, voxelization and 3D DFT. After finding the canonical position and orientation of a model (for the detailed process, readers can refer to Chapter 4), the feature extraction is performed in two steps: (1) voxelization using the bounding cube; (2) application of the 3D-DFT.

The bounding cube (BC) of a 3D model is defined to be the tightest cube in the canonical coordinate frame that encloses the model, with the center in the origin and the edges parallel to the coordinate axes. After determining the BC, voxelization is performed in the following manner: the BC is subdivided into N^3 (N is a power of 2) equally sized cubes and calculates the proportion of the total surface area of the mesh inside each of the new cubes (cells). The cell with the attributed value is regarded as the voxel at the given position. Obviously, with the increase in N, the fraction of all voxels inside BC having values greater than zero decreases. Therefore, a suitable way of storing a voxel-based feature vector is an octree structure. Thus, an efficient hierarchical feature representation can be obtained.

The information contained in this octree can be used in several ways. Vranić and Soupe formerly [49] used a similar voxelization as a feature in the spatial domain with a reasonably small N. The feature vector had N^3 components and the L_1 or L_2 norms were engaged for calculating distances. While in [53], their modification is as follows: A greater value of N is selected and the feature is represented in the frequency domain by applying the 3D-DFT to the voxelized model (i.e., calculated values in the N^3 cells).

Let $Q = \{q_{ikl} \mid q_{ikl} \in R, -N/2 \leq i, k, l < N/2\}$ be the set of all voxels. The set Q is transformed into the set $G = \{g_{uvw} \mid g_{uvw} \in C, -N/2 \leq u, v, w < N/2\}$ by

$$g_{uvw} = \sum_{i=-N/2}^{N/2-1} \sum_{k=-N/2}^{N/2-1} \sum_{l=-N/2}^{N/2-1} q_{ikl} \mathrm{e}^{-\frac{2\pi j}{N}(iu+kv+lw)} . \tag{3.44}$$

Finally, we find the absolute values of the coefficients g_{uvw} with indices $-K \leq u, v, w \leq K$ (the lowest frequencies). Except for the coefficient g_{000}, all selected complex numbers are pairwise conjugated. Therefore, the feature vector consists of $((2K+1)^3+1)/2$ real-valued components. In Vranić and Soupe's experiments, they select $K = 1, 2, 3$, i.e., the descriptors possess 14, 63, and 172 components, respectively.

The value of parameter N (the resolution of voxelization) should be sufficiently large in order to capture spatial properties of a model by the 3D DFT. In practice, Vranić and Soupe selected $N = 128$ and on average about 20,000 voxels (out of 1,283 elements of the set Q) have values greater than zero. This makes the octree representation very efficient. During the 3D-DFT, they computed only those elements of the set G that are used in the feature vector (14, 63, or 172 out of 1283). The proposed descriptor shows better retrieval performance than the voxel-based feature presented in [49]. Having in mind that the ray-based descriptor [49] was improved by incorporating spherical harmonics [54], they inferred that if the L_1 or L_2 norm is engaged, representation of a feature in the frequency domain is more efficient than representation of the same feature in the spatial domain.

3.6.1.3 Other Schemes

In [55], the Fourier descriptor is extended to produce a set of normalized coefficients which are invariant under any affine transformation (translation, rotation, scaling, and shearing). The method is based on a parameterized boundary description which is transformed to the Fourier domain and normalized to eliminate dependencies on the affine transformation and on the starting point. Invariance to affine transforms allows considerable robustness when applied to images of objects which rotate in all three dimensions, as is demonstrated by processing silhouettes of aircraft maneuvering in three-dimensional space. Richard and Hemani [56] utilized the Fourier descriptor to compute the boundary curvature of the 3D model and obtain its feature. Zhang and Fiume [57] adopted the Fourier descriptor to describe the closed 3D contours. This method possesses rotation-invariance. In addition, Sijbers et al. [58] proposed an efficient method to calculate the 3D Fourier descriptor.

3.6.2 Spherical Harmonic Analysis

Vranić [54] first introduced harmonic analysis into the field of 3D model feature extraction, which is a rotation-relevant feature descriptor. Kazhdan et al. [59] improved this scheme, making it rotation irrelevant. The key idea of this approach is to describe a spherical function in terms of the amount of energy it contains at different frequencies. Since these values do not change when the function is rotated, the resulting descriptor is rotation invariant. This approach can be viewed as a generalization of the Fourier Descriptor method to the case of spherical functions. The detailed procedure can be described as follows.

3.6.2.1 Spherical Harmonics

In mathematics, spherical harmonics are the angular portion of a set of solutions to Laplace's equation. Represented in a system of spherical coordinates, Laplace's spherical harmonics are a specific set of spherical harmonics which forms an orthogonal system, first introduced by Laplace. Spherical harmonics are important in many theoretical and practical applications, particularly in the computation of atomic orbital electron configurations, representation of gravitational fields, geoids and the magnetic fields of planetary bodies and stars, and characterization of cosmic microwave background radiation. In 3D computer graphics, spherical harmonics play a special role in a wide variety of topics including indirect lighting (ambient occlusion, global illumination, pre-computed radiance transfer, etc.) and in recognition of 3D shapes.

In order to represent a function on a sphere in a rotation invariant manner, Kazhdan et al. [59] utilized the mathematical notion of spherical harmonics to describe the way that rotations act on a spherical function. The theory of spherical harmonics says that any spherical function $f(\theta, \phi)$ can be decomposed as the sum of its harmonics:

$$f(\theta, \phi) = \sum_{l=0}^{\infty} \sum_{m=-l}^{l} a_{lm} Y_l^m(\theta, \phi). \tag{3.45}$$

The harmonics are visualized in Fig. 3.20. The key property of this decomposition is that if we restrict it to some frequency l, and define the subspace of functions:

$$V_l = Span(Y_l^{-l}, Y_l^{-l+1}, \dots, Y_l^{l-1}, Y_l^{l}), \tag{3.46}$$

we then have the following two properties: (1) V_l is a representation for the rotation group: For any function $f \in V_l$ and any rotation R, we have $R(f) \in V_l$. This can also be expressed in the following manner: if π_l is the projection onto the subspace V_l then π_l commutes with rotations:

$$\pi_l(R(f)) = R(\pi_l(f)) \,. \tag{3.47}$$

(2) V_l is irreducible: V_l cannot be further decomposed as the direct sum $V_l = V_l' \oplus V_l''$, where V_l' and V_l'' are also (nontrivial) representations of the rotation group.

The first property presents a way for decomposing spherical functions into rotation invariant components, while the second property guarantees that, in a linear sense, this decomposition is optimal.

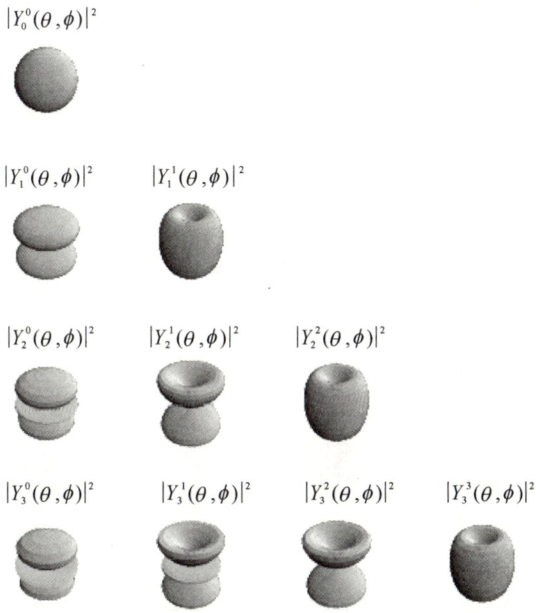

Fig. 3.20. Spherical harmonics

3.6.2.2 Rotation Invariant Descriptors

Using the properties of spherical harmonics and the observation that rotating a spherical function does not change its L_2-norm, we represent the energies of a spherical function $f(\theta, \phi)$ as:

$$SH(f) = \{\|f_0(\theta, \phi)\|, \|f_1(\theta, \phi)\|, \ldots\} \,, \tag{3.48}$$

where f_l is the frequency components of f as shown in steps (3) and (4) of Fig. 3.21:

$$f_l(\theta, \phi) = \pi_l(f) = \sum_{m=-l}^{l} a_{lm} Y_l^m(\theta, \phi) . \tag{3.49}$$

This representation has the property that it is independent of the orientation of the spherical function. To see this, we let R be any rotation and we have:

$$\begin{aligned} SH(R(f)) &= \{\|\pi_0(R(f))\|, \|\pi_1(R(f))\|, \ldots\} \\ &= \{\|R(\pi_0(f))\|, \|R(\pi_1(f))\|, \ldots\} \\ &= \{\|\pi_0(f)\|, \|\pi_1(f)\|, \ldots\} = SH(f), \end{aligned} \tag{3.50}$$

so that applying a rotation to a spherical function f does not change its energy representation.

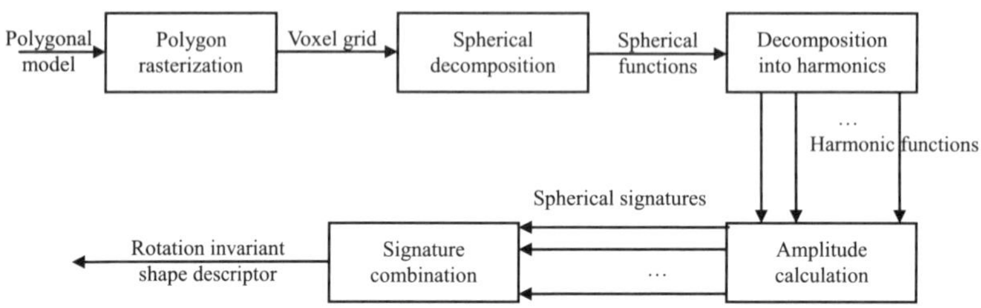

Fig. 3.21. Spherical harmonics analysis based feature extraction

3.6.2.3 Further Quadratic Invariance

Kazhdan et al. [59] made their representation still more discriminating by refining the case of the second order component. It can be proved that the L_2-difference between the quadratic components of two spherical functions is minimized when the two functions are aligned with their principal axes. Thus, instead of describing the constant and quadratic components by the two scalars $\|f_0\|$ and $\|f_2\|$, Kazhdan et al. [59] represented them by the three scalars a_1, a_2, and a_3, where after alignment to principal axes:

$$f_0 + f_2 = a_1 x^2 + a_2 y^2 + a_3 z^2 . \tag{3.51}$$

However, care must be taken because as functions on the unit sphere, x^2, y^2, and z^2 are not orthonormal. By fixing an orthonormal basis $\{\boldsymbol{v}_1, \boldsymbol{v}_2, \boldsymbol{v}_3\}$ for the span of $\{x^2, y^2, z^2\}$, the harmonic representation $SH(f)$ defined above can be replaced with the more discriminating representation:

$$SHQ(f) = \{\boldsymbol{R}^{-1}(a_1,\ a_2,\ a_3),\ \|f_1\|,\ \|f_3\|,\ \ldots\},\tag{3.52}$$

where $\boldsymbol{R}$ is the matrix whose columns are the orthonormal vectors $\boldsymbol{v}_i$.

3.6.2.4 Extensions to Voxel Descriptors

In order to obtain a rotation invariant representation of a voxel grid, Kazhdan et al. [59] used the observation that rotations fix the distance of a point from the origin. Thus, Kazhdan et al. [59] restricted the voxel grid to concentric spheres of different radii, and obtained the spherical harmonic representation of each spherical restriction independently. This process is shown in Fig. 3.21. First, Kazhdan et al. restricted the voxel grid to a collection of concentric spheres. Then they represented each spherical restriction in terms of its frequency decomposition. Finally, they computed the norm of each frequency component at each radius. The resultant rotation invariant representation is a 2D grid indexed by radius and frequency.

The method described above loses information as a result of the fact that the representation is invariant to independent rotations of the different spherical functions. For example, the plane in Fig. 3.22(b) is obtained from the one on the left by applying a rotation to the interior part of the model in Fig. 3.22(a). While the two models are not rotations of each other, the descriptors obtained are the same.

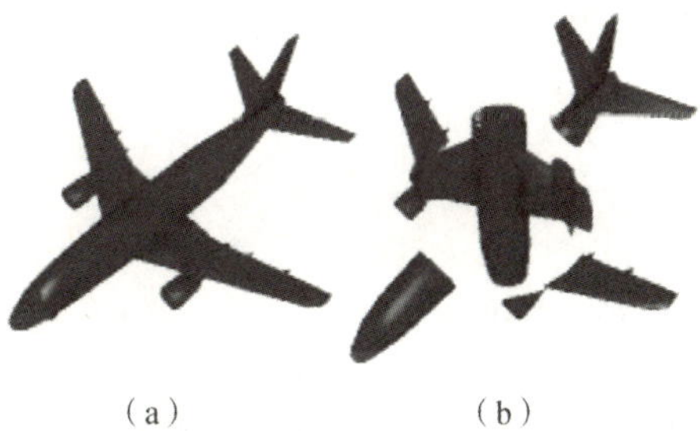

Fig. 3.22. The model (a) obtained by applying a rotation to the interior part of the model (b). While the models differ by more than a single rotation, their rotation invariant representations are the same [59] (With courtesy of Kazhdan et al.)

3.6.3 Wavelet Transform

A wavelet can also be used to describe the features of 3D models. Laga et al. [60] for the first time applied the spherical wavelet transform (SWT) to content-based 3D model retrieval. They proposed three new descriptors, i.e., spherical wavelet coefficients as feature vector (SWC_d), L_1 energy of the spherical wavelet coefficients ($SWEL_1$) and L_2 energy of the spherical wavelet coefficients ($SWEL_2$). They found that the sensitivity of the latitude-longitude parameterization to rotations of the North Pole affects the rotation invariance of the shape descriptors. Based on this fact, they proposed a new parameterization method based on regular octahedron sampling. Then they proposed three new spherical wavelet-based shape descriptors. The SWC_d takes into account the localization and local orientations of the shape features, while the $SWEL_1$ and $SWEL_2$ are compact and rotation invariant. The following is the detailed description of Laga et al.'s scheme.

3.6.3.1 Spherical Wavelets for 3D Shape Description

Let us first consider the problem of descriptor extraction from the spherical shape function. Wavelets are basis functions which represent a given signal at multiple levels of detail, called resolutions. They are suitable for sparse approximations of functions. In the Euclidean space, wavelets are defined by translating and dilating one function called mother wavelet. In the S^2 space, however, the metric is no longer Euclidean. Schröder and Sweldens [61] introduced the second generation wavelets. The idea behind this was to build wavelets with all desirable properties adapted to much more general settings than real lines and 2D images. The general wavelet transform of a function λ is constructed as follows.

Analysis (forward transform):

$$\begin{aligned} \lambda_{j,k} &= \sum\nolimits_{l \in K(j)} \tilde{h}_{j,k,l} \lambda_{j+1,l}, \\ \gamma_{j,m} &= \sum\nolimits_{l \in M(j)} \tilde{g}_{j,m,l} \lambda_{j+1,l}; \end{aligned} \tag{3.53}$$

Synthesis (backward transform):

$$\lambda_{j+1,l} = \sum\nolimits_{k \in K(j)} h_{j,k,l} \lambda_{j,k} + \sum\nolimits_{m \in M(j)} g_{j,m,l} \gamma_{j,m}, \tag{3.54}$$

where $\lambda_{j,\cdot}$ and $\gamma_{j,\cdot}$ are respectively the approximation and the wavelet coefficients of the function at resolution j. The decomposition filters $\tilde{h}$, $\tilde{g}$, and the synthesis filters h, g denote spherical wavelet basis functions. The forward transform is performed recursively starting from the shape function $\lambda = \lambda_{n,\cdot}$ at the finest resolution n, to get $\lambda_{j,\cdot}$ and $\gamma_{j,\cdot}$ at level j, $j = n-1, \ldots, 0$. The coarsest approximation $\lambda_{n-i,\cdot}$ is obtained after i iterations ($0 < i \le n$). The sets $M(j)$ and $K(j)$ are index sets on the sphere such that $K(j) \cup M(j) = K(j+1)$, and $K(n) = K$ is the index set at the finest resolution.

To analyze a 3D model, Laga et al. first applied spherical wavelet transform (SWT) to the spherical shape function and collected the coefficients to construct discriminative descriptors. The properties and behavior of the shape descriptors are therefore determined by the spherical wavelet basis functions used for transformation. Similar to 3D Zernike moments and spherical harmonics, the desired properties of a descriptor should be: (1) invariance to a group of transformations; (2) orthonormality of the decomposition; (3) completeness of the representation. The orthonormality ensures that the set of features will not contain redundant information. The completeness property implies that we are able to reconstruct approximations of the signal from the decomposition. The SW basis function should reflect these properties.

In Laga et al.'s work, they experimented with the second generation wavelets [61] including the linear and butterfly spherical wavelets with lifting scheme and image wavelets with spherical boundary extension rules. In their experiments on the Princeton Shape Benchmark, they found that the performance of both the linear and butterfly spherical wavelets is very low (comparable to shape distribution based descriptors). Therefore, they decided to use the image-based wavelet with spherical boundary extension rules to build their shape descriptors.

The image wavelet transform uses separable filters, so at each step it produces an approximation image A and three detail images HL, LH, and HH. The forward transformation algorithm, as illustrated in Fig. 3.23, is performed as follows: (1) initialization: (a) generate the geometry image I (the function f) of size $w \times h = 2^{n+1} \times 2^n$; (b) $A^{(n)} \leftarrow f$, $l \leftarrow n$. (2) forward transform: repeat the following steps until $l = 0$: (a) apply the forward spherical wavelet transform on $A^{(l)}$, obtaining the approximation $A^{(l-1)}$, and the detail coefficients $C^{(l-1)} = \{LH_{l-1}, HL_{l-1}, HH_{l-1}\}$ of size $2^l \times 2^{l-1}$. (b) $l \leftarrow l-1$. (3) collect the coefficients: the approximation $A^{(0)}$ and the coefficients $C^{(0)}, \ldots, C^{(n-1)}$ are collected into a vector $\boldsymbol{F}$.

Laga et al. experimented with the Haar wavelets, where the scaling function is designed to take the rolling average of the data, and the wavelet function is designed to take the difference between every two samples in the signal. They pointed out that another wavelet basis can also be used but requires a further investigation.

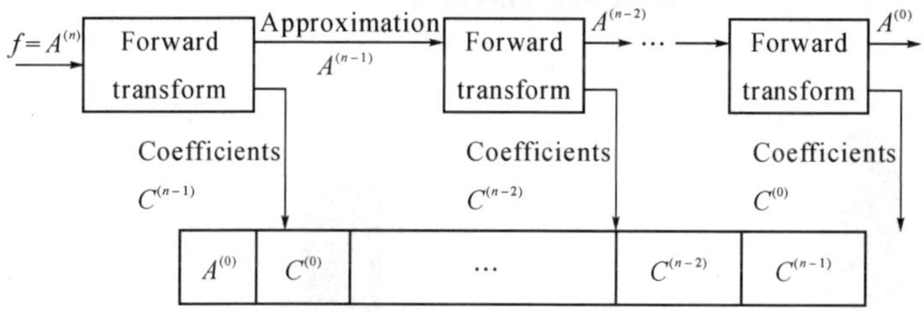

Fig. 3.23. Computation of spherical wavelet-based shape descriptors

3.6.3.2 Spherical Wavelet-Based Descriptors

Laga et al. proposed three methods to compare 3D shapes using their spherical wavelet transform: (1) wavelet coefficients as a shape descriptor (SWC_d) where the shape signature is built by considering directly the spherical wavelet coefficients, and (2) spherical wavelet energies: $SWEL_1$ based on the L_1 energy, and (3) $SWEL_2$ based on L_2 energy of the wavelet sub-bands. Fig. 3.24 shows an example model and its three different *SW* descriptors. The following parts detail each method.

(1) Wavelet coefficients as a shape descriptor. Once the spherical wavelet transform is performed, one may use the wavelet coefficients as the shape descriptor. Using the entire coefficients is computationally expensive. Instead, we can choose to keep the coefficients up to level d. Thus the obtained shape descriptor is called SWC_d, where $d = 0, \ldots, n-1$. In Laga et al.'s implementation, they used $d = 3$, therefore they obtained two dimensional feature vectors $\boldsymbol{F}$ of size $N = 2^{d+2} \times 2^{d+1} = 32 \times 16$.

Comparing directly wavelet coefficients requires efficient alignment of the 3D model prior to wavelet transform. A popular method for finding the reference coordinate frame is to pose normalization based on principal component analysis (PCA) as described in Section 3.2. During the preprocessing, they used the maximum area technique to resolve the positive and negative directions of the principal axis. Fig. 3.24 shows the SWC_3 descriptor extracted on the 3D "tree" model. Note that the vector $\boldsymbol{F}$ can provide an embedded multi-resolution representation for 3D shape features. This approach performs as a filtering of the 3D shape by removing outliers. A major difference with spherical harmonics is that SWT preserves the localization and orientation of local features. However, a feature space of dimension 512 is still computationally expensive.

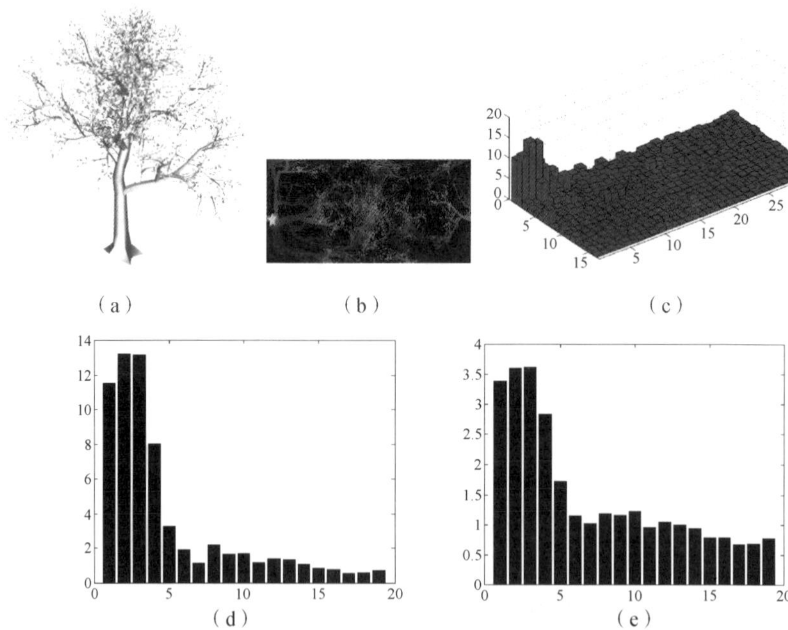

Fig. 3.24. Example of the "tree" model with its spherical wavelet-based descriptors [60]. (a) 3D shape; (b) Associated geometry image; (c) Spherical wavelet coefficients as descriptor (SWC_3); (d) L_2 energy descriptor ($SWEL_2$); (e) L_1 energy descriptor ($SWEL_1$) (©[2006]IEEE)

(2) Spherical wavelet energies. The wavelet energy signatures have been proven to be very powerful for texture characterization in [62]. Commonly the L_2 and L_1 norms are used as measures:

$$F_l^{(2)} = \left(\frac{1}{k_l} \sum_{j=1}^{k_l} x_{l,j}^2 \right)^{\frac{1}{2}}, \tag{3.55}$$

$$F_l^{(1)} = \frac{1}{k_l} \sum_{j=1}^{k_l} \| x_{l,j} \|, \tag{3.56}$$

where $x_{l,j}$ ($j = 1, 2, \ldots, k_l$) are the wavelet coefficients of the l-th wavelet sub-band. Using the observation that rotating a spherical function does not change its energy, Laga et al. proposed to adopt it to build general rotation invariant shape descriptors. For this purpose, they performed $n-1$ decompositions, then they computed the energy of the approximation $A^{(1)}$ and the energy of each detailed sub-band $HV^{(l)}$, $VH^{(l)}$ and $HH^{(l)}$ yielding a 1D shape descriptor $\boldsymbol{F} = \{F_l\}$, $l = 0, \ldots, 3\times(n-1)$ of size $N = 3\times(n-1)+1$. In Laga et al.'s case, they adopted $n = 7$, therefore $N = 19$. Laga et al. referred to L_1 energy and L_2 energy-based descriptors by $SWEL_1$ and $SWEL_2$ respectively.

The main benefits of this descriptor are its compactness and its rotation invariance. Therefore, the storage and computation time required for comparison are reduced. Since Laga et al. adopted the rotation invariant sampling method in [60], the shape descriptors invariant to general rotations can be obtained. However, similar to the power spectrum, information such as feature localization is lost in the energy spectrum.

Note that the above spherical wavelet analysis framework supports retrieval at different acuity levels. In some situations, only the main structures of the shapes are required for comparison while, in others, fine details are essential. In the former case, shape matching can be performed by considering only the wavelet coefficients on large scales while, in the later, coefficients on small scales are used. Hence, the flexibility of the developed method benefits different retrieval requirements. Finally, Table 3.1 summarizes [60] the length of the proposed descriptors. E-measure means the expected number of failures detected. discounted cumulative gain (DCG) measures the usefulness, or gain, of a document based on its position in the result list, and the gain is accumulated cumulatively from the top of the result list to the bottom, with the gain of each

Table 3.1 Performance of SW descriptors on the PSB base test classification [60] (©[2006]IEEE)

	Length	NN	1st-tier	2nd-tier	E-measure	DCG
SWC_d	512	46.9	31.4	39.7	20.5	65.4
$SWEL_1$	19	37.3	27.6	35.9	18.6	62.6
$SWEL_2$	19	30.3	24.9	31.5	16.1	59.4

Values of the length are in bytes, others are in (%). The length refers to the dimension of the feature space result discounted at lower ranks. From this table we can see that the $SWEL_1$ and $SWEL_2$ are more efficient in terms of storage requirement and comparison time, and they are also rotation invariant.

3.7 Visual-Image-Based Feature Extraction

Visual-image-based methods establish a functional mapping from the original 3D model to a predefined domain, typically several representative 2D planar views with reduced dimensions. This has long been studied in 3D engineering design and CAD communities, and has become one of the popular means to extract 3D shape signatures. The projections of a 3D model in all viewing directions are significant in 3D analysis. Visual-image-based feature extraction methods transform the complicated 3D problems into relatively mature image processing techniques to reduce the difficulty. At the same time, this kind of method is in accordance with the human visual system, thus the retrieval performance is better than other kinds of methods. However, for any single 3D model, it is necessary to extract features from several 2D images, thus a great deal of storage space and long executive time are required, and thus the retrieval efficiency is lower. Currently, many 3D model feature extraction methods based on projections have been proposed in the literature, where several 2D functional projections of a 3D model or 2D planar views from different perspectives are generated and combined as shape or silhouette feature descriptors. In this section we will introduce them in the following two categories.

3.7.1 Methods Based on 2D Functional Projection

The 2D functional projection reduces the 3D matching problem into a 2D case without computing multiple views of the object. The following are some typical methods in this category.

3.7.1.1 Spin Images

Johnson et al. [63] proposed a spin image representation, i.e., a 2D descriptive image associated with a sampling vertex set on a 3D surface, for which both the position and direction information are involved. The *x* and *y* coordinate values of the 2D spin image are defined as the accumulated values of two different distance functions of the 3D vertices, and the correlation coefficient between two spin images is computed as the similarity measure. However, since a 3D model usually consists of many surfaces, there is a large set of spin images generated for each 3D model. To achieve more concise and compact feature representation, the original set of spin images is compressed by the PCA method.

3.7.1.2 Geometry Images

Gu et al. and Praun et al. [64, 65] discussed the "geometry image" concept, a simple 2D array of quantized points with useful attributes, such as vertex positions, surface normals and textures. In fact, in Chapter 2 we have introduced the concept of geometry images. Laga et al. [66] applied this method to 3D shape matching by simplifying the 3D matching problem to measure similarities between parameterized 2D geometry images. All those methods make use of specific 3D geometry information from a 3D model in their 2D mapping process.

3.7.1.3 2D Slicing

Pu et al. [67] presented an approach based on 2D slices for measuring similarities between 3D models. The key idea is to represent the 3D model by a series of slices, as shown in Fig. 3.25, along certain directions so that the shape-matching problem between 3D models is transformed into similarity measuring between 2D slices. However, the following three problems should be solved: selection of cutting directions, cutting methods and similarity measuring. To solve these problems, some strategies and rules are proposed in [67]. Firstly, a maximum normal distribution method is presented to get three orthoaxes that coincide better with the human visual perception mechanism. Secondly, a cutting method is given which can be used to get a series of slices composed of a set of closed polygons. Thirdly, a 2D shape distribution method is developed to measure the similarity between the 2D slices. This scheme arises from such a fact as described in Fig. 3.25: Since 3D mesh models can be cut into a series of slices with polygon contours, why not the reverse? Could the reverse procedure be used to do shape matching? In this figure, the middle shape consists of 33 slices, while the right one consists of 100 slices. It is shown that a 3D model can be reconstructed precisely by overlapping a series of slices that represent the local contour of a 3D model. The larger the number of slices is, the more precise the final 3D model will be. For the detailed process, readers can refer to [67].

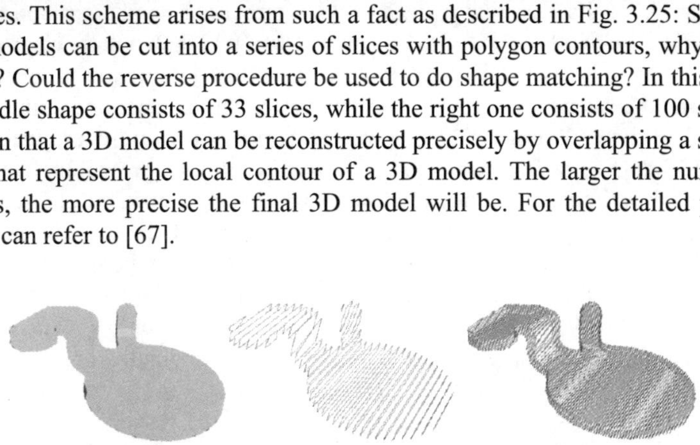

Fig. 3.25. Slice-based shape representation, where the shape on the right is reconstructed with more slices than the middle one [67] (©[2004]IEEE)

3.7.1.4 Harmonic Shape Images

Zhang et al. [68] reduced the 3D surface matching problem to a 2D image matching problem by employing the harmonic map theory [69], which studies the boundary mappings between different metric manifolds in terms of the energy-minimization principle. This representation scheme is called harmonic shape images that are used to represent the shape of 3D free-form surfaces that are originally represented by triangular meshes. Given a 3D surface patch with disc topology and a selected 2D planar domain, a harmonic map is constructed by a two-step process that includes boundary mapping and interior mapping. Under these mappings, there is one to one correspondence between the points on the 3D surface patch and the resultant harmonic image. Using this correspondence relationship, harmonic shape images are created by associating shape descriptors computed at each point of the surface patch at the corresponding point in the harmonic image. As a result, harmonic shape images are 2D shape representations of the 3D surface patch.

The detailed process to generate the harmonic shape images can be introduced as follows. Given a 3D surface S as shown in Fig. 3.26(a), let $\boldsymbol{v}$ denote an arbitrary vertex on S. Let $D(\boldsymbol{v}, R)$ denote the surface patch which has the central vertex $\boldsymbol{v}$ and radius R. R is measured by distance along the surface. $D(\boldsymbol{v}, R)$ is assumed to be a connected region without holes. $D(\boldsymbol{v}, R)$ consists of all the vertices in S whose surface distances are less than, or equal to, R. The overlaid region in Fig. 3.26(a) is an example of $D(\boldsymbol{v}, R)$. Its amplified version is shown in Fig. 3.26(b). The unit disc P on a 2D plane is selected to be the target domain. $D(\boldsymbol{v}, R)$ is mapped onto P by minimizing an energy functional. The resultant image $HI(D(\boldsymbol{v}, R))$ is called the harmonic image of $D(\boldsymbol{v}, R)$ as shown in Fig. 3.26(c). As can be seen in Figs. 3.26(a) and (c), for every vertex on the original surface patch $D(\boldsymbol{v}, R)$, one and only one vertex corresponds to it in the harmonic image $HI(D(\boldsymbol{v}, R))$. Furthermore, the connectivities among the vertices in $HI(D(\boldsymbol{v}, R))$ are the same as that of $D(\boldsymbol{v}, R)$. This means that the continuity of $D(\boldsymbol{v}, R)$ is preserved on the harmonic image $HI(D(\boldsymbol{v}, R))$. The preservation of the shape of $D(\boldsymbol{v}, R)$ is shown more clearly on the harmonic shape image $HSI(D(\boldsymbol{v}, R))$ (Fig. 3.26(d)) which is generated by associating the shape descriptor at every vertex on the harmonic image Fig. 3.26(c). The shape descriptor is computed at every vertex on the original surface patch Fig. 3.26(b). On $HSI(D(\boldsymbol{v}, R))$, high intensity values represent high curvature values and low intensity values represent low curvature values. The reason for harmonic shape images' ability to preserve the shape of the underlying surface patches lies in the energy function which is used to construct the mapping between a surface patch $D(\boldsymbol{v}, R)$ and the 2D target domain P. This energy function is defined to be the shape distortion when mapping $D(\boldsymbol{v}, R)$ onto P. Therefore, by minimizing the function, the shape of $D(\boldsymbol{v}, R)$ is maximally preserved on P. Another surface patch is shown in Figs. 3.26(e) and (f). Its harmonic image and harmonic shape image are shown in Figs. 3.26(g) and (h), respectively. In this case, there is occlusion in the surface patch as shown in Fig. 3.26(f). The occlusion is captured by its harmonic image and the harmonic shape image as shown in Figs.

3.26(g) and (h). The latter's ability to handle occlusion comes from the way the boundary mapping is constructed when mapping the boundary of $D(\boldsymbol{v}, R)$ onto the boundary of P. Because of the boundary mapping, the images remain approximately the same in the presence of occlusion. From the above generation process, it can be seen that the only requirement imposed on creating harmonic

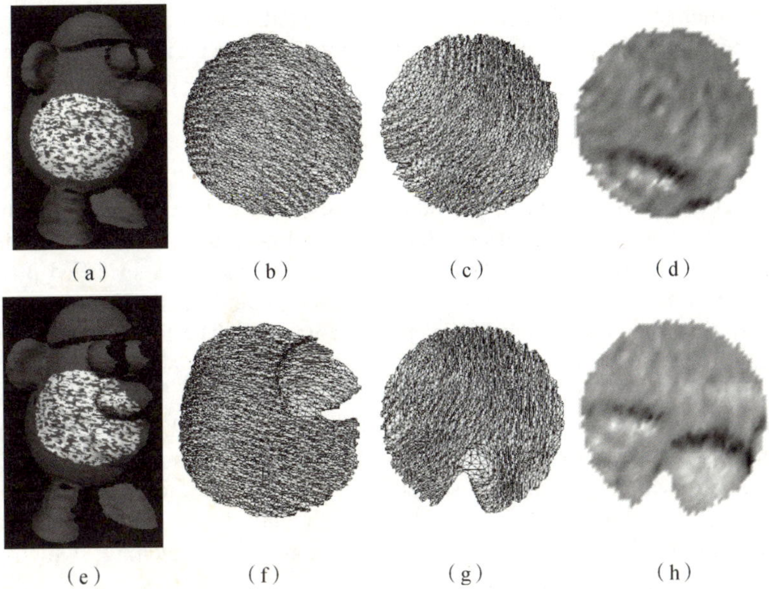

Fig. 3.26. Examples of surface patches and harmonic shape images [68]. (a), (e) Surface patches on a given surface; (b), (f) The surface patches in wireframe; (c), (g) Their harmonic images; (d), (h) Their harmonic shape images (With courtesy of Zhang and Hebert)

shape images is that the underlying surface patch is connected and without holes. This requirement is called the topology constraint.

Harmonic shape images have some properties that are important for surface matching. They are unique and their existence is guaranteed for any valid surface patches. More importantly, those images preserve both the shape and the continuity of the underlying surfaces. Furthermore, harmonic shape images are not designed specifically for representing surface shapes. Instead, they provide a general framework to represent surface attributes such as surface normal, color, texture and material. Harmonic shape images are discriminative and stable, and they are robust with respect to surface sampling resolution and occlusion. Extensive experiments have been conducted to analyze and demonstrate the properties of harmonic shape images in [68].

3.7.2 Methods Based on 2D Planar View Mapping

Compared with the methods in Subsection 3.7.1, the 2D mapping methods that establish mappings from a 3D view to a set of specific 2D planar views from different angles are much more natural and simple. The basic idea is that if two 3D shapes are similar, they should be similar from many different views. Thus, 2D shapes, such as 2D silhouettes, can be extracted and adopted for 3D shape matching. There is a prolific amount of literature on these particular techniques.

3.7.2.1 2D Boundary Information Based

Vranić et al. [70] presented a feature representation based on 2D boundary information, after having projected the 3D model onto three standard coordinate planes, i.e., *XY*, *XZ*, and *YZ* planes. For each projection on a specified plane, a silhouette is acquired by selecting contour points, equidistantly or equiangularly spanned; then the Fourier power spectrum is computed. The first n coefficients of the power spectrum are finally extracted as the feature. The drawback is the incapability of properly reflecting the 3D spatial information, since the 3D model is only viewed as a simple combination of three standard 2D projections, losing too much structure information. To solve this problem, Vranić et al. added depth information, which encoded the spatial distance difference of 3D surfaces into different gray values of their 2D projection images [70]. Also, they replaced contour-based 2D shape matching with region-based matching, which also increased the retrieval precision.

3.7.2.2 Aspect Graph

Cyr et al. [71] proposed an aspect-graph approach to represent 3D shapes, as shown in Fig. 3.27. First, 2D projection views are computed according to the view angles achieved after partitioning the viewing sphere by every 5°. Then, similar 2D projection views are clustered into the same group so as to generate a number of clusters called "aspect", from which the shape representation is created by selecting a representative view for each "aspect".

Similarly, Min et al. [72] projected each 3D model into several 2D silhouette images from m different viewpoints and then matched all their combinations with n ($m > n$) 2D sketches drawn by the user or the counterpart combinations of other 3D models. The similarity is measured as the minimal sum of all the pairwise sketch-to-image (or image-to-image) similar scores.

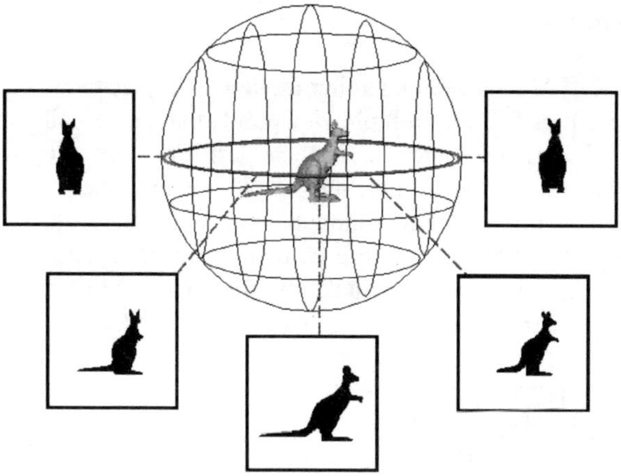

Fig. 3.27. Aspect graph [71] (©[2001]IEEE)

3.7.2.3 Light Field Descriptor

Chen et al. [73] proposed a light field descriptor representing the 4D light field of a 3D model with a collection of 2D images, which are captured by a set of uniformly distributed cameras by borrowing the concept of "light field" from image-based rendering. The cameras are controlled to rotate many times when measuring the similarity between descriptors of two 3D models, as shown in Fig. 3.28, so as to be switched onto their different vertices. The final 3D model retrieval results are combined from the matching results of all those acquired 2D images by integrating 2D Zernike moment and Fourier descriptors.

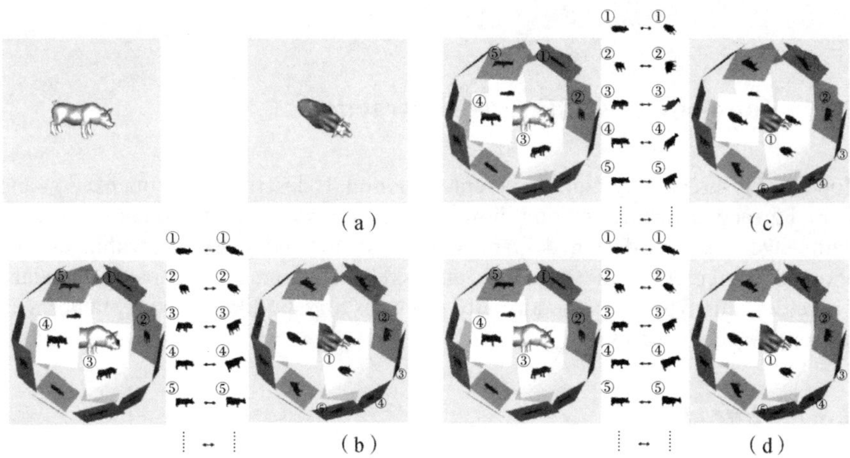

Fig. 3.28. (a)–(d) showing rotation and comparison in a light field [73] (With permission of Chen)

3.7.2.4 Depth Image

Ohbuchi et al. [74] presented a similar method. They generated a depth or z-value image of a 3D model from multiple viewpoints that are equally spaced on the unit sphere. The 3D model matching is then performed by adopting a 2D Fourier descriptor [70] for similarity matching of 2D images. The main difference is that Chen's 2D image only contains silhouettes while Ohbuchi's has depth information. Fig. 3.29 depicts Ohbuchi's feature extraction process. The depth image is first mapped from the Cartesian coordinate into the polar coordinate to perform Fourier transformation before Fourier descriptors are computed.

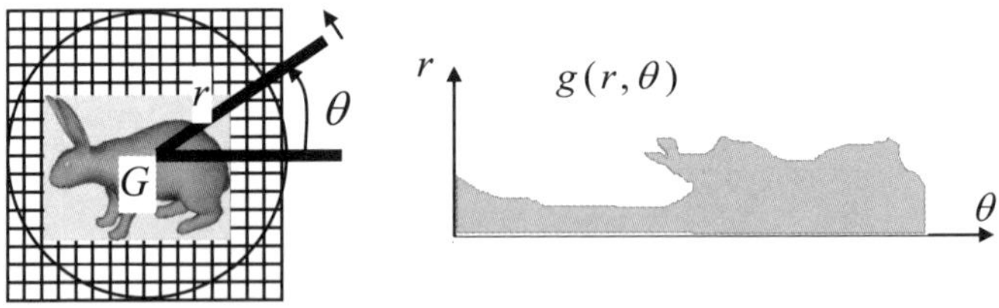

Fig. 3.29. Depth image

Since many more features can be extracted for a 2D shape, the function mapping methods make the retrieval process more flexible. They can also largely reduce the complexity of feature computation and make the feature descriptor more compact. However, this inevitably causes much loss of important 3D information, since the function mapping process is restricted by different constraints. Moreover, for 2D planar view mapping, how to decide the necessary number of 2D projection views is another problem in practice [71].

3.8 Topology-Based Feature Extraction

Topology is a relatively high-level representation. It describes the organization and spatial arrangement information: how vertices are connected to compose surfaces with edges. A well-designed graph data structure and graph algorithm can be adopted to represent the topology and skeleton characteristic of 3D models. Therefore, this type of method usually produces a graph-like structure, rather than numeric feature descriptors.

3.8.1 Introduction

Bardinet et al. [75] presented a structured 3D shape representation based on a 3D skeleton and medial axes, as an extension for the concept of 2D medial axis transform (MAT) [76]. First, adequate attributed relational graphs (ARGs), consisting of a set of nodes with attributes and a set of links are generated and the topological features are then extracted from the node and link structures of those graphs. Hilaga et al. [77] represented the topology of a 3D object as a Reeb graph using a function of "geodesic distance" [78] between points on the mesh. The Reeb graph is a skeleton representation using a continuous scalar function defined on an object with arbitrary dimensions [79].

Topology analysis can also be carried out by decomposing a 3D model as a parametric model of a set of simple elementary regular shapes. The topology is depicted as the spatial relationships and arrangements of those basic shapes, such as generalized cylinders [80], deformable regions [81], shock scaffold [82] and superquadrics [83]. Ma et al. [84] even presented a practical approach, using a model based on radial basis functions (RBFs) to extract 3D skeletons. For a 3D polygonal object, the vertices are treated as centers for RBF-level set construction and a gradient descent algorithm is employed on each vertex to locate the local maxima in the RBF. Finally, all the connected maxima pairs are handled using the Snake method and the final positions of the Snake sequences are extracted as the skeleton features. Tal et al. [85] first decomposed a mesh into elements called "watersheds" using a Watershed decomposition algorithm [86], then fit and classified them into four kinds of basic shapes: spherical surfaces, cylindrical surfaces, cone surfaces and planar surfaces. Next, the shape signature, an attributed decomposition graph, is constructed.

The topological and skeletal shape features are attractive for 3D retrieval because they are able to capture the significant shape structures of a 3D object. Meanwhile, they are relatively high-level and close to human intuitive perception, which makes them useful for defining more natural 3D query representation. They can also perform part-matching tasks for containing both local and global structural properties. For some kinds of topological representations, they are also robust against the LOD structure of 3D models, due to their multiresolution properties. However, 3D models are not always defined well enough to be easily and naturally decomposed into a canonical set of features or basic shapes. In addition, the decomposition process is usually computationally expensive. Moreover, model decomposition processes are quite noise-sensitive to small perturbations of the model. Thus, extra effort is, in turn, required to handle them. Finally, compared with the comparatively straightforward indexing and similarity matching algorithms based on numeric feature vectors, the indexing and matching algorithms for graph-like representations are relatively complex and time-consuming, due to the necessary graph searching processes. And, since there is currently no universal general-purpose graph matching solution, different graph matching algorithms need to be designed to accommodate different graph-like representations.

Here, we briefly introduce two typical methods, i.e., multi-resolution Reeb graph and skeleton graph.

3.8.2 Multi-resolution Reeb Graph

Hilaga et al. [77] proposed a novel technique, called topology matching, in which similarity between polyhedral models is quickly, accurately and automatically calculated by comparing multi-resolution Reeb graphs (MRGs). The basic idea of MRGs can be introduced as follows.

3.8.2.1 Reeb Graph

A Reeb graph is a topological and skeletal structure for an object of arbitrary dimensions. In topology matching, the Reeb graph is used as a search key that represents the features of a 3D shape. The definition of a Reeb graph is as follows:

Definition 3.1 (Reeb graph) *Let μ: $C \rightarrow R$ be a continuous function defined on an object C. The Reeb graph is the quotient space of the graph of μ in $C \times R$ by the equivalent relation $(X_1, \mu(X_1)) \sim (X_2, \mu(X_2))$ that holds if, and only if, (1) $\mu(X_1) = \mu(X_2)$ and (2) X_1 and X_2 are in the same connected component of $\mu^{-1}(\mu(X_1))$.*

When the function μ is defined on a manifold and critical points that are not degenerate, the function μ is referred to as a Morse function, as defined by Morse theory [87]. However, topology matching is not subject to this restriction.

It is clear that if the function μ changes, the corresponding Reeb graph also changes. Among the various types of μ and related Reeb graphs, one of the simplest examples is a height function on a 2D manifold. That is, the function μ returns the value of the z-coordinate (height) of the point $\boldsymbol{v}$ on a 2D manifold:

$$\mu(\boldsymbol{v}(x, y, z)) = z. \tag{3.57}$$

Most existing studies have used the height function as the function μ for generating the Reeb graph. Fig. 3.30 shows the distribution of the height function on the surface of a torus and the corresponding Reeb graph. In the left figure, the red and blue coloring represents minimum and maximum values, respectively, and the black lines represent the isovalued contours. The Reeb graph in the right figure corresponds to connectivity information for these isovalued contours.

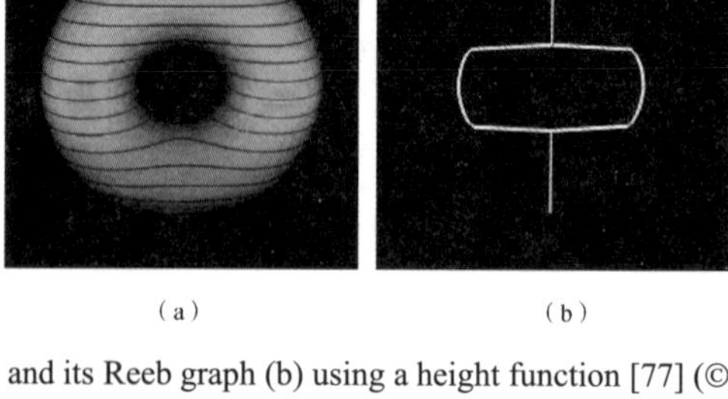

(a) (b)

Fig. 3.30. Torus (a) and its Reeb graph (b) using a height function [77] (©2001, Association for Computing Machinery, Inc. Reprinted by permission)

3.8.2.2 Multi-Resolution Reeb Graph

The basic idea of the MRG is to develop a series of Reeb graphs for an object at various levels of detail. To construct a Reeb graph for a certain level, the object is partitioned into regions based on the function μ. A node of the Reeb graph represents a connected component in a particular region, and adjacent nodes are linked by an edge if the corresponding connected components of the object contact each other. The Reeb graph for a finer level is constructed by re-partitioning each region. In topology matching, the re-partitioning is done in a binary manner for simplicity. Fig. 3.31 shows an example where a height function is employed as the function μ for convenience. In Fig. 3.31(a), there is only one region r_0 and one connected component s_0. Therefore, the Reeb graph consists of one node n_0 that corresponds to s_0. In Fig. 3.31(b), the region r_0 is re-partitioned into r_1 and r_2, producing connected components s_1 and s_2 in r_1, and s_3 in r_2. The corresponding nodes are n_1, n_2 and n_3 respectively. According to the connectivities of s_1, s_2 and s_3, edges are generated between n_1 and n_3, and also between n_2 and n_3. Finer levels of the Reeb graph are constructed in the same manner, as shown in Fig. 3.31(c). The MRG has the following properties:

Property 1 *There are parent-child relationships between nodes of adjacent levels. In Fig. 3.31, the node n_0 is the parent of n_1, n_2 and n_3, and the node n_1 is the parent of n_4 and n_6, etc.*

Property 2 *By repeating the re-partitioning, the MRG converges to the original Reeb graph as defined by Reeb. That is, finer levels approximate the original object more exactly.*

Property 3 *A Reeb graph of a certain level implicitly contains all of the information of the coarser levels. Once a Reeb graph is generated at a certain resolution level, a coarser Reeb graph can be constructed by unifying adjacent nodes. Consider the construction of the Reeb graph shown in Fig. 3.31(b) from that shown in Fig. 3.31(c) as an example. The nodes $\{n_4, n_6\}$ are unified to n_1, $\{n_5, n_7, n_8\}$ to n_2, and $\{n_9, n_{10}, n_{11}\}$ to n_3. Note that the unified nodes satisfy the parent-child relationship.*

Using the above three properties, MRGs are easily constructed and a similarity between objects can then be calculated using a coarse-to-fine strategy of different resolution levels as described in [77].

3.8.2.3 MRG Feature Extraction

MRG uses a continuous function based on the distribution of the geodesic distance, which is defined as follows:

$$\mu(\boldsymbol{v}) = \int_{p \in S} g(\boldsymbol{v}, p)\, \mathrm{d}S, \tag{3.58}$$

where $\boldsymbol{v}$ is a point on a surface S, and $g(\boldsymbol{v}, \boldsymbol{p})$ represents the geodesic distance between $\boldsymbol{v}$ and another point $\boldsymbol{p}$ on S, which is the length of the shortest path from $\boldsymbol{v}$

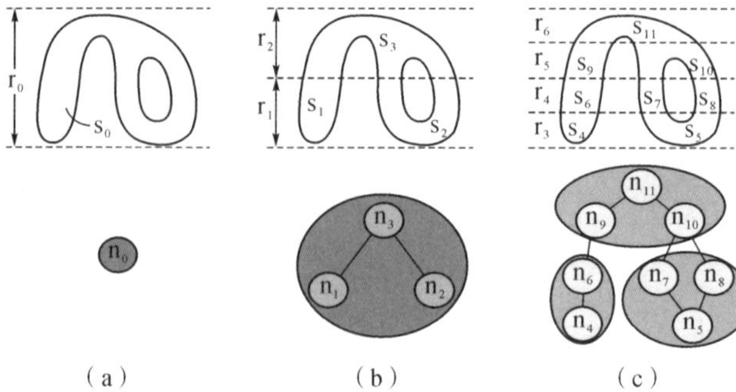

Fig. 3.31. Multi-resolution Reeb graph [77]. (a) With one node; (b) With three nodes; (c) With finer levels (©2001, Association for Computing Machinery, Inc. Reprinted by permission)

to $\boldsymbol{p}$. To produce the scaling invariance, a normalization step is also used and represented as follows:

$$\mu_n(\boldsymbol{v}) = \frac{\mu(\boldsymbol{v}) - \min_{p \in S} \mu(\boldsymbol{p})}{\max_{p \in S} \mu(\boldsymbol{p})}. \tag{3.59}$$

The MRG feature is invariant to translation and rotation and robust against changes in topology structure caused by a mesh simplification or subdivision. In consequence, it is discriminative of different levels of detail. However, MRG lacks the ability to correctly distinguish the corresponding parts of 3D models.

3.8.3 *Skeleton Graph*

In [88], Sundar et al. encoded the geometric and topological information in the form of a skeletal graph and used graph matching techniques to match the skeletons and to compare them. The skeletal graphs can be manually annotated to refine or restructure the search. This is a directed graph structure adopted to represent the skeleton of a 3D volumetric model [88], where an edge is directional according to a principle similar to a shock graph [89]. The skeleton is a nice shape descriptor because it can be utilized in the following ways:

(1) Part/Component matching. In contrast to a global shape measure, skeleton-matching can accommodate part-matching, i.e. whether the object to be matched can be found as part of a larger object, or vice versa. This feature can potentially give the users flexibility towards the matching algorithm, allowing them to specify what part of the object they would like to match or whether the matching algorithm should weight one part of the object more than another.

(2) Visualization. The skeleton can be used to register one object to another and visualize the result. This is very important in scientific applications where one is interested in both finding a similar object and understanding the extent of the

similarity.

(3) Intuitiveness. The skeleton is an intuitive representation of shape and can be understood by the user, allowing the user more control in the matching process.

(4) Articulation. The method can be used for articulated object matching, because the skeleton topology does not change during articulated motion.

(5) Indexing. We can index the skeletal graph for restricting the search space for the graph matching process.

The steps in the skeletal graph matching process include: obtaining a volume, computing a set of skeletal nodes, connecting the nodes into a graph, and then indexing into a database and/or verification with one of more objects. The results of the match are then visualized. Here we focus on the construction of the skeleton and preliminary results of using the graph matching in conjunction with skeletonization.

The term skeleton has many meanings. It generally refers to a "central-spine" or "stick-figure" like the representation of an object. The line is centered within the 3D/2D object. For 2D objects, the skeleton is related to the medial-axis of the 2D picture. For 3D objects a medial surface is computed. To use graph matching what is needed is a medial core/skeleton also known as a curve-skeleton which can be represented as a graph. The method utilized in [88] is a parameter-based thinning algorithm. This algorithm thins the volumes to a desired threshold based on a parameter given by the user. A family of different point sets can be obtained, each one thinner than its parent. This point set, termed skeletal voxels, is unconnected and must be connected to form an appropriate stick-figure representation. In what follows, we describe the various steps necessary to compute the skeleton/graph representation.

First, a volumetric cube is thinned into a skeletal-graph, a line-like sketch composed of the points on the medial axis of the medial surface planes. Then a clustering algorithm is implemented on the thinned voxels to increase the robustness against small perturbations on the surface and to reduce the number of graph nodes. An undirected acyclic graph is first generated out of the skeletal points by applying the minimum spanning tree (MST) algorithm. After that, the directed graph is finally constructed by directing the edge from a voxel with the higher distance to the one with the lower distance. Here the distance means the minimum distance from a voxel to the boundary of the volumetric object. Fig. 3.32 shows two examples of skeletal graphs.

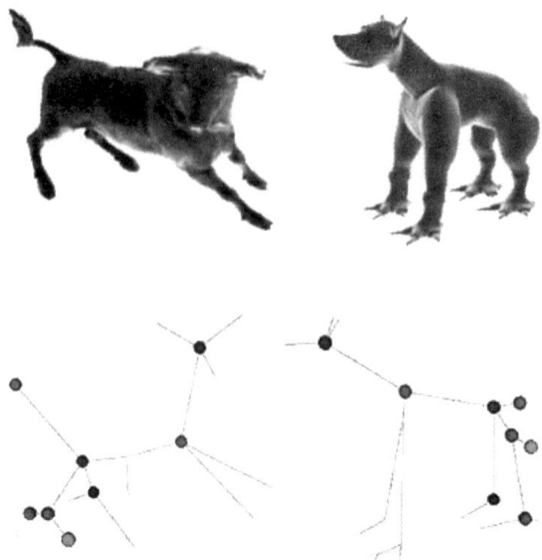

Fig. 3.32. Sample skeletal graphs: In the upper row, different volumes are shown. At the bottom are the resulting skeletal graphs [88] (©[2003]IEEE)

3.9 Appearance-Based Feature Extraction

The 3D models usually possess multimodal feature descriptors. Besides the shape features, the appearance attributes of 3D models such as material color, color distribution and texture, are also an important part of content-based 3D model retrieval. In particular, color and texture databases are necessary to render 3D models.

3.9.1 Introduction

In many practical applications, 3D appearance features, such as smoothness, roughness and distribution of light, might also be of interest, so that 3D model databases may also need to be searched according to the selected appearance properties. Besides, the visual perception of the geometry of the human being is indeed influenced by color, and separate colors are often analyzed as distinct entities in a human's visual system.

However, there are still insufficient research data on the appearance representation and extraction methodologies, compared with the abundant literature of 3D shape representation and extraction. This is partly due to the diversity and complexity of appearance attributes. For example, the distribution and spatial relationship of colors in 2D images or videos can be successfully defined and represented, whereas in 3D models this is not the case. Therefore, the issues of appearance representation and measurement in 3D situations, particularly how to integrate appearance information into the shape descriptor, or how to directly derive feature descriptors from appearance information that can then be combined with traditional shape descriptors to comprehensively feature 3D models, and similarity measurements suitable for these appearance-contained feature descriptors, are necessary and require intensive study. Although some shape features also contain partial appearance information such as color and texture, for example geometry image and histograms, they are still too superficial to depict the 3D appearance attributes properly. Here we briefly introduce several color and texture feature extraction methods for 3D models as follows.

3.9.2 Color Feature Extraction

To date, the appearance representations adopted in 3D model retrieval are mostly related to surface colors or surface textures. Paquet et al. [31] presented a color feature extraction method by separately taking into account the material color and its luminosity: on the one hand, the material color attribute is described with a color histogram of each component of red, green, blue (RGB) color space; on the other hand, luminosity attributes are represented by employing seven different histograms of diffuse reflection coefficients, specular reflection coefficients and textures.

Suzuki et al. [90] presented a color feature extraction method from a different perspective, which is based on material colors. This method can retrieve 3D polygonal models according to colors by reflecting the user's preferences from some material color databases. It is believed that material colors greatly influence the appearance of 3D models. In a simple rendering model, material colors of a 3D model can be specified by ambient color, diffuse color, specular color, emissive color, shininess and transparency. Each material color item contains several light values. Since the shading model is given by equations with these light parameters, switching these values can generate a large number of different colors and change the appearance of objects. Hence, Suzuki et al. proposed a color extraction and matching method to handle the material color databases efficiently, based on the user's subjective evaluation scales. First, users are asked to evaluate and describe material colors for some portions of the database as a study dataset. User inputs are then analyzed and a multidimensional space is created, which reflects the user's personalized evaluations of material colors. To create a complete, personalized search space, a set of non-studied data is mapped into the multidimensional space. Since the light characteristics of each material color are known, coordinates of each material color can be predicted by using multiple-regression analysis. In that way, each material color can be represented and matched.

3.9.3 Texture Feature Extraction

Suziki et al. evaluated another appearance feature representation using the surface textures of 3D models where the higher order local autocorrelation (HLAC) masks are extracted as texture features [91].

2D HLAC has been used as a feature descriptor for various 2D image pattern recognition applications. It is well known that the autocorrelation function is shift-invariant. The N-th-order autocorrelation functions with N displacements $a_1, \ldots, a_N$ are defined by

$$x^N(m) = \int P^m_{<r>} P^m_{<r+a_1>} \cdots P^m_{<r+a_N>} \mathrm{d}r, \tag{3.60}$$

where the function $P^m_{<r>}$ denotes the m-th order PARCOR coefficient of pixel $<r> = <x, y>$.

Since the number of these autocorrelation functions obtained by the combination of the displacements over the PARCOR images P^m is enormous, we must reduce them for practical applications. First, we restrict the order N up to the second, i.e., $N = 0, 1, 2$. We also restrict the range of displacements within a local 3×3 window, the center of which is the reference point. By eliminating the displacements which are equivalent to the shift, the number of patterns of the displacements is reduced to 25.

Although the HLAC mask patterns were previously applied to 2D images, they have not been applied to 3D models or volume data. Suziki et al. extended 2D HLAC mask patterns to 3D HLAC mask patterns, and this method enables masks to extract features from 3D models. 3D HLAC mask patterns are generated by using a simulation program, and 251 patterns have been found that are about 10 times more than 2D HLAC mask patterns. By using these 3D HLAC mask patterns, the search system can perform efficient retrieval.

3.10 Summary

In this chapter, we have discussed six types of feature extraction methods for 3D models. It should be borne in mind that these methods are not absolutely independent and isolated. In fact, many of them are quite interdependent. The purpose of our taxonomy is to provide a rational and comprehensible classification and summarization of the existing research literature. Currently, most of the work on shape feature extraction places emphasis on geometrical and surface topological properties of 3D shape features, based on surfaces, voxels, vertex sets, and structural shape models. Generally, geometrical features usually represent the specific shape and spatial position of surfaces, edges and vertices, while topological features maintain the linking relationship between surfaces, edges and vertices.

The common characteristic of global-geometrical-analysis-based methods is that they are almost all derived directly from the elementary unit of a 3D model, that is the vertex, polygon, or voxel, and a 3D model is viewed and handled as a vertex set, a polygon mesh set or a voxel set. Their advantages lie in their easy and direct derivation from 3D data structures, together with their relatively good representation power. However, the computation processes are usually too time-consuming and sensitive for small features. Also, the storage requirements are too high due to the difficulties in building a concise and efficient indexing mechanism for them in large model databases.

The spherical mapping based methods produce invariant shape features, which avoids the time-consuming canonical coordinate normalization process in feature extraction. However, they also have some shortcomings. Firstly, it is generally assumed that a 3D model will have valid topology (for meshes), or explicit volume (for volumetric models), which cannot be guaranteed in practice. Secondly, the spherical function mapping process is complicated and time-consuming. Since many more features can be extracted for a 2D shape, the function mapping methods make the retrieval process more flexible. They can also largely reduce the complexity of feature computation and make the feature descriptor more compact. However, this inevitably causes much loss of important 3D information, since the function mapping process is restricted by different constraints. Moreover, for 2D planar view mapping, how to decide the necessary number of 2D projection views is another problem in practice.

Many statistical shape feature descriptors are simple to compute and useful for keeping invariant properties. In many cases they are also robust against noise, or the small cracks and holes that exist in a 3D model. Unfortunately, as an inherent drawback of a histogram representation, they provide only limited discrimination between objects: They neither preserve nor construct spatial information. Thus, they are often not discriminating enough to make small differences between dissimilar 3D shapes and usually fail to distinguish different shapes having the same histogram.

The topological and skeletal shape features are attractive for 3D retrieval because they are able to capture the significant shape structures of a 3D object. Meanwhile, they are relatively high-level and close to human intuitive perception, which makes them useful for defining more natural 3D query representation. They can also perform part-matching tasks by containing both local and global structural properties. For some kinds of topological representations, they are also robust against the LOD structure of 3D models due to their multiresolution properties. However, 3D models are not always defined well enough to be easily and naturally decomposed into a canonical set of features or basic shapes. In addition, the decomposition process is usually computationally expensive. Moreover, model decomposition processes are quite noise-sensitive to small perturbations of the model. Thus, extra effort is, in turn, required to handle them. Finally, compared with the comparatively straightforward indexing and similarity matching algorithms based on numeric feature vectors, the indexing and matching algorithms of graph-like representations are relatively more complex and time-consuming, due to the necessary graph searching processes. And, since there is currently no universal general-purpose graph matching solution, different graph matching algorithms need to be designed to accommodate different graph-like representations.

Finally, further development of non-shape descriptors of 3D models, such as material color and texture, is very important. Furthermore, extraction of high-level semantic features and similarity measurements, combined with semantic information, will also raise important research issues and challenges.

References

[1] Y. K. Lai, Q. Y. Zhou, S. M. Hu, et al. Robust feature classification and editing. IEEE Transactions on Visualization and Computer Graphics, 2007, 13(1):34-45.

[2] H. T. Ho and D. Gibbins. Multi-scale feature extraction for 3D models using local surface curvature. In: Digital Image Computing: Techniques and Applications (DICTA'2008), 2008, pp. 16-23.

[3] C. B. Akgül, B. Sankur, Y. Yemez, et al. Density-based 3D shape descriptors. EURASIP Journal on Advances in Signal Processing, 2007, pp. 1-16.

[4] C. Cui, D. Wang and X. Yuan. Feature extraction of 3D model based on fuzzy clustering. In: Proceedings of the SPIE, 2005, Vol. 5637, pp. 559-566.

[5] Y. Yang, H. Lin and Y. Zhang. Content-based 3-D model retrieval: A survey. IEEE Transactions on Systems, Man and Cybernetics-Part C: Appliactions and Reviews, 2007, 37(6):1081-1098.

[6] J. L. Martínez, A. Reina and A. Mandow. Spherical laser point sampling with application to 3D scene genetic registration. In: 2007 IEEE International Conference on Robotics and Automation, 2007, pp. 1104-1109.

[7] T. Hlavaty and V. Skala. A survey of methods for 3D model feature extraction. Bulletin of IV Seminar Geometry and Graphics in Teaching Contemporary Engineer, 2003, 13(3):5-8.

[8] K. Siddiqi, A. Shokoufandeh, S. J. Dickinson, et al. Shock graphs and shape matching. In: Proceedings of the IEEE International Conference on Computer Vision (ICCV'98), 1998, pp. 222-229.

[9] T. Tung and F. Schmitt. The augmented multiresolution Reeb graph approach for content-based retrieval of 3D shapes. International Journal of Shape Modeling, 2005, 11(1):91-120.

[10] H. Sundar, D. Silver, N. Gagvani, et al. Skeleton based shape matching and retrieval. In: Proceedings of the International Conference on Shape Modeling and Applications (SMI'03), 2003, pp. 130-139.

[11] S. Kang and K. Ikeuchi. The complex EGI: a new representation for 3-D pose determination. IEEE Transactions on Pattern Analysis and Machine Intelligence, 1993, 15(7):707-721.

[12] E. Paquet and M. Rioux. Nefertiti: a query by content software for three-dimensional models databases management. In: Proceedings of the 1st International Conference on Recent Advances in 3-D Digital Imaging and

Modeling (3DIM '97), 1997, pp. 345-352.
[13] M. Ankerst, G. Kastenmüller, H. P. Kriegel, et al. 3D shape histograms for similarity search and classification in spatial databases. In: Proceedings of the 6th International Symposium on Advances in Spatial Databases (SSD'99), 1999, Vol. 1651, pp. 207-226.
[14] T. Funkhouser, P. Min, M. Kazhdan, et al. A search engine for 3D models. ACM Transactions on Graphics, 2003, 22(1):83-105.
[15] R. Osada, T. Funkhouser, B. Chazelle, et al. Shape distributions. ACM Transactions on Graphics, 2002, 21(4):807-832.
[16] J. Daniels II, L. K. Ha, T. Ochotta, et al. Robust smooth feature extraction from point clouds. Paper presented at The IEEE International Conference on Shape Modeling and Applications (SMI'07), 2007, pp. 123-136.
[17] M. Pauly, R. Keiser and M. Gross. Multi-scale feature extraction on point-sampled surfaces. Computer Graphics Forum, 2003, 22(3):281-290.
[18] S. Gumhold, X. Wang and R. McLeod. Feature extraction from point clouds. Paper presented at The 10th International Meshing Roundtable, Sandia National Laboratories, 2001.
[19] K. Demarsin, D. Vanderstraeten, T. Volodine, et al. Detection of closed sharp feature lines in point clouds for reverse engineering applications. Report TW 458, Department of Computer Science, K.U. Leuven, Belgium, 2006.
[20] M. K. Hu. Visual pattern recognition by moment invariants. IRE Trans. Inf. Theory, 1962, 8(2):179-187.
[21] A. P. Ashbrook, N. A. Thacker, P. I. Rockett, et al. Robust recognition of scaled shapes using pairwise geometric histograms. In: Proc. BMVC, 1995, pp. 503-512.
[22] M. Elad, A. Tal and S. Ar. Content based retrieval of VRML objects-An iterative and interactive approach. In: Proc. 6th Eurograph. Workshop Multimedia, 2001, pp. 97-108.
[23] R. M. Duda and P. E. Hart. Pattern Classification and Scene Analysis. Wiley, 1973.
[24] N. Canterakis. 3-D Zernike moments and Zernike affine invariants for 3D image analysis and recognition. Paper presented at The 11th Scand. Conf. Image Anal., 1999.
[25] M. Novotni and R. Klein. 3D Zernike descriptors for content based shape retrieval. Paper presented at The 8th ACM Symp. Solid Model. Appl., 2003.
[26] M. Ankerst, G. Kastenmuller, H. Kriegel, et al. 3D shape histograms for similarity search and classification in spatial databases. In: Proc. Symp. Large Spatial Databases, 1999, pp. 207-226.
[27] P. Besl. Triangles as a primary representation. Object recognition in computer vision. Lecture Notes in Computer Science, Springer-Verleg, 1994, Vol. 1929, pp. 1191-1206.
[28] M. Novotni and R. Klein. A geometric approach to 3D object comparison. In: Proc. Int. Conf. Shape Model. Appl., 2001, pp. 167-175.
[29] R. Ohbuchi and T. Takei. Shape-similarity comparison of 3D models using alpha shapes. In: Proc. 11th Pacific Conf. Comput Graph. Appl. (PG 2003), 2003, pp. 293-302.
[30] H. Edelsbrunner and E. P. Mücke. Three-dimensional alpha shapes. ACM Trans. Graph., 1994, 13(1):43-72.

[31] E. Paquet and M. Rioux. A content-based search engine for VRML databases. In: Proc. IEEE Int. Conf. Comput. Vis. and Pattern Recognit., Santa Barbara, CA, USA, 1998, pp. 541-546.
[32] MPEG Video Group. MPEG-7 Visual Part of eXperimentation Model (version 9.0 ed.). Pisa, Italy, 2001.
[33] M. T. Suzuki, T. Kato and N. Otsu. A similarity retrieval of 3D polygonal models using rotation invariant shape descriptors. Paper presented at The IEEE International Conference on Systems, Man, and Cybernetics, 2000, pp. 2946-2952.
[34] R. Osada, T. Funkhouser, B. Chazelle, et al. Shape distributions. ACM Transactions on Graphics, 2002, 21(4):807-832.
[35] J. L. Shih, C. H. Lee and J. T. Wang. 3D object retrieval system based on grid D2. Electronics Letters, 2005, 41(4):179-181.
[36] J. J. Song and F. Golshani. Shape-based 3D model retrieval. In: Proc. 15th IEEE Int. Conf. Tools Artif. Intell., 2003, pp. 636-640.
[37] B. K. P. Horn. Extended Gaussian Image. In: Proc. of IEEE, 1984, 72(12):1671-1686.
[38] H. Luo, J. S. Pan, Z. M. Lu, et al. A new 3D shape descriptor based on rotation. Paper presented at The Sixth International Conference on Intelligent Systems Design and Applications (ISDA2006), 2006.
[39] R. Ohbuchi, T. Minamitani and T. Takei. Shape-similarity search of 3D models by using enhanced shape functions. International Journal of Computer Applications in Technology, 2005, 23(2/3/4):70-85.
[40] Z. M. Lu, H. Luo and J. S. Pan. 3D model retrieval based on vector quantization index histograms. Paper presented at The 4th International Symposium on Instrumentation Science and Technology (ISIST'2006), 2006.
[41] Y. Linde, A. Buzo and R. M. Gray. An algorithm for vector quantizer design. IEEE Trans. Communications, 1980, 28(1):84-95.
[42] L. Kolonias, D. Tzovaras, S. Malassiotis, et al. Fast content based search of VRML models based on shape descriptors. In: Proc. IEEE Int. Conf. Image Process., 2001, Vol. 2, pp. 133-136.
[43] D. V. Vranić and D. Saupe. 3D model retrieval. Paper presented at The Spring Conf. Comput. Graph. (SCCG 2000), 2000.
[44] MPEG Requirements Group. Overview of the MPEG-7 Standard. Doc. ISO/MPEG N3158, Maui, Hawaii, 1999.
[45] M. Yu, I. Atmosukarto, W. K. Leow, et al. 3D model retrieval with morphing-based geometric and topological feature maps. In: Proc. IEEE Conf. Comput. Vis. Pattern Recognit., 2003, pp. 656-661.
[46] J. Tangelder and R. Veltkamp. Polyhedral model retrieval using weighted point sets. Int. J. Image Graph., 2003, 3:1-21.
[47] Y. Rubner, C. Tomasi and L. J. Guibas. A metric for distributions with applications to image databases. Paper presented at The IEEE Int. Conf. on Computer Vision, 1998, pp. 59-66.
[48] J. Rossignac and P. Borrel. Multi-resolution 3D approximation for rendering complex scenes. Geometric Modeling in Computer Graphics, 1993, pp. 455-465.
[49] M. Heczko, D. Keim, D. Saupe, et al. A method for similarity search of 3D objects (in German). In: Proc. BTW, 2001, pp. 384-401.
[50] V. Cicirello and W. Regli. Machining feature-based comparisons of mechanical

parts. In: Proc. Int. Conf. Shape Model. Appl., 2001, pp. 176-185.
[51] D. McWherter, M. Peabody, W. Regli, et al. Transformation invariant shape similarity comparison of solid models. Paper presented at The ASME DETC, Pittsburgh, PA, 2001.
[52] C. Zhang and T. Chen. Efficient feature extraction for 2D/3D objects in mesh representation. Paper presented at The ICIP, 2001.
[53] D. Vranić and D. Saupe. 3D shape descriptor based on 3D Fourier transform. In: The EURASIP Conference on Digital Signal Processing for Multimedia Communications and Services, 2001, pp. 271-274.
[54] D. Vranić, D. Saupe and J. Richter. Tools for 3D-object retrieval: Karhunen-Loeve transform and spherical harmonics. In: Proc. IEEE 2001 Workshop Multimedia Signal Process, Cannes, France, 2001, pp. 293-298.
[55] K. Arbter, W. E. Snyder, H. Burkhardt, et al. Application of affine invariant fourier descriptors to recognition of 3-D objects. IEEE Trans. on Pattern Analysis and Machine Intelligence, 1990, 12(7):640-647.
[56] C. W. Richard and H. Hemami. Identification of 3D objects using Fourier descriptors of the boundary curve. IEEE Transactions on Systems, Man, and Cybernetics, 1974, 4(4):371-378.
[57] H. Zhang and E. Fiume. Shape matching of 3D contours using normalized Fourier descriptors. Paper presented at International Conference on Shape Modeling and Applications, 2002, pp. 261-271.
[58] J. Sijbers, T. Ceulemans and D. van Dyck. Efficient algorithm for the computation of 3D Fourier descriptors. Paper presented at The 1st International Symposium on 3D Data Processing Visualization and Transmission, 2002, pp. 640-643.
[59] M. Kazhdan, T. Funkhouser and S. Rusinkiewicz. Rotation invariant spherical harmonic representation of 3D shape descriptors. Paper presented at The Eurographics/ACM Siggraph Symposium on Geometry Processing, 2003, pp. 156-164.
[60] H. Laga, H. Takahashi and M. Nakajima. Spherical wavelet descriptors for content-based 3D model retrieval. Paper presented at The IEEE International Conference on Shape Modeling and Applications, 2006, pp. 15-25.
[61] P. Schroder and W. Sweldens. Spherical wavelets: efficiently representing functions on the sphere. In: SIGGRAPH'95: Proceedings of the 22nd Annual Conference on Computer Graphics and Interactive Techniques, 1995, pp. 161-172.
[62] G. van de Wouwer, P. Scheunders and D. van Dyck. Statistical texture characterization from discrete wavelet representations. IEEE Transactions on Image Processing, 1999, 8(4):592-598.
[63] A. Johnson and M. Hebert. Using spin images for efficient object recognition in cluttered 3D scenes. IEEE Trans. Pattern Anal. Mach. Intell., 1999, 21(5):433-449.
[64] X. Gu, S. Gortler and H. Hoppe. Geometry images. In: Proc. ACM Siggraph, 2002, pp. 355-361.
[65] E. Praun and H. Hoppe. Spherical parametrization and remeshing. In: Proc. SIGGRAPH, 2003, pp. 340-349.
[66] H. Laga, H. Takahashi and M. Nakajima. Geometry image matching for similarity estimation of 3D shapes. In: Proc. Comput. Graph. Int., Crete, Greece,

2004, pp. 490-496.
[67] J. Pu, Y. Liu, G. Xin, et al. 3D model retrieval based on 2D slice similarity measurements. Paper presented at The 2nd International Symposium on 3D Data Processing, Visualization and Transmission, 2004, pp. 95-101.
[68] D. Zhang and M. Hebert. Harmonic shape images: A 3D free-form surface representation and its applications in surface matching. In: Proc. Energy Minimization Methods Comput. Vis. Pattern Recognit., 1999, pp. 30-43.
[69] J. Eells and L. H. Sampson. Harmonic mappings of Riemannian manifolds. Amer. J. Math., 1964, 86:109-160.
[70] D. Vranić. 3D model retrieval. Ph.D Dissertation, Univ. Leipzig, Leipzig, Germany, 2004.
[71] C. Cyr and B. Kimia. 3D object recognition using shape similarity-based aspect graph. In: Proc. 8th IEEE Int. Conf. Comput. Vision., Vancouver, 2001, pp. 254-261.
[72] P. Min, A. Halderman, M. Kazhdan, et al. Early experiences with a 3D model search engine. In: Proc. Web3D Symp., 2003, pp. 7-18.
[73] D. Y. Chen. Three-dimensional model shape description and retrieval based on lightfield. Ph.D Dissertation, Dept. Compute. Sci. Inf. Eng., National Taiwan Univ., Taipei, Taiwan, 2003.
[74] R. Ohbuchi, M. Nakazawa and T. Takei. Retrieving 3D shapes based on their appearance. In: Proc. 5th ACM SIGMM, Int. Workshop Multimedia Inf. Retrieval, 2003, pp. 39-45.
[75] E. Bardinet, S. Vidal, S. Arroyo, et al. Structural object matching. Paper presented at The Adv. Concepts Intell. Vision Syst. (ACIVS 2000), 2000.
[76] H. Blum. Biological shape and visual science. J. Theoret. Biol., 1973, 38:205-287.
[77] M. Hilaga, Y. Shinagawa, T. Kohmura, et al. Topology matching for fully automatic similarity estimation of 3D shapes. In: Proc. SIGGRAPH, 2001.
[78] M. Sharir and A. Schorr. On shortest paths in polyhedral spaces. SIAM J. Comput., 1986, 15(1):193-215.
[79] G. Reeb. On the singular points of a completely integrable PfAFF form or of a numerical function (in French). Comptes Randus Acad. Sci., 1946, 222:847-849.
[80] T. Binford. Visual perception by computer. In: Proc. IEEE Conf. Syst. Sci., 1971.
[81] R. Basri, L. Costa, D. Geiger, et al. Determining the similarity of deformable shapes. Vis. Res., 1998, 38:2365-2385.
[82] F. Leymarie and B. Kimia. The shock scaffold for representing 3D shape. In: Proc. 4th Int. Workshop Visual Form, 2001, pp. 216-228.
[83] Y. Zhang, A. Koschan and M. Abidi. Superquadrics based 3D object representation of automotive parts utilizing part decomposition. In: Proc. SPIE 6th Int. Conf. Qual. Control Artif. Vis., 2003, Vol. 5132, pp. 241-251.
[84] W. Ma, F. Wu and M. Ouhyoung. Skeleton extraction of 3D objects with radial basis functions. In: Proc. Shape Model. Int., 2003, pp. 207-216.
[85] A. Tal and E. Zuckerberger. Mesh retrieval by components. Paper presented at The Int. Conf. Comput. Graph. Theory Appl., 2006.
[86] J. C. Serra. Image Analysis and Mathematical Morphology (1st ed.). Academic, 1982.
[87] Y. Shinagawa, T. L. Kunii and Y. L. Kergosien. Surface coding based on Morse theory. IEEE Computer Graphics and Applications, 1991, 11(5):66-78.

[88] H. Sundar, D. Silver, N. Gagvani, et al. Skeleton based shape matching and retrieval. In: Proc. Shape Model. Int., 2003, pp. 130-139.
[89] K. Siddiqi, A. Shokoufandeh, S. Dickinson, et al. Shock graphs and shape Matching. Comput. Vis., 1998, pp. 222-229.
[90] M. Suzuki. A web-based retrieval system for 3D polygonal models. In: Proc. Joint 9th IFSA World Congr. 20th NAFIPS (IFSA/NAFIPS 2001), 2001, pp. 2271-2276.
[91] M. Suzuki, Y. Yaginuma and Y. Shimizu. A texture similarity evaluation method for 3D models. In: Proc. Int. Conf. Internet Multimedia Syst. Appl. (IMSA 2005), 2005, pp. 185-190.

4

Content-Based 3D Model Retrieval

Rapid development in computer graphics and 3D modeling tools has resulted in an increasing number of 3D models. Furthermore, the rapid development of the Internet enables us to have access to 3D models created by people everywhere. As the number of available 3D models grows, people have an increasing demand to index and retrieve them based on their contents. This chapter discusses the steps and techniques involved in content-based 3D model retrieval systems.

4.1 Introduction

First, we introduce the background, performance evaluation criteria, the basic framework, challenges and several important issues related to content-based 3D model retrieval systems.

4.1.1 Background

If we view audio as the first wave of multimedia, images as the second wave of multimedia and video as the third wave of multimedia, then we can regard 3D digital models and 3D scenes as the fourth wave of multimedia. Unlike 2D images, 3D models are capable of overcoming the illusion problem caused by the human eye, and therefore object segmentation becomes less error-prone and easier to achieve. Modern computer technology and powerful computing capacity, together with new acquisition and new modeling tools, make it much easier and cheaper to create and process 3D models with basic hardware, resulting in an increasing number of 3D models from various sources, such as those over the Internet and those from professional 3D model databases in the areas of biology, medicine, chemistry, archaeology and geography, and so on. In the past two decades, tools for retrieving and visualizing complex 3D models have become an integrated part of data processing in the fields of medicine, chemistry, architecture and entertainment, and so on. This naturally results in an increasing demand for powerful retrieval tools, by which these large-scale and complicated new generation media can be easily organized and searched by users. In addition, modeling highly realistic 3D models is still a very laborious, high-cost and time-consuming process. If the currently available 3D models can be efficiently retrieved and reused, much less time and much less effort will be required to complete the modeling task. Thus, the request for retrieving the expected 3D models from a huge database is increasingly urgent.

"Content-based processing" is a preferred and popular scheme for processing multimedia data efficiently [1]. However, compared with the booming achievements in search engines or retrieval systems for 1D and 2D multimedia, the research and development of 3D model retrieval systems lag behind. Many websites only allow users to retrieve 3D models in quite a limited and primitive way, such as browsing a directory structure, performing a keyword-based search engine [2], or retrieving based on file types or file sizes [3]. Those traditional text-based search techniques are no longer effective for 3D models, as they suffer from problems such as low efficiency, low accuracy and high ambiguity. In addition, the most significant issue is that 3D models embody both shape and appearance information, which are hard to represent and query based merely on text keywords.

To address the above issues, the idea of retrieving 3D models in a "content-based" manner has already attracted considerable attention as a new hotspot in several research areas, such as computer vision, computer graphics, geometric modeling, pattern recognition, mechanical computer-aided design and molecular biology. This "content-based" scheme is now developing into a "content-based 3D model retrieval" methodology, concentrating on the representation, recognition and matching of 3D models on the basis of extraction and comparison of their intrinsic representative features, such as shapes, colors, texture and light distribution. A complete content-based 3D model retrieval system involves several aspects, i.e., preprocessing, feature extraction, similarity measures, query interface, model classification, indexing and retrieval quality evaluation. A large number of researchers have been absorbed in this area and have already made much progress. Many algorithms have been proposed and reported, and there has been a subsequent increase in the publication of academic papers and books related to this topic in a wide range of international journals and conferences. The new international standard MPEG-7 has also covered some 3D shape descriptors as one of its feature sets. In fact, content-based 3D model retrieval can be applied widely in many fields, such as CAD, cultural heritage applications, robotics, molecular biology, the virtual geography environment (VGE), 3D spatial terrain, medicine, chemistry, military and industrial manufacturing. It can also be potentially applied in e-business and web search engines in distributed data environments. There have been several survey papers on 3D model retrieval [4-6]. The core of a content-based 3D model retrieval system includes query interface, feature extraction and similarity measures. Designing algorithms for geometry similarity comparison is one of the most significant research aspects in 3D model retrieval systems, and has become one aspect of MPEG-7 standards [7]. The key problem in similarity comparison between two 3D models is to generate shape descriptors that can form an index conveniently and achieve geometry shape matching effectively. In general, 3D descriptors should hold the following four characteristics: transformation invariance, high-speed computing, convenient index structures and easy storage.

4.1.2 Performance Evaluation Criteria

To compare and evaluate the effectiveness of 3D model retrieval algorithms, i.e., how well the system meets the users' demand, the investigation of retrieval performance evaluation is essential in content-based 3D model retrieval.

4.1.2.1 3D Model Benchmark Databases

Since there are many kinds of specialized 3D models in different domains, the relevant research work, including versatile shape representations and similarity measures, may also affect the retrieval task in different ways. As a result, when considering the performance evaluation issue, the first response will be to define a relatively common and general-purpose 3D model collection as a benchmark database, in order to define a common method to provide relevance judgments.

Currently, there are several representative 3D model databases for the purpose of performance evaluation, among which the Princeton Shape Benchmark (PSB) [8] is maybe the most popular and well-organized one. PSB is a publicly available 3D model benchmark database containing 1,814 classified 3D models, which have been collected from the Internet and organized into hierarchical semantic classifications by experts. PSB provides us with separate training and test sets, and each 3D model has a set of annotations. Fig. 4.1 shows some samples from the PSB.

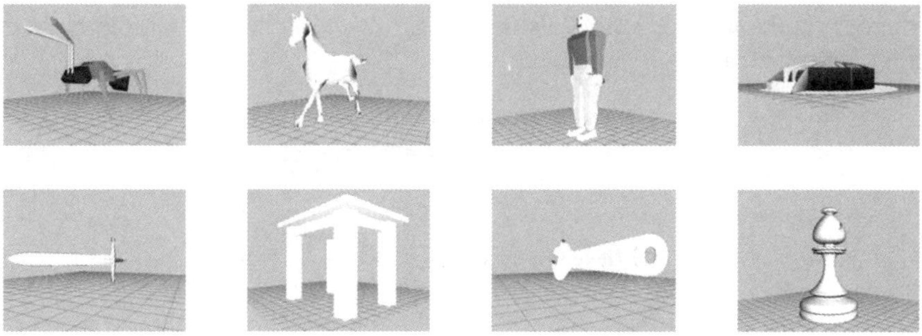

Fig. 4.1. Samples selected from the PSB

Besides the PSB, some other 3D model databases, which contain a wide variety of 3D objects that have been independently gathered by different research groups, can also be employed as standard benchmarks. These include the Utrecht databases [9], MPEG-7 databases [10] and Taiwan databases [11]. In addition, there are also several benchmark databases constructed for specific domains, e.g., CAD models [12] and 3D protein structures [13]. To know more detailed statistics for most currently available 3D model databases, readers can refer to [8].

Unfortunately, since most 3D model databases primarily focus on 3D shapes, there are currently no standard benchmark databases constructed for appearance attributes, such as color, texture and light distribution. Although the PSB can partially perform this function, it is still neither ideal nor optimal.

4.1.2.2 Performance Evaluation Methods

The two most common evaluation measures adopted in 3D model retrieval are precision and recall, which were introduced from the information retrieval (IR) community and have been widely employed to evaluate image retrieval systems [14]. Given a query model belonging to the category *C*, precision measures the ability of the system to retrieve models from *C*, thus precision can be defined as follows:

$$precision = \frac{N_{rc}}{N_r}, \tag{4.1}$$

where N_{rc} is the number of retrieved models belonging to *C* and N_r is the number of retrieved models. On the other hand, recall measures how many relevant models are retrieved to answer a query, thus recall is defined as

$$recall = \frac{N_{rc}}{N_c}, \tag{4.2}$$

where N_c is the number of models in *C*. Fig. 4.2 shows the relationship between precision and recall.

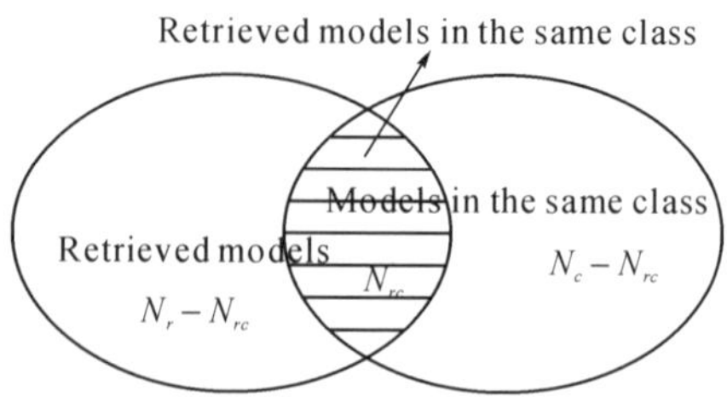

Fig. 4.2. Illustration of the relationship between precision and recall

In general, recall and precision are in a trade-off relationship. If one goes up, the other usually comes down. As the standard database is designed for similarity-based search, on the one hand if the similarity matching criteria are rather strict, then the precision value and the recall value go in opposite directions. On the other hand, if the matching criteria are too loose, most retrieved 3D models are useless.

Precision and recall can be separately used to evaluate the retrieval performance, e.g., the graph of precision vs. the number of retrieved models, or the graph of recall vs. the number of retrieved models. They can be also combined as a "precision-recall" (P-R) graph [15], which shows how precision falls and how recall rises as more and more 3D objects are retrieved. Fig. 4.3 gives a vivid example of achieving the P-R graph. Here, we assume that there are five 3D models in the same class as the query model, i.e. $N_c = 5$. With an increase in retrieved models, the precision value decreases but the recall value increases. The closer the precision value is to 1, the better the performance is obtained. Moreover, the performance can also be evaluated from some other aspects based on the P-R graph, such as effectiveness and robustness [16]. However, since "relevant" and "irrelevant" are both judged subjectively by users, this evaluation is naturally subjective.

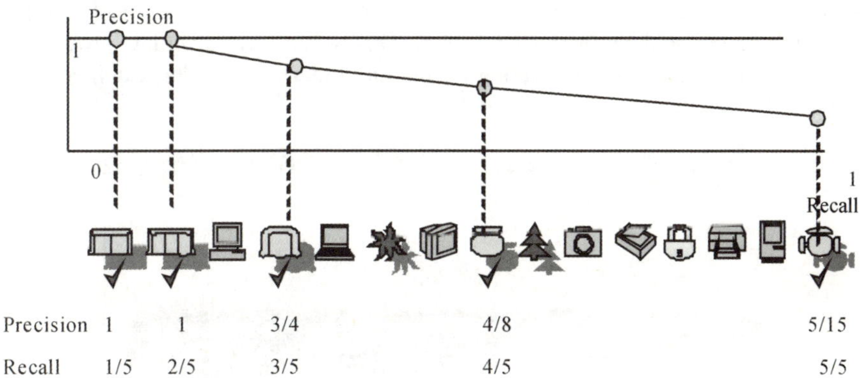

Fig. 4.3. Illustration of P-R graph calculation

Besides the P-R graph, to integrate the precision and recall criteria, another commonly-used criterion is called F_1 score [17]. In statistics, the F_1 score (also F-score or F-measure) is a measure of a test's accuracy, which considers both the precision and the recall of the test. The F_1 score can be interpreted as a weighted average of the precision and recall values, where an F_1 score reaches its best value at 1 and its worst score at 0. The traditional F-measure or balanced F-score (F_1 score) is the harmonic mean of precision and recall values, which can be defined as follows:

$$F_1 = \frac{2 \times precision \times recall}{precision + recall} \times 100\%. \tag{4.3}$$

In fact, the general formula for non-negative real β is given as

$$F_\beta = \frac{(1+\beta^2)\times precision \times recall}{\beta^2 \cdot precision + recall} \times 100\% \,. \tag{4.4}$$

The F-score is often used in the field of information retrieval to measure performance for search, document classification and query classification algorithms. Earlier works focused primarily on the F_1 score, but with the proliferation of large scale search engines, performance goals changed to lay heavy emphasis on either precision or recall and so F_β is seen in wide applications.

There are also some other performance evaluation methods used for 3D model retrieval. For example, a similarity matrix measurement is presented as a graphical performance evaluation, by which a matrix with higher contrast is usually rated either very similar or very dissimilar according to specifically designed criteria [18]. Many other types of evaluation measures, such as "best matches", "distance image" and "tier image" have also been proposed in [8] as follows.

The measure of "best matches" is a web page for each model displaying images of its best matches in rank order. Here, each image is the 2D effect picture of a 3D model from a certain view angle. The associated rank and distance value appears below each image, and images of models in the query model's class (hits) are highlighted with a thickened frame. This simple visualization provides a qualitative evaluation tool emulating the output of many 3D model search engines. A typical example is shown in Fig. 4.4.

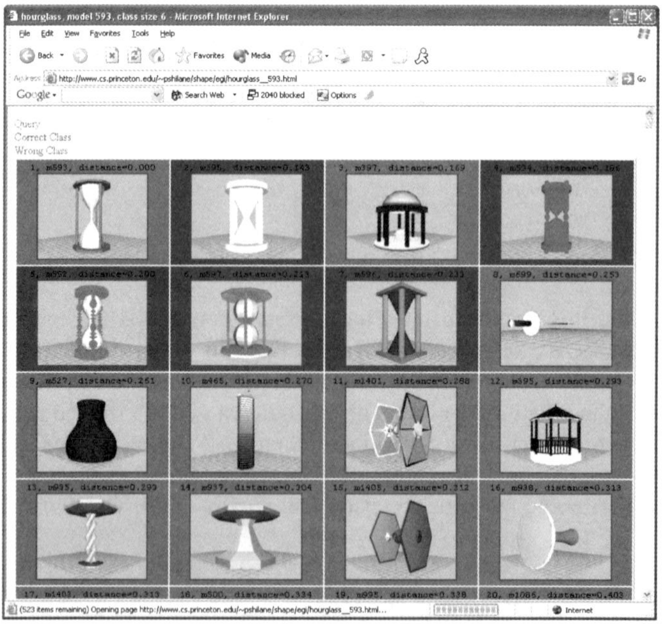

Fig. 4.4. A typical "best matches" evaluation measure (With courtesy of Shilane et al.)

The measure of "distance image" is an image of the distance matrix where the lightness of each pixel (i, j) is proportional to the magnitude of the distance between models $\boldsymbol{M}_i$ and $\boldsymbol{M}_j$ [19]. Models are grouped by class along each axis, and lines are added to separate classes, which makes it easy to evaluate patterns in the match results qualitatively. The optimal result is a set of darkest, class-sized blocks of pixels along the diagonal indicating that every model matches the models within its class better than those in other classes. Otherwise, the reasons for poor match results can often be seen in the image, e.g., off-diagonal blocks of dark pixels indicate that two classes match each other well. A typical example is shown in Fig. 4.5.

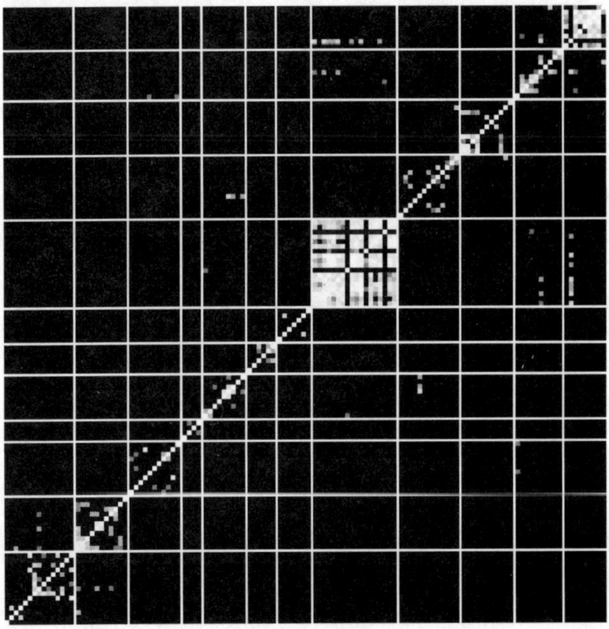

Fig. 4.5. A typical "distance image" evaluation measure (With courtesy of Shilane et al.)

The measure of "tier image" is an image visualizing nearest neighbor, first tier and second tier matches [19]. Specifically, for each row representing a query with the model $\boldsymbol{M}_j$ in a class with $|C|$ members, the pixel (i, j) is white if the model $\boldsymbol{M}_i$ is just the model $\boldsymbol{M}_j$ or its nearest neighbor, yellow if the model $\boldsymbol{M}_i$ is among the $|C| - 1$ top matches (i.e., the first tier) and golden if the model $\boldsymbol{M}_i$ is among the $2{\cdot}(|C| - 1)$ top matches (i.e., the second tier). Similar to the distance image, models are grouped by class along each axis, and lines are added to separate classes. However, this image is often more useful than the distance image because the best matches are clearly shown for every model, regardless of the magnitude of their distance values. The optimal result is a set of white/yellow, class-sized blocks of pixels along the diagonal indicating that every model matches the models within its class better than those in other classes. Otherwise, more colored pixels in the class-sized blocks along the diagonal represent a better result. A typical example is shown in Fig. 4.6.

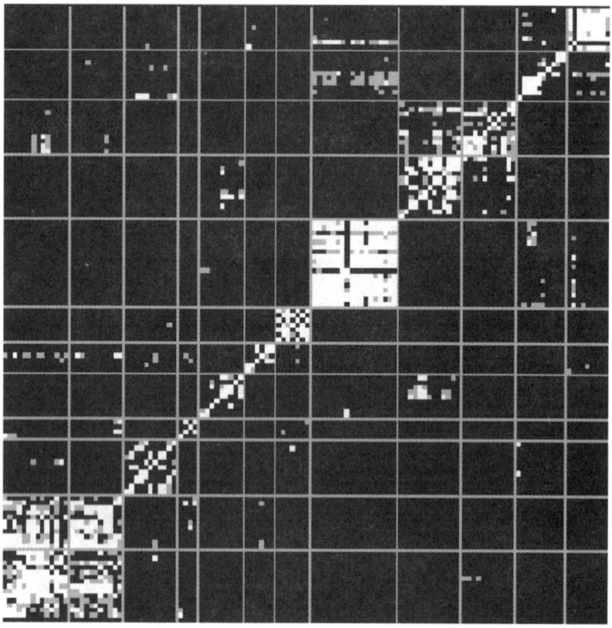

Fig. 4.6. A typical "tier image" evaluation measure (With courtesy of Shilane et al.)

4.2 Content-Based 3D Model Retrieval Framework

Then, we analyze and discuss several topics for content-based 3D model retrieval, including preprocessing, feature extraction, similarity matching and query interfaces.

4.2.1 Overview of Content-Based 3D Model Retrieval

The essential processing flow of a content-based 3D model retrieval system can be roughly described as follows: the compact and representative features, such as geometric shapes, spatial and topological relationships, statistical properties, textures and material attributes, are first computed and extracted automatically from 3D models to build their multidimensional indices. The similarity or dissimilarity measure between a query and each target model in the database is then defined and calculated in the multidimensional feature space. The similarity values are then sorted in descending order so that the models having the largest similarity values are returned as the matching results, on the basis of which browsing and retrieval in 3D model databases are finally implemented. Here, "content-based" means that the retriever utilizes the visual features of 3D models themselves, rather than relying on human-inputted metadata such as captions or keywords. The visual features of 3D models should be automatically or semi-automatically extracted and expected to characterize their contents.

The ultimate aim of content-based 3D model retrieval systems is to approximate human visual perception so that semantically similar 3D models can be correctly retrieved based on their looks. However, most of the existing types of 3D feature extraction methods which can be termed "low-level similarity-induced semantics," capture some, but not all, aspects of the content from a 3D model, and do not coincide with the high-level semantics it contains. As shown in Fig. 4.7, a sphere-like shape feature alone can be used to describe either a 3D ball or a 3D model of the globe. This is the well-known "semantic gap" issue [20] that indicates the relatively limited descriptive power of low-level visual features for approaching human subjective high-level perception. Therefore, high-level feature extraction methods that can derive semantics from low-level features should also be integrated as an important part in the content-based 3D model retrieval system. If 3D shape content is to be extracted in order to understand the meaning of a 3D model, the only available independent information is the low-level geometry data, connectivity data and surface appearance data. Annotations always depend on the knowledge, capability of expression and specific language of the annotator. They are therefore unreliable. To recognize the displayed scenes from the raw data of a model, the algorithms for selection and manipulation of vertices must be combined and parameterized in an adequate manner and finally linked with the natural description. Even the simple linguistic representation of shape or texture mapping, such as round or yellow, requires entirely different mathematical formalization methods, which are neither intuitive nor unique or sound.

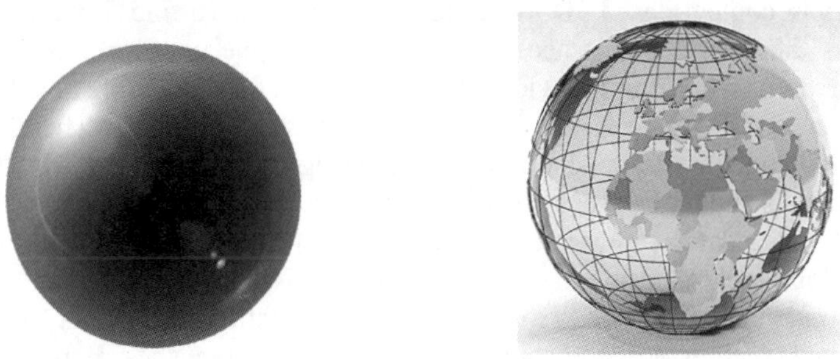

(a) (b)

Fig. 4.7. A sphere-like shape can be used to describe (a) a 3D ball or (b) a 3D model of the globe

4.2.2 Challenges in Content-Based 3D Model Retrieval

The new intricacies existing in 3D models have led to new challenges in content-based 3D model retrieval. These challenges can be listed as follows [4].

First, building accurate features for 3D models is more difficult and time-consuming than for other multimedia. 3D models embodying more complex and excessive poses than 2D media, i.e., with different translations, rotations, scales and reflections. This fact makes 3D models possess many more arbitrary and unpredictable positions, orientations and measurements and makes them difficult to be parameterized and searched. However, it is essential to search 3D models in an invariant manner with respect to translation, rotation, scaling and reflection. Therefore, in many cases, more additional alignment or pose registration processes may be required to align 3D models to their canonical coordinate systems. Otherwise, more complicated mappings or transformations may be performed to extract invariant features of a 3D model before a similarity match, which are time-consuming, computing-intensive and unstable.

Second, the diversity of 3D shape representations may obstruct the implementation of simple convenient and efficient 3D model retrieval systems. Up to now, there is no single common 3D shape format that serves as a standard. As we know, 3D models are usually represented with two types of data: geometric data and appearance attributes. Geometric data have a wide variety of representations including vertex data, surface data, volumetric data, solid structures, parametric surfaces, polygon meshes, implicit surfaces, volumetric arrays of voxel grids, or just unstructured "polygon-soup" and point clouds. Appearance attributes may contain material types, material colors, transparency, reflection coefficients and texture mapping. Due to the diversity of 3D representations, most of the currently available 3D model matching algorithms merely depend on 3D shape properties based on some specific data formats. How to hurdle the unnecessary complexity and ineffective matching induced by the format diversity issue is one of the major challenges in content-based 3D model retrieval systems. To find feasible solutions to address these issues, it is necessary to develop some new types of high-level descriptors to provide a unified view of the perceptual understanding of a 3D model. Nevertheless, a 3D model usually lacks high-level semantic clues. Therefore, it is also a challenge to establish an effective bridge between low-level 3D data representations and high-level semantic descriptions.

Third, 3D data representations have been well-designed for efficient visualization tasks, resulting in many problems for feature indexing and similarity comparison. For example, some 3D representations are not inherently well-defined, such as polygon soups and some unclosed meshes. Here, a polygon soup is just a list of triangles and has no inherent structure, like collision proxies or a height field. This makes it less efficient to collide against, but more easy for the users and content creators as they do not have to keep a certain structure in mind. However, it is difficult and ineffective to transform these into well-defined representations before feature extraction. Therefore, to accept "polygon soup" and other ill-defined 3D models is a further challenge for 3D model retrieval systems [21].

Finally, 3D models embody both considerable appearance attributes and complex geometric properties, which greatly increase the amount of information. In addition, the dimension of 3D data is also too high to be processed effectively and efficiently. Moreover, the multiresolution feature representations should be effectively generated in order that they are robust against different levels of detail of 3D model representations [22].

4.2.3 Framework of Content-Based 3D Model Retrieval

From the point of view of the conceptual level, a typical 3D model retrieval system framework as shown in Fig. 4.8 consists of a database with an index structure created offline and an online query engine [6]. This system generally consists of four main components: 1) the model preprocessing module for pose registration, noise removing and so on; 2) the feature extraction module for generating both low-level 3D shapes or appearance features and high-level semantic features; 3) the similarity matching phase, i.e., the relevance ranking procedure according to calculated similarity degrees; 4) the query interface, i.e., a practical online user interface designed to represent and process user queries. In general, a 3D model retrieval procedure is performed in four steps: indexing, querying, matching and visualizing. Except the first step that is done off-line, the remaining three steps are performed online to deal with each user query that supports input modes based on text, 3D sketches or 3D model examples, 2D projections and 2D sketches. For each of these input modes, the relevant shape descriptors are extracted from the 3D database models during the offline stage in order that they can be compared with the queries efficiently in the online phase. These shape descriptors provide a compact overall description of each 3D model.

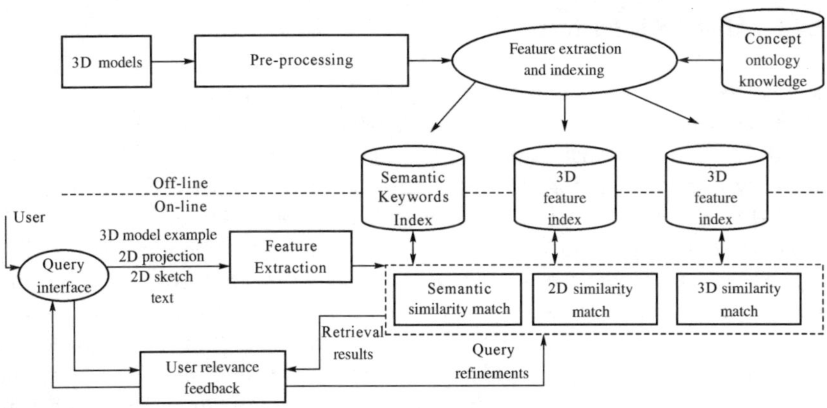

Fig. 4.8. Typical architecture framework of content-based 3-D model retrieval [6] (©[2008]IEEE)

To efficiently search large 3D model repositories online, an indexing data structure and an effective search algorithm should be well-designed. The online query engine computes the query descriptor, and then quantifies the similarity between the query descriptor and each shape descriptor in the database based on a specific similarity measure. The entire 3D model search engine allows a user to search for 3D models interactively, such as query methods based on text keywords, 2D sketching, 3D sketching, model matching and iterative refinement (i.e., relevance feedback). Min et al. [23] found that combining the results of text and shape matching can further improve the retrieval performance.

Different from the conventional 3D object recognition systems, which are usually performed at the cost of high computational complexity by establishing correspondences between a pair of 3D models and then comparing them, content-based 3D model retrieval systems are required to be performed on a "per-model" basis, which means that the feature used for matching should be calculated and stored independently of the target 3D models [24]. This also allows for the so-called "offline" feature extraction process because there is no demand to explicitly establish correspondences. Thus, during the genuine "online" retrieval phase, matching is performed by comparing the query's descriptor with each model's descriptor in the database. The feature of each 3D model from the database is extracted during the offline stage to enable comparison with online queries later on.

4.2.4 Important Issues in Content-Based 3D Model Retrieval

Here we would like to address five important issues in content-based 3D model retrieval as follows.

4.2.4.1 Model File Format

The first important issue is the type of model file format that a model retrieval system can accept. Most of the 3D models provided over the Internet are meshes defined in a file format supporting visual appearance [25]. Currently, the commonly-used formats for 3D model retrieval include VRML, 3D studio, PLY, AutoCAD, Wavefront, Lightwave objects, etc. These 3D model files over the Internet are both in plain links as well as in compressed archive files. As VRML is designed to be used over the Internet, it is often kept in a non-compressed format. Thus, the most commonly used format for retrieval is the VRML format. Most 3D models are represented by "polygon soups" consisting of unorganized and degenerate sets of polygons. They are rarely manifold, most are even not self-consistent and seldom have any solid modeling information. By contrast, for volume models, many retrieval techniques depending on a properly defined volume can be applied.

4.2.4.2 Normalization

Without prior knowledge, most 3D model search systems need the normalization step before feature extraction. Typically, this step is just a conversion of 3D models into their canonical representations to guarantee that the corresponding shape descriptors are invariant to rotation, translation and scaling operations. The Principal Component Analysis (PCA) algorithm for pose registration is fairly simple and efficient [26]. There are also some similarity measures which are invariant under the rotation operation [27-29]. We will discuss in detail the normalization step in the next section.

4.2.4.3 Dissimilarity Measures

How to define the "dissimilarity" or "similarity" measure is significant in implementing the whole retrieval process. To measure how similar two objects are, we need to adopt a dissimilarity measure to computer the distance between two descriptors. Typically, in information retrieval, a similarity metric is defined and applied to search similar files, such as documents, images, audio and videos. In fact, the reciprocal of the distance between two descriptors can be viewed as the similarity measure between two models, i.e., a small distance means a large similarity or a small dissimilarity. We will discuss the similarity matching problem in Section 4.5.

4.2.4.4 Criteria for Shape Representation

In general, the shape of a 3D object is described by a feature vector that serves as a search key in the database. If an unsuitable feature extraction method had been used, the whole retrieval system would be useless. The criteria for shape representation have been shown in Chapter 3. For more detailed information, readers can refer to Subsection 3.1.2. The shape representation method that would satisfy all requirements does not probably exist. Nevertheless, some methods that try to find a compromise among ideal properties exist. We will overview the feature extraction problem in Section 4.4 and, for more detailed information, readers can refer to Chapter 3.

4.2.4.5 Index for Highly Efficient Search

In general, the index structure is adopted to avoid the sequential scan that may be time-consuming during similarity matching. Researchers have presented many index structures and algorithms for efficient querying in the high-dimensional space. For example, metric access methods are index structures that utilize the metric properties of the distance function (especially triangle inequality) to filter out zones of the search space [30], while spatial access methods are index structures especially designed for vector spaces that, together with the metric properties of the distance function, use geometric information to discard unlikely points from the space [31].

4.3 Preprocessing of 3D Models

Finally, advantages and disadvantages of several typical 3D model retrieval systems are compared and some future works are proposed.

4.3.1 Overview

In general, 3D models have arbitrary scales, orientations and positions in the 3D space. In many situations, we are required to normalize the size and orientation of a 3D model before feature extraction in order to represent it in a canonical coordinate system. The aim of the normalization step is to guarantee that the same feature representation can be properly extracted from the same 3D object with any different scale, position and orientation. This enables us to perform search and retrieval tasks on a "per-model" basis, without further alignment of 3D models to each other. At present, there are two schemes to realize such a "per-model"-based normalization [32]:

(1) The normalization technique to find a canonical coordinate frame based on methods similar to the Principle Component Analysis (PCA), also referred to as pose estimation or pose registration.

(2) The invariance-based technique to define and extract feature descriptors that possess the inherent invariance characteristics, so as not to change under any rigid transformations. The invariance-based approaches have been accorded increasing weight in recent research because of their robustness and simplicity.

However, invariance characteristics are not always complete and all-sided to represent a 3D model. Moreover, the computation of these feature descriptors is necessarily performed over a unit coordinate frame. Thus, to guarantee the descriptive power and robustness of the feature representations, canonical coordinate normalization, such as alignment and scaling, is also a necessary step before invariant feature extraction.

Besides the normalization process, performing some other preprocessing steps [27, 33, 34] on 3D models before feature extraction is also inevitable. These steps include the transformation between different 3D data representations (e.g., to transform polygon meshes into voxel grids), the partition of model units and vertex clustering, etc. In some 3D model retrieval systems, at the preprocessing stage, a set of reference models is selected from the database based on cluster analysis, and distances between database models and reference models are computed and stored. In the following sections, we would like to introduce four typical preprocessing steps, i.e., pose normalization, polygon triangulation, mesh segmentation and vertex clustering.

4.3.2 Pose Normalization

In the absence of prior knowledge, 3D models have arbitrary scales, orientations and positions in the 3D space. Consequently, a normalization stage is required to achieve invariance characteristics of feature descriptors, which corresponds to placing the 3D model into a canonical coordinate system. The following attributes provide useful data for normalizing 3D models for differences in translation, scale and orientation:

(1) Center of mass: the average (x, y, z) coordinates for all points on the surfaces of all polygons. These values can be used to normalize the models for translation-invariance.

(2) Scale: the average distance from all points on the surfaces of all polygons to the center of mass. This value can be used to normalize the models for isotropic scaling-invariance.

(3) Principal axes: the eigenvectors and associated eigenvalues of the covariance matrix obtained by integrating the quadratic polynomials $\boldsymbol{v}_i \cdot \boldsymbol{v}_j$, with $\boldsymbol{v}_i \in \{x, y, z\}$, over all points on the surfaces of all polygons. These axes can be used to normalize the models for rotation-invariance.

Here we introduce two typical pose normalization methods. One is the Principal Component Analysis (PCA) based method, which makes the resulting shape feature vector independent of translations and rotations as much as possible. The other is to find the only bounding box of a 3D model.

4.3.2.1 PCA-Based Pose Normalization

Principal component analysis involves a mathematical procedure that transforms a number of possibly correlated variables into a smaller number of uncorrelated variables called principal components. The first principal component accounts for as much of the variability in the data as possible, and each succeeding component accounts for as much of the remaining variability as possible. Depending on the application field, it is also named the discrete Karhunen–Loève transform (KLT), the Hotelling transform or proper orthogonal decomposition (POD). PCA was invented in 1901 by Karl Pearson [35]. Now it is mostly used as a tool in exploratory data analysis and for making predictive models. PCA involves the calculation of the eigenvalue decomposition of a data covariance matrix or singular value decomposition of a data matrix, usually after mean centering the data for each attribute. The results of a PCA are usually discussed in terms of component scores and loadings. PCA is the simplest kind of true eigenvector-based multivariate analysis. In general, its operation can be viewed as revealing the internal structure of the data in a way which best explains the variance in the data. If a multivariate dataset is visualized as a set of coordinates in a high-dimensional data space (1 axis per variable), PCA provides the user with a lower-dimensional picture, i.e., a "shadow" of this object when viewed from its (in some sense) most informative viewpoint. PCA is closely related to factor analysis, and indeed, some statistical packages deliberately conflate the two techniques. Actually, true factor analysis makes different assumptions about the underlying structure and solves eigenvectors from a slightly different matrix.

In 3D model normalization, the aim of PCA is to change the coordinate system axes to new ones which coincide with the directions of the three largest spreads of the point (i.e. vertex) distribution. The detailed steps can be described as follows:

Step 1: Translation. First, the model's center of mass should be shifted to the coordinate origin as follows:

$$I_1 = I - c = \{u \mid u = v - c, v \in I\}, \tag{4.5}$$

where I is the original 3D model's coordinate frame, I_1 is the new coordinate frame after translation and $\boldsymbol{c}$ is the model's centroid [32].

Step 2: PCA-based rotation. Next, PCA is used to determine the canonical coordinate axes of a 3D model, based on calculating the corresponding eigenvectors and the resulting diagonal matrix $\boldsymbol{R}$ of eigenvalues, decreasingly ordered by their values. The rotation transformation is represented as:

$$I_2 = \boldsymbol{R} \cdot I_1 = \{x \mid x = \boldsymbol{R} \cdot u, u \in I_1\}, \tag{4.6}$$

where I_1 is the 3D model's coordinate frame before rotation and I_2 is the new coordinate frame after rotation, which are identical to the directions having the top three largest variances of the point distribution.

The general PCA transformation in 3D model retrieval is defined on the given set of representative points of a 3D model, such as vertices, centroids of each surface, or even randomly selected locations on each surface using statistical techniques, e.g., the Monte Carlo approach [36]. In considering the different sizes of triangles or meshes of a 3D model, some appropriate weighting factors, proportional to their surface areas can be accommodated, so as to make the transformation more robust and improve the reliability and veracity of feature representation [32, 37, 38]. However, the point-based PCA transformation may cause an inaccurate normalization result that will seriously affect the retrieval precision if the chosen vertices do not have an even distribution on the surface. Therefore, a more thorough improvement, termed CPCA (continuous PCA), which performs PCA transformation based on the whole 3D polygon mesh, is proposed in [39]. CPCA generalizes the PCA transformation by using the sums of integrals over surfaces instead of the sums over selective vertices. Assume that the whole size of all the surfaces in a 3D model is represented as

$$S = \sum_{i=1}^{N_f} S_i = \int_I \mathrm{d}v, \tag{4.7}$$

where $\boldsymbol{v} \in I$ is the point on the surface, N_f is the number of surfaces on the 3D model and I is the point set of the 3D model as follows:

$$I = \bigcup_{i=1}^{N_v} v_i, \tag{4.8}$$

where N_v is the number of points, $\boldsymbol{v}_i$ is the i-th point. Similarly, the triangle set T can be denoted as

$$T = \bigcup_{i=1}^{N_t} \Delta_i, \quad \Delta_i = (a_i, b_i, c_i). \tag{4.9}$$

where N_t is the number of triangles, Δ_i means the i-th triangle. The covariance matrix $\boldsymbol{R}$ is then defined as

$$\boldsymbol{R} = \frac{1}{S} \cdot \int_I v \cdot v^{\mathrm{T}} \mathrm{d}v \,. \tag{4.10}$$

After finding the covariance matrix, we then compute the matrix of eigenvectors which diagonalizes the covariance matrix. This step typically involves the use of a computer-based algorithm for calculating eigenvectors and eigenvalues. The eigenvalues and eigenvectors are then ordered and paired. The i-th eigenvalue corresponds to the i-th eigenvector. We then sort the columns of the eigenvector matrix and eigenvalue matrix in the descending order of eigenvalues. Finally, we select a subset of the eigenvectors as basis vectors.

The PCA algorithm is fairly simple and efficient. However, it may erroneously assign the principal axes and produce inaccurate normalization results, especially when the eigenvalues are equal or close to each other, which usually happens to different models within the same category [27, 40]. A typical example of PCA [32] is depicted in Fig. 4.9, where the axes of the original coordinate system are denoted with x, y, z, while the principal components are marked with p_1, p_2, and p_3.

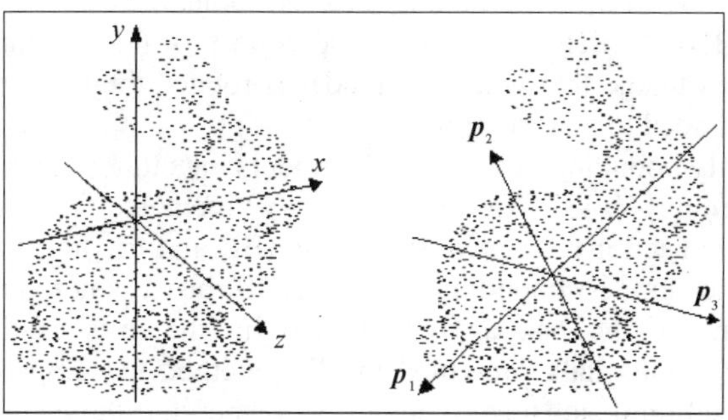

Fig. 4.9. Principal component analysis [32] (With courtesy of Vraníc and Saupe)

Step 3: Reflection. A diagonal flipping matrix $\boldsymbol{F}$ is designed to accomplish the reflection invariance, which ensures a model and its reflection will have the same feature descriptor.

Step 4: Scaling. Finally, the 3D model should also be scaled by multiplying a proper scaling coefficient s to a certain unit size to guarantee the scaling invariance.

The definition of the flipping matrix and scaling coefficient can be found in [39]. Consequently, the whole normalization process can be described as follows [32]:

$$\tau(I) = s^{-1} \cdot \boldsymbol{F} \cdot \boldsymbol{R} \cdot (I - c)\,. \tag{4.11}$$

4.3.2.2 Finding the Only Bounding Box of the 3D Model

In computer graphics and computational geometry, a bounding volume for a set of objects is a closed volume that completely contains the union of the objects in the set. Bounding volumes are used to improve the efficiency of geometrical operations by using simple volumes to contain more complex objects. Normally, simpler volumes have simpler ways to test for overlap. A bounding volume for a set of objects is also a bounding volume for the single object consisting of their union, and the other way around. Therefore, it is possible to confine the description to the case of a single object, which is assumed to be non-empty and bounded.

A bounding box is a cuboid, or in 2D a rectangle, containing the object. In dynamical simulation, bounding boxes are preferred to other shapes of bounding volumes such as bounding spheres or cylinders for objects that are roughly cuboid in shape when the intersection test needs to be fairly accurate. The benefit is obvious, for example, for objects that rest upon one another, such as a car resting on the ground: a bounding sphere would show the car as possibly intersecting with the ground, which then would need to be rejected by a more expensive test of the actual model of the car. A bounding box immediately shows the car as not intersecting with the ground, saving the more expensive test. In many applications the bounding box is aligned with the axes of the co-ordinate system, and it is then known as an axis-aligned bounding box (AABB). To distinguish the general case from an AABB, an arbitrary bounding box is sometimes called an oriented bounding box (OBB). AABBs are much simpler to test for intersection than OBBs, but have the disadvantage that when the model is rotated they cannot be simply rotated with it, but need to be recomputed.

Finding the only bounding box of the 3D model is another popular method for pose standardization [41-43]. To date, a lot of methods for constructing a bounding box have been investigated, such as AABB and Inertial Principal Axes (IPA) [41, 42]. The simplest bounding box is AABB, but it is not unique because the side directions of the box are determined by the axes of the universal coordinate system. Gottschalk presented the IPA method to compute a good fit bounding box based on a statistical method. By computing the eigenvectors of a 3×3 covariance matrix, the direction vectors for a good-fit box can be obtained. In Fig. 4.10, the bounding boxes shown in (a), (b) and (c) are some examples obtained by this method. Maximum Normal Distribution (MND), which is also a potent method to compute the only bounding box of a 3D model, has been provided by Pu et al. [43]. MND based 3D model standardization establishes the coordinate orientation of a bounding box according to normal distribution, and thereby obtains the intrinsic coordinate of a 3D object. The main idea of the maximum normal distribution method is to get three ortho-axes that coincide better with the human visual perception mechanism. Despite the fact that the IPA method can obtain three ortho-axes uniquely, they are still not ideal for the three directions that are not in accordance with our visual perception mechanism. Therefore, Pu et al. proposed adopting the maximum normal distribution as one of the principal axis. This method can be introduced as follows:

Firstly, we should compute the normal direction $\boldsymbol{N}_d$ for each triangle $\Delta\boldsymbol{pqr}$ and normalize it. It is the cross product of any two edges as

$$N_d = \frac{pq \times qr}{\|pq \times qr\|}. \tag{4.12}$$

Secondly, the area of each triangle Δ_i is calculated and the areas of all triangles with the same or opposite normals are added. Here Pu et al. thought the normals that are located in the same direction belong to a similar distribution.

The next step is to determine the three principal axes. From all normal distributions, the normal with the maximum area is selected as the first principal axis $\boldsymbol{b}_u$. To get the next principal axis $\boldsymbol{b}_v$, we can search from the remaining normal distributions and find out the normal that satisfies two conditions: (1) with the maximum area; (2) orthogonal to the first normal. Naturally, the third axis can be obtained by doing a cross product between $\boldsymbol{b}_u$ and $\boldsymbol{b}_v$:

$$b_w = b_u \times b_v. \tag{4.13}$$

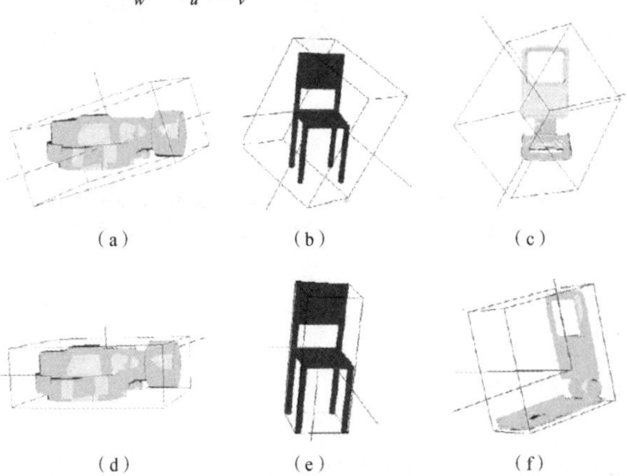

Fig. 4.10. Bounding box examples [41]. The bounding boxes shown in (a), (b) and (c) are obtained by the IPA method, while the boxes shown in (d), (e) and (f) by the MND method (With courtesy of Gottschalk)

To find the center and the half-length of the bounding box, Pu et al. projected the points of the polygon mesh onto the direction vector and find the minimum and maximum along each direction. Finally, the positive direction for each principal axis has to be decided. For this purpose, Pu et al. proposed a rule: the farthest side from the centroid is the positive direction. In Fig. 4.10, the boxes shown in (d), (e) and (f) are obtained by the maximum normal distribution method, and they look much better than Figs. 4.10(a), (b) and (c).

For models with obvious normal distributions, such as CAD models, the MND method outperforms the IPA method. However, for models without obvious normal distributions, as shown in Fig. 4.11, the former method will fail because the normal distribution has a random property for this case. From Fig. 4.11, we can observe that the IPA method is good in describing the mass distribution of 3D models, and it can find out the symmetric axes according to the mass distributions. Therefore, to overcome this limitation and make full use of the merits of the two methods, Pu et al. proposed a rule to combine the two methods: select the bounding box with smaller volume as the final box. Its validity has been proved by a large amount of models in their 3D library consisting of more than 2 700 models.

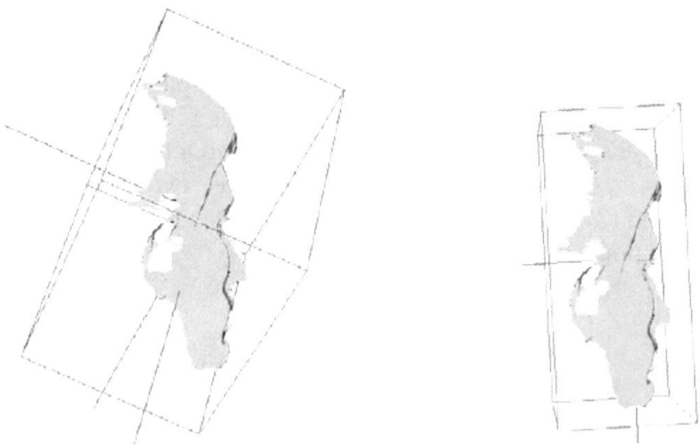

(a) (b)

Fig. 4.11. An example for the bounding box of a mesh model, in which the MND method fails [41]. (a) The bounding box obtained by the MND method; (b) The bounding box obtained by the IPA method (With courtesy of Gottschalk)

4.3.3 Polygon Triangulation

Transformation between different 3D data representations is often required before feature extraction, for the feature extraction method is often designed for only certain types of 3D data representations. For example, sometimes we may require extracting features based on triangles, thus a preprocessing step is commonly required to triangulate the polygons of the mesh. Here we introduce the polygon triangulation problem and algorithms.

In computational geometry, polygon triangulation [44] is the decomposition of a polygon into a set of triangles. A triangulation of a polygon ***P*** is its partition into non-overlapping triangles whose union is ***P***. In a strict sense, these triangles may have vertices only at the vertices of ***P***. In a less strict sense, points can be added anywhere on or inside the polygon to serve as vertices of triangles.

Triangulations are special cases of planar straight-line graphs. It is trivial to triangulate a convex polygon in linear time, by adding edges from one vertex to all other vertices. A monotone polygon can easily be triangulated in linear time as described by Fournier and Montuno [45].

For a long time there has been an open problem in computational geometry, whether a simple polygon may be triangulated faster than $O(N_v \log N_v)$ time [44], where N_v is the number of vertices of the polygon. In 1990, researchers discovered an $O(N_v \log \log N_v)$ algorithm for triangulation. Eventually, Chazelle showed in 1991 that any simple polygon can be triangulated in linear time. This algorithm is very complex though, so Chazelle and others are still looking for easier algorithms [46]. Although a practical linear time algorithm has yet to be found, simple randomized methods such as Seidel's [47] or Clarkson et al.'s have $O(N_v \log^* N_v)$ behavior which, in practice, are indistinguishable from $O(N_v)$. The time complexity of the triangulation of a polygon with holes has $O(N_v \log N_v)$ lower bound [44]. Over time, a number of algorithms have been proposed to triangulate a polygon. The following are two typical ones.

4.3.3.1 Ear Subtraction Method

One way to triangulate a simple polygon is to use the assertion that any simple polygon without holes has at least two so-called "ears". As shown in Fig. 4.12, an ear is a triangle with two sides on the edge of the polygon and the other one completely inside it. The algorithm then consists of finding such an ear, removing it from the polygon (which results in a new polygon that still meets the conditions) and repeating this until there is only one triangle left. This algorithm is easy to implement, but suboptimal, and it only works on polygons without holes. An implementation that keeps separate lists of convex and reflex vertices will run in $O(N_v^2)$ time. This method is also known as ear clipping and sometimes ear trimming.

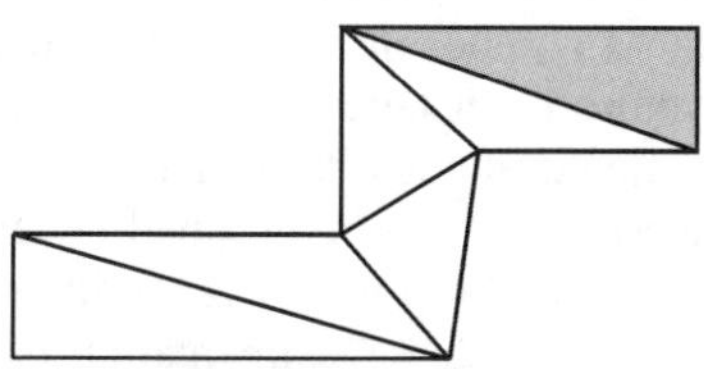

Fig. 4.12. A polygon ear

4.3.3.2 Monotone-Polygons-Based Method

A simple polygon may be decomposed into monotone polygons as follows [44]. For each point, we check if the vertices are both on the same side of the "sweep line", a horizontal or vertical line. If they are, we check the next sweep line on the other side. Break the polygon on the line between the original point and one of the points on this one. Note that if we are moving downwards, the points where both of the vertices are below the sweep line are "split points". Fig. 4.13 shows an example of breaking a polygon into monotone polygons. They mark a split in the polygon. From there we have to consider both sides separately. Using this algorithm to triangulate a simple polygon takes $O(N_v \log N_v)$ time.

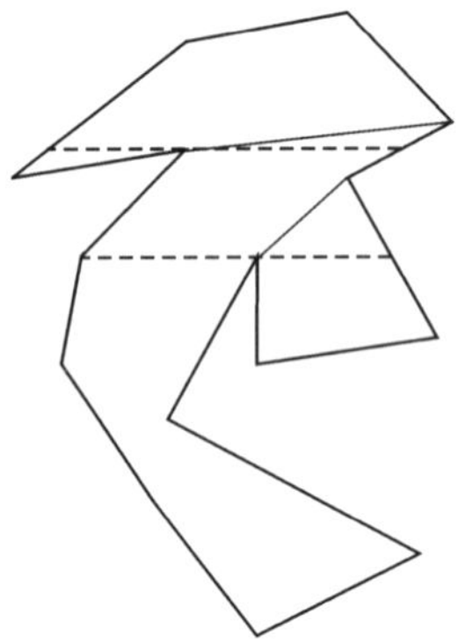

Fig. 4.13. Breaking a polygon into monotone polygons

4.3.4 Mesh Segmentation

The partition of model units is also required if we extract features from various parts of the 3D models. It is a segmentation problem. Mesh segmentation has become an important and challenging problem in computer graphics, with applications in areas as diverse as modeling, metamorphosis, compression, simplification, 3D shape retrieval, collision detection, texture mapping and skeleton extraction.

Mesh, and more generally shape, segmentation can be interpreted either in a purely geometric sense or in a more semantics-oriented manner. In the first case, the mesh is segmented into a number of patches that are uniform with respect to some property (e.g., curvature or distance to a fitting plane), while in the latter case the segmentation is aimed at identifying parts that correspond to relevant features of the shape. Methods that can be grouped under the first category have been presented as a pre-processing step for the recognition of meaningful features. Semantics-oriented approaches to shape segmentation have gained great interest recently in the research community, because they can support parameterization or re-meshing schemes, metamorphosis, 3D shape retrieval, skeleton extraction as well as modeling by composition paradigm that is based on natural shape decompositions.

It is rather difficult, however, to evaluate the performance of the different methods with respect to their ability to segment shapes into meaningful parts. This is due to the fact that the majority of the methods used in computer graphics are not devised for detecting specific features within a specific context. Also, the shape classes handled in the generic computer graphics context are a broadly varying category: from virtual humans to scanned artifacts, from highly complex free-form shapes to very smooth and featureless objects. Moreover, it is not easy to formally define the meaningful features of complex shapes in a non-engineering context and therefore the comparison of the different methods is mainly qualitative. Finally, shape segmentation methods are usually devised to solve a specific application problem, for example retrieval or parameterization, and therefore it is not easy to compare the efficacy of different methods for the shape segmentation itself.

The following are some typical mesh segmentation methods, and Fig. 4.14 shows some segmentation effects by these methods.

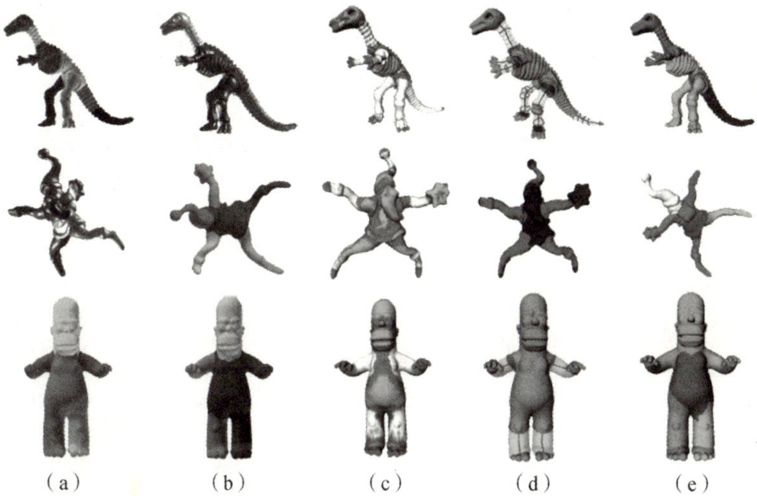

Fig. 4.14. Segmentations of miscellaneous models by various methods [48]. (a) Fuzzy clustering and cuts based; (b) Feature point and core extraction based; (c) Tailor; (d) Plumber; (e) Fitting primitives based (©[2006]IEEE)

(1) Mesh decomposition using fuzzy clustering and cuts [49]. The key idea of this algorithm is to first find the meaningful components using a clustering algorithm, while keeping the boundaries between the components fuzzy. Then, the algorithm focuses on the small fuzzy areas and finds the exact boundaries which go along the features of the object.

(2) Mesh segmentation using feature point and core extraction [50]. This approach is based on three key ideas. First, multi-dimensional scaling (MDS) is used to transform the mesh vertices into a pose insensitive representation. Second,

prominent feature points are extracted using the MDS representation. Third, the core component of the mesh is found. The core, along with the feature points, provides sufficient information for meaningful segmentation.

(3) Tailor: multi-scale mesh analysis using blowing bubbles [51]. This method provides a segmentation of a shape into clusters of vertices that have a uniform behavior from the point of view of the shape morphology, analyzed on different scales. The main idea is to analyze the shape by using a set of spheres of increasing radius, placed at the vertices of the mesh. The type and length of the sphere-mesh intersection curve are good descriptors of the shape and can be used to provide a multi-scale analysis of the surface.

(4) Plumber: mesh segmentation into tubular parts [52]. Based on the Tailor shape analysis, the Plumber method decomposes the shape into tubular features and body components and extracts, simultaneously, the skeletal axis of the features. Tubular features capture the elongated parts of the shape, protrusions or wells, and are well suited for articulated objects.

(5) Hierarchical mesh segmentation based on fitting primitives (HFP) [53]. Based on a hierarchal face clustering algorithm, the mesh is segmented into patches that best fit a pre-defined set of primitives. In the current prototype, these primitives are planes, spheres, and cylinders. Initially, each triangle represents a single cluster. At each iteration, all the pairs of adjacent clusters are considered, and the one that can be better approximated with one of the primitives forms a new single cluster. The approximation error is evaluated using the same metric for all the primitives, so that it makes sense to choose the most suitable primitive to approximate the set of triangles in a cluster.

4.3.5 Vertex Clustering

Some retrieval systems may require the mesh simplification step before feature extraction. Vertex clustering [54] is a practical technique to automatically compute approximations of polygonal representations of 3D objects. It is based on a previously developed model simplification technique which applies vertex-clustering. Major advantages of the vertex-clustering technique are its low computational cost and high data reduction rate, and thus suitable for interactive applications.

As we know, in a synthetic scene, when an object is far away from the viewpoint, its image size is small. Due to the discreteness of the image space, many points on the object are mapped onto the same pixels, and this happens often when the object's model is complex and the image size is relatively small. For points mapped to the same pixel, only one point appears on the image at the pixel, and the others are eliminated by hidden-surface removal. This is wastage in rendering as many such points are processed but never make their way to the final image. A potential solution to cut down this wasteful processing is to find out which are the points that are going to fall onto the same pixel and use a new point to represent them. Only this new point is sent for rendering.

The vertex-clustering method applies the above principle. The clustering process determines the closeness of the vertices in the object space and, for those vertices found to be close to one another (which are likely to be mapped onto the same pixel), a new representative vertex is created to replace them. Indirectly, determining the closeness of the vertices also helps to determine the closeness of the polygons. For example, two rectangles are close together if their corresponding vertices are close to each other. When each pair of the corresponding vertices is represented by a new vertex, the two rectangles are indirectly fused to become one rectangle (after removal of the duplicate). By using different clustering-cell sizes, we will have a different definition of "closeness", and this allows us to simplify the original model to models of different levels of detail (LODs).

Specifically, the process has the following steps: (1) Grading. A weight is computed for each vertex according to its visual importance. (2) Triangulation. Polygons are divided into triangles. (3) Clustering. Vertices are grouped into clusters based on geometric proximity. (4) Synthesis. A vertex representative is computed to replace the vertices in each cluster and thus simplifies some triangles into edges and points. (5) Elimination. Duplicated triangles, edges and points are removed. (6) Adjustment of normals. Normals of resulting edges and triangles are reconstructed.

4.4 Feature Extraction

In fact, feature extraction techniques have been discussed in detail in the last chapter. In this section, we would like to briefly introduce them with another categorical method. Here, methods addressing retrieval by global similarity of 3D models are classified according to the principles under which shape representations are derived. This section discusses feature extraction methods in the following four categories, i.e., primitive-based, statistics-based, geometry-based and view-based.

4.4.1 Primitive-Based Feature Extraction

Primitive-based approaches represent 3D objects with reference to a basic set of parameterized primitive elements. Parameter values are used to control the shape of each primitive element and are determined so as to best fit each primitive element with a part of the model. An example of this class of solutions has been proposed by Kriegel and Seidl in [55] where surface segments are used to model the potential docking sites of molecular structures. This approach develops on the approximation error of the surface. However, assumptions about the form of the function to be approximated limit the applicability of the approach only to special contexts. The main concept of Kriegel and Seidl's method is the approximation of

3D surface segments to provide comparable representations of shapes. Kriegel and Seidl presented a generic method based on modeling 3D shapes by a multi-parametric surface function. They called this function the approximation model. The similarity of 3D segments is measured by their mutual approximation error (and their extensions). The better the chosen multi-parametric surface function fits the characteristics of the application, the more powerful is the distance function in distinguishing between shapes that differ only slightly. This approach can be described as follows.

4.4.1.1 Approximation Models

The basic component of any approximation technique is the approximation model. Kriegel et al. adopted surface functions since they fit the 2D character of the 3D surface segments. Whereas any multi-parametric 2D surface function f: $R^2 \to R$ can be employed as an approximation model, we focus on a particular class of functions for which efficient algorithms to compute the approximation of a 3D segment are available. The class is characterized by the following definition.

Definition 4.1 (Surface Approximation Model) *The class of multi-parametric 2D surface functions* $f_{\mathbf{app}}$: $R^2 \to R$ *is called a d-dimensional surface approximation model, if it is the scalar product of a vector* **app** = $(a_1, ..., a_d) \in R^d$ *of d approximation parameters and a vector* $(f_1, ..., f_d)$ *of d 2D base functions* f_i : $R^2 \to R$:

$$f_{app}(x, y) = a_1 f_1(x, y) + ... + a_d f_d(x, y) = (a_1, \ ..., \ a_d) \cdot (f_1, \ ..., \ f_d)(x, y). \tag{4.14}$$

As we can see, surface approximation models are linear combinations of the base functions. The base functions themselves, however, may be as simple or complex as it is useful for the particular application. Examples for multi-parametric surface functions are paraboloids and trigonometric polynomials of various degrees.

4.4.1.2 Approximation of a 3D Segment

The notion by which Kriegel and Seidl related 3D surface segments and multi-parametric approximation models is the approximation error. For any arbitrary 3D surface segment s and any instance ***app*** of approximation parameters, the approximation error indicates the deviation of the surface function f_{app} from the points of the segment s:

Definition 4.2 (Approximation Error) *Let the 3D surface segment s be represented by a set of n surface points. Given an approximation model f and a vector* **app** *of approximation parameters, the approximation error of* **app** *and s is defined as*

$$d_s(\boldsymbol{app}) = \frac{1}{n}\sum_{p \in s}(f_{app}(p_x, p_y) - p_z)^2 , \tag{4.15}$$

where $\boldsymbol{p} = (p_x, p_y, p_z)$ is a 3D point in s. Given this definition, from all possible choices, Kriegel and Seidl selected the parameter vector ***app*** which yields the minimum approximation error for a given 3D segment s.

Definition 4.3 (Approximation of a Segment) *Given an approximation model f and a 3D surface segment s, the (unique) approximation of s is given by the parameter set* $\boldsymbol{app}_s$ *for which the approximation error is minimum*:

$$\boldsymbol{app}_s \text{ is approximation of } s \Leftrightarrow \forall \boldsymbol{app} : d_s^2(\boldsymbol{app}) \geq d_s^2(\boldsymbol{app}_s) . \tag{4.16}$$

The approximation $\boldsymbol{app}_s$ of s is required to be unique. Theoretically, it is possible that the approximation parameters vary without affecting the approximation error (in which case $\boldsymbol{app}_s$ would not be well defined). This indicates that the approximation model has been chosen inappropriately for the application domain and has to be changed. The algorithm will detect this situation and notify the user. Note that in all Kriegel and Seidl's experiments this situation never occurred.

In general, even the approximation error $d_s^2(\boldsymbol{app}_s)$ will be greater than zero. In order to obtain a similarity function that characterizes the similarity of an object to itself by the value zero, the relative approximation error is introduced as follows.

Definition 4.4 (Relative Approximation Error) *Given an approximation model f, a 3D surface segment s, and an arbitrary vector* ***app'*** *of approximation parameters, the relative approximation error* $\Delta d_s^2(\boldsymbol{app}')$ *of* ***app'*** *and s is defined as*

$$\Delta d_s^2(\boldsymbol{app}') = d_s^2(\boldsymbol{app}') - d_s^2(\boldsymbol{app}_s) . \tag{4.17}$$

The (unique) approximation $\boldsymbol{app}_s$ is closest to the original surface points and may be used as a more or less coarse representation of the shape of s, whereas the other surface functions do not fit the shape of the segment s very well.

Kriegel and Seidl focused on two immediate implications of this definition: First, the relative approximation never evaluates to a negative value, and it reaches zero for the (unique) approximation of a segment.

Lemma 4.1 *(1) For any 3D surface segment s and any approximation parameter set* ***app'****, the relative approximation error is non-negative:* $\Delta d_s^2(\boldsymbol{app}') \geq 0$. *(2) The relative approximation error reaches zero. In particular,* $\Delta d_s^2(\boldsymbol{app}') = 0$ *for all segments s.*

Two different segments s and q may share the same approximation $\boldsymbol{app}_s = \boldsymbol{app}_q$. Consequently, they cannot be distinguished by a simple comparison of their approximation parameters. The approximation error, however, provides additional information, and the segments may be discriminated if they differ in their approximation errors.

If too many 3D segments share the same approximation or even the same approximation error for a particular application, it is recommended to modify the approximation model, since it does not reflect the differences between the shapes very well. Another parametric surface function may be better suited to describe the variety of shapes that occur in the application.

4.4.1.3 Computation by Singular Value Decomposition

For Kriegel and Seidl's approximation models, they restrict themselves to the class of linear combinations of non-parameterized base functions as introduced in Definition 4.1. According to Definitions 4.2 and 4.3, finding an approximation is a least squares minimization problem for which an efficient numerical computation method is required. For linearly parameterized functions in particular, it is recommended to perform least-squares approximation by Singular Value Decomposition (SVD) [56].

Besides the d approximation parameters $\boldsymbol{app}_s = (a_1, \ldots, a_d)$, the SVD also returns a d-dimensional vector $\boldsymbol{w}_s$ of confidence or condition factors, and an orthogonal $d \times d$-matrix $\boldsymbol{V}_s$. Using $\boldsymbol{V}_s$, we can compute the relative approximation error for any approximation parameter vector $\boldsymbol{app}'$ with respect to the segment s. Let $\boldsymbol{A}_s = \boldsymbol{V}_s \cdot \mathrm{diag}(w_s)^2 \cdot \boldsymbol{V}_s^T$ and let us denote the rows of $\boldsymbol{V}_s$ by V_{si}. Now the error formula can be written as:

$$\Delta d_s^2(\boldsymbol{app}') = \sum_{i=1,\ \ldots,\ d} w_{si}^2((\boldsymbol{app}_i' - \boldsymbol{app}_{si}) \cdot V_{si})^2 = (\boldsymbol{app}' - \boldsymbol{app}_s) \cdot \boldsymbol{A}_s \cdot (\boldsymbol{app}' - \boldsymbol{app}_s)^{\mathrm{T}} . \quad (4.18)$$

4.4.1.4 Normalization in the 3D Space

In general, the points of a segment s are located anywhere in the 3D space and are oriented arbitrarily. Since we are only interested in the shape of the segment s, but not in its location and orientation in the 3D space, we transform s by a rigid 3D transformation into a normalized representation. There are two ways to integrate normalization into Kriegel and Seidl's method: (1) Separate. We first normalize the segment s, and then compute the approximation $\boldsymbol{app}_s$ by least-squares minimization. (2) Combined. We minimize the approximation error simultaneously over all the normalization and approximation parameters. In Kriegel and Seidl's experiments, they used the combined normalization approach. For similarity search purposes, only the resulting approximation parameters are used. However, the normalization parameters may be required later for superimposing segments.

4.4.2 Statistics-Based Feature Extraction

Shape descriptions based on statistical models consider the distribution of local features measured at the vertices of the 3D object mesh. The simplest approach approximates a feature distribution with its histogram. Any metric can be used to compute the similarity between the distributions of two models.

4.4.2.1 Overview

Vandeborre et al. [57] captured the representation of 3D objects by using histograms of the curvature of mesh vertices. As introduced in Chapter 3, Osada et al. [29] introduced shape functions as distributions of shape properties. Each distribution is approximated through the histogram of the values of the shape function. Local features such as the distance of mesh vertices to the centroid, the distance between random pairs of vertices of the mesh, and the area of triangles between three random vertices of the mesh are considered. Ohbuchi et al. [58] defined shape functions suited for objects with rotational symmetry. They have considered the principal axes of inertia of the object and used as shape functions three histograms: the moment of inertia about the axis, the average and the variance of the distance from the surface to the axis.

A limitation to statistical approaches is that they do not consider how local features are spatially distributed over the model surface. For this purpose, spatial map representations have been presented to capture either the spatial location of an object or the spatial distribution of relevant features on the object surface. Map entries correspond to locations or sections of the object and are arranged so as to preserve the relative positions of the object features. Vranic et al. [59] presented a solution in which a surface is described by associating with each ray from the origin, the value of the distance to the last point of intersection of the model with the ray and then extracting spherical harmonics for this spherical extent function. Assfalg et al. [60] proposed a method for the description of shapes for 3D objects whose surface is a simply connected region. The 3D object is deformed until it is a function on the sphere. Then, information about surface curvature is projected onto a 2D map that is used as the descriptor of the object shape.

4.4.2.2 Antini et al.'s Method

Recently, Antini et al. [61] proposed curvature correlograms to capture the spatial distribution of curvature values on the object surface. Previously, correlograms have been successfully used for image retrieval based on color content [62]. In particular, with respect to a description based on histograms of local features, correlograms also enable us to encode the information about the relative localization of local features. In [63], histograms of surface curvature have been used to support the description and retrieval of 3D objects. However, since histograms do not include any spatial information, the system is liable to false positives. Therefore, Antini et al. presented a model for representation and retrieval of 3D objects based on curvature correlograms. Correlograms are used to encode the information about curvature values and their localization on the object surface. For this peculiarity, description of 3D objects based on correlograms of curvature proves to be very effective for the purpose of content based retrieval of 3D objects.

High resolution 3D models obtained through scanning real world objects are often affected by high frequency noise, due to either the scanning device or the subsequent registration process. Hence, smoothing is required to deal with such models for the purpose of extracting their salient features. This is especially true if salient features are related to differential properties of the mesh surface, e.g. surface curvature. Selection of a smoothing filter is a critical step, as application of some filters entails changes in the model's shape. In the proposed solution, Antini et al. adopted the filter first proposed by Taubin [64]. This filter, also known as $\lambda|\mu$ filter, operates iteratively and interleaves a Laplacian smoothing weighed by λ with a second smoothing weighed with a negative factor μ ($\lambda > 0$, $\mu < -\lambda < 0$). This second step is introduced such that the model's original shape can be preserved.

Let $\boldsymbol{M}$ be a mesh. We denote with E, V and F the sets of all edges, vertices and faces of the mesh. We denote the cardinality of sets V, E and F with N_v, N_e and N_f, respectively. Given a vertex $\boldsymbol{v} \in \boldsymbol{M}$, the principal curvatures of $\boldsymbol{M}$ at the vertex $\boldsymbol{v}$ are indicated as $k_1(\boldsymbol{v})$ and $k_2(\boldsymbol{v})$, respectively. The mean curvature $\bar{k}_v$ is related to the principal curvatures $k_1(\boldsymbol{v})$ and $k_2(\boldsymbol{v})$ by the equation:

$$\bar{k}_v = \frac{k_1(\boldsymbol{v}) + k_2(\boldsymbol{v})}{2}. \tag{4.19}$$

Details about the computation of the principal curvatures for a mesh can be found in [65].

Values of the mean curvature are quantized into $2N+1$ intervals of discrete values. For this purpose, a quantization module processes the mean curvature value through a stairstep function so that many neighboring values are mapped to one output value as follows:

$$Q(\bar{k}) = \begin{cases} N\Delta, & \text{if } \bar{k} > N\Delta; \\ i\Delta, & \text{if } \bar{k} \in [i\Delta, (i+1)\Delta); \\ -i\Delta, & \text{if } \bar{k} \in (-(i+1)\Delta, \ -i\Delta]; \\ -N\Delta, & \text{if } \bar{k} < -N\Delta, \end{cases} \tag{4.20}$$

with $i \in \{0, ..., N-1\}$ and Δ is a suitable quantization parameter. The function $Q(\cdot)$ quantizes values of $\bar{k}$ into $2N+1$ distinct classes $\{c_i\}_{i=-N}^{N}$.

To simplify notations, $\boldsymbol{v} \in \boldsymbol{M}_i$ is synonymous with $\boldsymbol{v} \in \boldsymbol{M}$ and $Q(\bar{k}) = c_i$ in the following descriptions.

Definition 4.5 (Histogram of Curvature) *Given a quantization scheme to quantize curvature values into 2N+1 intervals* $\{c_i\}_{i=-N}^{N}$, *the histogram of curvature* $h_{ci}(\boldsymbol{M})$ *of the mesh* $\boldsymbol{M}$ *is defined as*:

$$h_{ci}(M) = N_v \cdot \Pr_{v_i \in M}[v_i \in M_i], \tag{4.21}$$

where N_v is the number of mesh vertices. $h_{ci}(\boldsymbol{M})/N_v$ is the probability that the quantized curvature of a generic vertex of the mesh belongs to the interval c_i.

The correlogram of curvatures is defined with respect to a predefined distance value δ. In particular, the curvature correlogram $\gamma_{c_i,c_j}^{(\delta)}$ of a mesh $\boldsymbol{M}$ is defined as:

$$\gamma_{c_i,c_j}^{(\delta)}(M) = \Pr_{v_1,v_2 \in M}[v_1 \in M_{c_i}, v_2 \in M_{c_j} \mid \|v_1 - v_2\| = \delta], \tag{4.22}$$

where $\gamma_{c_i,c_j}^{(\delta)}(\boldsymbol{M})$ means the probability that two vertices that are δ far away from each other have curvatures belonging to intervals c_i and c_j, respectively. Ideally, $\|\boldsymbol{v}_1 - \boldsymbol{v}_2\|$ should be the geodesic distance between two vertices $\boldsymbol{v}_1$ and $\boldsymbol{v}_2$. However, it can be approximated with the k-ring distance if the mesh $\boldsymbol{M}$ is regular and triangulated [66].

Definition 4.6 (1-ring) *Given a generic vertex* $\boldsymbol{v}_i \in \boldsymbol{M}$, *the neighborhood or 1-ring of* $\boldsymbol{v}_i$ *is the set*:

$$V^{v_i} = \{v_j \in M : \exists e_{ij} \in \mathrm{E}\}. \tag{4.23}$$

E is the set of all mesh edges (if $e_{ij} \in E$, there is an edge that links vertices $\boldsymbol{v}_i$ and $\boldsymbol{v}_j$). The set V^{v_i} can be easily computed using the morphological operator *dilate* [67] as follows:

$$V^{v_i} = dilate(v_i). \tag{4.24}$$

Through the *dilate* operator, the concept of 1-ring can be used to define, recursively, the generic k-th order neighborhood:

$$ring_k = dilate(\mathrm{e}^k) \cap dilate(\mathrm{e}^{k-1}). \tag{4.25}$$

Definition of the k-th order neighborhood enables the definition of a true metric between vertices in a mesh. This metric can be used for the purpose of computing curvature correlograms as an approximation of the usual geodesic distance (That is computationally much more demanding). According to this, the k-ring distance between two mesh vertices is defined as $d_{ring}(\boldsymbol{v}_1, \boldsymbol{v}_2) = k$ if $\boldsymbol{v}_2 \in ring_k(\boldsymbol{v}_1)$. Function $d_{ring}(\boldsymbol{v}_1, \boldsymbol{v}_2) = k$ is a true metric, in fact:

(i). $d_{ring}(\boldsymbol{u}, \boldsymbol{v}) \geq 0$, and $d_{ring}(\boldsymbol{u}, \boldsymbol{v}) = 0$ if and only if $\boldsymbol{u} = \boldsymbol{v}$;
(ii). $d_{ring}(\boldsymbol{u}, \boldsymbol{v}) = d_{ring}(\boldsymbol{v}, \boldsymbol{u})$;
(iii). $\forall \boldsymbol{w} \in \boldsymbol{M}$, $d_{ring}(\boldsymbol{u}, \boldsymbol{v}) \leq d_{ring}(\boldsymbol{u}, \boldsymbol{w}) + d_{ring}(\boldsymbol{w}, \boldsymbol{v})$.

Based on the above $d_{ring}(\cdot)$ distance, the curvature correlogram can be redefined as follows:

$$\gamma_{c_i,c_j}^{(k)}(M) = \Pr_{v_1,v_2 \in M}[v_1 \in M_{c_i}, v_2 \in M_{c_j} \mid d_{ring}(v_1, v_2) = k]. \tag{4.26}$$

4.4.3 Geometry-Based Feature Extraction

Geometry-based methods use geometric properties of the 3D object and their measures as global shape descriptors. Many geometry-based approaches have been proposed. Kolonias et al. [68] used dimensions of the object bounding box (i.e., its aspect ratios), a binary voxel-based representation of geometry and "set of paths", that outline the shape (i.e., model routes). In [69], each point where Gaussian and median curvatures are maxima and the torsion is maximum has been considered as a representative of the object shape. Elad et al. [70] used moments (up to the 7th order) of surface points, according to the fact that, different from the case of 2D images, the computation of moments for 3D models is not affected by self-occlusions. In [71], a representation based on moment invariants and Fourier transform coefficients has been combined with active learning to take into account user relevance feedback and improve the effectiveness of retrieval. In [72], a method has been presented to compute 3D Zernike descriptors from voxelized models. 3D Zernike descriptors capture object coherence in the radial direction and in the direction along a sphere. However, the effectiveness of the approach is strongly dependent on the quality of the voxelization process.

Here, we would like to introduce the system developed within the Nefertiti project, supporting retrieval of 3D models based on both shape geometry and appearance (i.e., color and texture) [73]. The detailed description for shape geometry is as follows:

The global analysis in [73] is performed in order to define a reference frame that shall be used by the other algorithms. The reference frame is defined as the principal axes of the tensor of inertia which is defined as

$$I = [I_{qr}] = \left[\frac{1}{n} \sum_{i=1}^{n} [S_i (q_i - q_{CM})(r_i - r_{CM})] \right], \tag{4.27}$$

where S_i is the surface of a triangular face (assuming a triangular decomposition of the object), CM is the center of mass of the object and q and r are equal to x, y or z. If the model is not made out of triangles, the triangulation is generated automatically by the software based on the Open Inventor Library (SGI). The principal axes are obtained by computing the eigen vectors of the tensor

$$\left[\boldsymbol{I}\boldsymbol{a}_i = \lambda_i \boldsymbol{a}_i \right]_{i=1,2,3}. \tag{4.28}$$

The identification of the axes is performed by comparing the eigen values. The eigen vector with the highest eigen value is labeled one, the second highest is labeled two and the remaining axis is labeled three. The tensor of inertia has a mirror symmetry problem which can be handled by computing the statistical distribution of the mass in the positive and negative direction in order to identify the positive direction. For each axis, the points are divided between "North" and "South": a point belongs to the North group if the angle between the corresponding cord and a given axis is smaller than 90°, and to the South group if it is greater than 90°. A cord is defined as a vector that goes from the center of mass of the model to the center of mass of the triangle. The standard deviation for the length of the cords is calculated for each group of each axis and it is defined as

$$s = \frac{\sqrt{n\sum_{i=1}^{n} d_i^2 - \left(\sum_{i=1}^{n} d_i\right)^2}}{n(n-1)}, \tag{4.29}$$

where d is the length of a cord and n the number of points. If the standard deviation of the North group is higher than the standard deviation of the South group, then the direction of the corresponding eigen vector is not changed while, in the other case, the direction is flipped by 180°. This technique is also applied to the first and second axes. Then the outer product between them is calculated. If the third axis does not have the same direction, then the resulting vector is flipped by 180° in order to have a direct orthogonal system.

The scale is simply handled by a bounding box which is the smallest box that can contain the model. The axes of the box are parallel to the principal axes of the tensor of inertia. A rough description of the mass distribution inside the box is obtained by using the eigen values of the tensor of inertia (i.e., moment description).

In [73], the shape is analyzed at three levels. The local level is defined by the normals. Assuming a triangular decomposition of the object and a normal for each triangle, the angles between the normals and the first two principal axes are computed using

$$\alpha_q = \cos^{-1}\left(\frac{\boldsymbol{n} \cdot \boldsymbol{a}_q}{\|\boldsymbol{n}\| \|\boldsymbol{a}_q\|} \right), \tag{4.30}$$

where

$$\boldsymbol{n} = \frac{[(\boldsymbol{r}_2 - \boldsymbol{r}_1) \times (\boldsymbol{r}_3 - \boldsymbol{r}_1)]}{\|(\boldsymbol{r}_2 - \boldsymbol{r}_1) \times (\boldsymbol{r}_3 - \boldsymbol{r}_1)\|}. \tag{4.31}$$

The statistic of this description is then presented in the form of a histogram. Reference [73] used three kinds of histograms, called histogram of the first, second and third kinds depending on the complexity of the description. The histogram of the first kind is defined as $h(\alpha_q)$ where q equals 1 and 2. This histogram does not distinguish between the two angles and does not take into account the relation between them. Because of that, it has a low discrimination capability. The histogram of the second kind is made out of two histograms: one for each angle. Thus it can distinguish the angles but it does not establish any relation between them. The histogram of the third kind is a bidimensional histogram defined as $h(\alpha_1, \alpha_2)$. Not only does it distinguish between the angles but it also maps the relation between them.

In general, normals are very sensitive to local variation in shape. In some cases, this may cause severe drawbacks. Let us consider an example: a pyramid and a step pyramid. In the case of the pyramid, the orientations of the normals are the same for all the triangles belonging to a given face, while in the other case they have two orientations corresponding to those of the step. The histograms corresponding to these pyramids are very distinct although both models have a very similar global shape. In order to solve this problem, reference [73] introduced the concept of a cord measurement. A cord is defined as a vector that goes from the center of mass of the model to the center of mass of a given triangle. The cord is not a unit vector since it has a length. As opposed to a normal, a cord can be considered as a regional characteristic. If we take the pyramid and the step pyramid as an example, we can see that the cord orientation changes slowly in a given region, while the normal orientation can have significant variations. As for normals, the statistical distribution of the cord orientations can be represented by three histograms, namely histograms of the first, second and third kinds. Since the cord has a length, it is also possible to describe the statistical distribution of the length of the cords by a histogram. This histogram is scale-dependent but it can be made scale-independent, by normalizing the scale, e.g., zero corresponding to the shortest cord and one to the longest.

Explicitly or implicitly, we are used to considering 3D models made out of surfaces. From a certain point of view this is right, but at the same time we should not forget that a 3D object is also a volume and consequently it might be interesting to analyze it as such. In a 3D discrete representation, the building blocks are called voxels. Using such a representation, it is possible to binarize a 3D model by losing a small amount of information. The idea is simply to map the model's coordinates to the discrete voxel coordinates as follows:

$$\begin{bmatrix} x \\ y \\ z \end{bmatrix} \Rightarrow \begin{bmatrix} i\Delta_x \\ j\Delta_y \\ k\Delta_z \end{bmatrix}. \tag{4.32}$$

where Δ_x, Δ_y and Δ_z are the dimensions of the voxel and i, j and k are the discrete coordinates. If the density of points in the original model is not high enough, it may be necessary to interpolate the original model so as to generate more points and to achieve a better description in the voxel space.

Reference [73] chose to analyze the voxel representation with a wavelet transform. Recent experiments tended to demonstrate that the human eye would perform a kind of wavelet transform. This would also mean that the brain would perform a part of its analysis based on such a transform. The wavelet transform performs a multi-scale analysis. By multi-scale we mean that the model is analyzed at different levels of detail. There is a fast implementation of the wavelet transform that makes it possible to perform the calculation rapidly. The fast wavelet transform is an orthogonal transformation, meaning that its base is orthogonal. The elements of the above are characterized by their scale and position. Each element of the base is bounded in space, which means that it occupies a well-defined region. This means that the analysis performed by the wavelet transform is local and that the size of the analyzed region depends on the scale of the wavelet. As an example, the 1D wavelet is defined as

$$\sqrt{2^j}\, w(2^j q - n)_{n,j \in Z}. \tag{4.33}$$

Reference [73] used DAU4 (Daubechies) wavelets which have two vanishing moments. The $N \times N$ (N being a multiple of two) matrix corresponding to the 1D transform is

$$\boldsymbol{W} - \begin{bmatrix} c_0 & c_1 & c_2 & c_3 & & & & & & \\ c_3 & -c_2 & c_1 & -c_0 & & & & & & \\ & & c_0 & c_1 & c_2 & c_3 & & & & \\ & & c_3 & -c_2 & c_1 & -c_0 & & & & \\ & & & & & & \ddots & & & \\ & & & & & & c_0 & c_1 & c_2 & c_3 \\ & & & & & & c_3 & -c_2 & c_1 & -c_0 \\ c_2 & c_3 & & & & & & & c_0 & c_1 \\ c_1 & -c_0 & & & & & & & c_3 & -c_2 \end{bmatrix}, \tag{4.34}$$

where

$$\begin{aligned} c_0 &= \frac{(1+\sqrt{3})}{4\sqrt{2}}, \quad c_1 = \frac{(3+\sqrt{3})}{4\sqrt{2}}, \\ c_2 &= \frac{(3-\sqrt{3})}{4\sqrt{2}}, \quad c_3 = \frac{(1-\sqrt{3})}{4\sqrt{2}}. \end{aligned} \tag{4.35}$$

Based on Eq.(4.35) we can define

$$\boldsymbol{H} = [c_0 \quad c_1 \quad c_2 \quad c_3], \tag{4.36}$$

$$\boldsymbol{G} = [c_3 \quad -c_2 \quad c_1 \quad -c_0]. \tag{4.37}$$

The doublet $\boldsymbol{H}$ and $\boldsymbol{G}$ is a quadrature mirror filter. $\boldsymbol{H}$ can be considered as a smoothing filter, while $\boldsymbol{G}$ is a filter with two vanishing moments. The 1D wavelet transform is computed by applying the wavelet transform matrix hierarchically, first on the full vector of length N, then to the $N/2$ values smoothed by $\boldsymbol{H}$, then the $N/4$ values smoothed again by $\boldsymbol{H}$, until two components remain. In order to compute the wavelet transform in three dimensions, the array is transformed sequentially on the first dimension (for all values of its other dimensions), then on its second dimension and finally on its third dimension. The final result of the wavelet transform is an array of the same dimension as the initial voxel array.

The set of wavelet coefficients represents a tremendous amount of information. In order to reduce it, Reference [73] computed the logarithm of base 2 of the coefficients in order to enhance the coefficients corresponding to small details. These usually have a very low value compared to those that have a large value and Reference [73] integrated the signal for each scale. A histogram representing the distribution of the signal at different scales is then constructed: the vertical axis represents the total amount of signal at a given scale and the horizontal axis represents the "scale" or level of resolution. It is important to notice that each "scale" in the histogram represents in fact a triplet of scales corresponding to s_x, s_y and s_z.

4.4.4 View-Based Feature Extraction

View-based descriptions use a set of 2D views of the model and appropriate descriptors of their content to represent the 3D object shape. One problem with this approach concerns the need for representations that are computationally tractable. In [74], a number of views of the 3D object are taken and, for each view, the 2D profile is considered. Then the PCA is used to reduce all object views to a limited set of representative views that are used to represent the whole 3D object shape. In [75], signatures of spin images have been proposed. In their original formulation [76], spin images are 2D histograms of the surface locations around a point. Each mesh vertex defines a family of cylindrical coordinate systems, with the origin in $\boldsymbol{p}$, and the axis along $\boldsymbol{n}$. The spin image is obtained by projecting all the other vertices over the tangent plane, retaining for each vertex the radial distance and the elevation and discarding the polar angle. The 2D information of the spin image is reduced to a 1D feature vector, partitioning the image into a finite number of regions and considering the point density in each region. Signatures are hence derived by clustering all spin image vectors and taking the centers of the clusters as their representatives. In [77], 2D views (light fields) of the object are taken from observation points uniformly distributed on the surface of a sphere centered in the object's centroid. For each of these views, Zernike moments and Fourier descriptors are computed so as to reduce the 2D information to a 1D feature vector. Computational complexity of retrieval is reduced by a multistep approach supporting early rejection of non-relevant models. For the detailed description, readers can refer to Chapter 3.

4.5 Similarity Matching

After the feature extraction process, appropriate similarity measurements should be designed to measure the content similarity. The ideal goal of similarity measurement has two aspects: (1) to make the feature vectors of similar 3D models as close as possible in the feature space and (2) to maintain the largest possible distances for dissimilar 3D models. Therefore, the task of the similarity match is to compute the suitable distances or dissimilarities in the multidimensional feature space between the user query and all the 3D models in the database and rank them in the descending order of similarities as well. A variable number of models are then retrieved by listing the top-ranking items.

At present, the available similarity matching methods in content-based 3D model retrieval can be categorized into four classes: (1) distance metrics; (2) graph matching; (3) machine learning; (4) semantic measures. The following are detailed descriptions for these four types of similarity matching methods.

4.5.1 Distance Metrics

Currently, distance metrics are perhaps the most popular and widely used similarity matching methods, most of which have already been used in content-based 2D media retrieval.

4.5.1.1 Minkowski Distances

A distance metric is a dissimilarity measurement with some particular properties, for which there is a comprehensive body of research. For content-based 3D model retrieval, the successfully used distance metrics include Manhattan distances [36], Euclidean distances [72] and Hausdorff distances [78]. The Manhattan and Euclidean measurements are both based on L_p distances (p=1, 2), meaning Minkowski distances. The L_p distance between two points $\boldsymbol{x}$, $\boldsymbol{y}$ in the N-dimensional space R^N is defined as

$$L_p = \left(\sum_{i=1}^{N} (x_i - y_i)^p \right)^{1/p} . \tag{4.38}$$

All distances are metrics when $p \geq 1$. The L_p distance itself can also be directly used as a similarity measurement. For example, Osada et al. [19] employed it to implement a similarity match on the probability density function of shape distribution features. In particular, to assign different impacts to different features or to allow relevance feedback, Euclidean distance is often modified into the weighted Euclidean distance with the weight matrix [19, 70, 79].

4.5.1.2 Hausdorff Distances

The Hausdorff distance, another frequently used metric, is defined for comparing two point sets of different sizes as follows:

$$h(A, B) = \min_{a \in A} \max_{b \in B} d(A, B), \tag{4.39}$$

where $d(A,B)$ is a distance metric, e.g., the Euclidean distance. However, it is very sensitive to noise since even a single outlier can change the Hausdorff distance [80].

4.5.1.3 Elastic-Matching Distances

Many other distance metrics have also been studied for the 3D model retrieval task. Ohbuchi et al. [36, 81] introduced an elastic-matching distance in order to compensate the "larger-than-wanted" effect caused by "rigid" distance metrics, e.g., the Euclidean distance, and the results were promising. Elastic matching has been used extensively in speech recognition. Ohbuchi et al. performed elastic matching along the distance axis, using the dynamic programming technique for its implementation to compute the distance $D_E(\boldsymbol{X},\boldsymbol{Y})$. It locally stretches and shrinks the distance axis of the histogram in order to find minimal distance matches. If the matching is too elastic, a pair of shapes having very different histograms could have a low distance value. Ohbuchi et al. implemented and experimentally compared the performance of the linear and the quadratic penalty functions, the latter of which is depicted in Eq.(4.42). Ohbuchi et al. used the better performing quadratic penalty function for their experiments:

$$D_E(\boldsymbol{X},\boldsymbol{Y}) = g(\boldsymbol{X}_n,\boldsymbol{Y}_n), \tag{4.40}$$

$$g(\boldsymbol{X}_n,\boldsymbol{Y}_n) = \min\begin{bmatrix} g(\boldsymbol{X}_n,\boldsymbol{Y}_{n-1}) + \Delta g(\boldsymbol{X}_n,\boldsymbol{Y}_n) \\ g(\boldsymbol{X}_{n-1},\boldsymbol{Y}_{n-1}) + 2\Delta g(\boldsymbol{X}_n,\boldsymbol{Y}_n) \\ g(\boldsymbol{X}_{n-1},\boldsymbol{Y}_n) + \Delta g(\boldsymbol{X}_n,\boldsymbol{Y}_n) \end{bmatrix}, \tag{4.41}$$

$$\Delta g(\boldsymbol{X}_i,\boldsymbol{Y}_j) = |i-j|\sqrt{\sum_{k=1}^{I_a}(x_{i,k} - y_{i,k})^2}, \tag{4.42}$$

where $\boldsymbol{X} = (x_{i,k})$ and $\boldsymbol{Y} = (y_{i,k})$ are the feature vectors (2D histograms having $I_d \times I_a$ elements) for the model A and B, respectively.

4.5.1.4 Improved Earthmover's Distances

Tangelder et al. [9] used an improved Earthmover's Distance (EMD) [82] as the distance measure. Intuitively, given two distributions, one can be seen as a mass of earth properly spread in space, the other as a collection of holes in that same space.

Then the EMD measures the least amount of work needed to fill the holes with earth. Here a unit of work corresponds to transporting a unit of earth by a unit of ground distance. Computing the EMD is based on a solution to the well-known transportation problem a.k.a. the Monge-Kantorovich problem. That is, signature matching can be naturally cast as a transportation problem by defining one signature as the supplier and the other as the consumer, and by setting the cost for a supplier-consumer pair to equal the ground distance between an element in the first signature and an element in the second signature. Intuitively, the solution is then the minimum amount of "work" required to transform one signature into the other. Thus, the EMD naturally extends the notion of a distance between single elements to that of a distance between sets or distributions of elements. The advantages of the EMD over previous definitions of distribution distances should now be apparent. First, the EMD applies to signatures, which subsume histograms. The greater compactness of signatures is in itself an advantage, and having a distance measure that can handle these variable-size structures is important. Second, the cost of moving "earth" reflects the notion of nearness properly, without the quantization problems in most current measures. Even for histograms, in fact, items from neighboring bins now contribute similar costs, as appropriate. Third, the EMD allows for partial matches in a very natural way. This is important, for instance, in order to deal with occlusions and clutter in image retrieval applications and when matching only parts of an image. Fourth, if the ground distance is a metric and the total weights of two signatures are equal, the EMD is a true metric, which allows endowing image spaces with a metric structure. Of course, it is important that the EMD can be computed efficiently, especially if it is used for image retrieval systems where a quick response is required. In addition, retrieval speed can be increased if lower bounds to the EMD can be computed at low cost. These bounds can significantly reduce the number of EMDs that actually need to be computed by pre filtering the database and ignoring images that are too far from the query. Fortunately, efficient algorithms for the transportation problem are available. For example, we can use the transportation-simplex method [12], a streamlined simplex algorithm that exploits the special structure of the transportation problem. A good initial basic feasible solution can drastically decrease the number of iterations needed. We can compute the initial basic feasible solution by Russell's method [23].

4.5.2 Graph-Matching Algorithms

When two 3D models to be compared are represented by graph-like structures, specific graph matching algorithms should be designed for similarity matching between them. However, matching two graphs is generally regarded as the largest isomorphic subgraph problem, which is almost impossible to solve in the general sense. Therefore, the currently available 3D shape similarity measures for graph matching are all customized to the given 3D topological features.

To compare two 3D models based on their skeleton-based Attributed

Relational Graphs (ARGs), we need to solve a graph matching problem. Bardinet et al. [83] compared two graphs by finding their optimal association matrix $\boldsymbol{P}$ so that an objective function E involving all types of nodes, links and attributes in the graph is minimized. Some heuristic constraints are also exploited in the objective function to guarantee the correctness of graph matching. They proposed an error-correcting consistent-labeling graph matching algorithm suitable to treat ARGs and adopted a nonlinear optimization method called graduated assignment. Given two ARGs $\boldsymbol{G}$ and $\boldsymbol{H}$, with I and J nodes respectively, assume there are R link types and S attribute types. The problem is to find the association matrix $\boldsymbol{P}$ such that the following objective function is minimized:

$$E_{ARG} = -\frac{1}{2}\sum_{i=1}^{I}\sum_{j=1}^{J}\sum_{k=1}^{I}\sum_{l=1}^{J} P_{ij}P_{kl}\sum_{r=1}^{R} C_{ijkl}^{(r)} + \alpha\sum_{i=1}^{I}\sum_{j=1}^{J} P_{ij}\sum_{s=1}^{S} C_{ij}^{(s)}, \tag{4.43}$$

subject to:

$$\begin{cases} \forall i, & \sum_{j=1}^{J} P_{ij} \le 1; \\ \forall j, & \sum_{i=1}^{I} P_{ij} \le 1; \\ \forall i,j, & P_{ij} \in \{0,1\}, \end{cases} \tag{4.44}$$

where $\{C_{ijkl}^{(r)}\}$ is the compatibility matrix for a link of type r, whose components are defined as $C_{ijkl}^{(r)} = cl^{(r)}(G_{ij}^{(r)}, H_{kl}^{(r)})$ (0 if either $G_{ij}^{(r)}$ or $H_{kl}^{(r)}$ is NULL); $\{C_{ij}^{(s)}\}$ is the similarity matrix for an attribute of type s, whose components are defined as: $C_{ij}^{(s)} = cn^{(s)}(G_i^{(s)}, H_j^{(s)})$; $\{G_{ij}^{(r)}\}$ and $\{H_{kl}^{(r)}\}$ are the adjacency matrices for the r-link; $cl^{(r)}(\cdot,\cdot)$ is a compatibility measure between a r-link in $\boldsymbol{G}$ and a r-link in $\boldsymbol{H}$; $\{G_i^{(s)}\}$ and $\{H_j^{(s)}\}$ are vectors corresponding to the s-attribute of the nodes of $\boldsymbol{G}$ and $\boldsymbol{H}$; $cn^{(s)}(\cdot,\cdot)$ is a measure of similarity between a node in $\boldsymbol{G}$ and a node in $\boldsymbol{H}$, with respect to the same attribute s. $\boldsymbol{P}$ is an $I \times J$ association matrix that at the end of the minimization process provides the correspondences between one set of primitives and the other: $P_{ij}=1$ if Node i in G corresponds to Node j in $\boldsymbol{H}$, 0 otherwise. Note that the approach does not always converge to an exact permutation matrix, thus a clean-up heuristic should be defined. Bardinet et al. set in each column of the association matrix $\boldsymbol{P}$ the maximum element to 1 and others to 0. In this specific case, $\boldsymbol{P}$ provides the correspondences between the skeleton parts of the two objects to be compared. Above constraints adopted in the objective function guarantee that two graph nodes, or two object skeleton parts, will be matched only if they are similar and if they share the same type of relations with their neighboring primitives in their respective graphs. Fig. 4.15 gives an example of skeleton-based ARG matching.

(a) (b) (c)

Fig. 4.15. Example of graph matching [82]. (a) Original object with superimposed skeleton and labeled object partition; (b) Deformed object obtained by occlusion with a polygonal shape and scaling, rotation and translation, with superimposed skeleton and labeled object partition; (c) Original object labeled by propagating labels of the deformed object through the skeleton-based ARG matching (With courtesy of Bardine et al.)

In [84-86], a graph matching algorithm for 2D shock graphs was proposed. The shock graph is an emerging shape representation for object recognition, where a 2D silhouette is decomposed into a set of qualitative parts, captured in a directed acyclic graph. A structural "signature" is defined for each graph node, which characterized the node's underlying subgraph structure, whose components are based on the eigenvalues of the subgraph's adjacency matrix. All the edges in the graph are discarded and the problem is transformed to find the maximum cardinality and minimum weight matching in bidirectional graphs. However, this approach cannot be guaranteed to conform to the hierarchical structures of two graphs. To solve this problem, a recursive depth-first search should be combined in order to exploit the matching at higher levels to constrain the matching at lower levels [87]. The graph matching algorithm typically outputs a number of parameters that can be used to determine the "goodness" of the similarity matching results, such as the number of nodes matched and information about which nodes are matched to other nodes. Furthermore, a coarse-to-fine graph matching strategy can also be easily adopted.

In addition, Hilaga et al. [88] associated each graph node with several attributes and defined the similarity between two nodes as the similarity between their attributes. Then, the similarity for a given set of node pairs was computed as a whole similarity measure.

4.5.3 Machine-Learning Methods

The main idea of similarity matching based on machine learning is to train a specific learning classifier for computing and ranking similarity degrees on a preselected training sample set with a specific scale by utilizing machine-learning methods such as artificial neural networks (ANNs) and support vector machines (SVMs), and so on. This is particularly proper in cases where no suitable distance metric can effectively measure the similarity, e.g., between two high-dimensional feature vectors. In those cases, some appropriate similarity measures can be approximated by learning the hidden correlations and mappings from a number of result-known training samples, which allows for great flexibility in the retrieval process.

4.5.3.1 SVM

Support vector machines (SVMs) [89] are a set of related supervised learning methods used for classification and regression. Viewing the input data as two sets of vectors in the n-dimensional space, an SVM will generate a separating hyperplane that maximizes the margin between the two data sets. To compute this margin, two parallel hyperplanes are constructed, one on each side of the separating hyperplane, which are "pushed up against" the two data sets. Intuitively, a good separation can be achieved by the hyperplane with the largest distance to the neighboring data points of both classes since, in general, the larger the margin, the lower the generalization error of the obtained classifier. The basic idea of the SVM approach can be described as follows.

Given some training data, a set of points with the following form

$$D = \{(\boldsymbol{x}_i, c_i) \mid \boldsymbol{x}_i \in R^p, c_i \in \{-1, 1\}\}_{i=1}^{n}, \tag{4.45}$$

where c_i is either 1 or -1, indicating one of two classes to which the point $\boldsymbol{x}_i$ belongs. Each $\boldsymbol{x}_i$ is a p-dimensional real vector. Our goal is to find the maximum-margin hyperplane which divides the points with $c_i = 1$ from those with $c_i = -1$. In fact, any hyperplane can be written as the set of points $\boldsymbol{x}$ satisfying

$$\boldsymbol{w} \cdot \boldsymbol{x} - b = 0, \tag{4.46}$$

where $\cdot$ denotes the dot product between two vectors. The vector $\boldsymbol{w}$ is a normal vector that is perpendicular to the hyperplane. The parameter $b / \|\boldsymbol{w}\|$ is the offset of the hyperplane from the origin along the normal vector $\boldsymbol{w}$. Our aim is to choose the $\boldsymbol{w}$ and b to maximize the margin, namely the distance between the two parallel hyperplanes that are as far apart as possible while still separating the data into two classes. These hyperplanes can be described by the equations

$$\boldsymbol{w} \cdot \boldsymbol{x} - b = 1, \tag{4.47}$$

and

$$\boldsymbol{w} \cdot \boldsymbol{x} - b = -1, \tag{4.48}$$

Note that if the training data are linearly separable, we can select the two hyperplanes of the margin in such a way that there are no points between them and then try to maximize their distance. According to geometry, we can find that the distance between these two hyperplanes equals $2/\|\boldsymbol{w}\|$, thus our goal is transformed to minimize $\|\boldsymbol{w}\|$. As we should also prevent data points from falling into the margin, we may add the following constraint: for each i, either $\boldsymbol{w} \cdot \boldsymbol{x}_i - b \geq 1$ for $\boldsymbol{x}_i$ in the first class or $\boldsymbol{w}\cdot\boldsymbol{x}_i - b \le -1$ for $\boldsymbol{x}_i$ in the second class. Then we have

$$c_i\left(\boldsymbol{w}\cdot\boldsymbol{x}_i - b\right) \ge 1 \text{ for } \forall 1 \le i \le n. \tag{4.49}$$

Based on the above descriptions, we obtain the following optimization problem:
Minimize (in $\boldsymbol{w}$, b): $||\boldsymbol{w}||$,

Subject to (for $\forall 1 \le i \le n$): $c_i\left(\boldsymbol{w}\cdot\boldsymbol{x}_i - b\right) \ge 1$. (4.50)

The above optimization problem is very hard to solve because it depends on $||\boldsymbol{w}||$, the norm of $\boldsymbol{w}$, which involves a square root. Luckily, it is possible to modify the equation by replacing $||\boldsymbol{w}||$ with $\frac{1}{2}\|\boldsymbol{w}\|^2$ without changing the optimal solution, since the minimum of the original and the modified equation have the same $\boldsymbol{w}$ and b. This is a quadratic programming (QP) optimization problem. More clearly:

Minimize (in $\boldsymbol{w}$, b): $\frac{1}{2}\|\boldsymbol{w}\|^2$,

Subject to (for $\forall 1 \le i \le n$): $c_i\left(\boldsymbol{w}\cdot\boldsymbol{x}_i - b\right) \ge 1$. (4.51)

Note that the factor of 0.5 is used for mathematical convenience. This problem can now be solved by standard quadratic programming techniques and programs. A typical 2D case is shown in Fig. 4.16.

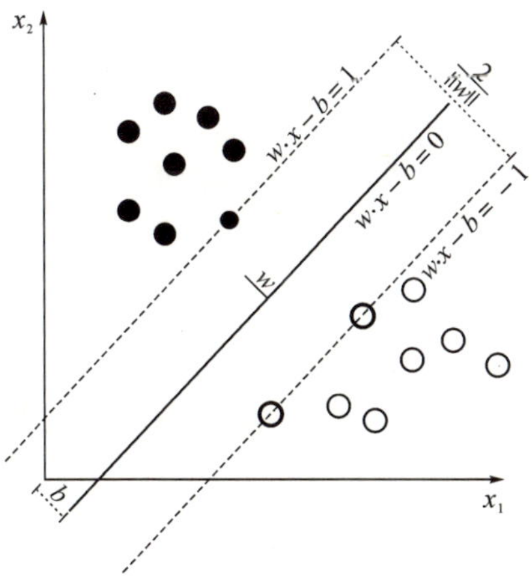

Fig. 4.16. 2D example to explain the SVM scheme

Ibato et al. [90] presented a shape-similarity search method that combines a 3D shape feature that is independent of the model's pose and size with the SVM-based learning classifier. This system is a human-oriented query-by-example system. By tagging similar and dissimilar models among the list of previous retrieval results, the system learns the models the user desires by using the SVM approach. Ibato et al. carried out many experiments by combining the transform-invariant D2 shape features [19] with the SVM, feeding the feature vector to an SVM to compute the dissimilarity. The experimental results show that, despite its simplicity, the system works well in retrieving shapes that a user feels "similar" to the given examples.

4.5.3.2 SOM

A self-organizing map (SOM) or self-organizing feature map (SOFM) is a type of artificial neural network that is trained by exploiting unsupervised learning methods to produce a low-dimensional (typically 2D), discretized representation of the input space of the training samples, called a map. Self-organizing maps are different to other artificial neural networks in the sense that they adopt a neighborhood function to preserve the topological properties of the input space. This makes SOM useful for visualizing low-dimensional views for high-dimensional data, akin to multidimensional scaling. The Finnish professor Teuvo Kohonen first described the model as an artificial neural network, sometimes called a Kohonen map. Similar to most artificial neural networks, SOMs operate in two modes: training and mapping. The training process builds the map based on input examples, which is a competitive process also called vector quantization, while the mapping process automatically classifies a new input vector.

A self-organizing map consists of components called nodes or neurons. Associated with each node is a weight vector of the same dimension as the input data vectors, and it is a point in the map space. The common arrangement of nodes is a regular spacing in a hexagonal or rectangular grid. The self-organizing map describes a mapping from a higher dimensional input space to a lower dimensional map space. The procedure for placing a vector from the data space onto the map space is to find the node with the closest weight vector to the vector taken from the data space and to assign the map coordinates of this node to our vector. While it is typical to regard this type of network structure as related to feedforward networks where the nodes are visualized as being attached, this type of architecture is fundamentally different in arrangement and motivation. Useful extensions include using toroidal grids where opposite edges are connected and use a large number of nodes. It has been shown that while self-organizing maps with a small number of nodes behave in a way that is similar to the K-means method, larger self-organizing maps rearrange data in a way that is fundamentally topological in character. It is also common to use the U-matrix. The U-matrix value of a particular node is the average distance between the node and its nearest neighbors. In a rectangular grid, for example, we might consider the nearest 4 or 8 nodes. Large SOMs display properties that are emergent. Therefore, large maps are preferable to smaller ones. If the self-organizing map consists of thousands of nodes, it is possible to perform clustering operations on the map itself.

The aim of SOM-based learning is to cause different parts of the network to respond similarly to certain input patterns. This is partly motivated by the way that the visual, auditory or other sensory information is handled in separate parts of the cerebral cortex in the human brain. The weights of the neurons are initialized either as small random values or sampled evenly from the subspace spanned by the two largest principal component eigenvectors. Obviously, with the latter alternative, learning is much faster since the initial weights already give good approximation of SOM weights. The network must be fed a large number of example vectors that represent, as closely as possible, the kinds of vectors expected during the mapping process. The examples are usually administered multiple times. The training utilizes competitive learning methods. When a training example is fed to the network, its Euclidean distance to all weight vectors is calculated. The neuron with its weight vector most similar to the input is called the best matching unit (BMU). The weights of the BMU and neurons close to the input in the SOM lattice are then adjusted towards the input vector. The magnitude of the modification decreases with both time and the distance from the BMU. In the simplest form, the magnitude is one for all neurons close enough to BMU and zero for others. A Gaussian function is also a common choice. Regardless of the functional form, the neighborhood function shrinks with time. At the beginning, when the neighborhood is broad, the self-organizing operation takes place on a global scale. When the neighborhood has shrunk to just a couple of neurons, the weights are converging to local estimates. This process is repeated for each input vector for a large number of cycles. The network winds up the associated output nodes with groups or patterns in the input data set. If these patterns can be named, the names can be attached to the associated nodes in the trained net. During the mapping process, there will be one single winning neuron, i.e., the neuron whose weight vector lies nearest to the input vector. This can be simply determined by computing the Euclidean distance between the input and weight vectors. It should be noted that any kind of object that can be represented digitally, and with which an appropriate distance measure is associated and in which the necessary operations for training are possible, can be used to construct a self-organizing map.

Pedro et al. [91] described a system for querying 3D model databases based on the spin image representation as a shape signature for objects depicted as triangular meshes. The spin image representation facilitates the task of aligning the query object with respect to matched models. The main contribution of this work is the introduction of a three-level indexing schema with artificial neural networks. The indexing schema improves greatly the efficiency in matching the query spin images against those stored in the database. Their results are suitable for content-based retrieval in 3D general object databases. Their method achieves both compression and indexing of the original set of spin images. Basically, a self-organized map is built from the stack of spin images of a given object. This is a way of "summarizing" the whole stack into a set of representative spin images. Then, the kernel *K*-means clustering algorithm is utilized in order to group representative views in the SOM map into a reduced set of clusters. At the query time, the input spin images will be first compared with the clusters' centers resulting from the kernel *K*-means method and subsequently with the SOM map if a finer answer is requested.

4.5.3.3 KNN Learning

In pattern recognition, the k-nearest neighbor (KNN) algorithm is a method to classify objects based on nearest training examples in the feature space. KNN is a kind of instance-based learning or lazy learning, where the function is only approximated locally and all calculations are deferred until classification. KNN can also be used for regression. KNN is one of the simplest machine-learning algorithms. An object is classified by a majority vote of its neighbors, with the object being assigned to the class most common amongst its k nearest neighbors. k is a positive integer, typically small. If $k = 1$, then the object is simply assigned to the class of its nearest neighbor. In binary (i.e., two-class) classification problems, it is helpful to choose k to be an odd number to avoid tied votes. The same method can be used for regression by simply assigning the property value of the object to be the average of the values of its k nearest neighbors. It is useful if we weigh the contributions of the neighbors, such that the nearer neighbors contribute more to the average than the more distant ones. The neighbors are taken from a set of objects for which the correct classification (or, in the case of regression, the value of the property) is known. This can be regarded as the training set for the algorithm, though no explicit training step is required. In order to identify neighbors, the objects are represented by position vectors in the multi-dimensional feature space. Usually the Euclidean distance is adopted, though other distance measures, such as the Manhattan distance, could in principle be used instead. The k-nearest neighbor algorithm is sensitive to the local structure of the data.

The training examples are vectors in the multi-dimensional feature space. The space is partitioned into regions by locations and labels of the training samples. A point in the space is assigned to the class c if it is the most frequent class label among the k nearest training samples. Usually the Euclidean distance is adopted as the distance metric, but this will only work for numerical values. In other cases, e.g., text classification, another metric, such as the overlap metric (or the Hamming distance) can be adopted. The training stage of the algorithm consists only of storing the feature vectors and class labels of the training samples. At the actual classification stage, the test sample (whose class is unknown) is represented as a vector in the feature space. Distances from the new vector to all stored vectors are calculated and k closest samples are selected. There are many ways to classify the new vector to a particular class, and one of the most frequently used techniques is to predict the new vector to the most common class amongst the k nearest neighbors. The major drawback of this technique is that the classes with the more frequent examples tend to dominate the prediction of the new vector, since they tend to come up in the k nearest neighbors when the neighbors are calculated due to their large number. One way to alleviate this problem is to consider the distance of each k nearest neighbors with the new vector that is to be classified and predict the class of the new vector based on these distances.

The best choice of k depends upon the data. In general, larger values of k reduce the effect of noise on the classification, but make boundaries between classes less distinct. A suitable k can be selected by various heuristic techniques, e.g., cross-validation. The special case where the class is predicted to be the class of the closest training sample (i.e. when k=1) is called the nearest neighbor algorithm. The accuracy of the KNN algorithm will be severely degraded if there are noisy or irrelevant features, or if the feature scales are not consistent with their importance. Many research efforts have been put into selecting or scaling features to improve classification. A particularly popular approach is to utilize evolutionary algorithms to optimize feature scaling. Another popular approach is to scale features by the mutual information of the training data with the training classes.

Ip et al. [92] proposed a weighted similarity function for CAD model classification based on an underlying shape distribution feature representation and a KNN learning algorithm. Given a set of CAD solid models and corresponding classes, the KNN learning method was used to extract the related patterns to automatically construct a model classifier and identify new or hidden classifications using the shape distribution feature, learning from the stored, correctly categorized training examples. In addition, probabilistic approaches, such as Bayes theorem, are also a practical way for similarity matching, in which specific probabilities of features are calculated and the 3D model having the highest probability will be identified as the closest matching result [93].

4.5.3.4 Relevance Feedback

Relevance feedback is a feature of some information retrieval systems. The idea behind relevance feedback is to take the results that are initially returned from a given query and to use information about whether or not those results are relevant to perform a new query. There are mainly three types of feedback, i.e., explicit feedback, implicit feedback and blind or "pseudo" feedback. Explicit feedback is obtained from assessors of relevance indicating the relevance of a document retrieved for a query. This type of feedback is defined as explicit only when the assessors (or other users of a system) know that the provided feedback is interpreted as relevance judgments. Users may indicate relevance explicitly using a binary or graded relevance system. Binary relevance feedback indicates that a document is either relevant or irrelevant for a given query. Graded relevance feedback indicates the relevance of a document to a query on a scale using numbers, letters or descriptions (such as "not relevant", "somewhat relevant", "relevant", or "very relevant"). Graded relevance may also take the form of a cardinal ordering of documents created by an assessor that places documents of a result set in order of (usually descending) relevance. And example of this would be the "SearchWiki" feature recently implemented by Google on their search website. SearchWiki is a Google search feature which allows logged-in users to annotate and re-order search results. The annotations and modified order only apply to the user's searches, but it is possible to view other users' annotations for a given search query. A performance metric which became popular around 2005 to measure the effectiveness of a ranking algorithm based on the explicit relevance feedback is normalized discounted cumulative gain (NDCG). Discounted cumulative gain (DCG) is a measure of the effectiveness of a Web search engine algorithm or related applications, often used in information retrieval. Using a graded relevance scale of documents in a search engine result set, DCG measures the usefulness, or gain, of a document based on its position in the result list. The gain is accumulated cumulatively from the top of the result list to the bottom, with the gain of each result discounted at lower ranks. Other measures include the precision at k (i.e., precision of top k results) and the mean average precision. Implicit feedback is inferred from the user behavior, such as noting which documents they do or do not select for viewing, the duration of time spent in viewing a document, or page browsing or scrolling actions. The key differences between implicit and explicit relevance feedback include the following: The user is not assessing relevance for the benefit of the IR system, but only satisfying their own needs and the user is not necessarily informed that their behavior (selected documents) will be used as relevance feedback. An example of this is the Surf Canyon browser extension, which advances search results from later pages of the result set based on both the user interaction (clicking an icon) and the time spent in viewing the page linked to a search result. Blind or "pseudo" relevance feedback is obtained by assuming that the top k documents in the result set containing n results (usually where $k << n$) are relevant. Blind feedback automates the manual part of relevance feedback and has the advantage that assessors are not required.

Actually, machine-learning methods can also be used to implement users' relevance feedback mechanism in 3D model retrieval to iteratively refine the retrieval results step by step, by making designed reactions to the user's interactive evaluations. This can also achieve a personalized retrieval, based on different user's preferences. A good example is Elad et al.'s work on relevance feedback [70, 94]. They made use of the SVM learning algorithm to derive the optimal weight combination for a weighted Euclidean distance metric, and made stepwise improvements to the similarity match, according to every iteration of the user's interactive evaluation. The detailed approach can be illustrated as follows.

Assuming that two feature vectors $\boldsymbol{X}$ and $\boldsymbol{Y}$ constitute partial descriptions of database objects D_X and D_Y respectively, we can measure the distance between the objects using the squared Euclidean distance

$$d(D_X,\ D_Y) = \|\boldsymbol{X} - \boldsymbol{Y}\|^2. \tag{4.52}$$

Using the Euclidean distance alone, the automatic search of the database will indeed produce objects that are geometrically close to the given one. However, these may not be what the human user has in mind when initiating the search. Therefore, Elad et al. employed further "parameterization" of this distance by adding weights and a bias value

$$d(D_X,\ D_Y) = (\boldsymbol{X} - \boldsymbol{Y})^{\mathrm{T}} \boldsymbol{W} (\boldsymbol{X} - \boldsymbol{Y}) + b, \tag{4.53}$$

where $\boldsymbol{W}$ may be any matrix, yet in the following we assume it is a diagonal matrix.

Given a set of search results, a human user may consider some of them relevant and some of them irrelevant, no matter that they are all geometrically close. The adaptation of the distance function can be done by re-computing distances, based on the user preferences. The additional requirement is that the new distance between the given object and the relevant results should be small and, obviously, the new distance between the given object and the irrelevant results should be large. In essence, this is a classification or a learning problem. One way of formulating the requirements is to define weights on the components of the distance function and writing a set of constraints. Denote the feature vector of the object for which the system is to search by $\boldsymbol{O}$, the feature vectors of the "relevant" results by $\{\boldsymbol{G}_k\}_{k=1}^{n_G}$, and the feature vectors of the "irrelevant" results by $\{\boldsymbol{B}_l\}_{l=1}^{n_B}$. The constraints posed on the weight function are as follows:

$$\begin{aligned} &k=1,\ 2,\ \ldots,\ n_G,\ d(D_O,D_{G_k})=[\boldsymbol{O}-\boldsymbol{G}_k]^{\mathrm{T}}\boldsymbol{W}[\boldsymbol{O}-\boldsymbol{G}_k]+b\le 1,\\ &l=1,\ 2,\ \ldots,\ n_B,\ d(D_O,D_{B_l})=[\boldsymbol{O}-\boldsymbol{B}_l]^{\mathrm{T}}\boldsymbol{W}[\boldsymbol{O}-\boldsymbol{B}_l]+b\ge 2. \end{aligned} \tag{4.54}$$

This generates a margin between the "relevant" and "irrelevant" results. The above inequalities are linear with respect to the entries of $\boldsymbol{W}$. Denoting the main diagonal of $\boldsymbol{W}$ by $\boldsymbol{\omega}$, we may rewrite the constraints as follows:

$$\begin{aligned} &k=1,\ 2,\ \ldots,\ n_G,\ d(D_O,D_{G_k})=[\boldsymbol{O}-\boldsymbol{G}_k]^2\cdot\boldsymbol{\omega}+b\le 1,\\ &l=1,\ 2,\ \ldots,\ n_B,\ d(D_O,D_{B_l})=[\boldsymbol{O}-\boldsymbol{B}_l]^2\cdot\boldsymbol{\omega}+b\ge 2. \end{aligned} \tag{4.55}$$

where the notation $\boldsymbol{V}^2$ means multiplying each vector entry of $\boldsymbol{V}$ by itself. An additional constraint is that the entries of $\boldsymbol{W}$ are all non-negative. Note that we do not require b to be non-negative, which may therefore end up with a non-metric similarity measure.

It can be shown that the maximal margin of separation between the two sets of results is achieved by the $\boldsymbol{\omega}$ with the smallest squared norm, $\min_{\boldsymbol{\omega}}\left(\|\boldsymbol{\omega}\|\right)^2$. Choosing the $\boldsymbol{\omega}$ with the smallest norm also renders the solution to the constraint system robust to the number of examples from each of the two subsets, "relevant" and "irrelevant", and also the size of the rest of the database. This is good when the above constraints are insufficient, i.e., $n_G+n_B \ll U$, U being the arity of the feature vectors. That is, there are more unknowns than inequalities, and therefore multiple possible solutions $\{\boldsymbol{\omega},b\}$ all satisfy the constraints.

Thus, at each refinement iteration we essentially need to solve the following problem for $\boldsymbol{\omega}$:

Minimize $\|\boldsymbol{\omega}\|^2$,

Subject to:

$$\begin{aligned} &k=1,\ 2,\ \ldots,\ n_G, d(D_O,D_{G_k})=[\boldsymbol{O}-\boldsymbol{G}_k]^2\cdot\boldsymbol{\omega}+b\le 1,\\ &l=1,\ 2,\ \ldots,\ n_B,\ d(D_O,D_{B_l})=[\boldsymbol{O}-\boldsymbol{B}_l]^2\cdot\boldsymbol{\omega}+b\ge 2,\\ &\boldsymbol{\omega}\ge 0. \end{aligned} \tag{4.56}$$

This quadratic optimization problem may be solved either directly or through the dual problem which proves easier when the number of constraints is much lower than the number of unknowns, i.e., $n_G + n_B << M$. The use of the bias in the formulation is crucial since it frees us from considering the boundary values and therefore choosing these values to be 1 and 2 does not lose generality.

The system may use the new, refined distance function to perform a new search, offering the user a set of results to better suit personal preferences. The user may, on this new set of results, mark preferences as was done for the previous search results. The new "relevant" and "irrelevant" results sets may now be used to further refine the distance function. There is no limit imposed by the system on the number of refinement iterations allowed. However, practical experiments showed that very few iterations are required for any example before a human user is satisfied with the output search results.

4.5.4 Semantic Measurements

As the 3D model retrieval results achieved by low-level features have proven not to be as discriminative as people had expected, this raises another important issue, that is, subjective semantic measurement in similarity comparison. Furthermore, whether a retrieved 3D model is "relevant" or "irrelevant" to the query is also judged by the users according to their subjective perception, related to the semantic content. Consequently, it is highly significant to develop semantic similarity-matching methods that take human perception into account in content-based 3D model retrieval systems.

Many approaches that have been proposed in 2D media retrieval to reduce the "semantic gap" try to perform similarity measurement based on high-level semantics. One method is to learn the connections between a 3D model and a set of semantic descriptors, or the semantic meanings from those automatically extracted 3D model features. This approach is usually based on machine learning and statistical classification, which groups 3D models into semantically meaningful categories using low-level features so that semantically-adaptive searching methods can be applied to different categories. Examples are as follows. Suzuki et al. [78] constructed a multidimensional scaling mechanism so that semantic keyword descriptors used in the query and the shape features calculated from the 3D shapes were strongly correlated, based on a training data set. The multidimensional scaling mechanism can analyze matrices of similar or dissimilar data by representing the rows and the columns as a point in Euclidean space and then measure their similarities using Euclidean distances. They then created a special user preference space according to this principle, in which a function mapping from the 3D model space was constructed to integrate semantic keywords and 3D shapes as a representation of human subjective perception. Zhang et al. [95] introduced the concept of "hidden annotation" to construct a semantic tree of the whole 3D model database. They used an active learning method to calculate a list of probabilities for each 3D model, which indicated the model's probability of having a certain semantic attribute. The list of probabilities was then utilized to calculate the semantic distance between two models, or between the user query and a model in the database. The overall dissimilarity between two models was finally determined by combining the weighted sum of the semantic distance with the low-level feature distance. In [90], a novel semantic measurement that could simulate human visual perception was also presented. It was achieved by employing a well-trained SVM learning classifier constructed by performing SVM learning on the tagged similar and dissimilar models in the retrieval results of the current querying step. An SVM-based semantic clustering and retrieval method was also successfully implemented in the prototypical 3D engineering shape search system (3-DESS) designed by Purdue University [96].

In addition, some concept hierarchies, such as predefined domain ontology, can also be introduced into the semantic measuring process. There is some work involved in building a fundamental framework for representing and measuring the semantic information of 3D models, such as the "Aim@shape" project (http://www.aim-at-shape.net) launched by the European Commission in order to implement semantically capable digital representations of 3D shapes that are expected to acquire, build, transmit, retrieve and process shapes with their associated knowledge. This project is an attempt to formalize shape knowledge (in particular, metadata, used for knowledge-based shape modeling) and define shape ontologies in specific contexts used for linking semantic keywords to shape features.

Shape knowledge representation is built on three basic levels: geometric, structural and semantic, where, at the semantic level, the association of specific semantics to structured and geometric models is established through automatic annotation of shapes or shape parts according to the concepts formalized by the domain ontology. Furthermore, by introducing a common formalization framework, it is also possible to build a shared semantic conceptualization of a multilayered architecture for shape models.

Another effective method is to perform user relevance feedback after each search iteration in the database [70, 97, 98]. This is effective in narrowing the gap between the low-level feature similarity and the high-level semantic similarity [70] by which, what the user has in mind is able to be better captured. To some extent, it is also regarded as a method of semantic measurement and has been extensively used in 2D media retrieval [98, 99]. In the case of 3D retrieval, Leifman et al. [100] proposed a relevance feedback method combining query refinement and supervised feature extraction at each step, which tried to find an optimal linear transformation that reweighs the low-level feature components so as to achieve the maximal separation of the original result set. They found that this projection by maximizing a cost function is defined as Fisher's Linear Discriminant Criterion. Atmosukarto et al. [101] also presented a subjective similarity-measurement-based relevance feedback process by combining various distances measured for different feature representations. This was implemented by computing the integer rank $r_k(O_i|O_j)$ of the 3D object O_i with respect to the 3D object O_j based on a probability estimation method in the feature space of the "relevant" and "irrelevant" result sets.

4.6 Query Style and User Interface

A content-based 3D model retrieval system is expected to allow users to submit their query in a natural and interactive way. Giving what kind of query interface to a user is a key problem in a 3D model retrieval system that has significant application. The query interface should be convenient while searching for models whose functions include, on the one hand, how users express features of the promising model in a retrieval system, while the descriptors may present different formats, such as text, draft or use case query. On the other hand, since the evaluation of retrieval results is finally completed by users, the system should be able to carry out optimization operations according to users' feedback. Due to the abundance of content descriptions of 3D models, there should be a variety of query specifications to be supported as follows.

4.6.1 Query by Example

In traditional information retrieval, Query by example (QBE) is a database query language for relational databases. It was devised by Moshé M. Zloof at IBM Research during the mid 1970s, in parallel to the development of SQL. It is the first graphical query language based on visual tables where the user would enter commands, example elements and conditions. Many graphical front-ends for databases use the ideas from QBE today. Based on the notion of domain relational calculus, QBE can be used as a search tool as well. A QBE parser parses the search query and looks for the keywords while eliminating words like "a", "an" or "the". A more formal query string, in languages such as SQL, is then generated and which is finally executed. However, when compared with a formal query, the results in the QBE system will be more variable. The user can also search for similar documents based on the text of a full document that he or she may have. This is accomplished by the user's submission of documents (or numerous documents) to the QBE result template. The analysis of these documents the user has inputted via the QBE parser will generate the required query. QBE is a seminal work in end-user development, frequently cited in research papers as an early example of this topic. Currently, QBE is supported in different object-oriented databases.

In content-based 3D model retrieval, QBE means that a 3D model example is directly provided as a query, which is also called the Use Case interface. Three categories should be mentioned [33, 102, 103]: first, the example model is a user-owned model or an existing model in a certain URL address; second, a certain model fetched from the return of the last retrieval process is provided, i.e., secondary retrieval; third, we can directly choose the model in the database to commit a query, which is called bank retrieval. QBE is the most common query interface up to now. Fig. 4.17 gives a typical example of the QBE-based 3D model retrieval system developed by the authors of this book, where the "car" model in the upper-left corner of the interface is the query model inputted by the users, while the returned 16 similar models are listed below.

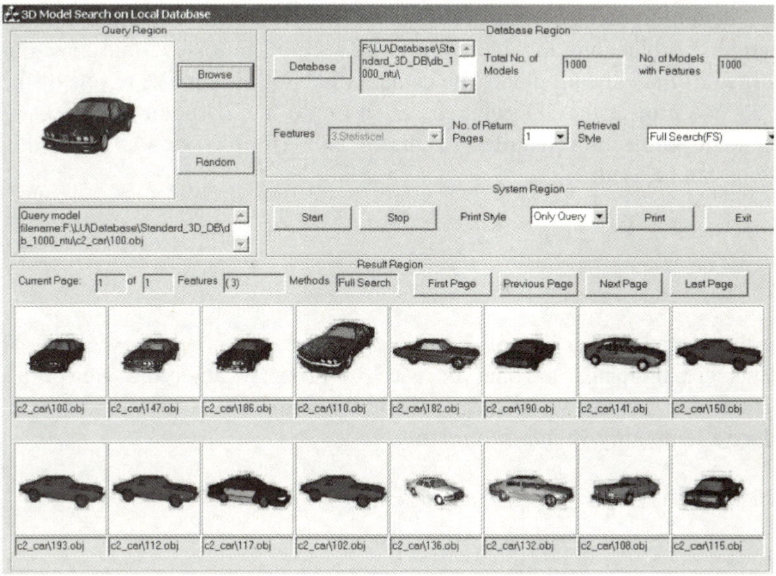

Fig. 4.17. The QBE-based 3D model retrieval demo system developed by the authors of this book

4.6.2 Query by 2D Projections

Draft or sketch is the most extensively applied query interface in practice. Since users paint basic features of a 3D model based on conception, the system extracts shape features from the drafts to match and retrieve in the database. The 2D draft is currently very attractive in image retrieval and afterwards can be extended to view based 3D model retrieval. In such a manner, with a number of drafts drawn by users as query request, the matching operation is conducted according to 2D projections of the 3D object from different view angles. Apart from the 2D sketches interface, there also exist 3D draft query interfaces. Teddy is a very typical 3D draft editing environment. For 2D-stroke-based users' input, it can construct 3D shape in accordance with certain rules. The technology has been adopted by the 3D search engine in Princeton University as a user input interface. In the subsequent three subsections, we will introduce query by 2D projections, query by 2D sketches and query by 3D sketches, respectively.

3D to 2D projection denotes any method of mapping 3D points to a 2D plane. Since most of the current methods for displaying graphical data are based on planar 2D media, the use of this type of projection is widespread, especially in computer graphics, engineering and drafting. There are two typical projection methods, i.e., orthographic projection and perspective projection, which can be described as follows:

(1) Orthographic projections are a small set of transforms often used to show profile, detail or precise measurements of a 3D object. Common names for orthographic projections include plan, cross-section, bird's-eye and elevation. If the normal of the viewing plane (the camera direction) is parallel to one of the 3D axes, e.g., to project the 3D point (a_x, a_y, a_z) onto the 2D point (b_x, b_y) using an orthographic projection parallel to the y axis (profile view), the following equations can be used:

$$\begin{aligned} b_x &= s_x a_x + c_x, \\ b_y &= s_z a_z + c_z, \end{aligned} \tag{4.57}$$

where the vector $\boldsymbol{s}$ is an arbitrary scale factor and $\boldsymbol{c}$ is an arbitrary offset. These constants are optional, and can be used to properly align the viewport. The projection can be shown through the following matrix notation, where we introduce a temporary vector $\boldsymbol{d}$ for clarity.

$$\begin{aligned} \begin{bmatrix} d_x \\ d_y \end{bmatrix} &= \begin{bmatrix} 1 & 0 & 0 \\ 0 & 0 & 1 \end{bmatrix} \begin{bmatrix} a_x \\ a_y \\ a_z \end{bmatrix}, \\ \begin{bmatrix} b_x \\ b_y \end{bmatrix} &= \begin{bmatrix} s_x & 0 \\ 0 & s_z \end{bmatrix} \begin{bmatrix} d_x \\ d_y \end{bmatrix} + \begin{bmatrix} c_x \\ c_z \end{bmatrix}. \end{aligned} \tag{4.58}$$

While orthographically projected images represent the 3D nature of the object projected, they do not represent the object as it would be recorded photographically or perceived by a viewer, who observes it directly. In particular, parallel lengths at all points in an orthographically projected image are of the same scale regardless of whether they are far away or near to the virtual viewer. As a result, lengths close to the viewer appear foreshortened.

(2) The perspective projection requires greater definition. A conceptual aid in understanding the mechanics of this projection involves treating the 2D projection as being viewed through a camera viewfinder. The camera's position, orientation and field of view control the behavior of the projection transformation. The following variables are defined to describe this transformation:

$\boldsymbol{a}_{x,y,z}$: the point in the 3D space that is to be projected;

$\boldsymbol{c}_{x,y,z}$: the location of the camera;

$\boldsymbol{\theta}_{x,y,z}$: the rotation of the camera. When $\boldsymbol{c}_{x,y,z}$ = (0, 0, 0), and $\boldsymbol{\theta}_{x,y,z}$ =(0, 0, 0), the 3D vector (1, 2, 0) is projected to the 2D vector (1, 2);

$\boldsymbol{e}_{x,y,z}$: the viewer's position relative to the display surface.

which results in:

$\boldsymbol{b}_{x,y}$: the 2D projection of $\boldsymbol{a}$.

First, we define a point $\boldsymbol{d}_{x,y,z}$ as a translation of Point $\boldsymbol{a}$ into a coordinate system defined by $\boldsymbol{c}$. This is achieved by subtracting $\boldsymbol{c}$ from $\boldsymbol{a}$, and then applying a vector rotation matrix using $-\boldsymbol{\theta}$ to the result. This transformation is often called a camera transform (note that these calculations assume a left-handed system of axes):

$$\begin{bmatrix} d_x \\ d_y \\ d_z \end{bmatrix} = \begin{bmatrix} 1 & 0 & 0 \\ 0 & \cos\theta_x & -\sin\theta_x \\ 0 & \sin\theta_x & \cos\theta_x \end{bmatrix} \begin{bmatrix} \cos\theta_y & 0 & \sin\theta_y \\ 0 & 1 & 0 \\ -\sin\theta_y & 0 & \cos\theta_y \end{bmatrix} \begin{bmatrix} \cos\theta_z & -\sin\theta_z & 0 \\ \sin\theta_z & \cos\theta_z & 0 \\ 0 & 0 & 1 \end{bmatrix} \left(\begin{bmatrix} a_x \\ a_y \\ a_z \end{bmatrix} - \begin{bmatrix} c_x \\ c_y \\ c_z \end{bmatrix} \right). \tag{4.59}$$

This transformed point can then be projected onto the 2D plane using the formula (here *x*-*y* is used as the projection plane, though other literatures may also use *x*-*z*):

$$\begin{aligned} b_x &= (d_x - e_x)(e_z / d_z), \\ b_y &= (d_y - e_y)(e_z / d_z). \end{aligned} \tag{4.60}$$

The distance of the viewer from the display surface, e_z, directly relates to the field of view, where $\alpha = 2\tan^{-1}(1/e_z)$ is the viewed angle. Note that this assumes that you map the points $(-1, -1)$ and $(1,1)$ to the corners of your viewing surface. Subsequent clipping and scaling operations may be necessary to map the 2D plane onto any particular display media.

In content-based 3D model retrieval, 2D projection views themselves can be adopted as features of a 3D model [104], while the query by 2D projections means representing a query with a set of 2D projection images of a 3D example model from different viewpoints [33]. Since both 2D projection and 2D sketch are 2D images, readers can refer to Fig. 4.18 as a similar demo system of query by 2D projections-based 3D model retrieval.

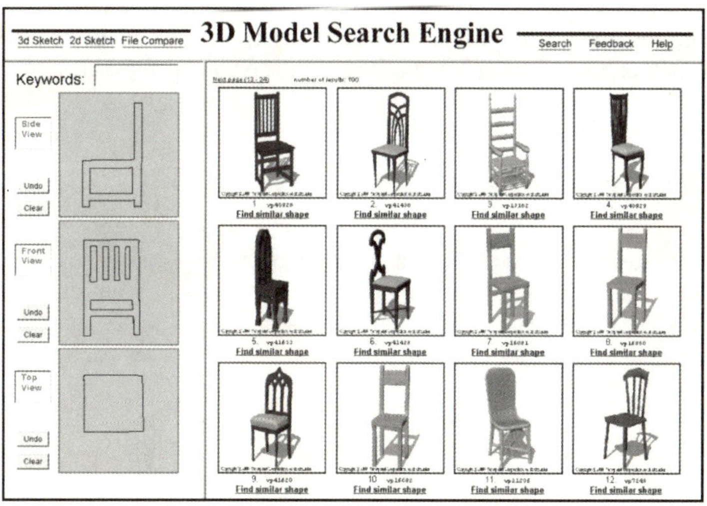

Fig. 4.18. Query by 2D sketch [105] (With courtesy of Min et al.)

4.6.3 Query by 2D Sketches

In content-based 3D model retrieval systems, query by 2D sketches means using 2D shapes sketched interactively by users as queries. Min et al. [33, 105] designed an interactive 2D sketch online interface, as shown in Fig. 4.18. The key problem is how to match 2D sketches to 3D objects, which is significantly different from classical problems in computer vision: the 2D input is hand-drawn rather than photographic and the interface is interactive. We must consider several new questions: How do people draw shapes? Which viewpoints do they select? How should the interface guide the user's input? What algorithms are robust enough to recognize human-drawn sketches?

To investigate these questions, Min et al. ran a pilot study in which 32 students were asked to draw three views of 8 different objects, with a time limit of 15 seconds per object. Min et al. found that people tend to sketch objects with fragmented boundary contours and few other lines, they are not very geometrically accurate, and they use a remarkably consistent set of view directions. Interestingly, the most frequently chosen views were not the characteristic views predicted by perceptual psychology, but instead ones that were simpler to draw (i.e. front, side and top views). Min et al. matched the n user sketches with projected 2D images of each 3D model in the database rendered from m different viewpoints ($m>n$). A model's similarity score is the minimal sum of n pairwise sketch-to-image similarity scores, subject to the constraint that no image can be matched to more than one sketch. These pairwise scores are calculated by comparing their shape signatures. These signatures are based on the amplitudes of the Fourier coefficients of a set of functions obtained by intersecting the 2D Euclidian distance transform of the image with a set of concentric circles. By taking the amplitude of each coefficient, we discard phase information, thereby making the signature rotation invariant. The distance transform of the image helps make Min et al.'s method robust to small variations in the positions of lines.

4.6.4 Query by 3D Sketches

In content-based 3D model retrieval systems, query by 3D sketches means using a 3D shape sketched interactively by users as queries. Min et al. [18] also implemented a 3D sketch online interface based on a 3D sketch tool Teddy, which was designed by Igarashi et al. [27, 106]. Fig. 4.19 gives an example of query by 3D sketches.

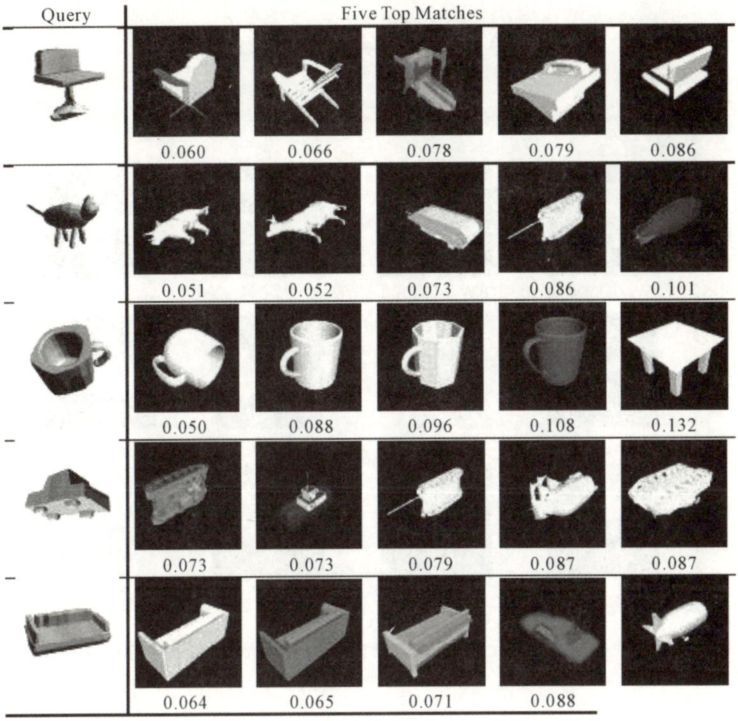

Fig. 4.19. Query by 3D sketches [105] (With courtesy of Min et al.)

4.6.5 Query by Text

Query by text means that the query interface is based on text keywords [33] and/or semantic descriptions [95]. Attempting to find a 3D model using just text keywords suffers from the same problems as any text search: a text description may be too limited, incorrect, ambiguous, or in a different language. Furthermore, 3D models contain shape and appearance information which is hard to query just based on text. In many cases, a shape query is able to describe a property of a 3D model that is hard to specify only adopting text. As shown in Fig. 4.20, query by the too common keyword "plane" will produce worse retrieval results. Thus, we often combine the text-based query with the sketch-based query, as discussed in the subsection below.

Fig. 4.20. The retrieval results for the query by the text keyword "plane" [105] (With courtesy of Min et al.)

4.6.6 Multimodal Queries and Relevance Feedback

Multimodal queries stand for combinations of multiple query representations mentioned earlier. In general, a query that is simultaneously done by integrating multiple-query specifications is more likely to produce better results than using any individual one. Moreover, the user interface of 3D model retrieval is responsible for displaying retrieval results to users in a visual and interactive way as well, in order to make the users browse them or pursue the next retrieval iteration easily. Fig. 4.21 shows an example of retrieval-based multimodal queries, query by text and query by 2D sketch.

Some 3D model retrieval systems also introduced an interactive user relevance feedback mechanism into their query interface. For example, a simple user relevance feedback interface is to give users a chance to mark a subset of the initial retrieval results as "relevant" or "irrelevant", using a "+" or a "–" symbol, as shown in Fig. 4.22. Zhang et al. [107] extended this kind of feedback interface by adding a way to mark the extent of "relevant" and "irrelevant," providing both qualitative and quantitative adjustments. Similar work can also be found in [100, 101]. The iterative refinement can automatically narrow the perception gap between the retrieval system and the users, which is expected to enhance the retrieval performance.

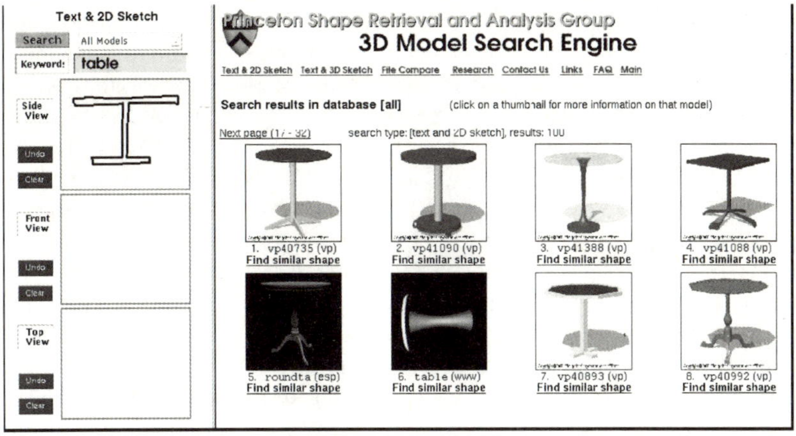

Fig. 4.21. The retrieval results for the query by the text keyword "table" and 2D sketch [105] (With courtesy of Min et al.)

Fig. 4.22. Relevance feedback interface developed by the authors of this book

4.7 Summary

From the beginning, content-based 3D model retrieval has witnessed much development and many achievements in both theory and application. There have been already a number of prototypes, standalone systems and Internet-based search engines implemented and publicized for the purpose of research. For example, "Nefertiti" [102] is the first content-based 3D model retrieval system for general use, where tensor of inertia, distribution of normals, distribution of cords and multiresolution analysis are used to describe each model. The database can be searched by scale, shape or color or any combination of these parameters. A user friendly interface makes the retrieval operation simple and intuitive and allows the editing of reference models according to the specifications of the user. The web-based 3D search engine [18] designed by Princeton University provides multi-modal query types. Available search types are Text & 2D Sketch, Text & 3D Sketch, File Compare and Find Similar Shape. The National Taiwan University [77] provides a web-based 3D model retrieval system in which features are represented using MPEG-7 Shape 3D descriptor and MPEG-7 Multi-view descriptor, so that it is also available for PC users.

Moreover, there are also some professional 3D model retrieval systems. For example, Ankerst et al. [34] developed a content-based retrieval system for 3D protein databases, while Heriot-Watt University implemented a web-based search engine, ShapeSifter (URL: http://www.shapesearch.net) and Drexel University (URL: http://edge.mcs.drexel.edu/repository/frameset.html) built a digital library for 3D CAD models and 3D engineering designs [108, 109]. Another noticeable trend is the 3D model retrieval service for handsets such as mobile phones and personal digital assistants. For example, Suzuki et al. [110] developed a 3D model retrieval prototype for mobile phone users. Moreover, a 3D model retrieval system adopting the MPEG-7 mechanism can also be easily tailored to Pocket-PC users.

Nevertheless, the accuracy of content information in 3D models, as a consequence of its versatile aspects and subjective cognition, is very much in question. Much work still needs to be undertaken to remedy this situation. The following are just some of the crucial issues and challenges deserving further investigation.

Research on a unified 3D model retrieval framework should be carried out urgently. The 3D data representations are very diversiform while the content of a 3D model remains independent of them, and thus the unified 3D model retrieval framework has been the main focus of attention. A practical unified 3D retrieval framework should be capable of accommodating most 3D data representations adaptively by extracting representation-independent features or performing standard transformations on the fly. Moreover, considering the efficiency of transmitting and retrieving 3D models over the Internet, performing the feature extraction and similarity matching operations directly from compressed 3D data are also meaningful.

It is important to develop more discriminative 3D shape features, especially those that are normalization-free and possess strong discriminative power. They must also be natural and simple for effective index mechanisms. Secondly, local partial shape feature extraction is required to achieve the feature vectors that are suitable for partial matching inside a 3D model. In practice, partial shape features that can describe the local details are often needed for more precise multiresolution and flexible retrieval. Further, multiple features need to be combined for effective similarity matching. Some work has already been undertaken toward this goal [23, 81, 111]. However, when a large number of feature descriptors are used for the query, the system may not be able to respond quickly because of the high computational complexity. Therefore, feature descriptor selection or reduction techniques must be designed and applied.

Consequently, how to select and weigh those feature descriptors is also an important and promising future research direction. In addition, it is essential to further develop non-shape descriptors of 3D models based on material colors and texture. Furthermore, extraction of high-level semantic features and similarity measurements combined with semantic information will also be important research issues and challenges.

With respect to user interfaces and query styles, it is significant to carry out research on the mechanism of relevance feedbacks and personalized retrieval integrating user preferences, by which users are able to tune the search criteria by themselves toward more satisfactory search results. In addition, the development of simple but powerful query interfaces is always one of our main concerns. The 3D sketching tool currently used for 3D shape queries is not user-friendly for novices. A less complex way for users to build simple 3D objects and 3D sketches should be provided, for example an interface that allows users to form a complicated 3D object by connecting some basic shapes, just like using building blocks. Besides, a more effective query interface that is able to locate objects in a non-rigid-body transformation should also be designed.

Finally, we should face up to retrieval issues targeted at 3D scenes that contain multiple 3D models. Currently, retrieval methods are mostly limited to the single 3D model. However, in many applications, such as a virtual reality environment, 3D models are usually presented in complex 3D scenes. Therefore, 3D model retrieval technology should be extended to handle more complex 3D scenes. A novel hierarchical object structure of 3D scenes may need to be investigated, to localize and recognize the 3D objects in a 3D scene.

References

[1] Y. X. Chen and J. Z. Wang. Machine Learning and Statistical Modeling Approaches to Image Retrieval. Kluwer, 2004.

[2] 3D Cafe Free 3D Models Meshes [Online]. Available: http://www.3dcafe.com. 2003.

[3] National Design Repository [Online]. Available: http://www.deepfx.com/meshnose. 2003.

[4] Y. Yang, H. Lin and Y. Zhang. Content-based 3D model retrieval: a survey. IEEE Transactions on Systems, Man and Cybernetics—Part C: Applications and Reviews, 2007, 37(6):1081-1098.

[5] J. Jia, Z. Qin, Q. Zhang, et al. An overview of content-based three-dimensional model retrieval methods. Paper presented at The IEEE International Conference on Systems Engineering, 2008, pp. 1-6.

[6] Z. Qin, J. Jia and J. Qin. Content based 3D model retrieval: a survey. Paper presented at The International Workshop on Content-Based Multimedia Indexing, 2008, pp. 249-256.

[7] E. Paquet and M. Rioux. The MPEG-7 standard and the content-based management of three-dimensional data: a case study. In: Proceedings of 1999 IEEE International Conference on Multimedia Computing and Systems, 1999,

pp. 375-380.
[8] P. Shilane, M. Kazhdan, P. Min, et al. The Princeton shape benchmark. In: Proceedings of Shape Modeling International, 2004.
[9] J. Tangelder and R. Veltkamp. Polyhedral model retrieval using weighted point sets. Int. J. Image Graph., 2003, 3:1-21.
[10] T. Zaharia and F. Prěteux. 3D versus 2D/3D shape descriptors: A comparative study. In: Proc. SPIE Conf. Image Process.: Algorithms Syst. III—SPIE Symp. Electron. Imaging, Sci. Technol., 2004, Vol. 5298, pp. 47-58.
[11] Meshnose, the 3D Objects Search Engine. [Online]. Available: http://www.deepfx.com/meshnose. 2003.
[12] National Design Repository. [Online]. Available: http://www.deepfx.com/meshnose. 2003.
[13] H. Berman, J. Westbrook, Z. Feng, et al. The protein data bank. Nucleic Acids Res., 2000, 28:235-242.
[14] K. Jarvelin and J. Kekalainen. IR evaluation methods for retrieving highly relevant documents. In: Proc. 23rd ACM SIGIR Conf. Res. Dev. Inf. Retrieval, 2000, pp. 41-48.
[15] J. Rocchio. Relevance feedback in information retrieval. The SMART Retrieval System: Experiments in Automatic Document Processing, Prentice-Hall, Englewood Cliffs, NJ, 1971, pp. 313-323.
[16] B. Bustos, D. Keim, D. Saupe, et al. An experimental comparison of feature-based 3D retrieval methods. Paper presented at The Int. Symp. 3D Data Process., Vis., Transmiss., 2004, pp. 215-222.
[17] S. M. Beitzel. On Understanding and Classifying Web Queries. Ph.D Thesis, 2006.
[18] P. Min. A 3D model search engine. Ph.D Dissertation. Dept. Comput. Sci. Princeton Univ., Princeton, NJ, 2004.
[19] R. Osada, T. Funkhouser, B. Chazelle, et al. Matching 3D models with shape distributions. Shape Modeling International, 2001, pp. 154-166.
[20] A. W. M. Smeulders, M. Worring, S. Santini, et al. Content-based image retrieval in the early years. IEEE Trans. Pattern Anal. Mach. Intell., 2000, 22(12):1349-1380.
[21] R. Ohbuchi, M. Nakazawa and T. Takei. Retrieving 3D shapes based on their appearance. In: Proc. 5th ACM SIGMM, Int. Workshop Multimedia Inf. Retrieval, Berkeley, CA, 2003, pp. 39-45.
[22] R. Ohbuchi and T. Takei. Shape-similarity comparison of 3D models using alpha shapes. In: Proc. 11th Pacific Conf. Comput Graph. Appl. (PG 2003), 2003, pp. 293-302.
[23] P. Min, M. Kazhdan and T. Funkhouser. A comparison of text and shape matching for retrieval of online 3D models. In: Proceedings of the 8th European Conference on Digital Libraries (ECDL 2004), 2004, pp. 209-220.
[24] M. Kazhdan, T. Funkhouser and S. Rusinkiewicz. Shape matching and anisotropy. ACM Trans. Graph., 2004, 23(3):623-629.
[25] J. W. H. Tangelder and R. C. Veltkamp. A survey of content based 3D shape retrieval methods. In: Proceedings of the Shape Modeling International 2004 (SMI'04), 2004, pp. 145-156.
[26] D. Y. Chen and M. Ouhyoung. A 3D model alignment and retrieval system. In: Proceedings of International Computer Symposium, Workshop on Multimedia Technologies, 2002, pp. 1436-1443.

[27] T. Funkhouser, P. Min, M. Kazhdan, et al. A search engine for 3D models. ACM Transactions on Graphics (TOG), 2003, 22:83-105.

[28] M. Kazhdan, T. Funkhouser and S. Rusinkiewicz. Rotation invariant spherical harmonic representation of 3D shape descriptors. In: Proceedings of the Eurographics/ACM SIGGRAPH Symposium on Geometry Processing, 2003, pp. 156-164.

[29] R. Osada, T. Funkhouser, B. Chazelle, et al. Shape distributions. ACM Transactions on Graphics (TOG), 2002, 21:807-832.

[30] E. Chávez, G. Navarro, R. Baeza-Yates et al. Searching in metric spaces. ACM Computing Surveys (CSUR), 2001, 33:273-321.

[31] C. Böhm, S. Berchtold and D. A. Keim. Searching in highdimensional spaces: Index structures for improving the performance of multimedia databases. ACM Computing Surveys (CSUR), 2001, 33:322-373.

[32] D. V. Vraníc and D. Saupe. 3D model retrieval. Paper presented at The Spring Conf. Comput. Graph. (SCCG 2000), 2000.

[33] P. Min, A. Halderman, M. Kazhdan, et al. Early experiences with a 3D model search engine. In: Proc. Web3D Symp., 2003, pp. 7-18.

[34] M. Ankerst, G. Kastenmuller, H. Kriegel, et al. Nearest neighbor classification in 3D protein databases. In: Proc. ISMB, 1999, pp. 34-43.

[35] K. Pearson. On lines and planes of closest fit to systems of points in space. Philosophical Magazine, 1901, 2(6):559-572.

[36] R. Ohbuchi, T. Otagiri, M. Ibato, et al. Shape-similarity search of three-dimensional models using parameterized statistics. In: Proc. 10th Pacific Conf. Comput. Graph. Appl., 2002, pp. 265-275.

[37] E. Paquet, A. Murching, T. Naveen, et al. Description of shape information for 2-D and 3-D objects. Signal Process.: Image Commun., 2000, 16:103-122.

[38] M. Heczko, D. Keim, D. Saupe, et al. A method for similarity search of 3D objects (in German). In: Proc. BTW, 2001, pp. 384-401.

[39] D. Vraníc, D. Saupe and J. Richter. Tools for 3D-object retrieval: Karhunen-Loeve transform and spherical harmonics. In: Proc. IEEE, Workshop Multimedia Signal Process, 2001, pp. 293-298.

[40] M. Kazhdan. Shape representations and algorithms for 3D model retrieval. Ph.D Dissertation, Dept. Comput. Sci., Princeton University, Princeton, NJ, 2004.

[41] S. Gottschalk. Collision queries using oriented bounding boxes. Ph. D Dissertation, Department of Computer Science, University of North Carolina at Chapel Hill, 1999.

[42] A. Tomas and H. Eric. Real-time Rendering (2nd ed.). A K Peters, Ltd., 2002, pp.564-567.

[43] J. Pu, Y. Liu, G. Xin, et al. Yusuke. 3D model retrieval based on 2D slice similarity measurements. In: Proc. 2nd International Symposium on 3D Data Processing, Visualization and Transmission (3DPVT 2004), 2004, pp. 95-101.

[44] M. de Berg, M. van Kreveld, M. Overmars, et al. Computational Geometry (2nd revised ed.). Springer-Verlag, 2000, pp.45-61.

[45] A. Fournier and D. Y. Montuno. Triangulating simple polygons and equivalent problems. ACM Transactions on Graphics, 1984, 3(2):153-174.

[46] A. Chazelle. Triangulating a simple polygon in linear time. Discrete & Computational Geometry, 1991, 6:485-524.

[47] R. Seidel. A simple and fast incremental randomized algorithm for computing trapezoidal decompositions and for triangulating polygons. Computational

Geometry: Theory and Applications, 1991, 1:51-64.
[48] M. Attene, S. Katz, M. Mortara, et al. Mesh segmentation - A comparative study. In: Proceedings of the IEEE International Conference on Shape Modeling and Applications, 2006, pp. 7.
[49] S. Katz and A. Tal. Hierarchical mesh decomposition using fuzzy clustering and cuts. ACM Trans. Graph. (SIGGRAPH), 2003, 22(3):954-961.
[50] S. Katz, G. Leifman and A. Tal. Mesh segmentation using feature point and core extraction. The Visual Computer, 2005, 21(8-10):865-875.
[51] M. Mortara, G. Patanè, M. Spagnuolo, et al. Blowing bubbles for the multi-scale analysis and decomposition of triangle meshes. Algorithmica, Special Issues on Shape Algorithms, 2004, 38(2):227-248.
[52] M. Mortara, G. Patanè, M. Spagnuolo, et al. Plumber: A multi-scale decomposition of 3D shapes into tubular primitives and bodies. In: Proc. of Solid Modeling and Applications, 2004, pp. 139-158.
[53] M. Attene, B. Falcidieno and M. Spagnuolo. Hierarchical mesh segmentation based on fitting primitives. The Visual Computer, 2006, 22(3):181-193.
[54] K. L. Low and T. S. Tan. Model simplification using vertex-clustering. In: Proceedings of the 1997 Symposium on Interactive 3D Graphics, 1997, pp. 75-82.
[55] H. P. Kriegel and T. Seidl. Approximation-based similarity search for 3D surface segments. GeoInformatica. Kluwer Academic Publisher, 1998, pp. 113-147.
[56] W. H. Press, S. A. Teukolsky, W. T. Vetterling, et al. Numerical recipes in C (2nd edition). Cambridge University Press, 1992.
[57] J. P. H. Vandeborre, V. Couillet and M. Daoudi. A practical approach for 3D model indexing by combining local and global invariants. In: Proceedings of the 1st International Symposium on 3D Data Processing, Visualization, and Transmission, 2002, Vol. 1, pp. 644-647.
[58] R. Ohbuchi, M. Nakazawa and T. Takei. Retrieving 3D shapes based on their appearance. In: Proceedings of MIR'03, Berkeley, CA, 2003, pp. 39-46.
[59] D. V. Vraníc, D. Saupe and J. Richter. Tools for 3D-object retrieval: Karhunen-Loeve-transform and spherical harmonics. In: Proceedings of the IEEE Workshop on Multimedia Signal Processing, 2001.
[60] J. Assfalg, A. D. Bimbo and P. Pala. Curvature maps for 3D CBR. In: Proceedings of the International Conference on Multimedia and Expo (ICME'03), 2003.
[61] G. Antini, S. Berretti, A. D. Bimbo, et al. Retrieval of 3D objects using curvature correlograms. In: Proceedings of the International Conference on Multimedia and Expo (ICME'05), 2005.
[62] J. Huang, R. Kumar, M. Mitra, et al. Statial color indexing and application. International Journal of Computer Vision, 1999, 35:245-268.
[63] G. Hetzel, B. Leibe, P. Levi, et al. 3D object recognition from range images using local feature histograms. In: Proc. of Int. Conf. on Computer Vision and Pattern Recognition (CVPR'01), 2001.
[64] G. Taubin. A signal processing approach to fair surface design. Computer Graphics (Annual Conference Series), 1995, 29:351-358.
[65] G. Taubin. Estimating the tensor of curvature of a surface from a polyhedral approximation. In: Proc. of Fifth International Conference on Computer Vision (ICCV'95), 1995, pp. 902-907.
[66] M. Desbrun, M. Meyer, P. Schroder, et al. Discrete Differential-Geometry

Operators in nD. Caltech, 2000.

[67] C. Rössl, L. Kobbelt and H. P. Seidel. Extraction of feature lines on triangulated surfaces using morphological operators. In: Smart Graphics, Proceedings of the 2000 AAAI Symposium, 2000.

[68] Kolonias, D. Tzovaras, S. Malassiotis, et al. Content-based similarity search of VRML models using shape descriptors. In: Proc. International Workshop on Content-Based Multimedia Indexing, 2001, pp. 19-21.

[69] F. Mokhtarian, N. Khalili, and P. Yeun. Multi-scale free-form 3D object recognition using 3D models. Image Vision Comput., 2001, 19(5):271-281.

[70] M. Elad, A. Tal and S. Ar. Content based retrieval of VRML objects-An iterative and interactive approach. EG Multimedia, 2001, pp. 97-108.

[71] C. Zhang and T. Chen. Indexing and retrieval of 3D models aided by active learning. ACM Multimedia, 2001, pp. 615-616.

[72] M. Novotni and R. Klein. 3D Zernike descriptors for content based shape retrieval. Solid Modeling, 2003.

[73] E. Paquet and M. Rioux. Nefertiti: A query by content system for three-dimensional model and image database management. Image Vision Comput., 1999, 17(2):157-166.

[74] S. Mahmoudi and M. Daoudi. 3D models retrieval by using characteristic views. In: Proc. 16th International Conference on Pattern Recognition, 2002, pp. 457-460.

[75] Assfalg, A. D. Bimbo and P. Pala. Spin images for retrieval of 3D objects by local and global similarity. In: Proc. 17th International Conference on Pattern Recognition (ICPR-04), 2004, pp. 23-26.

[76] A. E. Johnson and M. Hebert. Using spin-images for efficient multiple model recognition in cluttered 3-D scenes. IEEE Trans. Patt. Analy. Machine Intell., 1999, 21(5):433-449.

[77] D. Y. Chen, X. P. Tian, Y. T. Shen, et al. On visual similarity based 3D model retrieval. In: Proc. Eurographics Computer Graphics Forum (EG'03), 2003.

[78] M. Suzuki, T. Kato and N. Otsu. A similarity retrieval of 3D polygonal models using rotation invariant shape descriptors. In: Proc. IEEE Int. Conf. Syst., Man, Cybern. (SMC 2000), 2000, pp. 2946-2952.

[79] M. Kazhdan, B. Chazelle and D. Dobkin, et al. A reflective symmetry descriptor. In: Proc. Eur. Conf. Comput. Vision (ECCV), 2002, pp. 642-656.

[80] R. C. Veltkamp and M. Hagedoorn. Shape similarity measures, properties and constructions. In: Proc. VISUAL 2000, Lyon, France: Lecture Notes in Computer Science, 2000, Vol. 1929, pp. 467-476.

[81] R. Ohbuchi, T. Minamitani and T. Takei. Shape-similarity search of 3D models by using enhanced shape functions. In: Proc. Theory Pract. Comput. Graph, 2003, pp. 97-105.

[82] Y. Rubner, C. Tomasi and L. J. Guibas. The earth mover's distance as a metric for image retrieval. Int. J. Comput. Vis., 2000, 40(2):99-121.

[83] E. Bardinet, S. Vidal, S. Arroyo, et al. Structural object matching. Paper presented at The Adv. Concepts Intell. Vision Syst. (ACIVS 2000), 2000.

[84] A. Shokoufandeh, S. J. Dickinson, K. Siddiqi, et al. Indexing using a spectral encoding of topological structure. In: Proc. Comput. Vis. Pattern Recognit., 1999, 2:491-497.

[85] A. Shokoufandeh, S. Dickinson, C. Jonsso, et al. On the representation and matching of qualitative shape at multiple scales. In: Proc. 7th Eur. Conf. Comput.

Vis., Copenhagen, Denmark, 2002, pp. 759-775.
[86] K. Siddiqi, A. Shokoufandeh, S. Dickinson, et al. Shock graphs and shape matching. Comput. Vis. 1998, pp. 222-229.
[87] H. Sundar, D. Silver, N. Gagvani, et al. Skeleton based shape matching and retrieval. In: Proc. Shape Model. Int., 2003, pp. 130-139.
[88] M. Hilaga, Y. Shinagawa, T. Kohmura, et al. Topology matching for fully automatic similarity estimation of 3D shapes. Paper presented at The SIGGRAPH 2001, 2001.
[89] V. Vapnik. The Nature of Statistical Learning Theory (2nd edition). Springer-Verlag, 1999.
[90] Ibato, T. Otagiri and R. Ohbuchi. Shape-similarity search of three-dimensional models based on subjective measures. IPSJ SIG Notes Graph. CAD, 2002, 16: 25-30.
[91] A. Pedro, D. Alberto and M. José. Spin images and neural networks for efficient content-based retrieval in 3D object databases. In: Proc. CIVR 2002, Lecture Notes in Computer Science, 2002, Vol. 2383, pp. 225-234.
[92] A. Ip, W. Regli, L. Sieger, et al. Automated learning of model classifications. In; Proc. ACM Symp. Solid Model. Appl. Archive, 2003, pp. 322-327.
[93] T. Ansary, J. Vandeborre, S. Mahmoudi, et al. A Bayesian framework for 3D models retrieval based on characteristic views. In: Proc. 2nd Int. Symp. 3D Data Process., Vis. Transmiss. (3DPVT 2004), 2004, pp. 139-146.
[94] M. Elad, A. Tal and S. Ar. Directed search in a 3D objects database using svm. HP Laboratories, Haifa, Israel, Tech. Rep. HPL-2000-20R1, 2000.
[95] C. Zhang and T. Chen. Active learning for information retrieval: Using 3D models as an example. Tech. Rep. AMP01-04, Carnegie Mellon Univ., Pittsburgh, PA, 2001.
[96] S. Hou, K. Lou and K. Ramani. SVM-based semantic clustering and retrieval of a 3D model database. Proc. CAD, 2005, Vol. 2, pp. 155-164.
[97] Y. Rui, T. S. Huang, M. Ortega, et al. Relevance feedback: A power tool in interactive content-based image retrieval. IEEE Trans. Circuits Syst. Video Technol., 1998, 8(5):644-655.
[98] Y. Ishikawa, R. Subramanya and C. Faloutsos. Mindreader: Query databases through multiple examples. Paper presented at The 24th VLDB Conf., 1998.
[99] Y. Rui, T. S. Huang and S. F. Chang. Image retrieval: Current techniques, promising directions, and open issues. J. Vis. Commun. Image Represent., 1999, 10(1):39-62.
[100] G. Leifman, R. Meir and A. Tal. Relevance feedback for 3D shape retrieval. In: Proc. Israel–Korea Bi-Nat. Conf. Geom. Model. Comput. Graph., 2004, pp. 15-19.
[101] I. Atmosukarto, W. K. Leow and Z. Huang. Feature combination and relevance feedback for 3D model retrieval. In: Proc. 11th Int. Multimedia Model. Conf. (MMM 2005), 2005, pp. 128-133.
[102] E. Paquet and M. Rioux. A content-based search engine for VRML databases. In: Proc. IEEE Int. Conf. Comput. Vis. and Pattern Recognit., 1998, pp. 541-546.
[103] B. David. Methods for content-based retrieval of 3D models. Paper presented at The 3rd Annual CM316 Conf. Multimedia Syst., Southampton, U.K., 2003.
[104] H. Xiao and X. Zhang. A method for content-based 3D model retrieval by 2D projection views. WSEAS Transactions on Circuits and Systems Archive, 2008, 7(5):445-449.

[105] P. Min, J. Chen and T. Funkhouser. A 2D sketch interface for a 3D model search engine. In: Proc. SIGGRAPH Tech. Sketches, 2002, p. 138.
[106] T. Igarashi, S. Matsuoka and H. Tanaka. Teddy: A sketching interface for 3D freeform design. In: Proc. SIG-GRAPH 1999, ACM, 1999, pp. 409-416.
[107] C. Zhang and T. Chen. Efficient feature extraction for 2D/3D objects in mesh representation. Paper presented at The ICIP, 2001.
[108] J. Corney, H. Rea, D. Clark, et al. Coarse filters for shape matching. IEEE Comput. Graph. Appl., 2002, 22(3):65-74.
[109] D. McWherter, M. Peabody, A. Shokoufandeh, et al. Solid model databases: Techniques and empirical results. ASME/ACM Trans., J. Comput. Inf. Sci. Eng., 2001, 1(4):300-310.
[110] M. Suzuki, Y. Yaginuma and Y. Sugimoto. A 3D model retrieval system for cellular phones. In: Proc. IEEE Int. Conf. Syst Man Cybern, 2003, pp. 3846-3851.
[111] M. Novotni and R. Klein. A geometric approach to 3D object comparison. In: Proc. Int. Conf. Shape Model. Appl., 2001, pp. 167-175.

5

3D Model Watermarking

5.1 Introduction

3D meshes have been used more and more widely in industrial, medical and entertainment applications during the last decade. Many researchers, from both the academic and industrial sectors, have become aware of intellectual property protection and authentication problems arising with their increasing use. Apart from in familiar multimedia combinations, such as images, text, audio and video, the issues of copyright protection and piracy detection are now emerging in the fields of CAD, CAM, computer aided education (CAE) and computer graphics (CG), etc. Scicntific visualization, computcr animation and virtual rcality (VR) arc three hot topics in the field of computer graphics. On the one hand, with the development of collaborative design and virtual products in the network environment, it is expected that consumers will prefer models consisting of points, lines and faces, rather than material objects or accessories. As a result, only the authorized user can replicate, modify or recreate the model. The models that we handle are all three dimensional and digital, which can be called 3D graphics, 3D objects or 3D models. The issue of how to protect and even manipulate and control 3D models and other CAD products is now involved. On the other hand, with the rapid development in communication and distribution technology, digital content creation sometimes requires the cooperation of many creators. In particular, the scale of 3D objects is large and special skills are needed for the creation of 3D objects. Therefore, to create good and complex 3D content, the cooperation of many creators may be necessary and important. In the scenario of the joint-creation of 3D objects in a manufacturing environment, the creatorship of the participating creators becomes a big issue. There are some concerns for participating creators during the creation process. Firstly, each participating creator wants to prove his/her creatorship. Secondly, all of the participating creators want to verify the joint-creatorship of the whole product. Thirdly, it is necessary to prevent some creators from neglecting other creators and asserting the whole creatorship of the final product and selling the product to a buyer. How we protect each creator's creatorship and how we account for his/her level of contribution are a major challenge.

Digital watermarking has been considered a potentially efficient solution for copyright protection of various multimedia content. This technique carefully hides some secret information in the functional part of the cover content. Compared with cryptography, the digital watermarking technique is able to protect digital works (assets) after the transmission phase and legal access. Thus, digital watermarking techniques can provide us with a very effective approach to embed digital watermarks in 3D model data, such that the copyright of 3D models and other CAD products can be effectively protected. Nowadays, this research area is becoming a new hot topic in the field of digital watermarking. 3D model digital watermarking technology is a branch of digital watermarking technology, and its main aim is to embed invisible watermarks in 3D models to authenticate 3D models or embed information to claim the model's ownership. Watermarking 3D objects has been performed from various perspectives. In [1], an optical-based system employing phase shift interferometry was devised for mixing holograms of 3D objects, representing the cover media and the hidden data, respectively. Watermarking of texture attributes has been attempted by Garcia and Dugelay [2]. Hartung et al. watermarked the stream of MPEG-4 animation parameters, representing information about shape, texture and motion, by using a spread spectrum approach [3].

Attributes of 3D graphical objects can be easily removed or replaced. This is why most of the 3D watermarking algorithms are applied on the 3D graphical object geometry. Authentication is concerned with the protection of the cover media and should indicate when it has been modified. Authentication of 3D graphical objects by means of fragile watermarking has been considered in [4, 5]. Ohbuchi et al. discussed three methods for embedding data into 3D polygonal models in [6]. Many approaches applied to 3D object geometry aim to ensure invariance at geometrical transformations. This can be realized by using ratios of various 2D or 3D geometrical measures [6-9]. Results provided by a watermarking algorithm for copyright protection, by employing modifications in histograms of surface normals, were reported by Benedens in [10]. Local statistics have been used for watermarking 3D objects in [11, 12]. Multiresolution filters for mesh watermarking have been considered in connection with interpolating surface basis functions [13] and with pyramid-based algorithms [14]. Benedens and Busch introduced three different algorithms, each of them having robustness to certain attacks while being suitable for specific applications [15]. Algorithms that embed data in surfaces described by NURBS use changes in control points [9] or re-parameterization [16]. Wavelet decomposition of polygons was used for 3D watermarking in [17, 18]. Watermarking algorithms that embed information in the mesh spectral domain using graph Laplacian have been proposed in [19-21].A few characteristics can be outlined for the existing 3D watermarking approaches. Some of the 3D watermarking algorithms are based on displacing vertex locations [9, 12, 13] or on changing the local mesh connectivity [14, 16]. Minimization of local norms has been considered in the context of 3D watermarking in [15, 22]. Localized embedding has been employed in [6, 9, 15]. Arrangements of embedding primitives have been classified according to their locality in: global, local and indexed [6]. Localized and repetitive embedding is used in order to increase the robustness to 3D object cropping [21].

Preferably, a watermarking system would require in the detection stage only the knowledge of the watermark given by a key and that of the stego object. However, most of the approaches developed for watermarking 3D graphical objects are nonblind and require the knowledge of the cover media in the detection stage [2, 3, 5, 10, 13, 14, 16-20, 22]. Some algorithms require complex registration procedures [13-15, 18, 22] or should be provided with additional information about the embedding process in the detection stage [8, 9, 17]. A nonlinear 3D watermarking methodology that employs perturbations in the 3D geometrical structure is described in [23]. The watermark embedding is performed by a processing algorithm in two steps. In the first step, a string of vertices and their neighborhoods are selected from the cover object. The selected vertices are ordered according to a minimal distortion criterion. This criterion relies on the calculation of the sum of Euclidean distances from a vertex to its connected neighbors. A study of the effects of perturbations in the surface structure is provided in this paper. The second step estimates first and second order moments and defines two regions: one for embedding the bit "1" and another one for embedding the bit "0". First- and second-order moments have desirable properties of invariance to various transformations [24, 25] and have been used for shape description [7]. These properties can ensure the detection of the watermark after the 3D graphical object is transformed by affine transformations. Two different approaches that produce controlled local geometrical perturbations are considered for data embedding, i.e., using parallel planes and bounding ellipsoids. The detection stage is completely blind in the sense that it does not require the cover object.

This chapter is organized as follows. The description of general requirements for 3D watermarking is provided in Section 5.2. Section 5.3 focuses on the classification of 3D model watermarking algorithms. Section 5.4 discusses typical spatial domain 3D mesh model watermarking schemes. Section 5.5 introduces the robust adaptive 3D mesh watermarking algorithm proposed by the authors of this book, and it belongs to the spatial domain techniques. Section 5.6 introduces typical transform-domain 3D mesh model watermarking schemes. Section 5.7 overviews watermarking algorithms for other types of 3D models. Finally, conclusions and summaries are given in Section 5.8.

5.2 3D Model Watermarking System and Its Requirements

We introduce the concepts of 3D model watermarking, its framework and requirements.

5.2.1 Digital Watermarking

Digital watermarking is the process of possibly irreversibly embedding information into a digital signal. The signal may be audios, pictures, videos or 3D models. If the signal is copied, then the information is also carried in the copy. A digital watermark can be visible or invisible. In visible watermarking, the information is visible in the picture or video. Typically, the information is text or a logo which identifies the owner of the media. The image shown in Fig. 5.1 has a visible watermark. When a television broadcaster adds its logo to the corner of the transmitted video, this is also a visible watermark. On the other hand, in invisible watermarking, information is added as digital data to audios, pictures, videos or 3D models, but it cannot be perceived as such (although it is possible to detect the hidden information). An important application of invisible watermarking is to copyright protection systems, which are intended to prevent or deter unauthorized copying of digital media. The existence of an invisible watermark can only be determined using an appropriate watermark extraction or detection algorithm. In this chapter we restrict our attention to invisible watermarks. Steganography is an application of digital watermarking, where two parties communicate a secret message embedded in the digital signal. Annotation of digital photographs with descriptive information is another application of invisible watermarking. While some file formats for digital media can contain additional information called metadata, digital watermarking is distinct in that the data is carried in the signal itself.

Fig. 5.1. A visible watermark embedded in the Lena image [27] (©[2003]IEEE)

An invisible watermarking technique, in general, consists of an encoding process and a decoding process. The watermark insertion step is represented as:

$$\hat{X} = E_K(X, W), \tag{5.1}$$

where X is the original product, $\hat{X}$ is the watermarked variant, W is the watermark information being embedded, K is the user's insertion key, and E represents the watermark insertion function. Depending on the way the watermark is inserted, and depending on the nature of the watermarking algorithm, the detection or extraction method can take on very distinct approaches. One major difference between watermarking techniques is whether or not the watermark detection or extraction step requires the original image. Watermarking techniques that do not require the original image during the extraction process are called oblivious (or public or blind) watermarking techniques. For oblivious watermarking techniques, watermark extraction works as follows:

$$\hat{W} = D_{K'}(X'), \tag{5.2}$$

where X' is a possibly corrupted watermarked image, K' is the extraction key, D represents the watermark extraction/detection function, and $\hat{W}$ is the extracted watermark information. Oblivious schemes are attractive for many applications where it is not feasible to require the original image to decode a watermark.

5.2.2 3D Model Watermarking Framework

A typical 3D model watermarking system [26] is shown in Fig. 5.2. During the watermark embedding process of this system, the watermark is embedded in some way in the spatial or transformed domains of the original 3D model (i.e., cover model), so that the watermarked 3D model (i.e., stego model) is acquired. For example, a watermark bit can be embedded into the original 3D NURBS model surface to get the watermarked NURBS model. The stego 3D model is transmitted or sent through various channels, during which the stego model may be subject to a variety of attacks, including unintentional attacks and intentional attacks. Here, unintentional modifications are applied to a data object during the course of its normal use, while intentional modifications are applied to the data object with the intention of modifying or destroying the watermark.

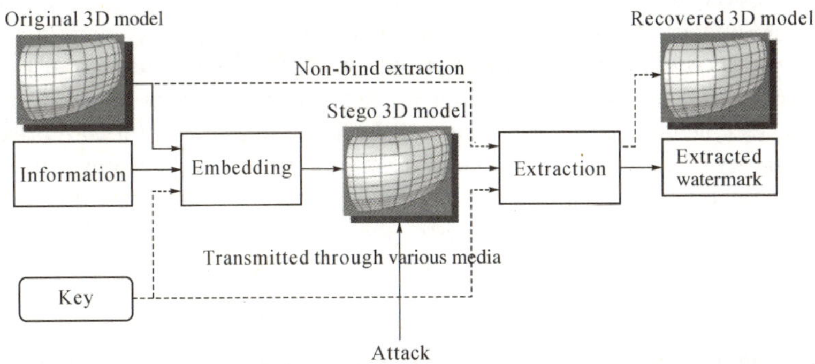

Fig. 5.2. Basic diagram of a typical 3D model watermarking system

In the detection end, we can extract the watermark from a suspect model through blind or non-blind detection methods. By comparison of the extracted watermark with the original watermark to calculate the similarity, the existence of the original watermark can be judged and the authenticity of the 3D model copyright source or content can be identified. On some special occasions, the original 3D model may also need to be restored in the watermark extraction, such as reversible watermarking applications.

5.2.3 Difficulties

There are still few watermarking methods for 3D meshes, in contrast with the relative maturity of the theory and practices of image, audio and video watermarking. This situation is mainly caused by the difficulties encountered while handling the arbitrary topology and irregular sampling of 3D meshes, as well as the complexity of the possible attacks on watermarked meshes.

A 3D mesh model can be very little, so the payload capacity can be low. Besides, there are multiple representations for exactly the same models and 3D model because of the lack of an inherent order. We can consider an image as a matrix, and each pixel as an element of this matrix. This means that all of these pixels have an intrinsic order in the image, for example, the order established by row or column scanning. This order is usually used to synchronize watermark bits (i.e. to know where the watermark bits are and in which order). On the contrary, there is no simple robust intrinsic ordering for mesh elements, which often constitute the watermark bit carriers (primitives). Some intuitive orders, such as the order of the vertices and facets in the mesh file, and the order of vertices obtained by ranking their projections on an axis of the objective Cartesian coordinate system, are easy to alter. In addition, because of their irregular sampling, it is very difficult to transform a 3D model into the frequency domain for further operation, and thus we still lack an effective spectral analysis tool for 3D meshes. This situation makes it difficult to apply existing successful spectral watermarking schemes on 3D meshes.

In addition to the above point, robust watermarks also have to face various intractable attacks. Many attacks on geometric or topology may undermine the watermark, such as mesh simplification and remeshing. The reordering of vertices and facets does not have any impact on the shape of the mesh, while it can seriously desynchronize the watermarks that rely on this straightforward ordering. The similarity transformations, including translation, rotation, uniform scaling and their combination, are supposed to be common operations through which a robust watermark should survive. Even worse, the original watermark primitives can disappear after a mesh simplification or remeshing. Such tools are available in many software packages, and they can completely destroy the connectivity information of the watermarked mesh while well conserving its shape. Usually, the possible attacks can be classified into two groups: the geometric attacks that only modify the positions of the vertices, and the connectivity attacks that also change the connectivity aspect. In addition, similar to the problems encountered by other digital watermarking technology, lossy compression will modify the 3D model geometry, so the synchronization problems must be resolved.

Watermarking 3D meshes in computer aided design applications has other difficulties caused by design constraints. For example, the symmetry of the object has to be conserved and the geometric modifications have to be within a tolerance for future assembly. In this situation, the watermarked mesh will no longer be evaluated just by the human visual system, which is quite subjective, but also by some strict objective metrics.

5.2.4 Requirements

The aim of digital watermarks not only lies in ensuring that the data will not be found and destroyed, but also to ensure that after the carrier, together with the embedded information, subject to intentional or unintentional operations (such as conversion, compression and simplification), information can be extracted correctly (from carriers) or some kind of measure is designed to estimate the existence possibility of the information. Therefore, the digital watermark normally should have the following characteristics: (1) Vindicability. The watermark should be able to provide complete and reliable evidence for the attribution of multimedia products that are copyright protected; (2) Imperceptivity. It is not visible and statistically irreparable; (3) Robustness. It should be able to bear a large number of different physical and geometric distortions, including intentional or unintentional attacks. The watermarking diagram for 3D mesh is basically similar to that for other media as shown in Fig. 5.2. However, in a 3D mesh, as points, lines and surface data are without a natural sequence, and 3D meshes usually subject to affine transformations such as translation, rotation and scaling, mesh compression and mesh simplification, 3D mesh watermarking methods are therefore distinguished greatly from other media watermarking methods. A brief description for all of the requirements for 3D model watermarking is given as follows.

5.2.4.1 Imperceptivity (Transparency)

Clearly, one of the most important requirements is the transparency of the watermark [26], i.e., the nonperceptibility of changes brought to the original model by the watermark. Due to the special nature of 3D models, two concepts of transparency need to be distinguished here, namely functional transparency and perceptual transparency. For the traditional carrier data, such as images and audio data, the transparency of the watermark can be recognized by the human eye and ear. In other words, the human perceptual system participates in identification of the difference between the cover data and the stego data, which is the issue of the perceptual transparency of the watermark. For 3D CAD geometry data, the transparency of the watermark should be judged according to the impact of the watermark accession to the 3D data, which is the issue of functional transparency. A perceptually transparent watermark may or may not be functionally transparent. Similarly, a functionally transparent watermark may or may not be perceptually transparent. For example, if a perceptually transparent watermark is embedded into the engine cylinder CAD data, the shape and even the impact of the engine cylinder may change. In another example, holes of 11 mm and 10 mm are, in normal circumstances, perceptually transparent, but may be completely different in the actual design of their functions. Therefore, for 3D mesh models used in production and design, not only perceptual transparency but also functional transparency should be satisfied.

5.2.4.2 Robustness and Security

The second important requirement for 3D watermarks is the ability to detect the watermark even after the object has undergone various transformations or attacks. In any watermarking or fingerprinting approach, there is a trade-off between being able to make the watermark survive a set of transformations and the actual visibility of the watermark. Such transformations can be inherent for 3D object manipulation in computer graphics or computer vision or they may be done intentionally with the malicious purpose of removing the watermark. Transformations of 3D meshes can be classified into geometrical and topological transformations. Geometrical transformations include affine transformations such as rotation, translation, scale normalization, vertex randomization, or their combinations, and can be local or applied to the entire object. Topological transformations consist of changing the order of vertices in the object description file, mesh simplification for the purpose of accelerating the rendering speed, mesh smoothing, insection operation, remeshing, partial deformation or cropping parts of the object. Other processing algorithms include object compression and encoding, such as MPEG-4. Smoothing and noise corruption algorithms can be mentioned in the category of intentional attacks. A large variety of attacks can be modeled generically by noise corruption. Noise corruption in 3D models amounts to a succession of small perturbations in the location of vertices.

Table 5.1 compares potential attacks of image watermarking algorithms and 3D objects watermarking algorithms [27]. Evident from the table, virtually every image watermarking algorithm attack method corresponds to their counterpart in 3D watermarking algorithms. However, an important distinction must not be ignored: Attack methods on 3D meshes are much more complicated. In fact, an image is 2D and is uniformly sampled, while a 3D mesh corresponds to 3D space points with a certain topology and non-uniform sampling. Therefore, many image processing methods cannot be directly extended to 3D geometric data. In Table 5.1, the remeshing operation is a unique attack on 3D models. Remeshing is actually a resampling operation on the geometric shape of a 3D model and usually causes topology alterations.

Table 5.1 Comparisons of image watermarking and 3D model watermarking

Attacks	Descriptions
Image attacks	Cropping
	2D translation/rotation/scaling
	Noise
	Compression
	Downsampling
	Upsampling
	2D free deformation
	Filtering
Mesh attacks	3D insection/decimation
	3D translation/rotation/scaling
	Noise
	Geometry compression
	Simplification
	Subdivision (e.g., subdivision surface)
	3D free deformation
	Mesh filtering (e.g., Taubin smoothing)
	Topology change (e.g., remeshing)
	Mesh optimization (e.g. topology compression)
	Reordering

In principle, watermarks should be able to withstand geometric or topology attacks that do not damage the visual effects of a model. In addition, some more complex geometry operations are likely to undermine visual effects and usability, so the basic objective of a watermarking system does not include robustness against uneven scale transformation along arbitrary axes, projection (e.g., a 2D projection), or overall deformation. The key issue of 3D mesh robust watermarking algorithm research is to find the location to embed watermarks so that the embedded watermark can withstand a series of attacks, and how to embed watermarks in the 3D mesh model as much as possible. After the attack, apparently, the robust watermarking algorithm should still be able to extract the embedded watermark or prove its existence in the 3D mesh. Most of the existing 3D watermarking algorithms are robust to certain attacks, but not to others. Usually, topology-based watermarking algorithms are not robust to affine transformations, while vertex displacement algorithms are not robust to mesh simplifications. It would be preferable to embed the watermark in regions displaying a great amount of variation in the 3D object. This is similar to the procedure of considering regions of high frequency for image watermarking.

From the security perspective, if we do not grasp all the relevant knowledge of the embedded watermark, it can hardly be forged. If the attacker tried to delete the watermark, the 3D mesh will be damaged. In theory, removal of any watermark is possible, so a watermark with high security should meet the requirement that the cost of removal is far greater than the 3D model value.

5.2.4.3 Payload Capacity

Watermarking systems should allow for a certain amount of embedded watermark information [28], not an insignificant amount of data. At least 32 bytes of payload capacity is required for embedding the sequence code that indicates the status of the purchaser or copyright owner. To prove the ownership, sufficient capacity to store a hash value is usually required (for example, the MD5 hash function requires 128 bytes, and the SHA hash algorithm needs 160 bytes). The watermarking system known as a statistical method can perform an arbitrarily long Hash transform and feed back some type of random number generator, which can obtain the data modification position based on the overall statistics, such as the mean and the variance. These systems allow the detection of the existence of the watermarks with prior knowledge, but they also have shortcomings, although these systems may claim there are no capacity constraints. For example, the identification of the registered authorization on the network model requires testing all of the identities, which requires a large amount of computation. The objective of a high-capacity watermark is simply to hide a large amount of secret information within the mesh object for applications such as content labeling and covert communication. High-capacity watermarks are often fragile (in sense that they are not robust), and some of them have the potential to be successful fragile watermarks with precise attack localization capability.

There is a classic problem, i.e., the trade-off between capacity, robustness and imperceptibility. These measures are often contradictory. For example, high watermarking intensity provides better robustness, but normally degrades the visual quality of the watermarked mesh and risks making the watermark perceptible. Redundant insertion can considerably strengthen the robustness, but unavoidably decreases the capacity. Local adaptive geometric analysis seems favorable for finding optimum watermarking parameters in order to achieve a sufficient compromise between these indicators. A valuable solution could lie in detecting rough (noised) regions where slight geometric distortions would be nearly invisible. As observed in [23], these regions are characterized by the presence of many short edges, and they are somewhat equivalent to highly textured or detailed image areas, which are often used by image watermarking algorithms to obtain a better invisibility.

In addition to the above requirements, an ideal 3D model watermarking system needs to meet the following additional requirements.

5.2.4.4 Space Utilization

Space utilization and robustness normally contradict each other. As a result, the most efficient use of space as possible is one important parameter for the evaluation of mesh watermarking algorithms, and this involves how to properly coordinate the relationship between the robustness of the watermark and space utilization.

5.2.4.5 Background Processing and Suitable Speed

Watermark embedding and extraction are better performed without user participation. Using the "robot" engine to automatically search for watermarks in websites and databases is very useful for monitoring the applications of legal and illegal copies. The ultimate goal of this application is real-time monitoring. But in that case it will bring pressure on the implementation speed and storage requirements of watermarking systems.

5.2.4.6 Embedding Multiple Watermarks

In practical applications, people may need multiple watermarks to be embedded. This usually occurs in the sales requirements of manufacturers and resellers: The manufacturer embeds his copyright information and secret information about resellers, while resellers embed users' information and the end-authorization information. Ideally, these watermarks cannot interfere with each other.

5.2.4.7 Minimum Knowledge of a Priori Data

An ideal watermarking system needs only 3D model data and a watermark extraction key. The key corresponds to the creator or the company making the model, the type of model, the model itself or authentication. All the necessary parameters, such as the seed, are all included in the key. In public watermarking systems, all the models of one creator may use the same key or the system uses the same key for models from different creators. Unfortunately, the extraction process may need more priori knowledge: knowledge of the model itself, especially specific embedding positions to meet synchronization; part of, or the whole of, the original model for registration.

In addition, an ideal watermarking system has a blind detection algorithm. A non-blind system would need the original cover media in the detection stage. Usually, it is expected that a nonblind approach can provide better robustness to various attacks. However, a nonblind watermarking approach is not suitable for most applications.

5.2.4.8 Minimum Preprocessing Overhead

An ideal watermarking system must allow immediate access to the embedded watermark, without the need for preprocessing the model data. Preprocessing may involve model data transformation, model identification, surface normal correction, model registration or scaling.

5.3 Classifications of 3D Model Watermarking Algorithms

There exist different classifications of 3D model watermarking algorithms. We can classify 3D model watermarking algorithms from angles such as robustness, redundancy utilization, 3D model types, embedding domains, obliviousness, reversibility, transparency and complexity. The following are detailed descriptions of different classifications.

5.3.1 Classification According to Redundancy Utilization

Usually, watermarking algorithms utilize the carrier's redundancy to embed additional information. For 3D CAD geometry data, there are three types of redundancy that can be used to embed watermarks [26].

5.3.1.1 Innate Redundancy

Innate redundancy is the redundancy that the 3D geometry data themselves possess. Without affecting the shape functions, information can be embedded into shapes through the revision of part shapes. The shape modification manner and embedding locations should be carefully controlled to meet the functional transparency. Each shape has a certain function in the CAD geometry. However, the vast majority of 3D CAD geometry data have a certain arbitrariness, based on which we can embed a watermark without affecting the shape function. For example, a method that has been used for many years is to inscribe names or partial figures of manufacturers in some parts of the machine.

5.3.1.2 Representation Redundancy

The description forms of 3D models may also be redundant. At this time, people may amend the description of the shape itself without altering the shape to embed information. For example, we can embed knots into NURBS surfaces without altering the geometry. Once embedded, it is very difficult for node removal if we force the model geometry to be maintained.

5.3.1.3 Encoding Redundancy

There will be also some redundancy in encoding the shape description, thus we can also embed watermarks without changing the geometry or shape descriptions. For example, suppose a CAD model coordinate of each control point has an accuracy of up to 6 bits, while the data format is up to 10 bits, thus 4 bits out of the 10 bits can be used to embed watermarks.

A watermark is usually embedded in parametric curves and surfaces if we utilize the second type of redundancy mentioned above. Such a method can be divided into four categories in accordance with its two characteristics, i.e., model shapes and sizes of the data model [26]: (1) maintaining the shape and the data size; (2) maintaining the shape unchanged but changing the data size; (3) changing the shape but maintaining the data size unchanged; (4) both the shape and size of data are changed by watermarking. Here, the same data size means that the number of parameters used to define the shape (such as control points and knots) is unchanged, while specific values of these parameters may be changed. For the effectiveness of communication and storage, keeping the data size unchanged is very useful.

5.3.2 Classification According to Robustness

Another very important classification of watermarking algorithms is by their robustness. Usually, one hopes to construct a robust watermark, which is able to withstand common malicious attacks, for copyright protection purposes. However, sometimes the watermark is intentionally designed to be fragile, even to very slight modifications, in order to be used in authentication applications. Thus, according to the robustness features of digital watermarking, 3D model digital watermarking technologies can be divided into two categories: robust digital watermarking and fragile digital watermarking technologies. Usually fragile watermarking systems can find applications in tamper detection, while robust watermarking systems are commonly designed for copyright protection and piracy detection and the majority of algorithms belong to the latter type. Robust digital watermarking technologies should have a strong anti-jamming capability, so that the embedded watermark is difficult to remove in all kinds of incidents or malicious attacks. Contrary to robust watermarking, fragile digital watermarking must have a high vulnerability to external operations, i.e., once the model data is tampered with, the embedded watermark must be changed or even removed. Yeung and Yeo's algorithms [29, 30], Benedens's vertex flood algorithm [31] and the four algorithms proposed by Ichikawa et al. in [32] all fall into this category.

A robust watermark should withstand both intentional and unintentional modifications of the stego-data. On the other hand, a fragile watermark must be affected by intentional (and some unintentional) modifications so that tampering and other damage to the data can be detected. Here, unintentional (incident) modifications are applied to an object during the course of its normal use, while intentional modifications are applied to the object with the intention of modifying or destroying the watermark. Both robust and fragile digital watermarking schemes demand transparency, in other words the watermark embedding process should not undermine the visual effects and the commercial value of the model will not be reduced. However, robustness and transparency of the watermark are often at odds, i.e., making a watermark more robust tends to make it less transparent.

5.3.3 Classification According to Complexity

According to the complexity, the 3D model watermarking algorithms can be divided into two categories, i.e., algorithms to embed information directly into the structure geometry and algorithms to embed information indirectly into the constructed geometry. Embedding information directly into the geometric body refers to embedding watermarks directly in the structure geometry such as vertex coordinates, edge lengths and polygon areas. For example, the watermark is embedded into the area ratio between two similar triangles or polygons, the length ratio between two straight line segments, the volume ratio between two tetrahedrons, and so on. The TSQ (triangle similarity quadruple) algorithm by Ohbuchi and the TVR (tetrahedral volume ratio) algorithm [33] belong to this category. Preprocessing is a necessary step before the construction of indirect and non-intuitionistic geometry primitives, e.g., algorithms proposed by Yeung and Yeo [29, 30] and by Wagner [7], as well as the watermarking algorithm, through adjusting the mesh surface normal distribution by Benedens [28]. All belong to this category.

5.3.4 Classification According to Embedding Domains

According to different embedding domains, watermarking technologies can be divided into spatial-domain-based watermarking algorithms and transform-domain based watermark embedding algorithms. With respect to spatial-domain-based watermarking algorithms, a watermark is directly embedded in the original mesh by modifying the mesh's geometry, connectivity or other attribute parameters. With respect to transform-domain-based watermarking algorithms, a watermark is embedded through modifying the coefficients obtained after a certain transformation. In this book, algorithms are illustrated in these two categories. Generally speaking, spatial-domain-based watermarking algorithms are simple, transparent and fast, but with poor robustness, while transform-domain-based watermarking algorithms possess opposite properties. In each category, it seems convenient to subdivide the members into two subclasses, robust and fragile watermarking techniques depending on the robustness.

5.3.5 Classification According to Obliviousness

We distinguish between non-blind and blind (oblivious) watermarking schemes depending on whether or not the original digital work should participate in the watermark detection or extraction phase. Sometimes a blind watermarking algorithm is also called a public watermarking algorithm, and a non-blind watermarking algorithm is referred to as a private watermarking algorithm. A public watermarking scheme extracts a message using stego-data only. Such extraction is called blind-extraction or blind-detection. A private watermarking scheme requires original cover-data as well as the watermarked stego-data for its non-blind extraction of the embedded message. While a private watermarking scheme with non-blind extraction enables more robust and accurate extraction (e.g., subtraction of the cover-data from the stego-data reveals the signal, that is the watermark), a public scheme usually is easier to adopt in an application scenario. Usually, people hope that the watermark detection algorithm is blind.

In addition, a watermarking scheme may employ a cryptographic approach to make embedded messages secure from a third party. Therefore, people also mention public key and private key watermarking algorithms. The former means that the watermark is encrypted using a public key encryption algorithm prior to embedding, while the latter means that the embedded watermark is encrypted using a symmetric key encryption algorithm before embedding. Furthermore, a cryptographic function may be embedded into the watermarking process itself, for example, to scramble the mapping from a message bit to the corresponding watermark structure.

5.3.6 Classification According to 3D Model Types

According to different types of 3D models, 3D model watermarking technologies can be divided into 3D mesh watermarking technology, the NURBS watermarking technology [16, 34], digital watermarking technologies for facial motion parameters [3] and voxel-based digital watermarking technologies [35-38]. Due to space limitations, this chapter mainly introduces 3D mesh watermarking technology.

5.3.7 Classification According to Reversibility

According to reversibility, 3D model watermarking technologies can be divided into irreversible watermarking techniques and reversible watermarking techniques. This chapter mainly focuses on the former, while the next chapter focuses on the latter. Most of the watermarking techniques are irreversible, the cover media cannot return to its original state after embedding. In the embedding procedure, the irreversible distortion of the original content is introduced. Although this distortion is imperceptible, we will never regain the original content. This may not be acceptable in some sensitive applications, such as military data, medical data and 2D vector data for the geographical information system (GIS). Reversible watermarking is a technique for embedding data in a digital host signal in such a manner that the original host signal can be restored in a bit-exact manner in the restoration process. Thus, reversible watermarking has become an interesting research topic in recent years. It is also called lossless watermarking, i.e., the original content can be complctely restored when decoding or during the watermark extraction.

5.3.8 Classification According to Transparency

Arguably the most important property of a watermark is its transparency as discussed in Subsection 5.2.4. Watermarks must be transparent to the intended applications. We distinguish two kinds of transparency, functional and perceptual. For most of the traditional data types, such as image and audio data, transparency of a watermark is to be judged by human beings. If the cover-data and stego-data are indistinguishable to human observers, the watermark is perceptually transparent. For other data types, such as 3D geometric CAD data, transparency of the watermark is judged according to whether the functionality of the data is altered or not. A perceptually transparent watermark may or may not be functionally transparent. Likewise, a functionally transparent watermark may or may not be perceptually transparent. For example, a perceptually transparent watermark added to CAD data of an engine cylinder may alter the shape of the cylinder enough to interfere with the function of the engine.

5.4 Spatial-Domain-Based 3D Model Watermarking

In 1997, Ohbuchi et al. published a pioneer paper on 3D mesh watermarking at the ACM International Conference on Multimedia 97 [33] when he was working at the IBM Tokyo Research Laboratory in Japan. The paper is generally acknowledged to be the first paper published in the international community about 3D mesh watermarking technology, which provided new ideas and methods for 3D mesh model watermarking and digital watermarking research. It was a significant milestone. Over the next few years, researchers in Japan, South Korea, Germany, the United States, China and other countries conducted a series of watermarking research experiments and have achieved many results. In the following three sections, 3D model watermarking methods will be described in the spatial domain (two sections, one is for the algorithm proposed by the authors of this book) and transform domains, respectively. In this section, 3D model watermarking algorithms are described with classifications according to different embedding primitives and embedding objects, the former 8 categories of algorithms are designed for 3D meshes while the last subsection introduces watermarking for other types of 3D models. In the next section, a robust spatial-domain 3D mesh watermarking method proposed by the authors of this book is introduced in detail.

In order to facilitate and unify the representation later in this book, the definition of a 3D mesh model is defined here with mathematical symbols before 3D model watermarking algorithms are introduced. An ordered polygon mesh consisting of k vertices can be defined as $\boldsymbol{M} = \{\boldsymbol{P}, \boldsymbol{C}\}$, where $\boldsymbol{P} = \{\boldsymbol{p}_i\}$ is the set of vertices, $I = 0, 1, 2, \ldots, k-1$, $\boldsymbol{p}_i = \{x_i, y_i, z_i\}$ is a 3D coordinate triple, which are connected according to a certain topology $\boldsymbol{C} = \{i_l, j_l\}$, $l=0, 1, 2, \ldots, m-1$, $0 \le i_l, j_l \le k-1$. Other attributes such as color, material and surface normal are optionally included by a mesh model. As a result, mesh watermarking can be performed through altering vertices, topology or even attributes.

5.4.1 Vertex Disturbance

In fact, in many 3D model watermarking algorithms, a watermark is embedded through altering the vertex coordinates. However, methods in which triangles, tetrahedrons or a certain kind of distance are regarded as primitives for watermarking are not included in this category. The following are several typical watermarking methods based on the idea of vertex disturbance, i.e., embedding watermarks by modifying the vertex coordinates slightly according to the corresponding watermark bits.

5.4.1.1 Spread-Spectrum Mechanism

In 1999, Praun from Princeton University and Hoppe from the Microsoft Research Institute applied the spread spectrum technology to triangle meshes, providing a robust mesh watermarking algorithm for arbitrary triangle meshes [39]. Spread-spectrum technology is a technical means used in information transmission, which makes the signal bandwidth much wider than the minimum requirements to send information. Spread spectrum is implemented with a code independent of the data to be sent. The spread spectrum code should be received by the receiver synchronously for the subsequent de-spread and data recovery processes. Spread-spectrum technology makes signal detection and removal more difficult, therefore the watermarking methods based on spread-spectrum technologies are quite robust. Considering that the representation of mesh surfaces lacks natural parametric methods based on frequency decomposition, Praun et al. constructed a group of scalar functions using multi-resolution analysis on the mesh vertex structure (Due to space limitations, the construction details are not illustrated here). During the watermark embedding process, the basic idea is to disturb vertex coordinates slightly along the direction of surface normals and weighted by the corresponding basis function. Suppose that the watermark is a Gaussian noise sequence with zero mean and unit variance, $\boldsymbol{w} = \{w_0, w_1, \ldots, w_{m-1}\}$. To guarantee irreversibility, the original 3D model and its related information are both encrypted with Hash functions, e.g., MD5 or SHA-1 algorithms, and the encrypted sequence is used as the seed for the pseudo-randomizer. Basis functions, multiplied by a coefficient, are added to the 3D vertex coordinates. Every basis function i has a scalar impact factor ϕ_j^i and a global displacement $\boldsymbol{d}_i$ for every vertex j, $0 \le I \le m-1$, $0 \le j \le k-1$. For each direction of X, Y and Z, the embedding formula is as follows (take X for example):

$$\begin{bmatrix} x_0^w \\ x_1^w \\ \cdots \\ x_{k-1}^w \end{bmatrix} = \begin{bmatrix} x_0 \\ x_1 \\ \cdots \\ x_{k-1} \end{bmatrix} + \varepsilon \cdot \begin{bmatrix} \phi_0^0 & \phi_0^1 & \cdots & \phi_0^{m-1} \\ \phi_1^0 & \phi_1^1 & \cdots & \phi_1^{m-1} \\ \vdots & \vdots & \ddots & \vdots \\ \phi_{k-1}^0 & \phi_{k-1}^1 & \cdots & \phi_{k-1}^{m-1} \end{bmatrix} \begin{bmatrix} h_0 d_{0x} & 0 & \cdots & 0 \\ 0 & h_1 d_{1x} & \cdots & 0 \\ \vdots & \vdots & \ddots & \vdots \\ 0 & 0 & \cdots & h_{m-1} d_{(m-1)x} \end{bmatrix} \begin{bmatrix} w_0 \\ w_1 \\ \cdots \\ w_{m-1} \end{bmatrix}, \quad (5.3)$$

where x_j^w and x_j are the coordinate along X for the watermarked vertex $\boldsymbol{p}_j^w$ and the original vertex $\boldsymbol{p}_i$ respectively, $0 \le j \le k-1$, ε is the parameter for embedding, d_{ix} is the X component of the global displacement $\boldsymbol{d}_i$, and h_i is the amplitude of the i-th basis function. To countermine the topology attacks such as mesh simplification, an optimization method is used in this algorithm to remesh the attacked mesh model based on the connectivity of the original mesh model. Simulation results show that this watermarking method is rather robust to such operations as translation, rotation, uniform scaling, insection, smoothing, simplification and remeshing and can also resist attacks of added noise, least significant bits alteration and so on.

5.4.1.2 Masking Based on Connected Vertices

In 2003, a novel spatial domain 3D model watermarking algorithm was proposed in [40], in which the masking factor for additive embedding is acquired from connected vertices. Suppose $S_i = \{j | \{i, j\} \in \boldsymbol{C}\}$ represent the set of suffixes of vertices connecting to $\boldsymbol{p}_i$. For simplicity, the set S_i consisting of all vertices $\boldsymbol{p}_i$ is assumed not to be null, and the binary watermark sequences along the three axes are $W_{\boldsymbol{x}} = \{w_{x0}, w_{x1}, \ldots, w_{x(k-1)}\}$, $W_{\boldsymbol{y}} = \{w_{y0}, w_{y1}, \ldots, w_{y(k-1)}\}$ and $W_z = \{w_{z0}, w_{z1}, \ldots, w_{z(k-1)}\}$. The three watermark embedding formulas are as follows:

$$x_i^w = x_i + \alpha \Lambda_x(\boldsymbol{p}_i) w_{xi}, \quad (5.4)$$

$$y_i^w = y_i + \alpha \Lambda_y(\boldsymbol{p}_i) w_{yi}, \quad (5.5)$$

$$z_i^w = z_i + \alpha \Lambda_z(\boldsymbol{p}_i) w_{zi}, \quad (5.6)$$

where $\Lambda(p_i)=\{\Lambda_x(\boldsymbol{p}_i),\ \Lambda_y(\boldsymbol{p}_i),\ \Lambda_z(\boldsymbol{p}_i)\}$ is the mask function of the vertex $\boldsymbol{p}_i$, α is the embedding factor that is set to be 0.2 in [40]. The construction of the mask function is introduced here. First, a vector $\boldsymbol{n}_i$ is defined as follows:

$$\boldsymbol{n}_i=\frac{1}{|S_i|}\sum_{j\in S_i}(\boldsymbol{p}_j-\boldsymbol{p}_i)=(n_{ix},\ n_{iy},\ n_{iz}), \tag{5.7}$$

where $|S_i|$ represents the set cardinality, and the vector $\boldsymbol{n}_i$ is in essential a "discrete normal vector" that represents the change of the coordinates around $\boldsymbol{p}_i$. Thus the mask function can be defined as follows:

$$\Lambda(p_i)=\{|n_{ix}|,\ |n_{iy}|,\ |n_{iz}|\}. \tag{5.8}$$

5.4.1.3 Dithered Modulation in the Ellipsoid Derived from Connected Vertices

In [41, 42], the embedding location is first confirmed and then a dithering embedding method is performed in the ellipsoid that is derived from the vertices connected to the selected location (vertex). The selection of embedding locations is based on a geometry criterion. First, every "discrete normal vector" $\boldsymbol{n}_i$ for every vertex $\boldsymbol{p}_i$ is computed according to Eq.(5.7). Then an ellipsoid is defined for each vertex, which encloses all the connected vertices to $\boldsymbol{p}_i$. Obviously, the centroid of the ellipsoid is calculated as follows:

$$\boldsymbol{\mu}_i=\frac{1}{|S_i|}\sum_{j\in S_i}\boldsymbol{p}_j, \tag{5.9}$$

while the shape of the ellipsoid is determined by the variance (2-order statistics) as follows:

$$\boldsymbol{U}_i=K\frac{\sum_{j\in S_i}(\boldsymbol{p}_j-\boldsymbol{\mu}_i)(\boldsymbol{p}_j-\boldsymbol{\mu}_i)^{\mathrm{T}}}{|S_i|}, \tag{5.10}$$

where K is a normalized factor. In general, $\boldsymbol{U}_i$ is not singular unless all the vertices connected to $\boldsymbol{p}_i$ are coplanar. Obviously, we should avoid the vertex $\boldsymbol{p}_i$ that produces a singular matrix $\boldsymbol{U}_i$. In the case that $\boldsymbol{U}_i$ is non-singular, any vector $\boldsymbol{q}$ on the ellipsoid surface should satisfy the following condition:

$$(\boldsymbol{q}-\boldsymbol{\mu}_i)^{\mathrm{T}}U_i^{-1}(\boldsymbol{q}-\boldsymbol{\mu}_i)=1. \tag{5.11}$$

Consequently, an ellipsoid can be represented by ($\boldsymbol{\mu}_i$, $\boldsymbol{U}_i$). After every ellipsoid corresponding to $\boldsymbol{p}_i$ is calculated, the sum distance from the vertex $\boldsymbol{p}_i$ to its neighborhoods can be computed as follows:

$$D_i = \sum_{j \in S_i} \left\| \boldsymbol{p}_j - \boldsymbol{p}_i \right\| \cdot \tag{5.12}$$

Now we can safely select vertices that satisfy $D_i<T$ as the embedding primitives. T is a predefined threshold. It should be noticed that once a vertex is selected as an embedding location, all the connected vertices to this vertex should be excluded for embedding watermark bits in order not to interfere with each other.

All the vertices that satisfy the condition $D_i<T$ are divided into several groups, each group consisting of m vertices, and then a binary watermark sequence with the length of m, is embedded repeatedly. For any vertex in each group, two embedding methods are adopted in [42]. In the first method, two parallel planes that are of the same distance from the centroid $\boldsymbol{\mu}_i$ are defined, and the normal vector $\boldsymbol{Q}_i$ of the parallel planes and their distance e_i from the centroid are calculated respectively as below:

$$\boldsymbol{Q}_i = \frac{1}{|S_i|} \sum_{j \in S_i} \boldsymbol{n}_j \, , \tag{5.13}$$

$$e_i = \frac{1}{|S_i|} \sum_{j \in S_i} [(\boldsymbol{p}_j - \boldsymbol{\mu})^{\mathrm{T}} \boldsymbol{Q}]^2 \cdot \tag{5.14}$$

If the watermark bit is “1”, we should make the following formula come into existence

$$(\boldsymbol{p}_i^w - \boldsymbol{\mu}_i)^{\mathrm{T}} \boldsymbol{Q}_i < e_i , \tag{5.15}$$

where $\boldsymbol{p}_i^w$ is the watermarked vertex. If the watermark bit is “0”, then the following formula should come into existence:

$$(\boldsymbol{p}_i^w - \boldsymbol{\mu}_i)^{\mathrm{T}} \boldsymbol{Q}_i > e_i . \tag{5.16}$$

In the second method, a watermark is embedded with the ellipsoid surface defined above as the boundary, meaning that if we want to embed a bit “1”, we modify $\boldsymbol{p}_i$ along $\boldsymbol{p}_i - \boldsymbol{\mu}_i$ until the final $\boldsymbol{p}_i^w$ falls in the ellipsoid such that

$$(\boldsymbol{p}_i^w - \boldsymbol{\mu}_i)^{\mathrm{T}} \boldsymbol{U}_i^{-1} (\boldsymbol{p}_i^w - \boldsymbol{\mu}_i) < 1. \tag{5.17}$$

Otherwise, we can make $\boldsymbol{p}_i^w$ outside the ellipsoid so that a watermark bit “0” can be embedded, making $\boldsymbol{p}_i^w$ satisfy the following formula,

$$(\boldsymbol{p}_i^w - \boldsymbol{\mu}_i)^{\mathrm{T}} \boldsymbol{U}_i^{-1} (\boldsymbol{p}_i^w - \boldsymbol{\mu}_i) > 1. \tag{5.18}$$

5.4.1.4 Fragile Watermarking

Besides the above several algorithms, Yeung and Yeo from Intel presented a fragile 3D mesh watermarking algorithm for verification for the first time in 1999. The proposed algorithm can be used to verify whether or not the change on a 3D polygon mesh is authentic [29, 30]. As we know, in order to achieve this purpose, the embedded watermark should be very sensitive for even minor changes, and any mesh change will be immediately detected and located, and then presented in an intuitive way. The basic process is as follows: Firstly, the centroid $\boldsymbol{\mu}_i$ of all the vertices connected to the vertex $\boldsymbol{p}_i$ is computed according to Eq.(5.9). Then the floating vector $\boldsymbol{\mu}_i$ is converted to an integer vector $\boldsymbol{t}_i = (t_{ix}, t_{iy}, t_{iz})$ using a certain function. Finally, another function is utilized to convert $\boldsymbol{t}_i = (t_{ix}, t_{iy}, t_{iz})$ into two integers L_{ix} and L_{iy}, thus the mapping from the centroid to a 2D mesh is acquired, where (L_{ix}, L_{iy}) is the corresponding position in the 2D mesh. In fact, a 3D vertex coordinate can be converted into an integer using a certain function, where the integer can be regarded as a pixel value while (L_{ix}, L_{iy}) is the pixel's corresponding coordinate. As a result, the watermark can be embedded through slightly altering the coordinates in the image. The study of fragile watermarking is an important branch of watermarking and can be widely used in 3D model authentication and multi-level user management in collaborative design.

5.4.2 Modifying Distances or Lengths

We introduce a 3D mesh watermarking technique, a vertex flood algorithm, and a robust watermarking algorithm for polygon meshes.

5.4.2.1 Modifying the Distances from the Centroid to Vertices

A 3D mesh watermarking technique [43, 44] that utilizes the distances from the centroid to vertices to achieve watermarking is proposed by Yu et al. from Northwestern Polytechnical University. The watermark embedding process is as follows.

Step 1: Input the watermark to be embedded and/or the secret key into the pseudo-randomizer and the corresponding binary watermark sequence $\boldsymbol{w} = \{w_0, w_1, \ldots, w_{m-1}\}$ is generated, where m is the length of the watermark sequence, $\boldsymbol{w} = G(K)$ represents the watermark generation algorithm and K is a large enough set of keys.

Step 2: Use the function "*Permute*" to reorder the original vertex set $\boldsymbol{P} = \{\boldsymbol{p}_i\}$, $I = 0, 1, 2, \ldots, k-1$, with the key as the parameter: $\boldsymbol{P}' = Permute(\boldsymbol{P}, K)$, where

k is the number of vertices of a 3D model, K is the secret key for reordering and $\boldsymbol{P}' = \{\boldsymbol{p}'_i\}$ is the reordered vertex sequence of the 3D model.

Step 3: Select $L\times m$ vertices orderly from the reordered vertices $\boldsymbol{P}' = \{\boldsymbol{p}'_i\}$ and divide them into m groups, i.e. $\boldsymbol{P}' = \{\boldsymbol{P}'_0,\ \boldsymbol{P}'_1,\ \ldots,\ \boldsymbol{P}'_{m-1}\}$, where $\boldsymbol{P}'_i = \{\boldsymbol{p}'_{i0},\ \boldsymbol{p}'_{i1},\ \ldots,\ \boldsymbol{p}'_{i(L-1)}\}$, $0\le i\le m-1$, and L is the number of vertices in each group.

Step 4: Each group can be regarded as an embedding primitive $\boldsymbol{P}'_i$ and can be embedded with a watermark bit w_i. In [43], the watermark is embedded in the following manner:

$$\boldsymbol{L}^w_{ij} = \boldsymbol{L}_{ij} + \alpha w_i \boldsymbol{U}_{ij},\quad 0\le i\le m-1,\quad 0\le j\le L-1, \tag{5.19}$$

where $\boldsymbol{L}_{ij}$ denotes the vector from the center to the j-th vertex in the i-th group, $\boldsymbol{L}^w_{ij}$ represents the corresponding watermarked vertex, α is the embedding weight, w_i is the i-th bit of the watermark sequence and $\boldsymbol{U}_{ij}$ is the unit vector of $\boldsymbol{L}_{ij}$. To improve the transparency, the watermark can be embedded in the following manner:

$$\boldsymbol{L}^w_{ij} = \boldsymbol{L}_{ij} + \alpha\beta_{ij} w_i \boldsymbol{U}_{ij},\quad 0\le i\le m-1,\quad 0\le j\le L-1, \tag{5.20}$$

where α is the global embedding weight parameter that controls the overall energy of the embedded watermark, β_{ij} is the local embedding weight parameter that makes the embedding process adaptive to the local characteristic of the 3D model. In [44], the watermark is embedded in the following manner:

$$\boldsymbol{L}^w_{ij} = \boldsymbol{L}_{ij} + \beta_{ij}(\alpha) w_i \boldsymbol{U}_{ij},\quad 0\le i\le m-1,\quad 0\le j\le L-1, \tag{5.21}$$

where $\beta_{ij}(\alpha)$ shows that the local embedding weight is relevant to the global embedding weight α.

Step 5: Reorder the watermarked 3D model back to its original order.

The corresponding detection method for above-mentioned watermark embedding methods involves the original 3D model $\boldsymbol{M}$ and the detailed procedure is as follows:

Step 1: Some attackers may use simple translation, rotation and scaling operations to change the watermarked 3D model. Before the watermark extraction, the attacked 3D model must be registered to its original position, direction and scale. Usually, there is always a balance between computation complexity and accuracy, which affects the speed and accuracy of watermark extraction. As a result, we should make an appropriate trade-off between complexity and accuracy. The registration process should be performed between the model $\hat{\boldsymbol{M}}$ to be detected and the original model $\boldsymbol{M}$, because if the registration is performed between $\hat{\boldsymbol{M}}$ and the stego mesh $\boldsymbol{M}^w$, some additional information may be introduced to $\hat{\boldsymbol{M}}$.

Step 2: Since some attacks may alter the mesh topology, such as simplification, insection and remeshing, the watermark cannot be correctly extracted from the attacked model through a non-blind watermark detection method. In this case, resampling is required to recover the model with the original connectivity. The resampling process is as follows: a line from the center of the original model $\boldsymbol{M}$ to the vertex $\boldsymbol{p}_i$ is drawn and intersected with $\hat{\boldsymbol{M}}$. If there is more than one point of intersection that is closest to $\hat{\boldsymbol{p}}_i$, then $\hat{\boldsymbol{p}}_i$ is regarded as the match point of $\boldsymbol{p}_i$; Otherwise $\hat{\boldsymbol{p}}_i = \boldsymbol{p}_i$ is taken.

Step 3: This process is the same as Steps 2 and 3 in the embedding algorithm: reorder $\boldsymbol{M}$ and $\hat{\boldsymbol{M}}$ and group them to get $\boldsymbol{P}' = \{\boldsymbol{P}'_0, \boldsymbol{P}'_0, \ldots, \boldsymbol{P}'_{m-1}\}$ and $\hat{\boldsymbol{P}}' = \{\hat{\boldsymbol{P}}'_0, \hat{\boldsymbol{P}}'_1, \ldots, \hat{\boldsymbol{P}}'_{m-1}\}$.

Step 4: Regard the center of the original model as the center of the model to be detected. Compute the magnitude difference between the vector from the model center to original vertices and the vector from the model center to the vertices to be detected in each group:

$$D_{ij} = \hat{L}_{ij} - L_{ij}, \tag{5.22}$$

where L_{ij} is the vector magnitude from the center to the j-th vertex in the i-th group and $\hat{L}_{ij}$ is the corresponding vector magnitude for $\hat{\boldsymbol{M}}$.

Step 5: Sum the vector magnitude differences in each group:

$$D_i = \frac{1}{L}\sum_{j=0}^{L-1} D_{ij}, \tag{5.23}$$

where D_i is the sum of the differences in the i-th group.

Step 6: Extract the watermark as follows:

$$\hat{w}_i = \mathrm{sgn}(D_i). \tag{5.24}$$

Step 7: Verify whether or not the extracted watermark is identical to the original, according to the correlation between the extracted and the original watermarks. If the correlation is higher than the threshold T, then the extracted watermark is identical to the original, otherwise not. The correlation is defined as below:

$$Cor(\hat{\boldsymbol{w}}, \boldsymbol{w}) = \frac{\sum_{j=0}^{m-1}(\hat{w}_j - \hat{w}_{ave})(w_j - w_{ave})}{\sqrt{\sum_{j=0}^{m-1}(\hat{w}_j - \hat{w}_{ave})^2 \sum_{j=0}^{m-1}(w_j - w_{ave})^2}}, \tag{5.25}$$

where $\hat{\boldsymbol{w}}$ is the extracted watermark sequence, $\boldsymbol{w}$ is the original watermark sequence, $\hat{w}_{ave}$ is the mean of $\hat{\boldsymbol{w}}$, w_{ave} is the mean of $\boldsymbol{w}$ and m is the length of the watermark sequence.

The algorithms in [44] have the following characteristics: (1) They use the overall geometric features as primitives; (2) They distribute the watermark information throughout the model; (3) The watermark embedding strength is adaptive to local geometric features. Experiments show that this watermarking algorithm can resist ordinary attacks for a 3D model, such as simplification, adding noise, insection and their combinations. In addition, a progressive transmission method of 3D models is introduced in [45]. This literature has also proposed a watermarking algorithm based on the distance from the vertices to the mean of the base. This algorithm adopts the simple additive embedding mechanism. Due to space limitations, it will not be illustrated here.

5.4.2.2 Vertex Flood Algorithm

Benedens proposed two oblivious watermarking algorithms for polygon meshes in [31] and one of them is called a vertex flood algorithm. In the vertex flood algorithm, one or more triangles are first chosen as the initial triangles and then data can be embedded through adjusting the distances from the initial triangles' gravity to all vertices. This algorithm is a kind of fragile watermarking algorithm that can be used in model verification. Due to space limitations, this algorithm will not be elaborated here.

5.4.2.3 Altering the Length of Specific Vectors

A robust watermarking algorithm for polygon meshes with an arbitrary topology [7] was proposed by Wagner from Arizona State University in the USA. In this algorithm, a watermark is embedded in the coordinates of mesh data. Since the embedding is independent of the order of vertices, it shows high robustness to similarity transforms, but is less robust to remeshing and simplification operations. The basic procedure is as follows: First, compute the vector $\boldsymbol{n}_i$ according to Eq.(5.7). Then the relative vector magnitudes are regarded as the embedding primitives. Since the Euclidean norm $\|\boldsymbol{n}_i\|$ is invariant to affine transforms, the algorithm is robust to affine transforms. Let

$$d = \frac{1}{k}\sum_{i=0}^{k-1}\|\boldsymbol{n}_i\|, \tag{5.26}$$

and according to

$$n_i = \text{round}\left(\frac{c}{d}\|\boldsymbol{n}_i\|\right), \tag{5.27}$$

we can convert each vector $\boldsymbol{n}_i$ to an integer, where c is the primary parameter that is a fixed real value. The value of $\boldsymbol{n}_i$ remains unchanged during the geometry transform of 3D models. The watermark data are defined as a function $f(v)$ on the sphere surface, e.g. $f(v)$ = constant. Similarly, according to

$$w_i = \text{round}\left(2^b f\left(\frac{\boldsymbol{n}_i}{\|\boldsymbol{n}_i\|}\right)\right), \tag{5.28}$$

the value of $f(v)$ can be converted to an integer w_i. From the binary representation of $\boldsymbol{n}_i$, b bits can be selected to be replaced by the watermark data w_i (for each $\boldsymbol{n}_i$, the embedding location is fixed), so the modified vector $\boldsymbol{n}_i^w$ is acquired. With the above formulae, the watermarked vertex $\boldsymbol{p}_i^w$ can be calculated according to $\boldsymbol{n}_i^w$. The watermark extraction process is relatively simple, only requiring the calculation of $\hat{\boldsymbol{n}}_i$ and an appropriate position for extraction.

5.4.3 *Adopting Triangle/Strip as Embedding Primitives*

We introduce the triangle similarity quadruple method, mesh density pattern algorithm, quantization index modulation, and triangle flood algorithm.

5.4.3.1 Triangle Similarity Quadruple (TSQ)

In 1997, Ohbuchi et al. proposed several 3D model watermarking algorithms for triangle meshes based on the concepts of mesh displacement, topology displacement and visual pattern, the most representative and most historically significant algorithm of which is the triangle similarity quadruple (TSQ) method [6, 33, 46-48]. Just as its name implies, this algorithm utilizes the concept of similar triangles. A set of similar triangles can be defined as a two-tuple (b/a, h/c), as shown in Fig. 5.3. In addition, 4 neighboring triangles can form a macro-embedding primitive (MEP), as shown in Fig. 5.4. Each MEP can store a quadruple {Marker, Subscript, Data1, Data2}, where "Marker" is to uniquely mark the MEP, "Subscript" is the index, "Data1" and "Data2" are symbols to be embedded. In an MEP, the 4 triangles are denoted by M, S, D_1 and D_2, and store the values of "Marker", "Subscript", "Data1" and "Data2", respectively.

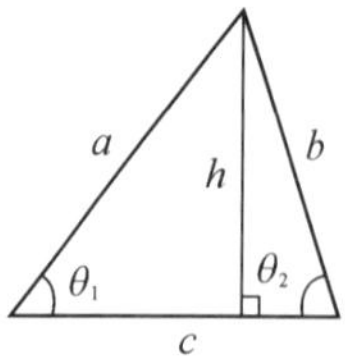

Fig. 5.3. Two-tuple $\{b/a, h/c\}$

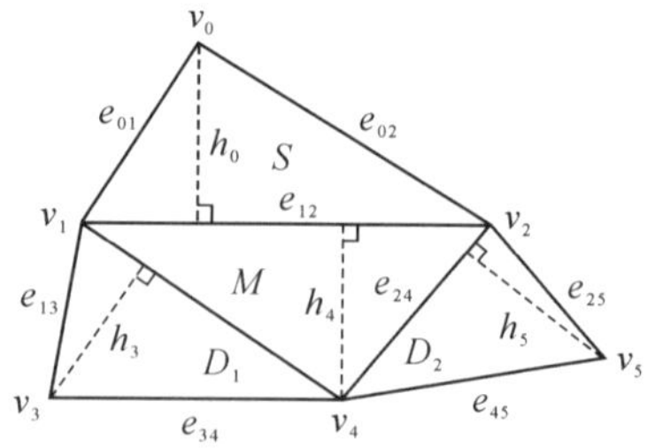

Fig. 5.4. The 4 triangles in MEP

The watermark embedding process is as follows: First we traverse the whole mesh and seek for a proper MEP. We make the middle triangle M similar to the given triangle through slightly altering the three vertices of M, so that the value of Marker can be embedded. Then, by changing the coordinates of $\boldsymbol{v}_0$, $\boldsymbol{v}_3$ and $\boldsymbol{v}_5$, the values of "Subscript", "Data1" and "Data2" can be embedded into two-tuples $\{e_{02}/e_{01}, h_0/e_{12}\}$, $\{e_{13}/e_{34}, h_3/e_{14}\}$ and $\{e_{45}/e_{25}, h_5/e_{24}\}$ respectively, repeating the above process until all the data are embedded.

The watermark extraction process is as follows: (1) Seek for the matched MEP in the stego mesh according to a given two-tuple, i.e., "Marker"; (2) Extract values of "Subscript", "Data1" and "Data2"; (3) Repeat the above process until all the data are extracted; (4) Reorder all "Data1" and "Data2" values according to the value of "Subscript", and combine them to acquire the final extracted watermark.

The above idea is simple and the corresponding design is elegant with easy implementation, so the algorithm can be used for copyright information reminder in a collaborative design process. The 3D model of Beethoven's bust with 4,889 triangles and 2,655 vertices embedded with 132 bytes information (embedded for 6 times with redundancy) is depicted in Fig. 5.5(a). The embedded information is lost gradually with the insection becoming heavier. As shown in Table 5.2, all the 132 bytes hidden information can be retrieved when the left side is cut, while only 102 bytes can be retrieved when the model is decimated by three quarters.

(a) (b) (c) (d)

Fig. 5.5. Watermarked insections of the 3D model of Beethoven bust [19]. (a) With 4,889 triangles; (b) With 2,443 triangles; (c) With 1,192 triangles; (d) With 399 triangles. (©1997, Association for Computing Machinery, Inc. Reprinted by permission)

Table 5.2 Information lost caused by insections

Subgraph of Fig. 5.6	Number of triangles	Information included
(a)	4,889	Embedded 6 times with redundancy, 132 bytes for each embedding
(b)	2,443	132/132 bytes
(c)	1,192	102/132 bytes
(d)	399	85/132 bytes

5.4.3.2 Mesh Density Pattern Algorithm

Another representative algorithm proposed by Ohbuchi is the mesh density pattern (MDP) algorithm in [33]. In this algorithm, a pattern that is visible in the wire-frame rendering mode can be embedded in the given triangle mesh by adjusting the triangle mesh size. The algorithm can resist certain geometric transformations, but is fragile to mesh topology attacks such as simplification. A visible watermark "IBM" is embedded in a mesh model (in the wire-frame rendering mode) in Fig. 5.6 while the simplified stego mesh is depicted in Fig. 5.7.

Fig. 5.6. A mesh model with a visible watermark [19] (©1997, Association for Computing Machinery, Inc. Reprinted by permission)

5.4.3.3 Quantization Index Modulation

In 2003, a mesh model watermarking algorithm based on quantization index modulation (QIM) was proposed [49], in which a certain edge in a triangle is regarded as the entry edge and the other two are exit edges, as shown in Fig. 5.8(a), where *AB* is the entry edge, *AC* and *BC* are exit edges. There are two steps in the algorithm:

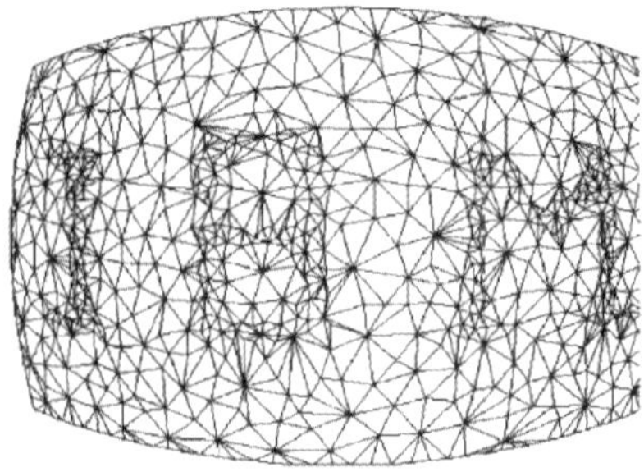

Fig. 5.7. Simplified stego mesh [19] (©1997, Association for Computing Machinery, Inc. Reprinted by permission)

First, a triangle strip peeling sequence is established based on a secret key, and the process is shown in Fig. 5.8. The initial triangle is determined by the specific geometry characteristic. The next triangle in the sequence should either be the first triangle (Its new entry edge is AC) or the second triangle (Its new triangle is BC), which is determined by the bits of the secret key. Here, the length of the secret key is allowed to be the same as the number of triangles. The path of the accessed triangles is called "Stencil" in [49].

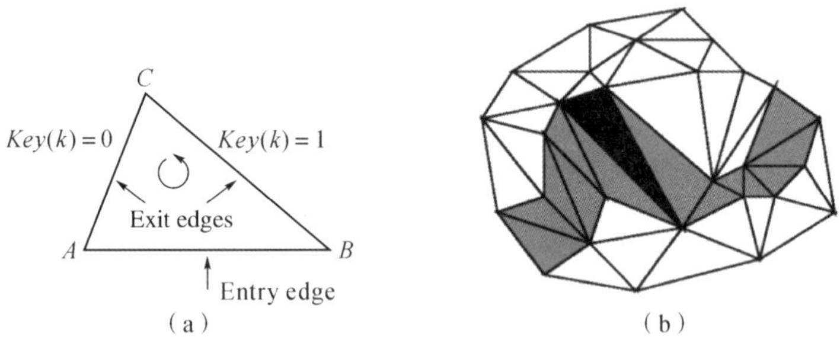

Fig. 5.8. Construction of the triangle strip peeling sequence (TSPS) [8]. (a) Two types of triangle edges; (b) TSPS is gray and the embedded location is black (©[2003]IEEE)

Second, the selected triangle is judged whether or not to be changed according to the binary watermark information, which is called the macro embedding procedure (MEP). Every triangle is regarded as a two-state object, where the state of a triangle is defined as being determined by the interval of Vertex C's perpendicular projection on the entry edge AB, and $P(C)$ represents the projection location. In order to describe the state of an interval, we can use a method similar to quantization index modulation to segment the entry edge AB into a series of intervals, two intervals forming a group, as shown in Fig. 5.9. The ends of these intervals are denoted by D_0 to D_{2n}, and divided into two subsets: S_0 and S_1, by allocating the ends into groups of two ends. If the length of these intervals is fixed to be $1/(2n)$, then the state of the triangle is invariant to geometry transforms. If the projection $P(C)$ belongs to the subset S_0, then the state of the triangle is "0"; otherwise, if $P(C)$ belongs to the subset S_1, the state of the triangle is "1". If the watermark bit to be embedded is w, then the embedding rule is as follows: If $P(C)$ belongs to the subset S_w, then no modification is needed; otherwise, move C to C' to make $P(C')$ belong to the subset S_w. The mapping from C to C' must satisfy affine transform invariance, and the distance between them should be short enough to satisfy the imperceptivity while being long enough to satisfy the robustness. As a result, the boundaries of intervals are viewed as the symmetry axes, in order to make the mapping symmetric with respect to the nearest axis, as shown in Fig. 5.10 (an example of $n = 2$).

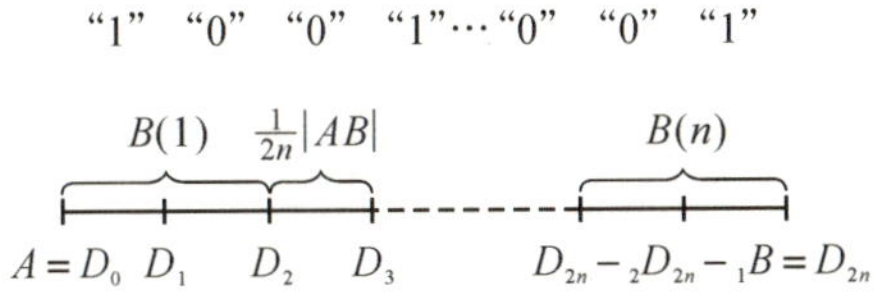

Fig. 5.9. Interval segmentation of the entry edge AB [8] (©[2003]IEEE)

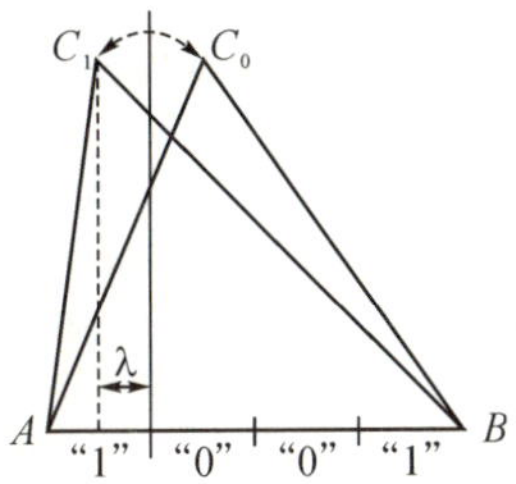

Fig. 5.10. Dithering of vertex C [8] (©[2003]IEEE)

5.4.3.4 Triangle Flood Algorithm

Besides the vertex flood algorithm, another oblivious mesh watermarking algorithm was also proposed by Benedens [31]. Similar to the vertex flood algorithm, one or two initial triangles are needed to be selected. A unique traverse order is generated according to the initial triangles and the watermark information is embedded through altering the vertex coordinates along the path and recording the triangle order along this path in order to ensure it to be unique. Due to space limitation, the algorithm will not be elaborated here.

5.4.4 Using a Tetrahedron as the Embedding Primitive

In addition to TSQ and MDP algorithms, Ohbuchi also proposed another representative and historically significant algorithm—tetrahedral volume ratio (TVR). This algorithm utilizes an affine transform invariant, i.e. the tetrahedral volume ratio, to embed the watermark in a mesh: Set an initial condition, that is, the initial vertex and the initial spanning direction are given and seek for a vertex spanning tree Vt according to the triangle mesh. At a given vertex, scan the connecting edges counterclockwise until an edge that is not available in Vt and is not connected to any vertex scanned in Vt. If an edge satisfying the above conditions is found, append it to Vt. And then a certain edge to be the initial edge is sought in order that the volume of the enclosed tetrahedron is maximal. A triangle bounding edge (TBE) list is required to be constructed before Vt is converted into a triangle list, where the initial list consists of the edges of a series of vertices in Vt. The list can be constructed as follows: scan Vt from the root node and then span all the vertices, and scan all connected edges clockwise at each vertex. If the scanned edge is not in TBE, then append it. If the three edges of a triangle are found in TBE for the first time, and the triangle is not available in the triangle sequence "Tris", then append the triangle in the "Tris", as shown in Fig. 8 in [19]. Convert Tris into a tetrahedron sequence "Tets", and regard the first tetrahedron of Tets as the denominator. Converting the "Tets" to a volume ratio sequence "Vrs", the data symbol can be embedded into each volume ratio through replacing the vertices of the numerator tetrahedrons. The embedded locations are depicted in Fig. 11 in [19], where the dark gray parts represent the embedded locations.

The watermark extraction process involves testing the candidate edges to find proper initial edges using pre-embedded symbols. However, because of factors such as noise, usually it is not accurate if the initial edge is determined only in accordance with the largest tetrahedron volume. The algorithm is highly robust to affine transforms (such as projection transformation), but is fragile to topology changes (such as remeshing and randomization of the vertex order) and geometry transformation. The stego mesh and the attacked stego mesh with an affine transform and an insection are rendered in Fig. 5.11. Simulation results show that the TVR algorithm can resist these two attacks.

In addition to TVR, another mesh watermarking algorithm based on Affine Invariant Embedding (AIE) was proposed by Benedens and Busch [50, 51]. Inspired by TVR, AIE uses tetrahedrons as embedding primitives as follows: A triangle with vertices $V=\{\boldsymbol{v}_1, \boldsymbol{v}_2, \boldsymbol{v}_3\}$ is selected and then an edge with an end in V is selected. The other end of the selected edge is denoted as $\boldsymbol{v}_4$, where the distance from $\boldsymbol{v}_4$ to $\{\boldsymbol{v}_1, \boldsymbol{v}_2, \boldsymbol{v}_3\}$ is large enough. Thus two initial triangles $\{\boldsymbol{v}_1, \boldsymbol{v}_2, \boldsymbol{v}_3\}$ and $\{\boldsymbol{v}_2, \boldsymbol{v}_3, \boldsymbol{v}_4\}$ are acquired, as shown in Fig. 5.12. Two sets G_1 and G_2 are constructed: G_1 consists of all vertices that only have one neighboring vertex in $V=\{\boldsymbol{v}_1, \boldsymbol{v}_2, \boldsymbol{v}_3, \boldsymbol{v}_4\}$, i.e. ***a***, ***b***, ***c***, ***d***, ***e*** in Fig. 5.12; G_2 is comprised of all vertices that are neighboring to the initial triangle through an edge and locate in a certain triangle, i.e. ***A***, ***B***, ***C***, ***D*** in Fig. 5.12. A set G is constructed based on G_1 and G_2: If $|G_1|<4$ (meaning that the cardinality is less than 4) and $|G_2|<4$, then set $G=G_2 \cup G_1$. Otherwise, let $G=\min_{i=1,2}\{G_i \mid |G_i| \geq 4\}$. If $|G|<4$, then abandon this primitive. The case of $G=G_2$ is shown in Fig. 5.12. Finally, divide G into 4 subsets g_1, g_2, g_3, g_4 (The number of elements should be similar) and record the watermark information and the control information in the vertices that formed g_1, g_2, g_3, g_4, as shown in Table 5.3, where the former 2 bits are the flag for groups, I_5–I_0 are index bits, D_9–D_0 are imbedded information bits, so the embedding capacity is deduced to be 640 bits. The GEOMARK system has been developed by Benedens et al. based on the above-mentioned algorithms and can be applied to watermarking for 3D models and virtual scenes.

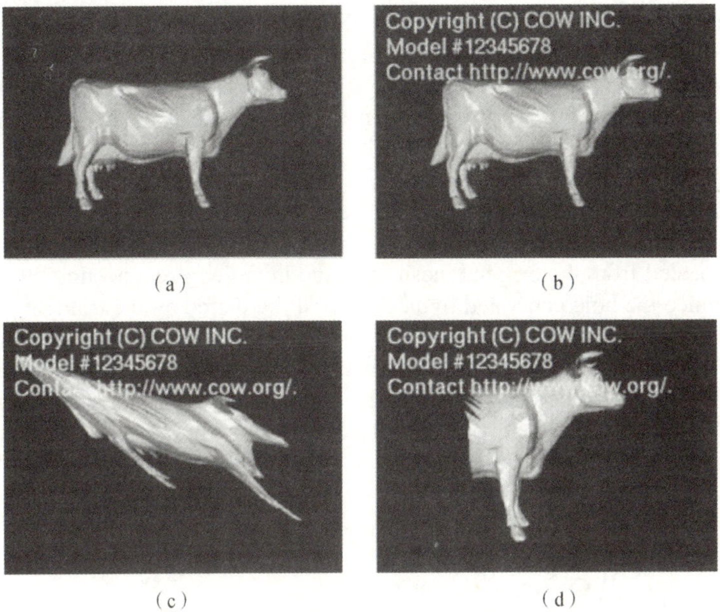

Fig. 5.11. Results of watermarking and attacks by TVR [19]. (a) Cover model; (b) Stego model; (c) Affine transform; (d) Insection. (©1997, Association for Computing Machinery, Inc. Reprinted by permission)

Table 5.3 Distribution of embedded information bits in each group

Group	Embedded information bits				
g_1	00	I_5	I_4	I_3	I_2
g_2	01	I_1	I_0	D_9	D_8
g_3	10	D_7	D_6	D_5	D_4
g_4	11	D_3	D_2	D_1	D_0

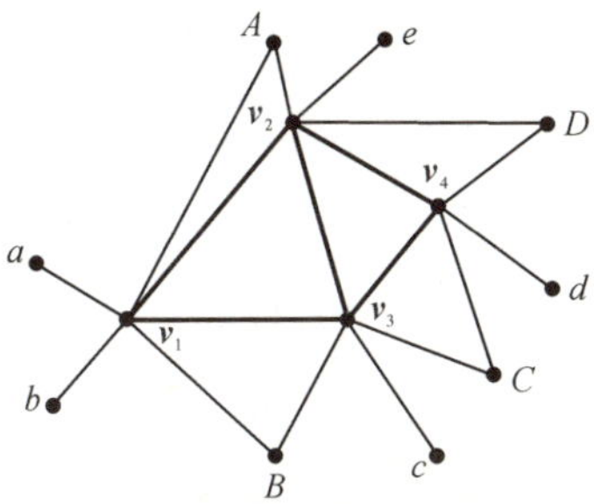

Fig. 5.12. The two initial triangles are the embedding primitive in AIE, denoted as $V = \{v_1, v_2, v_3, v_4\}$

5.4.5 Topology Structure Adjustment

Another representative watermarking algorithm proposed by Ohbuchi et al. in the early years is triangle strip peeling symbol sequence (TSPS) [33, 46, 47]. The algorithm is oblivious based on alteration of the mesh topology, with the relationship between a triangle pair in the triangle sequence as the embedding primitive. Each of these elements can be embedded with 1-bit ("0" or "1") information. The linear arrangement relations between primitives can be derived from the adjacency of every triangle in the triangle strip peeling sequence. An example is shown in Fig. 5.13, where 12 adjacent triangles (in real lines) form a triangle strip peeling sequence and represent 11 bits of information "10101001011". If the last bit of the bit sequence is not "1", then the last triangle is drawn in dashed lines. In essence, during the embedding process, the triangle strip is peeled from the original mesh, with the initial edge connecting the original mesh. Since the hole generated by peeling is still covered by the triangle strip, the operation is invisible. Because the algorithm is based on alteration of the topology, it can resist the attacks of geometry transforms. The algorithm can also resist insection through embedding redundantly. However, this algorithm is not robust to topology attacks such as polygon simplification. In addition, the space utilization is relatively low.

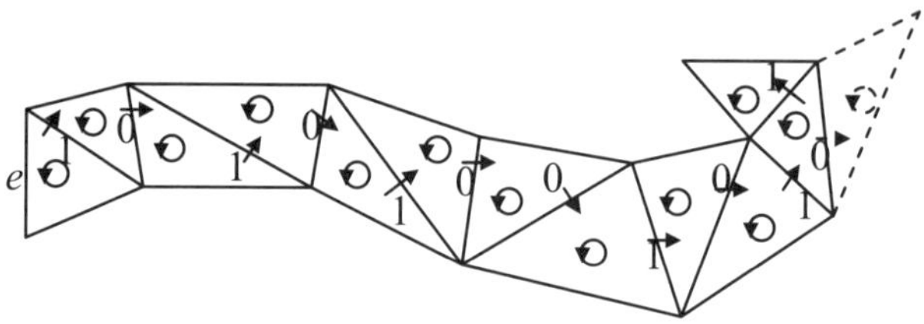

Fig. 5.13. The selection of the triangle strip according to the watermark bits

5.4.6 Modification of Surface Normal Distribution

Inspired by the works of Ohbuchi et al., Benedens proposed a mesh watermarking algorithm by modifying surface normal distribution [28, 52]. As we know, a 3D object can be regarded as a set of surfaces with different sizes and a certain direction, while surfaces can be represented or approached by a mesh with a series of planes or, in some cases, triangles. The distribution of mesh surface normals will be changed after the watermark is embedded, without any change in the mesh topology. In this algorithm, surface normal vectors from the centroid to the centers of triangles are constructed first, then the basic geometry unit—bin normal vector (in the pre-processing, the normal vectors of a mesh are divided into several sets—bins, each of which is a set of surface normal vectors and can be embedded with 1 bit watermark) is calculated, and finally the watermark is embedded by moving the centroid of bins, i.e. average normal vectors. n centroids of bins should be moved in order to embed a watermark with n bits. The replacement process is through substituting the mesh vertices, resulting in the changes of normal vectors of triangles and then the centroids of the corresponding bins. Simulation results show that the algorithm is robust to vertex randomization, remeshing and simplification. The embedded watermark can still survive when the stego mesh is simplified to 36% of the cover mesh. In addition, another mesh watermarking algorithm based on alteration of surface normal vectors is available in [53]. Due to space limitations, the details are not elaborated here.

5.4.7 *Attribute Modification*

A representative mesh watermarking algorithm that is based on shape attribute (e.g. texture mapping coordinates) adjustment was proposed in [33, 46, 48, 54]. For meshes with texture mapping, the watermark can be embedded through altering the coordinates of the texture mapping or the attributes (e.g. vertex color) of every vertex. The basic idea is to modulate each bit of the watermark to the coordinate displacement in the texture mapping. Similarly, a mesh watermarking based on alteration of line colors and widths was proposed in [55]. Due to space limitation, the methods of this category are not elaborated.

5.4.8 *Redundancy-Based Methods*

Apart from the above algorithms, several algorithms [32] based on the redundant data in a polygon mesh have been proposed by Ichikawa et al. from Japan's Toyohashi University in 2002. The algorithms, which maintain the original geometry and topology, are as follows: (1) Full permutation scheme (FPS) and partial permutation scheme (PPS) that permute the order of mesh vertices and polygons; (2) Polygon vertex rotation scheme (PVR), packet PVR, full PVR (FPVR) and partial PVR (PPVR) that embed watermarks through rotating vertices. Due to the low embedding capacity of these methods, they are only supplementary methods to those methods based on alteration of geometry and topology, and will not be detailed here.

5.5 A Robust Adaptive 3D Mesh Watermarking Scheme

Protection of intellectual properties is one of the most important problems in the production and consumption of digital multimedia data. The problem is gaining more and more attention as multimedia data are increasing, and thus there have been intensive efforts focused on securing the multimedia by encryption and watermarking. In the last decades, as more and more 3D models have been produced, distributed and consumed, they are confronted by the protection problem of intellectual property. Only recently, works focusing on watermarking of 3D model data begin to appear in the literature. As we know, since 1997, Ohbuchi et al. have published a series of literatures with respect to 3D mesh watermarking, which no doubt expand the territory of 3D mesh watermarking techniques. Subsequently, Benedens proposed several robust watermarking algorithms. However, the aforementioned algorithms are either short of robustness or with relatively high computation complexities. In this section, we introduce a robust watermarking scheme [56] proposed by the authors of this book. In the proposed algorithm, watermarks are embedded into a 3D model by altering model vertices with weights and along directions that are all adaptive to the local geometry. Thus, we can watermark the model imperceptibly with maximum possible energy of the watermark. Experimental results and the attack analysis demonstrate that the proposed watermarking algorithm is transparent and robust against a combination of various attacks, and it is time-saving and effective.

5.5.1 *Watermarking Scheme*

The basic flows of watermark embedding and extraction processes are outlined, including watermark embedding process and watermark extraction process.

5.5.1.1 Watermark Embedding Process

The detailed watermark embedding process can be shown in Fig. 5.14. Firstly, we adopt the non-adaptive watermark generation algorithm participating with copyright information. Copyright information and the secret key are inputted to a pseudo-random sequence generator G and the output is the permuted binary watermark as follows:

$$W = G(m, K), \tag{5.29}$$

where m denotes the original copyright information, K is the secret key and

$$W = \{w_i \mid w_i \in \{-1, 1\}, i = 0, 1, \cdots, N-1\}, \tag{5.30}$$

where N denotes the length of the watermark sequence.

Secondly, we disturb the order of vertices of the original model according to the key K:

$$V_o' = P(V_o, K), \tag{5.31}$$

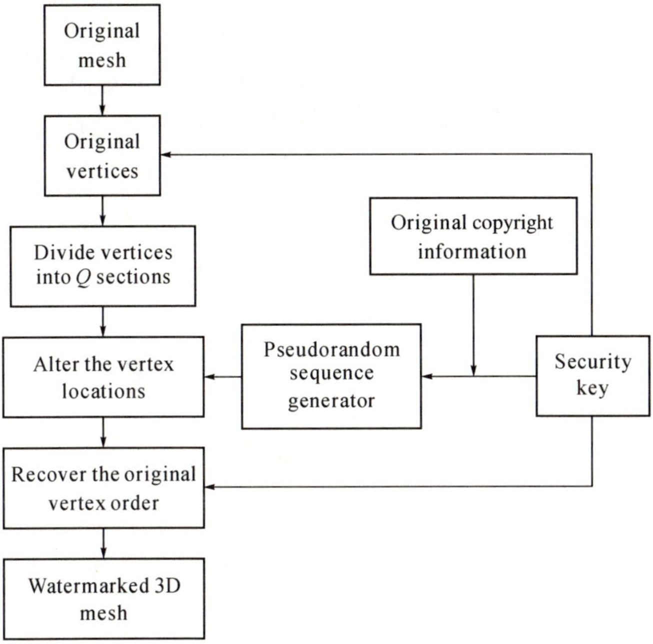

Fig.5.14. The watermark embedding procedure

where $V_o = \{\boldsymbol{v}_i^{(o)}\}$ and $V_p = \{\boldsymbol{v}_i^{(p)}\}$ are the sets of vertices of the original model and the permuted model respectively, $0 \le i \le L-1$ and L is the number of vertices.

Thirdly, we choose $N{\times}Q$ vertices from the vertex set V_p of the disturbed model and then divide these vertices into Q subsets $V_l^{(p)}$, $0 \le l \le Q-1$ as follows:

$$V_l^{(p)} = \{\boldsymbol{v}_{lj}^{(p)}\},\ 0 \le l \le Q-1,\ 0 \le j \le N-1, \tag{5.32}$$

where N is the number of vertices in each section, which equals the length of the watermark sequence.

Fourthly, we embed one watermark bit into each section by the following formula,

$$\boldsymbol{e}_{lj}^{(w)} = \boldsymbol{e}_{lj}^{(o)} + \beta \cdot \alpha_{lj} \cdot w_l \cdot \boldsymbol{n}_{lj}^{(o)}, \tag{5.33}$$

where $\boldsymbol{e}_{lj}^{(o)}$ denotes the original vector from the centroid of the model to the j-th vertex of the l-th section, $\boldsymbol{e}_{lj}^{(w)}$ denotes the watermarked vector from the centroid of the model to the j-th vertex of the l-th section, β is the watermarking coefficient to control the global energy of the embedded watermark sequence, w_l is the l-th bit of the watermark sequence, α_{lj} is the parameter controlling the local watermarking weight and is adaptive to the local geometry of the model, which will be detailed in Subsection 5.5.2. $\boldsymbol{n}_{lj}^{(o)}$ is the direction in which a watermark bit is embedded corresponding to the j-th vertex in the l-th section, which will be detailed in Subsection 5.5.2. The same watermark sequence is embedded into each section repeatedly in order to ensure robustness to local deformation. When a vertex is embedded with a watermark bit, its neighboring vertices cannot be used as embedding locations.

Finally, we permute reversely the order of the watermarked vertices by using the original key K.

5.5.1.2 Watermark Extraction Process

The detailed watermark extraction procedure is shown in Fig. 5.15. Note that an attack might change the 3D model via similarity transforms. We extract the potential watermark as follows.

Before extracting watermarks, we should firstly recover the object to its original location and scale via model registration. The annealing algorithm in [57] is adopted in our work. Secondly, we use the re-sampling scheme proposed in [58] in case attacks which change the mesh connectivity are applied to the watermarked mesh. Thirdly, for both the original model and the model to be detected, we disturb and divide their vertices to get their own Q sections as described in Eqs.(5.31) and (5.32) for the embedding procedure, respectively. We then compute the residual vectors between the vectors that connect the origin with the vertices in each section of the original model and those of the model to be detected as follows:

$$\boldsymbol{r}_{lj} = \boldsymbol{e}_{lj}^{(d)} - \boldsymbol{e}_{lj}^{(o)}, \tag{5.34}$$

where $\boldsymbol{e}_{lj}^{(o)}$ and $\boldsymbol{e}_{lj}^{(d)}$ are the vector from the origin to the j-th vertex in the l-th section of the original model and the vector from the origin to the j-th vertex in the l-th section of the model to be detected, respectively. Fourthly, we sum up the dot products of the residual vectors and their corresponding watermarking directions as follows:

$$s_j = \sum_{l=0}^{Q-1} \boldsymbol{r}_{lj} \cdot \boldsymbol{n}_{lj}^{(o)}, \tag{5.35}$$

where $\boldsymbol{n}_{lj}^{(o)}$ is the direction in which the watermark bit is embedded corresponding to the j-th vertex in the l-th section and $0 \le j \le N-1$. Finally, we extract the watermark sequence (based on the value of s_j, we can easily extract the l-th watermark bit $w_j^{(d)}$) and compute the correlation value between it and the

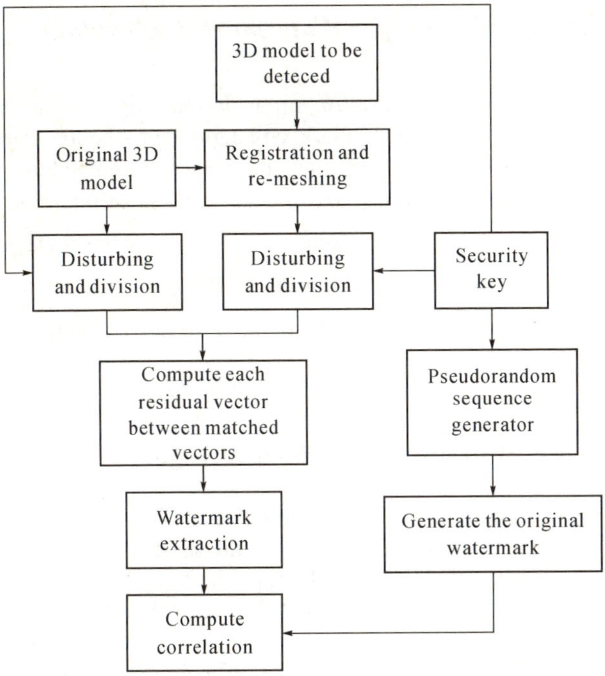

Fig.5.15. The watermark extraction procedure

original one to decide whether the watermark exists in the 3D model to be detected. If the correlation value is larger than the threshold, the watermark exists in the model to be detected, otherwise not. Here, the correlation function is defined as follows:

$$c(W^d,\ W)=\frac{\sum_{i=0}^{N-1}(w_i^{(d)}-\overline{w}^{(d)})(w_i-\overline{w})}{\sqrt{\sum_{i=0}^{N-1}(w_i^{(d)}-\overline{w}^{(d)})^2\sum_{i=0}^{N-1}(w_i-\overline{w})^2}}, \tag{5.36}$$

where $W^{(d)}$ is the extracted watermark sequence, W is the original watermark sequence, $\overline{w}^{(d)}$ is the mean value of $w_i^{(d)}$, $\overline{w}$ is the mean value of w_i and N is the length of the watermark sequence. We specify the threshold T for watermark detection to be 0.4 as a trade-off between the false-positive and false-negative possibilities.

5.5.2 Parameter Control for Watermark Embedding

In order to increase the robustness of our watermarking scheme, we introduce some novel features into the watermark embedding procedure. Particularly, we develop a weighting scheme that regulates the watermark embedding strength and direction. In this subsection, we discuss in detail how to compute parameters α_{lj} and $\boldsymbol{n}_{lj}^{(o)}$ respectively. Here, the parameter α_{lj} controls the local watermarking strength, and the watermark embedding direction $\boldsymbol{n}_{lj}^{(o)}$ is rather a novel feature proposed in this subsection as we present a criterion to determine in which direction to embed the watermark.

5.5.2.1 Control of Watermark Embedding Strength

For embedding watermarks into 3D models, we should regulate the watermark strength so that it adapts to the local feature of the mesh and the watermark can be embedded with high robustness and imperceptivity. Among the previous literatures concerning watermarking of 3D models, few have considered controlling the local watermarking strength. The watermarking algorithm in [39] uses the geometric magnitude of the vertex split operation to control the local watermarking strength while Reference [14] controls the watermark strength by the minimal length of vertices' 1-ring edge neighborhood. Both approaches have their limitations and potential defects. The watermarking algorithm in [59] controls the local watermarking strength using normals of triangle surfaces connecting to a vertex and the distances between the vertex and its subtenses. However, the computation is more complex than that of the algorithm proposed in this paper. In addition, the choice of the watermark embedding direction is not considered in [59], thus the algorithm in [59] cannot make full use of the local geometry of the 3D model either, which can be concluded from the simulation results.

We first compute the distance d_{ji} between the vertex $\boldsymbol{v}_i$ and each of its neighboring vertices, which is defined as

$$d_{ji} = \left\| \boldsymbol{v}_j - \boldsymbol{v}_i \right\|_2, \quad \boldsymbol{v}_j \in N(\boldsymbol{v}_i). \tag{5.37}$$

Regard the vertex $\boldsymbol{v}_i$ as a node in a circuit, the distances between it and its neighboring vertices as impedances between nodes $\boldsymbol{v}_i$ and its neighboring vertices, and the parallel connection impedance between $\boldsymbol{v}_i$ and its neighboring vertices as watermark embedding weights of the vertex $\boldsymbol{v}_i$. The computation formula is defined as follows:

$$wt_i = 1 / \sum_{\boldsymbol{v}_j \in N(\boldsymbol{v}_i)} (1 / d_{ji}). \tag{5.38}$$

As long as there is a relatively small value in the distances between a vertex and its neighboring vertices, the vertex cannot be modified considerably. Otherwise, the normal of the triangle surface containing the relatively shorter edge will be changed greatly, and thus the imperceptivity of the watermark cannot be satisfied [14]. However, if the distances between the vertex and its neighboring vertices are all long and relatively the same, the vertex can be greatly changed, while the triangle surface normals connecting to the vertex are only slightly changed and thus the imperceptivity of the watermark can be satisfied. The aforementioned characteristics mainly accord with those of parallel connection impedances in a circuit.

However, the weight of watermark embedding is slightly different from parallel connection impedance, as shown in Fig. 5.16, where point A represents the vertex $\boldsymbol{v}_i$. All lengths of the solid line segments in Fig. 5.16 are the same. Regard Fig. 5.16(a) and Fig. 5.16(b) as two circuits, each solid line segment representing a component whose impedance is R. It can be easily deducted that the parallel connection impedances between the node $\boldsymbol{v}_i$ and its neighboring nodes in Fig. 5.16(a) and Fig. 5.16(b) are $R/3$ and $R/6$, respectively. However, if we consider Fig. 5.16(a) or Fig. 5.16(b) as the connection of a vertex and its neighboring vertices of a 3D model in space, draw a dash line segment AH which is vertical to A's subtense as shown in Fig. 5.16. According to [59], given the watermark embedding direction, the local watermark embedding strength is determined by the length of the dash line segment AH. It is obvious that $AH = R/2$ in Fig. 5.16(a) while $AH = \sqrt{3}R/2$ in Fig. 5.16(b). The weights computed according to the above two methods are different because the neighboring vertices in Fig. 5.16(b) are larger than those in Fig. 5.16(a), which means that edges connecting to A are augmented and the angles between neighboring edges are decreased. According to the above discussion, the formula for computing watermark embedding weight can be modified as follows:

$$WT_i = wt_i \cdot q \cdot \sin\left(\frac{\pi}{2} - \frac{\pi}{q}\right) = \frac{1}{\sum\limits_{\boldsymbol{v}_j \in N(\boldsymbol{v}_i)} (1/d_{ji})} \cdot q \cdot \sin\left(\frac{\pi}{2} - \frac{\pi}{q}\right), \tag{5.39}$$

where q denotes the number of A's neighboring vertices. The first item of the right side of the above equation makes sure that the embedding weight is mainly determined by the minimal length between A and A's neighbors, the second one shows how the number of A's neighbors affects the embedding weight, and the last one is the effect of angles between the neighboring edges connecting with A to the embedding weight.

In our algorithm, a vertex and its neighbors can be regarded as a primitive where the watermark is embedded without computing the watermark embedding weight according to each triangle surface connecting to the vertex. Thus, the algorithm can make full use of the local geometry of the model with the imperceptivity of watermarks and is computationally timesaving, especially in the case where the number of surfaces of the model is considerable.

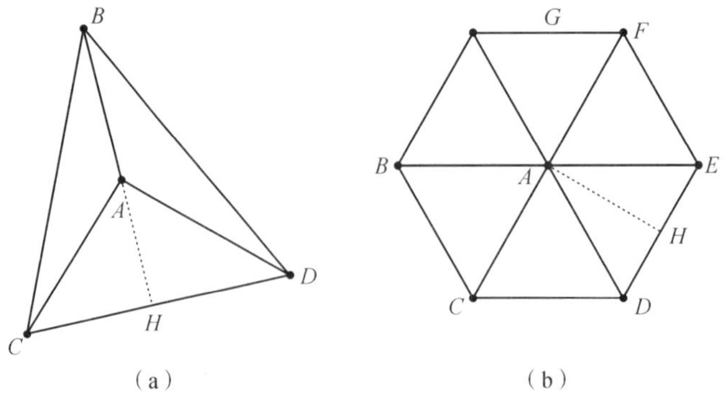

Fig.5.16. Two example cases of computing the locally adaptive watermarking strength with the local geometry. (a) Point A has 3 neighbors; (b) Point A has 6 neighbors

5.5.2.2 Control of the Watermark Embedding Direction

The local strength for embedding watermarks has been ascertained in the previous part. Now the direction in which the watermark should be embedded is to be confirmed. If two parameters are both acquired, the watermarking scheme is then fixed. By optimizing the watermark embedding direction, more energy of the watermark can be embedded with imperceptivity. Namely, the visual change in the model is relatively less if a watermark with fixed energy is embedded along the optimized direction. Enhancing the watermark strength and minimizing the visual change in the model supplement each other.

In most of the previous literature concerning 3D model watermarking techniques, the watermark is embedded along the vector that links the model centroid to a vertex, whose length is the embedding primitive. Though the primitive is a global geometry feature, it may not allow maximum possible watermark energy to be embedded. A rather novel method to ascertain the watermark embedding direction is proposed here, and the locally adaptive watermark embedding direction is not only a global geometry feature that is the primitive to be embedded with the watermark, but also makes sure that more (of the) energy of the watermark can be embedded under the precondition of imperceptivity.

The watermark energy that can be embedded lies on the watermark embedding direction under the precondition that the local geometry feature and the visual characteristic of the model are fixed. As shown in [59] and the example in Fig. 5.17, if the dot products of the unit vector of watermark embedding directions and normalized normals of triangle surfaces connecting to the vertex increase, the watermark energy that can be embedded decreases. The watermark energy that can be embedded is determined by the minimum value of the dot products to satisfy imperceptivity. Thus, the watermark energy that can be embedded is determined by

$$O = \max_i \{\boldsymbol{p}_i \cdot \boldsymbol{n}\}, \tag{5.40}$$

and the watermark energy is inversely proportional to *O*, where $\boldsymbol{p}_i$ is the *i*-th normalized normal of triangle surfaces, $i = 1, 2, \cdots, q$ and $\boldsymbol{n}$ is the watermark embedding direction. Now the embedding direction can be optimized by minimizing the object function *O*.

Let the unit normal of *A* be (0, 0, 1), where the vertex normal is defined as follows:

$$\boldsymbol{v} = \sum_{i=1}^{q} \boldsymbol{p}_i / q \cdot \tag{5.41}$$

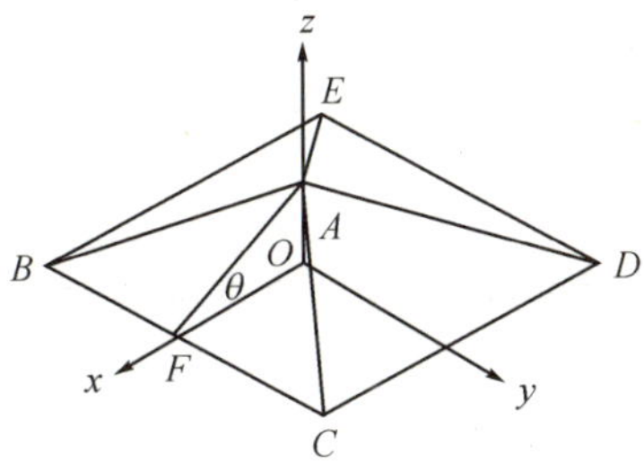

Fig. 5.17. An example of Vertex *A* connecting with 4 triangle surfaces

Let the angle between each of the *q* triangle surfaces and the underside of the polyhedron be

$$\theta = \arccos\left(S / \sum_{i=1}^{q} s_i \right), \ \theta > 0, \tag{5.42}$$

where *S* equals the area of the polyhedron underside, s_i denotes the area of a triangle surface connecting to *A*, *i* = 1, 2, …, *q* and, as a result, the normals of the surfaces are as follows:

$$\boldsymbol{p}_i = \left(\cos\left[(i-1) \cdot \frac{2\pi}{q} \right] \cdot \sin\theta, \ \sin\left[(i-1) \cdot \frac{2\pi}{q} \right] \cdot \sin\theta, \ \cos\theta \right). \tag{5.43}$$

If $\boldsymbol{n}$ is chosen as *A*'s unit normal (0, 0, 1), it is obvious that:

$$O_1 = \cos\theta. \tag{5.44}$$

Let the unit normal of the vector from the model centroid to A be $\boldsymbol{u} = \{x, \ y, \ z\}$ where

$$x^2 + y^2 + z^2 = 1, \ z > 0. \tag{5.45}$$

If $\boldsymbol{n}$ is chosen as $\boldsymbol{u}$, then

$$O_2 = \max_i \{\boldsymbol{p}_i \cdot \boldsymbol{u}\}. \tag{5.46}$$

Due to the rotation symmetry, we can let $O_2 = \boldsymbol{p}_1 \cdot \boldsymbol{u} = x\sin\theta + z\cos\theta$, and then we have

$$\boldsymbol{p}_1 \cdot \boldsymbol{u} > \boldsymbol{p}_k \cdot \boldsymbol{u}, \ k = 2, 3, \cdots, q. \tag{5.47}$$

It can be deducted from the above inequation that

$$(1-\cos\frac{2\pi}{q})x > y\sin\frac{2\pi}{q}, \ x > 0. \tag{5.48}$$

From the restriction conditions Eqs.(5.45) and (5.48), it can be deducted that

$$\left[\left(\frac{1-\cos\dfrac{2\pi}{q}}{\sin\dfrac{2\pi}{q}}\right)^2 + 1\right] \cdot x^2 > 1 - z^2. \tag{5.49}$$

In order to optimize the watermark embedding direction $\boldsymbol{n}$, O_1 and O_2 are compared as follows. From the restriction condition Eq.(5.49), it is known that if

$$\sqrt{\frac{1-z^2}{\left[\left(\left(1-\cos\dfrac{2\pi}{q}\right)\Big/\sin\dfrac{2\pi}{q}\right)^2 + 1\right]}} \cdot \sin\theta + z\cos\theta > \cos\theta, \tag{5.50}$$

then $x\sin\theta + z\cos\theta > \cos\theta$, namely $O_2 > O_1$. This conclusion demonstrates that if A satisfies the condition Eq.(5.50), less energy of the watermark can be embedded along the direction of the vector that links the model centroid to A than along the direction of A's normal, namely along the latter direction. The visual change in the model is relatively less than along the former direction under the precondition that the watermark embedding strength is fixed. Hence, if a vertex of a 3D model satisfies the condition Eq.(5.50), the direction along which the watermark is embedded should be chosen as the vertex's normal. Otherwise, it should be chosen as the direction of the vector that links the model centroid to the vertex.

5.5.3 Experimental Results

To test our watermarking technique in terms of robustness and imperceptibility, we perform some experiments on a triangle mesh model. This mesh consists of 2017 vertices and 3961 triangle surfaces. We embed a watermark with 256 bits into the model. To test the robustness of our technique, experimental results of our algorithm and the algorithm in [43] are compared here. In this subsection, our algorithm is referred to as Algorithm 1, while that in [43] is referred to as Algorithm 2. The two algorithms are compared under the same condition, namely with the watermark and model being the same and the watermark energy being the same value of 0.002133. The watermarked face models based on Algorithm 1 and Algorithm 2 are respectively shown in Fig. 5.18(b) and Fig. 5.18(c), while Fig. 5.18(a) is the original face model. Visually comparing Fig. 5.18(b) with Fig. 5.18(a), we can conclude that the embedded watermark is imperceptible. Fig. 5.18(d) shows the copyright information, which can be encrypted by a key into the watermark to be embedded. To evaluate the robustness of Algorithm 1 and Algorithm 2, we attack the watermarked face model with polygon simplification, adding noises, insection, rotation, translation, scaling, as well as some of their combined operations. Experimental results show that Algorithm 1 is more robust against these attacks than Algorithm 2, the detail is as follows.

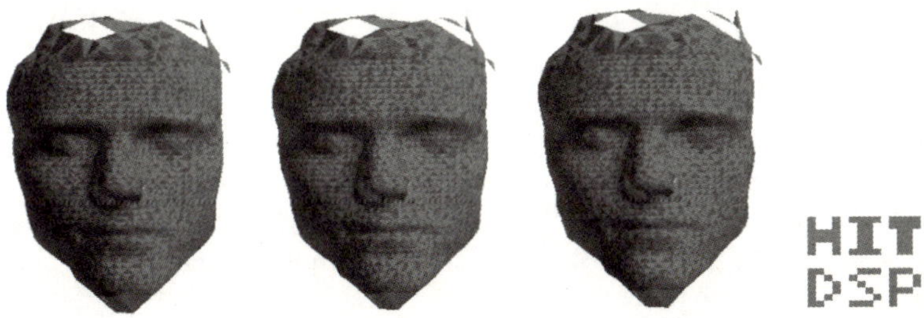

(a) (b) (c) (d)

Fig. 5.18. Face models and the watermark embedded. (a) Original face model; (b) Watermarked model by Algorithm 1; (c) Watermarked model by Algorithm 2; (d) Copyright information

5.5.3.1 Noise Attacks

To test the robustness against noise attacks, we add a noise vector to each vertex. We perform the test four times and the amplitude of the noise is 0.5%, 1.2% and 3.0%, respectively, of the length of the longest vector extended from the model centroid to a vertex. From Fig. 5.19, it can be visually seen that when the amplitude of the noise is 3.0% of the longest vector, the model is changed greatly. However, it can be seen from Table 5.4 that the watermark correlation is still 0.77 in Algorithm 1, which is better than that in Algorithm 2.

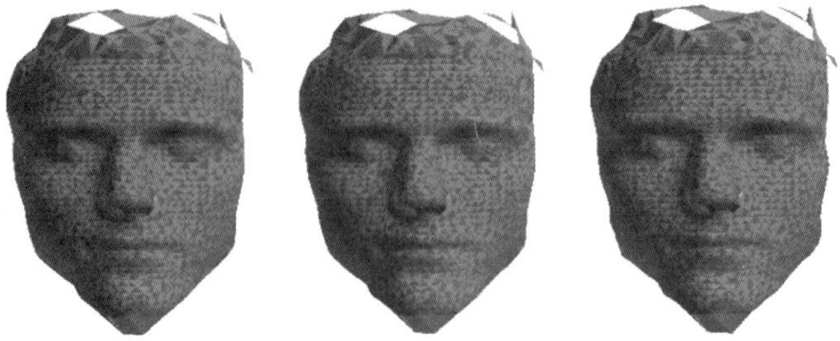

(a) (b) (c)

Fig. 5.19. Noise attacks on the watermarked model with different noise amplitudes. (a) 0.5%; (b) 1.2%; (c) 3.0%

Table 5.4 Results of noise attacks

Amplitude of noise/Max	Correlation 1	Correlation 2
0.5%	1.00	0.96
1.2%	0.98	0.64
3.0%	0.77	0.46

5.5.3.2 Similarity Transform Attacks

When the detected model is attacked by similarity transforms such as translation, rotation and uniform scaling, we must recover the attacked model back to its original location and scale via model registration. Because the registration is performed between the attacked model and the original model, registration errors may occur between the attacked model and the registered model. Hence, we should also test the robustness of our watermarking scheme against similarity transform as well as after registration. Since there are trade-offs between registration accuracy and speed for most registration techniques, it would be useful to investigate the robustness of our scheme against similarity transforms in order to test the registration technique. In Tables 5.5, 5.6 and 5.7, the experimental results show that our scheme has sufficient robustness to registration errors. Registration results are shown in Table 5.8. The watermarked face model subjected to similarity transforms such as rotation, translation and uniform scaling can be thoroughly recovered in few anneal registration times, as shown in the experimental results.

5.5.3.3 Simplification Attacks

The experimental results in Table 5.9 show high robustness of Algorithm 1, even if 20% of vertices are removed.

Table 5.5 Results of rotation attacks

Angle	Rotation axis	Correlation 1	Correlation 2
0.3°	*Z*	0.79	0.31
0.3°	*X*		
0.3°	*Y*	0.93	0.56
0.2°	*Z*		
0.6°	*X*		
0.5°	*Y*	0.85	0.24
0.5°	*Z*		
0.8°	*Y*	0.42	0.14

Table 5.6 Results of rotation and translation attacks

Angle	Rotation axis	Displacement	Translation direction	Correlation 1	Correlation 2
0.3°	*Z*	1%	(1, 1, 0)	0.83	0.72
0.5°	*X*				
0.2°	*Y*	0.5%	(1, 1, 1)	0.58	0.31
0.2°	*Z*				
0.5°	*Y*	2%	(0, 0, 1)	0.78	0.48

Table 5.7 Results of uniform scaling

Scaling (length from centroid to vertex)	Correlation	Correlation
10.99	0.68	0.23
21.005	1.00	0.61
1.01	0.83	0.37

Table 5.8 Results of registration

Rotation angle (round *X*, *Y* and *Z* axes)	Translation vector	Scaling (length from centroid to vertex)	Anneal registration times	Correlation 1	Correlation 2
(25°,50°,80°)	(2.0, 5.0, 4.0)	5.0	3	0.95	0.72
(25°,50°,80°)	(2.0, 5.0, 4.0)	0.2	5	1.00	0.86

Table 5.9 Results of simplification

Simplification rate	Correlation 1	Correlation
10%	0.93	0.92
215%	0.85	0.86
20%	0.51	0.53

5.5.3.4 Insection Attacks

It can be known from Table 5.10 that Algorithm 1 has high robustness against insection operations. Even if only 50% of vertices are left, the correlation value is still around 0.60.

Table 5.10 Results of insection

Insection rate	Correlation 1	Correlation 2
10%	0.97	0.96
20%	0.96	0.94
50%	0.60	0.59

5.5.3.5 Embedding with Two Watermarks

Two different watermarks can be embedded via our algorithm by using two different secret keys. The dual watermarked face model is shown in Fig. 5.20. Table 5.11 depicts the correlation value corresponding to each watermark. It can be known from the table that each watermark is well extracted via Algorithm 1.

Table 5.11 Results of extracting the two watermarks

	Cases	Correlation value
Algorithm 1	The primary watermark	0.82
	The secondary watermark	0.80
Algorithm 2	The primary watermark	0.78
	The secondary watermark	0.79

Fig.5.20. Dual watermarked face model

5.5.3.6 Combination Attacks

To test the robustness of our technique against combination attacks, the face model is subjected to combined attacks of simplification, insection, additional noise, translation, rotation and uniform scaling. Re-sampling operations are applied before the watermark is extracted. Experimental results are shown in Table 5.12. High robustness of Algorithm 1 against these combination attacks is demonstrated, while the watermark cannot be extracted via Algorithm 2, as shown in Table 5.12.

Table 5.12 Results of combined attacks

Insection rate	Simplification rate	Noise/Max	Translation vector/Max	Rotation angle (round X, Y and Z axes)	Scaling	Correlation 1	Correlation 2
10%	5%	0.1%	(0.1%, 0, 0)	(0.1°, 0°,0°)	0.995	0.69	0.34
15%	5%	0.3%	(0, 0, 0.1%)	(0°, 0.1°, 0.1°)	1.002	0.72	0.22
15%	10%	0.2%	(0.1%, 0, 0.1%)	(0.1°, 0°,0.1°)	1.005	0.64	0.16

From all the above experiments we can conclude that the proposed watermarking technique is highly robust against a lot of common attacks imposed on 3D mesh models in comparison with Algorithm 2. Experimental results of Algorithm 1 and Algorithm 2 against simplification and insection attacks are nearly the same because under such attacks, vertices are removed with some watermark information, while the remaining watermark information can be entirely extracted.

5.5.4 Conclusions

In this section, we introduce our robust watermarking scheme that embeds watermark information by altering the position of a vertex with a certain weight and along a certain direction which is (are all) adaptive with respect to the local geometry of the model. The watermark embedding weight is acquired from the local geometry of a vertex and its neighbors, other than the normal change of each face connecting to the vertex. In our method, the robustness is greatly enhanced due to the adaptive parameter control during the watermarking process. Moreover, the computation cost is rather low, especially in the case where considerable surfaces are in the model. Furthermore, not only is the locally adaptive watermark embedding direction a global geometry feature, but it also makes sure that more energy of the watermark can be embedded with imperceptivity.

Experimental results show that this approach is able to withstand common attacks such as polygon mesh simplifications, addition of Gaussian random noise, model insection, similarity transforms and some combined attacks and is applicable to all triangle mesh models. However, the main limitation of the proposed algorithm is that it is a public watermarking technique, namely the original cover signal is required during the detection process. It is necessary to investigate the blind-detection algorithm, which not only makes the watermark extracting process convenient, but intensifies the security of the original data.

5.6 3D Watermarking in Transformed Domains

According to our experiences in watermarking technologies for images, audio clips and video clips, we know that it is better to embed information in the spectral domain rather than in the spatial domain to achieve higher robustness. Since a watermark is embedded in the crucial position of the carrier in spectral domain based watermarking algorithms, the embedded watermark can resist attacks such as simplification. Most of the algorithms with high robustness are in the spectral domain. The principle of spectral domain based watermarking is to analyze the mesh spectrum which can be acquired by the mesh topology and graph theory [60]. Currently, there are few literatures related to transforming domain based 3D model watermarking algorithms, which can be mentioned in this section as follows.

5.6.1 Mesh Watermarking in Wavelet Transform Domains

In 1998, an oblivious mesh watermarking algorithm based on multi-resolution wavelet decomposition [61, 62], the first method for mesh watermarking in the spectral domain, was proposed by Kanai and Date from Japan's Hokkaido University. In this algorithm, the wavelet transform is used several times to decompose the original mesh $\boldsymbol{M}$ in its multi-resolution representation (MRR), and then a set of wavelet coefficient vectors $\boldsymbol{V}^1$, $\boldsymbol{V}^2$, …, $\boldsymbol{V}^d$ for different resolutions and a coarse approximation mesh $\boldsymbol{M}^d$ are acquired. The watermark is embedded by altering the norms of wavelet coefficient vectors, resulting in the watermarked wavelet coefficient vectors $\boldsymbol{V}^{1w}$, $\boldsymbol{V}^{2w}$, …, $\boldsymbol{V}^{dw}$ which can be inversely transformed to the stego mesh $\boldsymbol{M}^w$. The embedding process is illustrated in Fig. 5.21. The watermark extraction procedure is simple: The watermark can be extracted through the calculation of the difference between the wavelet coefficient vectors corresponding to the stego mesh and the cover mesh. The groundwork of the above method is the wavelet transform and multi-resolution representation, which were first developed by Lounsbery and Stollnitz [63, 64] and have been applied extensively in other 3D model processing areas.

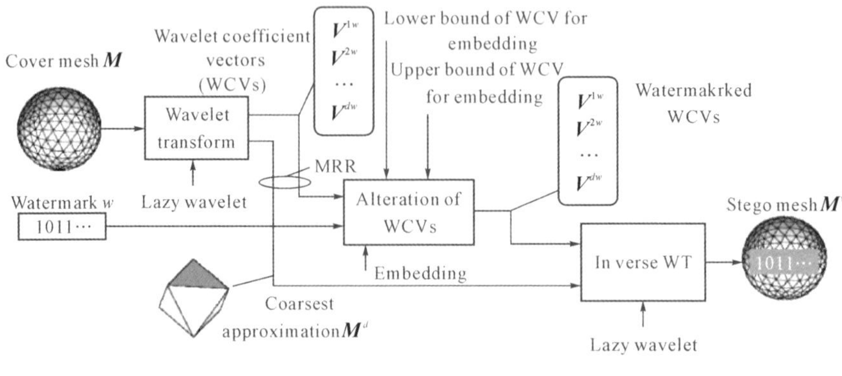

Fig. 5.21. The watermark embedding process [17] (With permission of ASME)

5.6.2 Mesh Watermarking in the RST Invariant Space

A mesh watermarking algorithm that is robust to rotation, translation and scaling is proposed in [65], in which a watermark sequence of 3 values is embedded in the 3D model vertices. Since the 3D model surface is transformed into a RST invariant space before the watermark embedding, this algorithm can be regarded as belonging to transformed domain methods. The detailed description is as follows.

5.6.2.1 3D Surface Transform

A 3D mesh model is composed of a set of vertices $\boldsymbol{P} = \{\boldsymbol{p}_i\}$ and their connectivity set $\boldsymbol{C}$. Every vertex $\boldsymbol{p}_i$ has its 3D coordinate $\boldsymbol{p}_i = (x_i, y_i, z_i)$. The goal of the transform is to convert the 3D data into a 1D signal in order to embed the watermark. The transform used here is invariant to rotation, scaling and translation as follows:

(1) Compute the centroid of all the vertices as follows: $\boldsymbol{\mu} = \frac{1}{k}\sum_{j=0}^{k-1}\boldsymbol{p}_j = (\mu_x, \mu_y, \mu_z)$.

(2) Translate the model. Subtract the centroid from each $\boldsymbol{p}_i=(x_i,y_i,z_i)$ and get $\boldsymbol{p}_i' = (x_i', y_i', z_i') = (x_i - \mu_x, y_i - \mu_y, z_i - \mu_z)$. The new vertices coordinates are invariant to translation.

(3) Principal component analysis. Denote the principal component of vertices as an eigenvector $\boldsymbol{T}$, which corresponds to the maximum eigenvalues of the covariance matrix of vertices. Here, the covariance matrix can be represented as follows:

$$\boldsymbol{H} = \begin{bmatrix} \sum_{i=0}^{k-1} x_i'^2 & \sum_{i=0}^{k-1} y_i' x_i' & \sum_{i=0}^{k-1} z_i' x_i' \\ \sum_{i=0}^{k-1} x_i' y_i' & \sum_{i=0}^{k-1} y_i'^2 & \sum_{i=0}^{k-1} z_i' y_i' \\ \sum_{i=0}^{k-1} x_i' z_i' & \sum_{i=0}^{k-1} y_i' z_i' & \sum_{i=0}^{k-1} z_i'^2 \end{bmatrix}. \tag{5.51}$$

(4) Model rotation. Rotate the model so that the eigenvector $\boldsymbol{T}$ is along the Z axis, so that the rotation invariance is achieved.

(5) Transform the mesh into spherical coordinates, in other words represent each vertex $\boldsymbol{p}_i''$ in the coordinates (r_i, θ_i, ϕ_i). The watermark is embedded in the r_i component, so the scaling invariance is also achieved.

5.6.2.2 Watermark Embedding and Detection

The watermark to be embedded is a 3-valued sequence: $\boldsymbol{w} = \{w_i | w_i \in \{-1, 0, 1\}\}$, which is adaptively generated by the secret key K and the above sequence $\boldsymbol{r} = \{r_i\}$.

$$r_i^w = \begin{cases} r_i, & w_i = 0; \\ g_1(r_i, n_i), & w_i = 1; \\ g_2(r_i, n_i), & w_i = -1, \end{cases} \tag{5.52}$$

where n_i denotes the function value determined by the neighborhoods of r_i, $g_1(r_i, n_i)$ and $g_2(r_i, n_i)$ are functions for embedding:

$$\begin{aligned} g_1(r_i, n_i) &= n_i + \alpha_1 r_i, \\ g_2(r_i, n_i) &= n_i + \alpha_2 r_i, \end{aligned} \tag{5.53}$$

where $\alpha_1>0$ and $\alpha_2<0$ are the embedding parameters. Accordingly, the detection formula is easily designed:

$$\hat{w}_i = \begin{cases} 1, & \hat{r}_i > r_i; \\ -1, & \hat{r}_i < r_i. \end{cases} \tag{5.54}$$

5.6.3 Mesh Watermarking Based on the Burt-Adelson Pyramid

Yin et al. from the CAD & CG State Key Laboratory of Zhejiang University addressed the two difficulties in mesh watermarking—mesh decomposition and topology recovery from the attacked mesh, by constructing a Burt-Adelson pyramid using a relaxation operator and embedding the watermark in the final coarsest approximation mesh [14]. This algorithm is integrated with the multi-resolution mesh processing toolbox of Guskov, and can embed watermarks in the low spectral coefficients without extra data structure or complex computation. In addition, the embedded watermark can survive the operations in the mesh processing toolbox. The mesh resampling algorithm described is simple but efficient, which enables watermark detection on simplified meshes and other meshes with topology changes. In this Subsection, the relaxation operator and the Burt-Adelson pyramid are firstly introduced and then the embedding algorithm, followed by the detection algorithm, is given.

5.6.3.1 Relaxation Operator and Burt-Adelson Pyramid

The neighborhood of a triangle mesh should be defined first. We denote a triangle mesh as $\boldsymbol{M} = (\boldsymbol{P}, \boldsymbol{C})$, where $\boldsymbol{P} = \{\boldsymbol{p}_i\}$ is the vertices set and $\boldsymbol{p}_i = (x_i, y_i, z_i)$. $\boldsymbol{C}$ consists of the topology information, i.e. the connectivity information. Given a vertex $\boldsymbol{p}_i$ and an edge e, then $V_1(i)$ is defined as a 1-ring vertex neighborhood of $\boldsymbol{p}_i$, $E_1(i)$ is a 1-ring edge neighborhood of $\boldsymbol{p}_i$, $V_2(i)$ is a 2-ring vertex neighborhood of $\boldsymbol{p}_i$, $E_2(i)$ is a 2-ring edge neighborhood of $\boldsymbol{p}_i$, and $U(e)$ is a vertex neighborhood of the edge e, as illustrated in Fig. 5.22, where the gray vertices are vertex neighborhoods and the thick lines are edge neighborhoods.

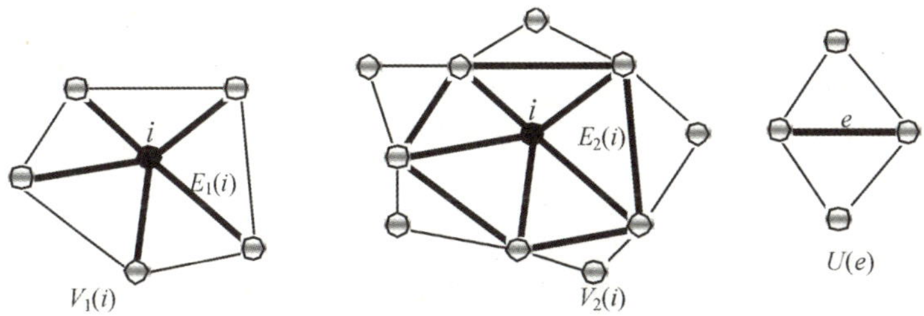

Fig. 5.22. Definition of neighborhoods

The definition of the relaxation operator [66] is given by Guskov et al. as below:

$$R_{p_i} = \sum_{j \in V_2(i)} \tau_{i,j} \boldsymbol{p}_j, \tag{5.55}$$

where $\tau_{i,j}$ is defined as

$$\tau_{i,j} = -\frac{\sum_{\{e \in V_2(i) | j \in U(e)\}} c_{e,i} c_{e,j}}{\sum_{\{e \in V_2(i)\}} c_{e,i}^2}. \tag{5.56}$$

According to the specific connectivity in Fig. 5.23, $c_{e,j}$ has the following 4 choices:

$$c_{e,l_1} = \frac{L_e}{A_{[l_1,s,j]}},\quad c_{e,l_2} = \frac{L_e}{A_{[l_2,j,s]}},\quad c_{e,j} = \frac{L_e A_{[s,l_2,l_1]}}{A_{[l_2,s,j]} A_{[l_2,j,s]}},\quad c_{e,s} = \frac{L_e A_{[j,l_1,l_2]}}{A_{[l_2,s,j]} A_{[l_2,j,s]}}, \tag{5.57}$$

where L_e is the length of the shared edge e, A represents the signed area of the triangle, $A_{[s,l_2,l_1]}$ and $A_{[j,l_1,l_2]}$ are areas of the rotated triangles of sl_2l_1 and jl_1l_2 on the same plane.

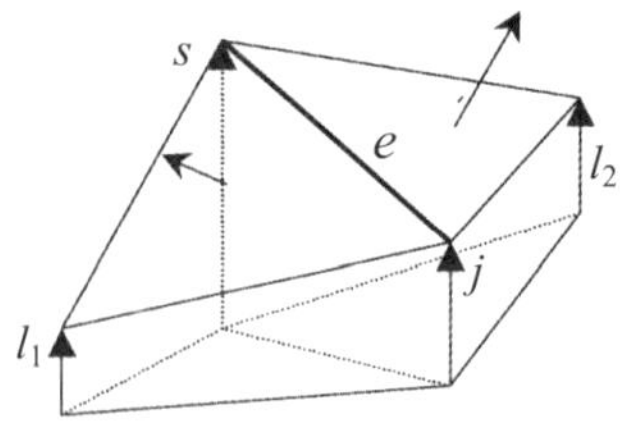

Fig. 5.23. Calculation of $c_{e,i}$, $i \in \{l_1, l_2, j, s\}$

According to the relaxation operator defined above, the Burt-Adelson (BA) pyramid [66] can be constructed. The pyramid algorithm belongs to mesh multi-resolution representation algorithms, of which a good multi-resolution method is the Hoppe progressive mesh [67] method. Usually, the second error metric by Garland [68] is used in constructing a progressive mesh and a vertex is removed each time using the half edge folding method. In this way, the mesh sequence ($\boldsymbol{P}^n$, $\boldsymbol{C}^n$) is constructed, $1 \le n \le N$, $\boldsymbol{P}^n = \{\boldsymbol{p}_i | 1 \le I \le n\}$. It is clear that the index of the removed vertex is n when $\boldsymbol{P}^n$ becomes $\boldsymbol{P}^{n-1}$.

A pure progressive mesh method only removes vertices, with the coordinates of the other vertices unchanged while, in the pyramid algorithms, the coordinates of the left vertices may be different from their counterparts in the finer mesh, so that differences at different levels come into being. Here the new coordinates of the left vertices are denoted as $\boldsymbol{q}_j^n$, the differences between different levels are represented as $\boldsymbol{d}_j^n$, which is also called the detail information. The detailed construction of the BA pyramid is illustrated in Fig. 5.24. The mesh sequence ($\boldsymbol{P}^n$, $\boldsymbol{C}^n$) can be constructed from the start of $\boldsymbol{P}^N = \boldsymbol{P}$, $1 \le n \le N$. There are 4 steps to construct $\boldsymbol{P}^{n-1}$ from $\boldsymbol{P}^n$ (i.e. removing vertex n) as follows:

(1) Pre-smoothing. Update the coordinate of the 1-ring vertex neighborhood $\forall j \in V_1^n(n)$ of vertex n: $\boldsymbol{p}_j^{n-1} = \sum_{k \in V_2^n(j)} \tau_{j,k}^n \boldsymbol{p}_k^n$; the other vertices $\forall j \in V_1^{n-1} \setminus V_1^n(n)$ of $\boldsymbol{P}^n$ are not changed and copied to $\boldsymbol{P}^{n-1}$, i.e., $\boldsymbol{p}_j^{n-1} = \boldsymbol{p}_j^n$.

(2) Downsampling. Remove n by half edge folding.

(3) Subdivision. Compute the coordinates of the vertex after subdivision, $\boldsymbol{q}_j^n$, according to the coordinates of $\boldsymbol{P}^{n-1}$. The coordinates of the newly removed vertex n are

$$\boldsymbol{q}_n^n = \sum_{j \in V_2^n(n)} \tau_{n,j}^n \boldsymbol{p}_j^{n-1} . \tag{5.58}$$

And the coordinate of a 1-ring vertex neighborhood of vertex n is as follows:

$$\forall j \in V_1^n(n): \ \boldsymbol{q}_j^n = \sum_{k \in V_2^n(j) \setminus \{n\}} \tau_{j,k}^n \boldsymbol{p}_k^{n-1} + \tau_{j,n}^n \boldsymbol{q}_n^n . \tag{5.59}$$

(4) Computation of details. Compute the details of the local structure F^{n-1} for the vertex n and its neighborhoods as follows:

$$\forall j \in V_1^n(n) \cup \{n\}: \ d_j^n = F_j^{n-1}(p_j^n - q_j^n), \tag{5.60}$$

where $Q^n = \{q_j^n\}$ and $D^n = \{d_j^n\}$.

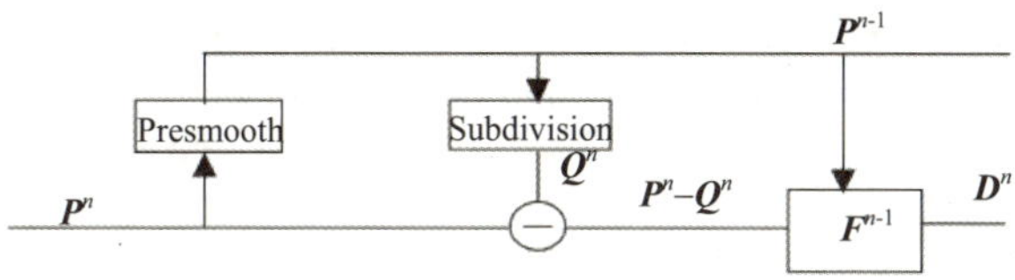

Fig. 5.24. BA pyramid scheme

In the construction of the lower level of the pyramid from the upper level, Q^n is first acquired by subdivision using vertices of P^{n-1}, and adding it to D^n so that P^n is recovered. At the same time, the pyramid data information is recorded in a proper data structure, such as the half edge folding sequence, the relaxation operator sequence τ^n and the details sequence D^n, which are all necessary for mesh multi-resolution processing as well as mesh watermark embedding and detection.

From the above pyramid structure construction process we can see that the coarser mesh in an upper level can be regarded as the low-frequency coefficients of the finer mesh in a lower level. From the point of view of signal processing, a vertex of a coarser mesh is the smoothed downsampled vertex of a finer mesh and corresponds to low-frequency. In the construction process, the most significant features are maintained while the details are abandoned. As a result, the process of embedding the watermark in a coarse mesh is analogous to watermarking in the low-frequency coefficients in still images.

5.6.3.2 Watermark Embedding

A bipolar sequence $\boldsymbol{w} = \{w_1, w_2, \ldots, w_m\}$ is used as the watermark and the embedding process is as follows:

(1) Construct a BA pyramid from the original mesh M and an appropriate level of coarse mesh M^c is the embedding object.

(2) Select $[m/3]$ vertices p_i randomly or according to some rules from M^c, $i = 1, 2, \ldots, [m/3]$; Compute the minimum length of the 1-ring edge neighborhood of p_i: $lm_i = \min\{length(e)|e \in E_1(i)\}$, then the watermark embedding equations are as follows:

$$\begin{cases} p_{ix}^{w} = p_{ix} + w_{3i+1} \cdot \alpha \cdot lm_i, \\ p_{iy}^{w} = p_{iy} + w_{3i+2} \cdot \alpha \cdot lm_i, \\ p_{iz}^{w} = p_{iz} + w_{3i+3} \cdot \alpha \cdot lm_i, \end{cases} \tag{5.61}$$

where p_{ix} is the x component of $\boldsymbol{p}_i$, p_{ix}^{w} is the corresponding watermarked x component and the others are defined in the same way; α is the watermark strength parameter which controls the energy of watermark; lm_i is a local watermark strength parameter which makes the embedding adaptive to local geometry features. In the real implementation, the threshold T is set and the watermark is embedded only when $lm_i > T$. The watermarked coarse mesh is finally acquired and denoted as $\boldsymbol{M}^{cw}$.

(3) Construct the watermarked fine mesh $\boldsymbol{M}^{w}$ according to the pyramid reconstruction method.

5.6.3.3 Watermark Detection

For a given suspect mesh $\hat{\boldsymbol{M}}$, a watermark detection method is needed to extract the potential watermark information in the mesh and compare it with a given watermark to judge if the watermark exists. Usually, this judgment is carried out by the holder of the original data, i.e. the person who embedded the watermark in the mesh. According to the embedding algorithm described above, the watermark detection algorithm can be described as follows: the watermark detector uses the pyramid of the original mesh $\boldsymbol{M}$ and of the suspect mesh $\hat{\boldsymbol{M}}$, respectively, to construct coarse meshes $\hat{\boldsymbol{M}}^{c}$ and $\boldsymbol{M}^{c}$. Compare $\hat{\boldsymbol{M}}^{c}$ and $\boldsymbol{M}^{c}$, and then the watermark can be calculated as follows:

$$\begin{cases} \hat{w}_{3i+1} = \mathrm{sgn}(\hat{p}_{ix} - p_{ix}), \\ \hat{w}_{3i+2} = \mathrm{sgn}(\hat{p}_{iy} - p_{iy}), \\ \hat{w}_{3i+3} = \mathrm{sgn}(\hat{p}_{iz} - p_{iz}), \end{cases} \tag{5.62}$$

where $\boldsymbol{p}_i$ belongs to $\boldsymbol{M}^{c}$, $\hat{\boldsymbol{p}}_i$ belongs to $\hat{\boldsymbol{M}}^{c}$ and "sgn" is the sign function.

In addition, when the stego mesh is attacked by operations such as simplification, the mesh topology will be changed and the above watermark detection method will have no effect. In order to address this issue, a resampling algorithm is also proposed in [14]. Due to space limitation, the resampling method is not elaborated.

5.6.4 Mesh Watermarking Based on Fourier Analysis

In 2001, Ohbuchi and Mukaivama developed a 3D mesh watermarking algorithm in the spectral domain [19]. In this algorithm, the Kirchhoff matrix is derived from the mesh connectivity first (The Kirchhoff matrix is used in this algorithm though various Laplacian matrices can be defined with different methods). The eigenvector decomposition is performed using the Kirchhoff matrix and then the frequency scope of the mesh can be calculated through projecting the spatial coordinates on a set of eigenvectors. The watermark is embedded by modifying the spectral coefficients, i.e. altering the mesh shape in the spectral domain based on mesh spectrum analysis. The watermark embedding algorithm is robust to affine transform, random noise on vertices, mesh smoothing (mesh low-pass filtering) and insection. In 2002, the above watermarking algorithm was extended by Ohbuchi [20], so that not only the embedding process is quicker, but the robustness to simplification and combined attacks is also improved. In 2003, Cayre et al. continued the research in this direction [21]. In this algorithm, the watermark is embedded based on relationship, instead of imbedding additively as in [19,20]. Below, a brief introduction to Fourier analysis of 3D meshes using the Laplacian operator and the watermarking algorithm in [21] is given.

5.6.4.1 Laplacian-Operator-Based Discrete Fourier Analysis for 3D meshes

First, a set of indices of the neighborhoods of $\boldsymbol{p}_i$ is collected as $\{i^*\}$:

$$\forall \boldsymbol{p}_j \in \boldsymbol{P},\quad j \in \{i^*\} \quad \Leftrightarrow (i,j) \in \boldsymbol{C}. \tag{5.63}$$

Define d_i as the degree of $\boldsymbol{p}_i$, i.e. $d_i = |\{i^*\}|$. Thus the $k\times k$ Laplacian matrix $\mathbf{L}$ defined by Taubin [69] is as follows:

$$L_{ij} = \begin{cases} 1, & i = j; \\ -d_i^{-1}, & j \in \{i^*\} \text{ and } d_i \neq 0; \\ 0, & \text{otherwise.} \end{cases} \tag{5.64}$$

The eigenvectors of $\boldsymbol{L}$ is a set of orthogonal basis of R^k, and the eigenvalues e_i, $0 \le I \le k-1$ can be regarded as the pseudo frequencies of the geometry, which is in a range from 0 to 2. Let $\boldsymbol{X}$ denote the set of all x coordinates, and $\boldsymbol{Y}$ and $\boldsymbol{Z}$ are defined in the same way for y coordinates and z coordinates, respectively. Define $\boldsymbol{B}$ as a matrix with each column as an eigenvector, and then we can get:

$$\begin{bmatrix} e_0 & 0 & \cdots & \cdots & 0 \\ 0 & \ddots & \ddots & & \vdots \\ \vdots & \ddots & e_i & \ddots & \vdots \\ \vdots & & \ddots & \ddots & 0 \\ 0 & \cdots & \cdots & 0 & e_{k-1} \end{bmatrix} = \boldsymbol{B}^{-1}\boldsymbol{L}\boldsymbol{B}. \tag{5.65}$$

Then we can perform the orthogonal transform on the three k-dimensional vectors $\boldsymbol{X}$, $\boldsymbol{Y}$ and $\boldsymbol{Z}$, thus the so-called spectrum or pseudo-frequency vectors $\boldsymbol{O}$, $\boldsymbol{Q}$ and $\boldsymbol{R}$ can be derived:

$$\boldsymbol{O} = \boldsymbol{B}\boldsymbol{X},\ \boldsymbol{Q} = \boldsymbol{B}\boldsymbol{Y},\ \boldsymbol{R} = \boldsymbol{B}\boldsymbol{Z}, \tag{5.66}$$

and the corresponding reconstruction formulae are:

$$\boldsymbol{X} = \boldsymbol{B}^{-1}\boldsymbol{O},\ \boldsymbol{Y} = \boldsymbol{B}^{-1}\boldsymbol{Q},\ \boldsymbol{Z} = \boldsymbol{B}^{-1}\boldsymbol{R}. \tag{5.67}$$

The Kirchhoff matrix (also called combinatorial Laplacian matrix) is suggested by Ohbuchi to compute the spectrum information. Characteristics of a Kirchhoff matrix are very similar to those of a Taubin matrix, and facilitate fast computing. As a result, the Laplacian power spectrum of the vertex sequence $\boldsymbol{P}$ can be represented by the sum of the power of the signal along the three pseudo-frequency axes as follows:

$$S_i = |O_i|^2 + |Q_i|^2 + |R_i|^2,\quad 0 \le i \le k-1. \tag{5.68}$$

5.6.4.2 Watermark Embedding

The watermark is embedded by randomly altering the relationship between $\boldsymbol{O}$, $\boldsymbol{Q}$ and $\boldsymbol{R}$ in [21]. The former i_0 low-frequency coefficients are kept unchanged to ensure imperceptivity. Every remaining S_i is embedded with one bit of watermark, i.e., in total $k-i_0$ bits can be embedded. Take a coefficient triple (O_i, Q_i, R_i) as an example and they are reordered as follows:

$$(O_i, Q_i, R_i) \to (C_{\min}, C_{\text{inter}}, C_{\max}), \tag{5.69}$$

where

$$C_{\min} = \min\{O_i, Q_i, R_i\},\ C_{\text{inter}} = \text{mid}\{O_i, Q_i, R_i\},\ C_{\max} = \max\{O_i, Q_i, R_i\}. \tag{5.70}$$

The interval $[C_{\min}, C_{\max}]$ with the length $\Delta = C_{\max} - C_{\min}$ is divided into two subintervals: $W_0 = [C_{\min}, C_{\min}+0.5\Delta]$ and $W_1 = [C_{\min}+0.5\Delta, C_{\max}]$. If the watermark bit to be embedded is "0", then alter C_{inter} to make it fall in the interval W_0;

otherwise, if the watermark bit is "1", then alter C_{inter} to make it fall in the interval W_1. Let $C_{\text{mean}} = 0.5(C_{\min} + C_{\max})$ and then the embedding can be formulized as

$$\begin{cases} C_{\text{inter}}^{w} = \dfrac{C_{\text{mean}} - |C_{\text{inter}} - C_{\text{mean}}|}{m}, & w = 0; \\ C_{\text{inter}}^{w} = \dfrac{C_{\text{mean}} + |C_{\text{inter}} - C_{\text{mean}}|}{m}, & w = 1, \end{cases} \tag{5.71}$$

where the parameter m is used to control the trade-off between the robustness and imperceptivity, and is set to be 10 in [21]. The watermark extraction is simple and blind, only requiring judging whether or not $\hat{C}_{\text{inter}}$ falls in the interval W_0.

5.6.5 *Other Algorithms*

In addition to the above mentioned algorithms, Reference [70] proposed an alternative transform domain mesh watermarking idea. The algorithm regards the virtual object to be embedded as an image generated by a 3D scanner. Principal component analysis is conducted on vertices so the object position in the scanner can be estimated. When we receive the 2D range image from the scanner, we can use traditional DCT image watermarking algorithms to embed a watermark. According to the altered 2D range image, we can modify 3D mesh vertices accordingly, thus completing the watermark embedding process. In the watermark detection phase, we can generate a 2D range image according to the 3D mesh to be detected, and then extract the watermark information from the range image. Experimental results show that the algorithm is robust to mesh simplification and Gaussian noise.

In addition, the literature [71] proposed a 3D polygon mesh robust watermark algorithm in the frequency domain based on singular spectrum analysis (SSA). The main idea is to regard all vertices as being in a vertex sequence, and then perform SSA on the trajectory matrix derived from the sequence in order to extract the spectrum of the vertices sequence. The embedded watermark in the spectrum can resist similarity transform and random noise. Due to space limitations, these algorithms are not illustrated.

5.7 Watermarking Schemes for Other Types of 3D Models

The above-mentioned algorithms are all designed for 3D polygon mesh models. In fact, not all 3D models are represented by polygons. As a result, watermarking algorithms for other types of 3D models are also available. Due to space limitations, they are briefly introduced here.

5.7.1 Watermarking Methods for NURBS Curves and Surfaces

3D models are usually represented by mesh, non-uniform rational B-spline (NURBS), or voxel. Among these models, mesh is quite widely used because many studies on the mesh have already been performed, and also because the scanned 3D data are naturally the sampling points of surfaces. However, the mesh representation has drawbacks in that it requires a large amount of data and it cannot represent mathematically rigorous curves and surfaces. Unlike mesh, the NURBS describes 3D models by using mathematical formulae. The data size for the NURBS is remarkably smaller than that for the mesh because the surface can be represented by only a few parameters. Also, the NURBS is smooth in nature so that the smoothness of NURBS is restricted only by hardware resolution. Hence, the NURBS is used in CAD and other areas where high precision is required, and it is also used in animation because the motion of an object can be realized by successively adjusting some of the parameters.

Although the amount of 3D multimedia data is dramatically increasing, there has not been much discussion on the watermarking of 3D models, especially on the 3D NURBS models. Currently, the vast majority of watermarking algorithms are directed at the 3D polygon mesh models. However, many 3D models are represented by parameterized curves and surfaces, such as non-uniform rational B-spline (NURBS) curves and surfaces. Therefore, 3D model watermarking algorithms based on NURBS curves and surfaces are available in [16, 34, 72]. Besides, many 3D model algorithms embed a watermark based on imperceptible change in geometry and/or topology, while such geometry/topology changes can be tolerated by few current CAD models. Therefore, a 3D model watermarking algorithm, without changing the NURBS curves and surface shapes, is presented in [16].

In [72], two watermarking algorithms are proposed for 3D NURBS, one is suitable for steganography (for secret communication between trusting parties) and the other for robust watermarking. In the proposed algorithm, a virtual NURBS model is first generated from the original one. Instead of embedding information into the parameters of NURBS data as in the existing algorithm, the proposed algorithms extract several 2D images from the 3D virtual model and apply the 2D watermarking methods. In the steganography algorithm, a 3D virtual model is first sampled in each of u and v directions, where u and v are parameters of NURBS. That means a sequence of $\{u, v\}$ is generated, where the number of elements is limited to be less than that of the control points. Then three 2D virtual images are extracted, the pixels of which are the distances from the sample points to the x, y, and z plane, respectively. The watermark is embedded into these 2D images, which leads to the modification of the control points of NURBS. As a result, the original model is changed by the watermark data as much as by the quantity of embedded data. But the data size of the NURBS model is preserved because there is no change in the number of knots and control points. For the extraction of embedded information, modified virtual sample points are first acquired by the matrix operation of basis functions in accordance with the $\{u, v\}$ sequence. Even if the third party has the original NURBS model, the embedded information cannot be acquired without $\{u, v\}$ sequence as a key, which is a good property for the steganography. The second algorithm is suitable for robust watermarking. This algorithm also samples the 3D virtual model. But the difference from the steganography algorithm is that the number of sampled points is not limited by the number of control points of the original NURBS model. Instead, the sequence $\{u, v\}$ is chosen so that the sampling interval in the physical space is kept constant. This makes the model robust against attacks on knot vectors, such as knot insertion, removal and so forth. The procedure for making 2D virtual images is the same as for the steganography algorithm. Then, the watermarking algorithms for 2D images are applied to these virtual images and a new NURBS model is made by the approximation of watermarked sample points. The watermarks in the coordinate of each sample point are distorted within the error bound by approximation. But such distortion can be controlled by the strength of embedded watermarks and the magnitude of error bound. Since the points are not sampled in the physical space (x-, y-, z-coordinate) but in the parametric space (u-, v-coordinate), the proposed algorithm for watermarking is also found to be robust against attacks on the control points that determine the model's transition, rotation, scaling and projection.

5.7.2 3D Volume Watermarking

Some 3D models are acquired using some special equipment (such as 3D laser scanners). Similar to 2D pixel-based images, the data unit of a 3D image is a voxel, which also has a color or gray-scale property. Watermarks can be embedded through altering the colors or gray properties in the spatial domain or transformed domains (e.g. 3D DCT, DFT, 3D DWT). Detailed descriptions of 3D image watermarking algorithms can be found in [35-38].

5.7.3 3D Animation Watermarking

Animation is the rapid display of a sequence of images of 2D or 3D artwork or model positions in order to create an illusion of movement. It is an optical illusion of motion due to the phenomenon of persistence of vision, and can be created and demonstrated in a number of ways. The most common method of presenting animation is as a motion picture or video program, although several other forms of presenting animation also exist. Computer animation (or CGI animation) is the art of creating moving images with the use of computers. It is a subfield of computer graphics and animation. Increasingly it is created by means of 3D computer graphics, though 2D computer graphics are still widely used for stylistic, low bandwidth and faster real-time rendering needs. Sometimes the target of the animation is the computer itself, but sometimes the target is another medium, such as film. It is also referred to as CGI (computer-generated imagery or computer-generated imaging), especially when used in films. For 3D animations, all frames must be rendered after modeling is complete. For 2D vector animations, the rendering process is the key frame illustration process, while in-between frames are rendered as needed. For pre-recorded presentations, the rendered frames are transferred to a different format or medium such as film or digital video. The frames may also be rendered in real time as they are presented to the end-user audience. Low bandwidth animations transmitted via the internet (e.g. 2D Flash, X3D) often use software on the end-users computer to render in real time as an alternative to streaming or pre-loaded high bandwidth animations.

3D animation watermarking technology is a brand new application of 3D animation data protection. Animation is referred to as a role continuously moving for a certain period of time. The role can be compactly represented by a skeleton formed by some key points with one or more degrees of freedom. The change of each degree of freedom in the time domain can be viewed as an independent signal, while the whole animation is a function of time. DCT can be used for a 3D animation oblivious watermarking algorithm by performing a slight quantization disturbance to mid-coefficients of DCT and combining the ideas of spread spectrum and quantization. Choosing a reasonable quantization step can ensure that the original movement is visually acceptable. At the same time, spreading every watermark bit over many frequency coefficients by spread spectrum can effectively increase the robustness. This algorithm exhibits high robustness to white Gaussian noise, resampling, movement smoothing and reordering.

In addition, Hartung et al. developed a watermarking algorithm [3] in the MPEG-4 facial animation parameters (FAP) sequence using spread spectrum technology. A remarkable aspect of this method is that not only can watermarks be extracted from parameters, but the facial animation parameter sequence (from which the watermark can be extracted) can also be generated from the real facial video sequence using the facial feature tracking system.

5.8 Summary

This chapter focuses on 3D model watermarking algorithms. Starting with a brief introduction, the 3D model watermarking system model, characteristics, requirements and classifications were discussed. Then several 3D mesh watermarking methods in the spatial domain were introduced. Next, a robust mesh watermarking scheme proposed by the authors of this book was introduced in detail. Then, according to different transformations when embedding information, we briefed some typical 3D model watermarking algorithms in the transform domain. Finally, watermarking algorithms for other types of 3D models were briefly introduced.

Through this chapter, we can see that 3D model watermarking is a new field of watermarking research, which has become the focus for domestic and foreign researchers who have done much exploratory work and provided a lot of new ideas for those working in CAD research and development. Thus a new research area has opened up. However, analysis shows that there is much unfinished work. There are many outstanding issues and thus a larger study space for 3D model watermarking. A number of issues need to be addressed by thorough studies-centered around 3D mesh watermarking:

Robust watermarking also needs improving. Robust watermarking research includes robustness against insection, non-uniform scaling and mesh simplification, as well as the introduction of geometric noise interference, and so on. In 3D mesh digital watermarking research, we can learn from the still image digital watermarking ideas and methods. In particular, we should introduce transform-domain methods into 3D mesh watermarking research, such as the pioneering work done by Kanai in this direction [61, 62]. With consideration of a balanced robustness-capacity relationship, improving the robustness of public watermarks is still a problem.

The applied research area of fragile watermarking is not yet mature. Visualization tools for detecting and locating the alteration should be further improved. In addition, research into authentication for VRML (virtual reality modeling language) models, along with multi-level verification of 3D meshes, has involved few people as yet.

It is necessary to develop watermarking methods for VRML files. VRML is widely used for creating a dynamic 3D virtual space over the Internet. VRML documents are text documents and send commands to Internet browsers about how to create 3D models for the virtual space. Research into watermarking methods for VRML files has a direct practical value.

Watermarking technology has extended to the CAD system and other forms of representation, mainly to the free surface and the solid model. There are many ways for describing object shapes, such as representation by voxels, CSG trees and borders. Border representation includes implicit function surfaces, parametric surfaces, subdivision surfaces and points, as well as polygonal meshes. Ohbuchi et al. and Mitsuhashi et al. have done exploratory work in the field of watermarking for interval curve surfaces and triangle domain curve surfaces. The solid model is far more extensively applied in the CAD field than mesh models, so it is more significant for copyright protection and product verification if we extend the watermarking technology to the CAD field.

Now, a potential application example of 3D watermark technology is given—the Virtual Museum. Although a museum exists for the collection, protection and use of important cultural relics, for various reasons most museums have the following drawbacks: (1) With limitations of technology, finance and space, cultural relics are being kept in poor conditions, and some are even facing problems of oxidation and mildew; (2) Heritage management methods are backward and, for safety reasons, museums are closed for long periods, resulting in a low utilization rate. In order to better protect our heritage, share our resources, disseminate knowledge of our civilization and fully realize the social and economic benefits of the museum, we can make use of digital tools and virtual reality technology to transform the museum into a digital and virtual museum. The digital museum can be represented as follows: The functions of a museum such as collection, display and exhibition are demonstrated in a digital way, so display and initiative can be emphasized, the knowledge and expertise of the designers can be reflected and the curiosity of users can be attracted. The digital museum is a typical example of virtual museums, which use digitally simulated artifacts and scenes of real 3D models to display the history. It is a combination of traditional archaeological technology and advanced virtual reality technology, in which the whole scene can be reproduced in the form of 3D interactive explorations. In a virtual museum, people can not only see the 3D model objects but also speculate in the computerized virtual world environment: Every detail in the virtual world looks exactly the same as the actual historical sites, without any restrictions, and 3D model objects can be displayed indefinitely, because of zero-risk of damage or theft to the artifacts. Digital technology will enable people to make better use of museums and the protection of cultural relics. Storage methods for artifacts should be diversified, such as text, images, sound, video and 3D models, etc. Reduction of the acidic gases exhaled by visitors will reduce the maintenance costs of the heritages. Valuable cultural relics will not fade or gather mildew as time goes on. Moreover, as digital technology had facilitated the spread of conditions for digital works, so the heritages can be easily demonstrated to online visitors and a better dissemination of history and culture is achieved. Our long history will be more widely known to people all over the world. While digital technology will bring about a series of benefits and convenience for museums, issues concerning heritage copyright protection come into being. Since digital products can be losslessly duplicated, stored or even re-generated, illegal acquisition of cultural relics also becomes easier, so there is an urgent need for effective protection of these digital heritages. A digital museum is a concentration of documents, images, audio, video and 3D models, so a comprehensive application of a variety of digital watermarking technologies is necessary for copyright protection and integrity verification for digitized cultural relics.

References

[1] S. Kishk and B. Javidi. 3D object watermarking by 3-D hidden object. Opt. Exp., 2003, 11(8):874-888.

[2] E. Garcia and J. L. Dugelay. Texture-based watermarking of 3-D video objects. IEEE Trans. Circuits Syst. Video Technol., 2003, 13(8):853-866.

[3] F. Hartung, P. Eisert and B. Girod. Digital watermarking of MPEG-4 facial animation parameters. Comput. Graph., 1998, 22(4):425-435.

[4] B. L. Yeo and M. M. Yeung. Watermarking 3-D objects for verification. IEEE Comput. Graph. Appl., 1999, 19(1):36-45.

[5] C. Fornaro and A. Sanna. Private key watermarking for authentication of CSG models. Comput. Aided Design., 2000, 32(12):727-735.

[6] R. Ohbuchi, H. Masuda and M. Aono. Watermarking three-dimensional polygonal models through geometric and topological modifications. IEEE J. Sel. Areas Commun., 1998, 16(4):551-560.

[7] M. G. Wagner. Robust watermarking of polygonal meshes. In: Proc. Geometric Modeling and Processing, 2000, pp. 201-208.

[8] F. Cayre and B. Macq. Data hiding on 3-D triangle meshes. IEEE Trans. Signal Process., 2003, 51(4):939-949.

[9] O. Benedens. Affine invariant watermarks for 3-D polygonal and NURBS based models. In: Proc. Int. Workshop Information Security, 2000, pp. 15-29.

[10] O. Benedens. Geometry based watermarking of 3-D models. IEEE Comput. Graph. Appl., 1999, 19(1):46-55.

[11] B. Koh and T. Chen. Progressive browsing of 3-D models. In: Proc. IEEE Workshop Multimedia Signal Processing, 1999, pp. 71-76.

[12] T. Harte and A. G. Bors. Watermarking 3-D Models. In: Proc. IEEE Int. Conf. Image Processing, 2002, Vol. III, pp. 661-664.

[13] E. Praun, H. Hoppe and A. Finkelstein. Robust mesh watermarking. In: Proc. Int. Conf. Computer Graphics and Interactive Techniques, 1999, Vol. 6, pp. 69-76.

[14] K. Yin, Z. Pan, J. Shi, et al. Robust mesh watermarking based on multiresolution processing. Comput. Graph., 2001, 25(3):409-420.

[15] O. Benedens and C. Busch. Toward blind detection of robust watermarks in polygonal models. In: Proc. EUROGRAPHICS, 2000, Vol. 19, pp. C199-C208.

[16] R. Ohbuchi, H. Masuda and M. Aono. A shape-preserving data embedding algorithm for NURBS curves and surfaces. In: Proc. Computer Graphics Int. Conf., Canmore, 1999, pp. 180-187.

[17] S. Kanai, H. Date and T. Kishinami. Digital watermarking for 3-D polygons using multiresolution wavelet decomposition. In: Proc. Int. Workshop Geometric Modeling: Fundamentals and Applications, 1998, pp. 296-307.

[18] S. H. Yang, C. Y. Liao and C. Y. Hsieh. Watermarking MPEG-4 2-D mesh animation in multiresolution analysis. In: Proc. Advances Multimedia Information Processing, 2002, pp. 66-73.

[19] R. Ohbuchi, S. Takahashi, T. Miyazawa, et al. Watermarking 3-D polygonal meshes in the mesh spectral domain. In: Proc. Graphics Interface, 2001, pp. 9-17.

[20] R. Ohbuchi, A. Mukaiyama and S. Takahashi. A frequency-domain approach to watermarking 3-D shapes. In: Proc. EUROGRAPHICS, 2002, Vol. 21, pp. 373-382.

[21] F. Cayre, P. Rondao-Alface, F. Schmitt, et al. Application of spectral decomposition to compression and watermarking of 3-D triangle mesh geometry. Signal Process.: Image Commun., 2003, 18(4): 309-319.

[22] O. Benedens. Robust watermarking and affine registration of 3-D meshes. In: Proc. Information Hiding, 2003, pp. 177-195.

[23] A. G. Bors. Watermarking mesh-based representations of 3-D objects using local moments. IEEE Transactions on Image Processing, 2006, 15(3):687-701.

[24] A. Papoulis. Probability, Random Variables, and Stochastic Processes. McGraw-Hill, 1965.
[25] R. M. Haralick and L. G. Shapiro. Computer and Robot Vision. Vol. I. Addison-Wesley, 1992.
[26] R. Ohbuchi and H. Masuda. Managing CAD data as a multimedia data type using digital watermarking. In: IFIP WG 5.2, Fourth International Workshop on Knowledge Intensive CAD (KIC-4), 2000.
[27] M. Corsini, M. Barni, F. Bartolini, et al. Towards 3D watermarking technology. In: The IEEE Region 8 Computer as a Tool (EUROCON'2003), Sept. 22-24, 2003, 2:393-396.
[28] O. Benedens. Geometry-based watermarking of 3D models. IEEE Computer Graphics and Applications, 1999, 19(1):46-55.
[29] M. Yeung and B. L. Yeo. Fragile watermarking of three-dimensional objects. Paper presented at The International Conference on Image Processing (ICIP'98), 1998, 2:442-446.
[30] B. L. Yeo and M. Yeung. Watermarking 3D objects for verification. IEEE Computer Graphics and Applications, 1999, 1:36-45.
[31] O. Benedens. Two high capacity methods for embedding public watermarks into 3D polygonal models. In: Proceedings of the Multimedia and Security-Workshop at ACM Multimedia 99, 1999, pp. 95-99.
[32] S. Ichikawa, H. Chiyama and K. Akabane1. Redundancy in 3D polygon models and its application to digital signature. Journal of WSCG, 2002, 10(1): 225-232.
[33] R. Ohbuchi, H. Masuda and M. Aono. Watermarking three-dimensional polygonal models. In: Proceedings of ACM International Conference on Multimedia, 1997, pp. 261-272.
[34] J. J. Lee, N. I. Cho and J. W. Kim. Watermarking for 3D NURBS graphic data. In: IEEE Workshop on Multimedia Signal Processing, 2002, pp. 304-307.
[35] A. Tefas, G. Louizis and I. Pitas. 3D image watermarking robust to geometric distortions. Paper presented at The IEEE International Conference on Acoustics, Speech, and Signal Processing (ICASSP'02), 2002, pp. IV-3465-IV-3468.
[36] G. Louizis, A. Tefas and I. Pitas. Copyright protection of 3D images using watermarks of specific spatial structure. Paper presented at The IEEE International Conference on Multimedia and Expo (ICME'02), 2002, 2:557-560.
[37] Y. H. Wu, X. Guan, M. S. Kankanhalli, et al. Robust invisible watermarking of volume data using the 3D DCT. Computer Graphics International, 2001, pp. 359-362.
[38] X. Peng, L. F. Yu and L. L. Cai. Digital watermarking in three-dimensional space with a virtual-optics imaging modality. Optics Communications, 2003, 226(1-6): 155-165.
[39] R. Praun, H. Hoppe and A. Finkelstein. Robust mesh watermarking. In: Annual Conference Series Computer Graphics Proceedings, ACM SIGGRAPH, New York, 1999, pp. 49-56.
[40] M. Ashourian and R. Enteshary. A new masking method for spatial domain watermarking of three-dimensional triangle meshes. Paper presented at The Conference on Convergent Technologies for Asia-Pacific Region (TENCON'2003), 2003, 1: 428-431.
[41] T. Harte and A. G. Bors. Watermarking 3D models. Paper presented at The International Conference on Image Processing, 2002, 3: 661-664.
[42] T. Harte and A. G. Bors. Watermarking graphical objects. Paper presented at The

14th International Conference on Digital Signal Processing (DSP'2002), 2002, 2:709-712.

[43] Z. Q. Yu, H. H. S. Ip and L. F. Kowk. Robust watermarking of 3D polygonal models based on vertex scrambling. In: Proceedings of Computer Graphics International, 2003, pp. 254-257.

[44] Z. Q. Yu, H. H. S. Ip and L .F. Kwok. A robust watermarking scheme for 3D triangular mesh models. Pattern Recognition, 2003, 36(11):2603-2614.

[45] L. Koh and T. H. Chen. Progressive browsing of 3D models. In: IEEE 3rd Workshop on Multimedia Signal Processing, 1999, pp. 71-76.

[46] R. Ohbuchi, H. Masuda and M. Aono. Data embedding algorithms for geometrical and non-geometrical targets in three-dimensional polygonal models. Computer Communications, 1998, 21(15):1344-1354.

[47] R. Ohbuchi, H. Masuda and M. Aono. Embedding data in 3D models. In; Proc. of European Workshop on Interactive Distributed Multimedia Systems and Telecommunication Services (IDMS'97), 1997.

[48] R. Ohbuchi, H. Masuda and M. Aono. Watermarking multiple object types in three-dimensional models. In; Multimedia and Security Workshop at ACM Multimedia'98, 1998.

[49] F. Cayre and B. Macq. Data hiding on 3-D triangle meshes. IEEE Transactions on Signal Processing, 2003, 51(4):939-949.

[50] O. Benedens. Affine invariant watermarks for 3D polygonal and NURBS based models. In: Information Security, Third International Workshop, 1975, pp.15-29.

[51] O. Benedens and C. Busch. Towards blind detection of robust watermarks in polygonal models. Computer Graphics Forum, 2000, 19(3).

[52] O. Benedens. Watermarking of 3D polygon based models with robustness against mesh simplification. In: Proc. SPIE: Security and Watermarking of Multimedia Contents, 1999, Vol. 3657, pp. 329-340.

[53] S. H. Lee, T. S. Kim, B. J. Kim, et al. 3D polygonal meshes watermarking using normal vector distributions. Paper presented at The International Conference on Multimedia and Expo (ICME'03), 2003, 3:105-108.

[54] L. J. Zhang, R. F. Tong, F. Q. Su, et al. A mesh watermarking approach for appearance attributes. Paper presented at The 10th Pacific Conference on Computer Graphics and Applications, 2002, pp. 450-451.

[55] H. Sonnet, T. Isenberg, J. Dittmann, et al. Illustration watermarks for vector. Paper presented at The 11th Pacific Conference on Graphics Computer Graphics and Applications, 2003, pp. 73-82.

[56] Z. Li, W. M. Zheng and Z. M. Lu. A robust geometry-based watermarking scheme for 3D meshes. Paper presented at The first International Conference on Innovative Computing, Information and Control (ICICIC-06), 2006, Vol. II, pp. 166-169.

[57] R. Otten and L. van Ginneken. The Annealing Algorithm. Kluwer Academic Publishers, 1989.

[58] J. Maillot, H. Yahia and A. Verroust. Interactive texture mapping. SIGGRAPH Proceedings on Computer Graphics, 1993, 27:27-34.

[59] Z. Q. Yu, H. S. I. Horace and L. F. Kowk. Robust watermarking of 3D polygonal models based on vertice scrambling. Computer Graphics International 2003 (CGI'03), 2003, p. 254.

[60] Z. Karni and C. Gotsman. Spectral compression of mesh geometry. In: Computer Graphics (Proceedings of SIGGRAPH), 2000, pp. 279-286.

[61] S. Kanai, H. Date and T. Kishinami. Digital watermarking for 3D polygons using multiresolution wavelet decomposition. In: Proc. Sixth IFIP WG 5.2 GEO-6, 1998, pp. 296-307.
[62] H. Date, S. Kanai and T. Kishinami. Digital watermarking for 3D polygonal model based on wavelet transform. In: Proceedings of DETC'99, 1999.
[63] J. M. Lounsbery. Multiresolution analysis for surfaces of arbitrary topological type. Ph.D Thesis, Department of Computer Science and Engineering, University of Washington, 1994.
[64] J. Stollnitz, T. D. Derose and D. H. Salesin. Wavelet for Computer Graphics. Morgan Kaufmann Publishers, 1996.
[65] A. Kalivas, A. Tefas and I. Pitas. Watermarking of 3D models using principal component analysis. In: IEEE International Conference on Acoustics, Speech, and Signal Processing (ICASSP'03), 2003, 5:676-679.
[66] Guskov, W. Sweldensy and P. Schroder. Multiresolution signal processing for meshes. In: SIGGRAPH'99 Conference Proceedings, 1999, pp. 325-334.
[67] H. Hoppe. Progressive Meshes. In: SIGGRAPH'96 Proceedings, 1996, pp. 99-108.
[68] M. Garland and P. S. Heckbert. Surface simplification using quadric error metrics. In: SIGGRAPH'97 Proceedings, 1997, pp. 119-128.
[69] G. Taubin, T. Zhang and G. Golub. Optimal surface smoothing as filter design. IBM Technical Report RC-20404, 1996.
[70] H. S. Song, N. I. Cho and J. W. Kim. Robust watermarking of 3D mesh models. In: IEEE Workshop on Multimedia Signal Processing, 2002, pp. 332-335.
[71] K. Muratani and K. Sugihara. Watermarking 3D polygonal meshes using the singular spectrum analysis. Paper presented at The IMA Conference on the Mathematics of Surfaces, 2003, pp. 85-98.
[72] J. Lee, N. I. Cho and S. U. Lee. Watermarking algorithms for 3D NURBS graphic data. EURASIP Journal on Applied Signal Processing, 2004, 14: 2142-2152.

6

Reversible Data Hiding in 3D Models

As mentioned in Chapter 5, 3D model watermarking techniques can be classified into irreversible watermarking techniques and reversible watermarking techniques. Chapter 5 focuses on irreversible watermarking techniques. Now we turn to reversible watermarking techniques in this chapter. In fact, reversible watermarking is a branch of reversible data hiding. Reversible watermarking schemes are designed mainly for copyright protection and content authentication, while reversible data hiding schemes are designed for more application areas, including covert communication, besides copyright protection and content authentication. Reversible data hiding is also called invertible data hiding, lossless data hiding, distortion-free data hiding or erasable data hiding. It was initially investigated and designed for digital images. Then reversible data hiding schemes were reported in the literature for other media such as video, audio, 2D vector data, motion data and 3D models. After the first work on 3D model data hiding was reported [1], most subsequent work focuses on the following four aspects: (1) to improve the robustness of the 3D model data hiding schemes [2, 3] against rotation, translation, scaling, mesh simplification, and so on; (2) to reduce the visual distortions introduced by data embedding [4]; (3) to achieve the goal of blind extraction of the hidden data [5]; (4) to enhance the embedding capacity of the confidential data [6]. Some of these methods are based on transform domains and/or multiresolution analysis [7-9].

Recently, 3D model reversible data hiding has drawn much attention among researchers. In this prototype, the marked model should be recovered as accurately as the original one after data exaction. This requirement is more restricted than the traditional 3D model data hiding paradigm. This chapter starts with introducing the background and performance evaluation metrics of 3D model reversible data hiding. As many available 3D model reversible data hiding techniques come from the counterpart ideas of digital image reversible data hiding schemes, some basic reversible data hiding schemes for digital images are briefly reviewed. Next, three kinds of 3D model reversible data hiding techniques are extensively introduced, i.e., spatial-domain-based, compressed-domain-based and transform domain based methods. Lastly, a summary is given.

6.1 Introduction

We first introduce the background and performance evaluation metrics of 3D model reversible data.

6.1.1 Background

Data hiding is a technique that embeds secret information called a mark into host media for various purposes such as copyright protection, broadcast monitoring and authentication. Although cryptography is another way to protect the digital content, it only protects the content in transit. Once the content is decrypted, it has no further protection. Moreover, cryptographic techniques cannot provide sufficient integrity for content authentication. Data hiding techniques can be used in a wide variety of applications, each of which has its own specific requirements: different payload, perceptual transparency, robustness and security [10-13].

Digital watermarking is a form of data hiding. From the application point of view, digital watermarking methods can be classified into two categories: robust watermarking and fragile watermarking [10]. On the one hand, robust watermarking aims at making a watermark robust to all possible distortions to preserve the contents. On the other hand, fragile watermarking makes a watermark invalid even after the slightest modification of the contents, so it is useful to control content integrity and authentication. Most multimedia data embedding techniques modify, and hence distort, the host signal in order to insert the additional information. Often, this embedding distortion is small, yet irreversible; i.e., it cannot be removed to recover the original host signal. In many applications, the loss of host signal fidelity is not prohibitive as long as original and modified signals are perceptually equivalent. However, in some cases, although some embedding distortion is admissible, permanent loss of signal fidelity is undesirable. For example, in quality-sensitive applications such as medical imaging, military imaging, law enforcement and remote sensing where a slight modification can lead to a significant difference in the final decision-making process, the original media without any modification is required during data analysis. Even if the modification is quite small and imperceptible to the human eye, it is not acceptable because it may affect the right decision and lead to legal problems. This highlights the need for reversible (lossless) data embedding techniques. These techniques, like their lossy counterparts, insert information bits by modifying the host signal, thus inducing an embedding distortion. Nevertheless, they also enable the removal of such distortions and the lossless restoration of the original host signal after extraction of embedded information. Most of the reversible data hiding schemes, or so-called lossless data hiding (invertible data hiding) schemes, belong to fragile watermarking. For content authentication and tamper proofing, this enables exact recovery of the original media from the watermarked image after watermark removal [14]. The hash value of the original content, as well as electronic patient records (EPRs) and metadata regarding the content can be represented as the watermark. In multimedia archives, content providers do not want to waste their storage space to store both the original media and the watermarked one, due to cost and maintenance problems [15].

In fact, reversible data hiding is mainly used for the content authentication of multimedia data such as images, video and electronic documents, because of its emerging demand in various fields such as law enforcement, medical imagery and astronomical research. One of the most important requirements in this field is to have the original media during judgment to take the right decision. Cryptographic techniques based on either symmetric key or asymmetric key methods cannot provide adequate security and integrity for content authentication, because the main problem within the cryptographic techniques is that they are irreversible. Some authors use synonyms distortion-free, lossless, invertible, erasable watermarking for reversible data hiding. Lossless watermarking, as a branch of fragile watermarking, is the process that allows exact recovery of the original media by extracting the embedded information from the watermarked media, if the watermarked media is deemed to be authentic. That means no single bit of the watermarked media is changed after embedding the payload to the original media. This technique embeds secret information with the media so that the embedded message is hidden, invisible and fragile. Any attempt to change the watermarked media will make the authentication fail.

6.1.2 Requirements and Performance Evaluation Criteria

The general principle of reversible data hiding is that for a digital object (say a JPEG image file) I, a subset J of I is chosen. J has the structural property that it can be easily randomized without changing the essential property of I, and it offers the lossless compression version of I enough space (at least 128 bits) to embed the authentication message (say hash of I). During embedding, J is replaced by the authentication message concatenated with the compressed J. If J is highly compressible, only a subset of J can be used. During the decoding process, authentication information together with compressed J is extracted. This extracted J (compressed) is decompressed to replace the modified features in the watermarked object; hence the exact copy of the original object is found. The decoding process is just the reverse of the embedding process.

Three basic requirements for reversible data hiding can be summarized as follows:

(1) Reversibility. Reversibility is defined as "one can remove his embedded data to restore the original media." It is the most important and essential property for reversible data hiding.

(2) Capacity. The data to be embedded should be as large as possible. A small capacity will restrict the range of applications. The capacity is one of the important factors for measuring the performance of the algorithm.

(3) Fidelity. Data hiding techniques with high capacity might lead to low

fidelity. The perceptual quality of the host media should not be degraded severely after data embedding, although the original content is supposed to be recovered completely.

In particular, the performance of a 3D model reversible data hiding algorithm is measured by the following aspects: (1) embedding capacity; (2) visual quality of the marked model; (3) computational complexity. Reversible data hiding aims at developing a method that increases the embedding capacity as much as possible while keeping the distortion and the computational complexity at a low level.

6.2 Reversible Data Hiding for Digital Images

Before introducing reversible data hiding schemes for 3D models, this section first introduces classifications, applications and typical schemes of reversible data hiding for images.

6.2.1 Classification of Reversible Data Hiding Schemes

According to the embedding strategies, available reversible data hiding can be classified into three types as follows.

6.2.1.1 Type-I Algorithms

The type-I algorithms are based on lossless data compression techniques. They losslessly compress selected features from the host media to obtain enough space, which is then filled up with the secret data to be hidden. For example, Fridrich et al. [16] used a JBIG lossless compression scheme for compressing a proper bit-plane that offers minimum redundancy and embedded the image hash by appending it to the compressed bit-stream. However, a noisy image may force us to embed the hash in the higher bit-plane, and hence it causes visual artifacts. Celik et al. [17] used a CALIC lossless compression algorithm and achieved high capacity by using a generalized least significant bit embedding (G-LSB) technique, but the capacity depends on image structures.

6.2.1.2 Type-II Algorithms

The type-II algorithms are performed in transform domains such as integer discrete cosine transform (DCT) or integer discrete wavelet transform (DWT) where message bits are embedded into the corresponding coefficients. In [18], Yang et al. proposed a reversible data hiding algorithm based on integer DCT coefficients of image blocks. The capacity and visual quality were adjusted by selecting different numbers of AC coefficients in different frequencies. In [19], an integer wavelet transform is employed. Secret bits are embedded into a middle bit-plane of the integer wavelet coefficients in the high frequency sub-band. In [15], Lee et al. applied the integer-to-integer wavelet transform to image blocks and embedded message bits into the high-frequency wavelet coefficients of each block.

6.1.2.3 Type-III Algorithms

The type-III algorithms can be grouped into two categories: difference expansion (DE) and histogram modification. The original difference expansion technique was proposed by Tian in [20]. It applies the integer Haar wavelet transform to obtain high-pass components considered as the differences of pixel pairs. Secret bits are embedded by expanding these differences. The main advantage is its high embedding capacity, but its disadvantages are the undesirable distortion at low capacities and lack of capacity control due to embedding of a location map which contains the location information of all selected expandable difference values. Alattar developed the DE technique for color images using triplets [21] and quads [22] of adjacent pixels and generalized DE for any integer transform [23]. Kamstra and Heijmans [24] improved the DE technique by employing low-pass components to predict which location will be expandable, so their scheme is capable of embedding small capacities at low distortions. To overcome the drawbacks of the DE technique, Thodi and Rodriguez [25] presented a histogram-shifting technique to embed a location map for capacity control and suggested a prediction error expansion approach utilizing the spatial correlation in the neighborhood of a pixel.

Histogram modification techniques use the image histogram to hide message bits and achieve reversibility. Since most histogram-based methods do not apply any transform, all processing is performed in the spatial domain, and thus the computational cost is moderately lower than type-I and type-II algorithms. Ni et al. [26] utilized a zero point and a peak point of a given image histogram where the amount of embedding capacity is the number of pixels in the peak point. Versaki et al. [27] also proposed a reversible scheme using peak and zero points. One drawback of these algorithms is that it requires the information of the histogram's peak or zero points to recover an original image. In [28] and [29], they extended Ni's scheme and applied the location map to reverse without the knowledge of the peak and zero points. Tsai et al. [30] achieved a higher embedding capacity than the previous histogram-based methods by using a residue image indicating a difference between a basic pixel and each pixel in a non-overlapping block. However, in their scheme, since the peak and zero point information per each block is required to be attached to message bits, it makes the actual embedding capacity lower. Lee et al. [31] explored the peak point in the difference image histogram and embedded data into locations where the values of the difference image are −1 and +1. In [32], Lin et al. divided the image into non-overlapping blocks and generated a difference image block by block. Then, message bits are embedded by modifying the difference image of each block after making an empty bin through histogram shifting. Although this technique is a high capacity reversible method using a multi-level hiding strategy, it is required to transmit the peak information of all blocks.

In the type-I algorithms, the embedding capacity varies according to the characteristic of the image and the performance highly depends on the adopted lossless compression algorithm. The type-II algorithms show satisfactory results, but require additional computational costs to convert the media into transform domains. The DE technique in type-III algorithms is required to control the capacity due to the embedding of the location map. Although histogram-based methods simply work through histogram modification, overhead information should be as little as possible. In the following two subsections, two typical reversible data hiding schemes for images are detailed.

6.2.2 Difference-Expansion-Based Reversible Data Hiding

In [20], Tian proposed a reversible data hiding method for images based on difference expansion. In this method, the secret data is embedded in the difference of image pixel values. For a pair of pixels (x, y) in a gray level image, their average l and difference h are defined as

$$\begin{cases} l = \left\lfloor \dfrac{x+y}{2} \right\rfloor, \\ h = x - y. \end{cases} \tag{6.1}$$

Then the message to be embedded is computed by $h' = 2 \times h + b$. Here b denotes one secret bit. The new marked pixels are given as

$$\begin{cases} x' = l + \left\lfloor \dfrac{h'+1}{2} \right\rfloor, \\ y' = l - \left\lfloor \dfrac{h'}{2} \right\rfloor. \end{cases} \tag{6.2}$$

During data extraction, the secret bit is extracted as $b = h' \bmod 2$ and the original difference is computed as

$$h = \left\lfloor \frac{x' - y'}{2} \right\rfloor. \tag{6.3}$$

The two original pixels are recovered as

$$\begin{cases} x = \left\lfloor \dfrac{x' + y'}{2} \right\rfloor + \left\lfloor \dfrac{h+1}{2} \right\rfloor, \\ y = \left\lfloor \dfrac{x' + y'}{2} \right\rfloor - \left\lfloor \dfrac{h}{2} \right\rfloor. \end{cases} \tag{6.4}$$

The major problem is that overflow and underflow might occur. The secret bit can be embedded only in the pixels which satisfy

$$0 \le l - \left\lfloor \frac{h'}{2} \right\rfloor, \ l + \left\lfloor \frac{h'+1}{2} \right\rfloor \le 255. \tag{6.5}$$

A pixel pair satisfying Eq.(6.5) is called the expandable pixel pair. In order to achieve lossless data embedding, a location map is employed to record the expandable pixel pair. The location map is then compressed by lossless compression methods and concatenated with the original secret message to be superimposed on the host signal later.

In [23], Alattar extended Tian's scheme using difference expansion of a vector instead of a pixel pair to hide message data for color images. In their scheme, a vector is formed by k non-overlapping pixels. Then they use a reversible integer transform function to transform the vector. If the transformed vector can be used to hide message data, then they use Tian's difference expansion algorithm to conceal the data. For restoring the host image, the algorithm needs a location map, as well as Tian's location map, to indicate whether the vector can be used to hide message bits or not.

For example, a vector with four pixels is used to embed three message bits. Let $\boldsymbol{p} = (p_1, p_2, p_3, p_4)$ be the vector and b_1, b_2, b_3 be the message bits. First, they use the reversible integer transformation function to compute the weighted average q_1, and the differences q_2, q_3 and q_4 of p_2, p_3, p_4 from p_1. The weighted average and the differences are calculated by

$$\begin{cases} q_1 = \left\lfloor \dfrac{a_1 p_1 + a_2 p_2 + a_3 p_3 + a_4 p_4}{a_1 + a_2 + a_3 + a_4} \right\rfloor, \\ q_2 = p_2 - p_1, \\ q_3 = p_3 - p_1, \\ q_4 = p_4 - p_1, \end{cases} \tag{6.6}$$

where a_1, a_2, a_3, a_4 are constant coefficients. Then, the weighted average and the differences are shifted according to the message bits to generate the one-bit left-shifted values q'_1, q'_2, q'_3 and q'_4. The shifted values are computed by

$$\begin{cases} q_1' = q_1, \\ q_2' = 2 \times q_2 + b, \\ q_3' = 2 \times q_3 + b, \\ q_4' = 2 \times q_4 + b. \end{cases} \tag{6.7}$$

Finally, the pixels containing the message bits p'_1, p'_2, p'_3 and p'_4 are calculated by

$$\begin{cases} p_1' = q_1 - \left\lfloor \dfrac{a_2 q_2' + a_3 q_3' + a_4 q_4'}{a_1 + a_2 + a_3 + a_4} \right\rfloor, \\ p_2' = q_2' + q_1 - \left\lfloor \dfrac{a_2 q_2' + a_3 q_3' + a_4 q_4'}{a_1 + a_2 + a_3 + a_4} \right\rfloor, \\ p_3' = q_3' + q_1 - \left\lfloor \dfrac{a_2 q_2' + a_3 q_3' + a_4 q_4'}{a_1 + a_2 + a_3 + a_4} \right\rfloor, \\ p_4' = q_4' + q_1 - \left\lfloor \dfrac{a_2 q_2' + a_3 q_3' + a_4 q_4'}{a_1 + a_2 + a_3 + a_4} \right\rfloor. \end{cases} \tag{6.8}$$

In the decoding phase, they compute the shifted values by using

$$\begin{cases} q_1'' = \left\lfloor \dfrac{a_1 p_1' + a_2 p_2' + a_3 p_3' + a_4 p_4'}{a_1 + a_2 + a_3 + a_4} \right\rfloor, \\ q_2'' = p_2' - p_1', \\ q_3'' = p_3' - p_1', \\ q_4'' = p_4' - p_1'. \end{cases} \tag{6.9}$$

The embedding data is inferred from the shifted values that are computed as

$$\begin{cases} b_1 = q_2'' - 2 \times \left\lfloor \dfrac{q_2''}{2} \right\rfloor, \\ b_2 = q_3'' - 2 \times \left\lfloor \dfrac{q_3''}{2} \right\rfloor, \\ b_3 = q_4'' - 2 \times \left\lfloor \dfrac{q_4''}{2} \right\rfloor. \end{cases} \tag{6.10}$$

The original q_1, q_2, q_3 and q_4 are given by

$$\begin{cases} q_1 = q''_1 \\ q_2 = \left\lfloor \dfrac{q''_2}{2} \right\rfloor, \\ q_3 = \left\lfloor \dfrac{q''_3}{2} \right\rfloor, \\ q_4 = \left\lfloor \dfrac{q''_4}{2} \right\rfloor. \end{cases} \tag{6.11}$$

Finally, the original pixels are restored by

$$\begin{cases} p_1 = q_1 - \left\lfloor \dfrac{a_2 q_2 + a_3 q_3 + a_4 q_4}{a_1 + a_2 + a_3 + a_4} \right\rfloor, \\ p_2 = q_2 + q_1, \\ p_3 = q_3 + q_1, \\ p_4 = q_4 + q_1. \end{cases} \tag{6.12}$$

In this way, the secret data is extracted and the host image is accurately recovered.

6.2.3 Histogram-Shifting-Based Reversible Data Hiding

In [33], Ni et al. proposed a reversible data-hiding method based on histogram shifting. It shifts part of the image histogram and then embeds data in the produced redundancy.

The basic principle is shown in Fig. 6.1. The left histogram is the original one computed, based on the host image. The center one is the shifted histogram and the right one is the version after data embedding.

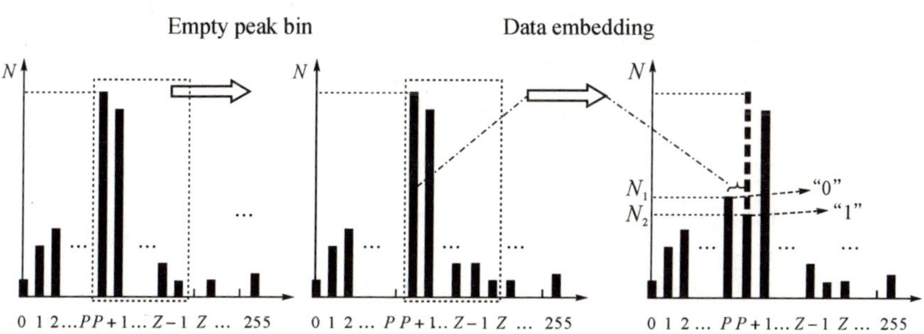

Fig. 6.1. Reversible watermark embedding based on histogram shifting

In these histograms, the horizontal axis denotes the pixel values in the range of [0, 255], while N on the vertical axis is the number of peak values corresponding to the pixel value P. In [33], P is called the peak point and the first one with magnitude 0 on the right side of P is called the zero point Z.

The peak and zero points must be found before shifting the histogram. Then all bins between $[P, Z-1]$ are shifted one gray level rightward. That is, to all pixel values between $[P, Z-1]$ add 1 and thus the original P is emptied. As a result, the magnitude in the original bin $P+1$ is changed as N.

Next, we can embed secret data by modulating 0 and 1 on P and $P+1$, respectively. In particular, the pixel values belonging to the bin $P+1$ are scanned one by one. If the bit "0" is to be embedded, the pixels with the value $P+1$ are modified as P, while they are kept unchanged when the bit "1" is to be embedded. In this way, the data embedding process is completed.

The data extraction and image recovery is the inverse process of data embedding. First, the peak point P and the zero point Z must be located accurately. Then we scan the whole image. If we come across a pixel with the value P, a secret bit "0" is extracted. If $P+1$ is encountered, a secret bit "1" is extracted. After the data is extracted, we only need to subtract 1 from all pixel values between $[P+1, Z]$ and thus the original image can be perfectly recovered.

6.2.4 Applications of Reversible Data Hiding for Images

There are many applications of reversible data embedding techniques, such as business, legislation and medical applications. Four typical applications can be expressed as follows.

6.2.4.1 Medical Diagnostic Images

Medical images require a high degree of restoration capability. The patient's information such as the personal data, medical history and results of diagnosis are suitable to be embedded. Because of the potential risk of medical lawsuits and of the physician misinterpreting an image, medical images are very sensitive and cannot be disturbed in any way. Reversible data hiding techniques are thus very useful in the medical imaging environment [34, 35].

6.2.4.2 Digital Photography as Legal Evidence

As establishing the integrity of evidence throughout the crime scene investigation is of paramount importance, if reversible secret data could be embedded by digital cameras, the picture evidence of a crime scene would be acceptable for law enforcement [36].

6.2.4.3 Remote Sensing Images for Military Imagery

Military images, such as satellite and reconnaissance images, might be inspected under special viewing conditions when typical assumptions about distortions apply. Those conditions include extreme zooming, iterative filtering and enhancement and so on. Reversible embedding techniques are appropriate for such applications because the original data can be restored without any loss of information [37].

6.2.4.4 Media Asset Management

Watermarking-based media asset management systems control the multimedia by embedding the catalog, index and annotation of the original content. As some people might be concerned about the quality degradation of an image as a result of watermark embedding, reversible data embedding could be a convenient method of embedding the description or control information without affecting the image quality [38].

6.3 Reversible Data Hiding for 3D Models

Although reversible data hiding was first introduced for digital images, it also has wide application scenarios for hiding data in 3D models. For example, suppose there is a column on a 3D mechanical model obtained by computed aided design. The diameter of this column is changed with a given data hiding scheme. In some applications, it is not enough that the hidden content is accurately extracted. This is because the remaining watermarked model is still distorted. Even if the column diameter is increased or decreased by 1 mm, it may cause a severe effect because this mechanical model cannot be assembled well with other mechanical accessories.Therefore, it also has significance in designing reversible data hiding methods for 3D models.

6.3.1 General System

As shown in Fig. 6.2, the general system for 3D model reversible data hiding can be deduced from that designed for images. In this typical system, $\boldsymbol{M}$ and W denote the host model and the original secret data, respectively. W is embedded in $\boldsymbol{M}$ with the key K and the marked model $\boldsymbol{M}_W$ is produced. Suppose the $\boldsymbol{M}_W$ is losslessly transmitted to the receiver and then the secret data is extracted as W_R with the same key K. Meanwhile, the original model is recovered as $\boldsymbol{M}_R$. The definition of 3D model reversible data hiding requires that both the secret data and the host model should be recovered accurately, i.e., $W_R = W$ and $\boldsymbol{M}_R = \boldsymbol{M}$. In a word, 3D model reversible data hiding schemes also satisfy the imperceptibility and inseparability properties that those general irreversible data hiding schemes do.

6.3.2 Challenges of 3D Model Reversible Data Hiding

According to the general model shown in Fig. 6.2, we can find that the requirements of 3D model reversible data hiding are more restricted than those of irreversible ones. Besides, as a special host media, 3D model reversible data hiding has several technical challenges as follows.

(1) Nowadays there are many types of 3D models such as 3D meshes and point cloud models. Most 3D models are represented as meshes, while point cloud models are stored and used in some specific applications such as 3D face recognition. Moreover, there exist many formats of meshes, such as .off and .obj. In practical applications, various types and formats of models are often interconverted. In contrast, most available reversible data hiding schemes are designed for one specific type or format. Thus, these schemes are usually not suitable for other types or formats. Therefore, developing a universal reversible data-hiding scheme is a challenging work.

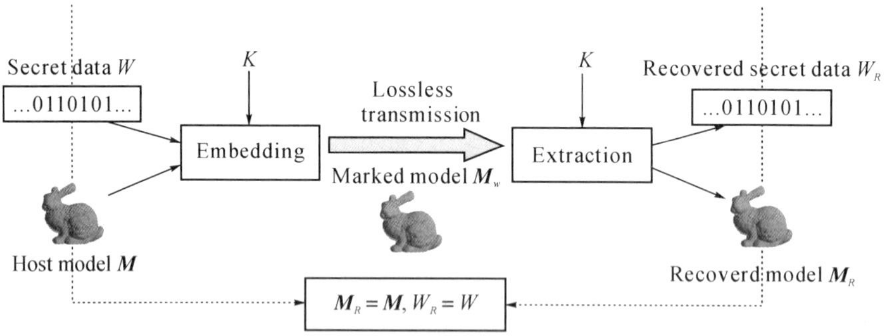

Fig. 6.2. A general system for 3D model reversible data hiding

(2) Various models may have different levels of detail. For example, a desk may only contain tens of vertices and faces, while a plane may have thousands of vertices and faces. This diversity of levels of detail should be considered in developing the reversible data hiding scheme for 3D models.

(3) The elements of data hiding in images are pixels, while in a 3D model the elements of data hiding are usually vertices and faces. In an image, each pixel has its fixed coordinates and data hiding is just to modify their pixel values. In contrast, the coordinates of the watermarked vertices of 3D models are usually changed before data extraction. For example, the watermarked model is rotated and translated. Thus, pose estimation is usually required. This causes a difficulty to extract data and recover the host model. Sometimes some affiliated knowledge must be used to assist the data extraction and model recovery. This affiliated

knowledge must be securely sent to the decoder along with the watermarked model. Thus researchers must try to reduce the amount of affiliated knowledge.

6.3.3 Algorithm Classification

Nowadays, some reversible data hiding schemes for 3D models are proposed in the literature [39-45]. According to different embedding domains, they can be classified into spatial-domain-based, compressed-domain-based and transform-domain-based methods. In spatial-domain-based methods [39, 42, 43], the task of data embedding is to modify the vertex coordinates, edge connections, face slopes and so on. These schemes usually have a low computational complexity. The compressed-domain-based methods [44, 45] are for embedding data with certain compression techniques involved, e.g., vector quantization. In addition, some of these methods are designed for compressed content of 3D models. Their advantage is to hide data without decompressing the host model. In transform domain-based methods [40, 41], the original model is transformed into a certain transform domain and then data are embedded in transform coefficients. In these schemes, the reversibility is guaranteed by that of the transforms.

6.4 Spatial Domain 3D Model Reversible Data Hiding

Most available 3D model reversible data hiding schemes belong to spatial domain methods. In [39], Chou et al. proposed a reversible data hiding scheme for 3D models. In this method, all of the 3D vertices are divided into a set of groups. Then they are transformed into the invariant space for resisting the attacks such as rotation, translation and scaling. The secret data are embedded in some carefully selected positions with unnoticeable distortions introduced. In this way, some parameters are generated for data extraction, and these parameters are also hidden in 3D models. In data extraction, these parameters are retrieved for data extraction and model recovery. In [42], a reversible data hiding scheme for 3D meshes is proposed based on prediction-error expansion. The principle is to predict a vertex's position by calculating the centroid of its traversed neighbors, and then the prediction error, i.e. the difference between the predicted and real positions, is expanded for data embedding. In this scheme, only the vertex coordinates are modified to embed data, and thus the mesh topology is unchanged. The visual distortion is reduced by adaptively choosing a threshold so that the prediction errors with too large a magnitude will not be expanded. The selected threshold value and the location information are saved in the mesh for model recovery. As the original mesh can be exactly recovered, this algorithm can be used for symmetric or public key authentication of 3D mesh models.

This section introduces another spatial-domain-based reversible data hiding method for 3D models [43]. It can be used to authenticate 3D meshes by modulating the distances from the mesh faces to the mesh centroid to embed a fragile watermark. It keeps the modulation information in the watermarked mesh so that the reversibility of the embedding process is achieved. Since the embedded watermark is sensitive to geometrical and topological processing operations, unauthorized modifications on the watermarked mesh can be therefore detected by retrieving and comparing the embedded watermark with the original one. Furthermore, as long as the watermarked mesh is intact, the original mesh can be recovered using some a priori knowledge.

6.4.1 3D Mesh Authentication

With the widespread use of polygonal meshes, how to authenticate them has become a real need, especially in the web environment. As an effective measure, data hiding for multimedia content (e.g. digital images, 3D models, video and audio streams) has been widely studied to prove the ownership of digital works, verify their integrity, convey additional information, and so forth. Depending on the applications, digital watermarking can be mainly classified into robust watermarking (e.g. [46-48]) and fragile watermarking. In this subsection, we concentrate on the latter only, in which the embedded watermark will change or even disappear if the watermarked object is tampered with. Therefore, fragile watermarking has been used to verify the integrity of digital works. In the literature, only a few fragile ones [5, 49-51] have been proposed to verify the integrity. Actually, the first fragile watermarking method for 3D object verification is addressed by Yeo and Yeung in [49], as a 3D version of the method for 2D image watermarking. In [52], invertible authentication of 3D meshes is first introduced by combining a public verifiable digital signature protocol with the embedding method in [53], which appends extra faces and vertices to the original mesh. After extracting the embedded signature, the appended faces and vertices can be removed on demand to reproduce the original mesh with a secret key. One of the algorithms proposed in [5] called Vertex Flood Algorithm can be used for model authentication with certain tolerances, e.g. truncation of mantissas of vertex coordinates. A fragile watermarking scheme for triangle meshes is presented by Cayre et al. in [50] to embed a watermark with robustness against translation, rotation and scaling transforms. Nevertheless, all those proposed algorithms are not reversible, i.e. the original mesh cannot be recovered from the watermarked mesh. Actually, it is advantageous to recover the original mesh from its watermarked version because the mesh distortion introduced by the encoding process can be compensated. In this subsection, a reversible data-hiding method is introduced to authenticate 3D meshes [43]. By keeping the modulation information in the watermarked mesh, the reversibility of the embedding process in [54] is achieved. Since the embedded watermark is sensitive to geometrical and topological processing, unauthorized modifications on the watermarked mesh can be detected by retrieving the embedded watermark and comparing it with the original one. Furthermore, as long as the watermarked mesh is intact, the original mesh can be recovered with some a priori knowledge.

6.4.2 Encoding Stage

In [54], the distance from the mesh faces to the mesh centroid is modulated to embed the fragile watermark to detect the modifications on the watermarked mesh. As a result, the original mesh is changed after the watermarking process. Nevertheless, we notice that the mesh topology is unchanged during the encoding process; the original mesh can be recovered by moving every vertex back to its original position. It can be achieved by keeping the modulation information in the watermarked mesh. Accordingly, the encoding and decoding processes will be shown as follows, respectively.

In the encoding process, a special case of quantization index modulation called dither modulation [55] is extended to the mesh. By modulating the distances from the mesh faces to the mesh centroid, a sequence of data bits is embedded into the original mesh.

Suppose $V = \{\boldsymbol{v}_1, \ldots, \boldsymbol{v}_U\}$ is the set of vertex positions in R^3, the position $\boldsymbol{v}_c$ of the mesh centroid is defined as

$$\boldsymbol{v}_c = \frac{1}{U}\sum_{i=1}^{U}\boldsymbol{v}_i. \tag{6.13}$$

Similarly, the face centroid position is defined as the mean of the vertex positions in the face. Subsequently, the distance d_{fi} from the face f_i to $\boldsymbol{v}_c$ can be defined as

$$d_{fi} = \sqrt{(v_{icx} - v_{cx})^2 + (v_{icy} - v_{cy})^2 + (v_{icz} - v_{cz})^2}, \tag{6.14}$$

where $(v_{icx}, v_{icy}, v_{icz})$ and (v_{cx}, v_{cy}, v_{cz}) are the coordinates of the face centroid $\boldsymbol{v}_{ic}$ and the mesh centroid $\boldsymbol{v}_c$ in R^3, respectively. It can be concluded that d_{fi} is sensitive to both geometrical and topological modifications made to the mesh model.

The distance d_i from a vertex with the position $\boldsymbol{v}_i$ to the mesh centroid is defined as

$$d_i = \sqrt{(v_{ix} - v_{cx})^2 + (v_{iy} - v_{cy})^2 + (v_{iz} - v_{cz})^2}, \tag{6.15}$$

where (v_{ix}, v_{iy}, v_{iz}) is the vertex coordinate in R^3. The quantization step S of the modulation is chosen as

$$S=D/N, \tag{6.16}$$

where N is a specified value and D is the distance from the furthest vertex to the mesh centroid. With the modulation step S, the integer quotient Q_i and the remainder R_i are obtained by

$$Q_i = \left\lfloor \frac{d_{fi}}{S} \right\rfloor, \tag{6.17}$$

$$R_i = d_{fi} \% S. \tag{6.18}$$

To embed one watermark bit w_i, Wu and Yiu [43] modulated the distance d_{fi} from f_i to the mesh centroid so that the modulated integer quotient Q'_i meets $Q'_i\%2 = w_i$. To keep the modulation information in the watermarked mesh, the modulated distance d'_{fi} is defined as

$$d'_{fi} = \begin{cases} Q_i \times S + S/2 + m_i, & \text{if } Q_i\%2 = w_i; \\ Q_i \times S - S/2 + m_i, & \text{if } Q_i\%2 = \overline{w}_i \text{ and } R_i < S/2; \\ Q_i \times S + 3S/2 + m_i, & \text{if } Q_i\%2 = \overline{w}_i \text{ and } R_i \geq S/2, \end{cases} \tag{6.19}$$

where $\overline{w}_i = 1 - w_i$ and m_i is the modulation component with the definition as follows: Suppose there are K faces used to embed the watermark information, for $I =$ 3, …, K, $m_i = \dfrac{d'_{f(i-1)} - d_{f(i-1)}}{4}$, while $m_1 = \dfrac{d'_{fK} - d_{fK}}{4}$ and $m_2 = \dfrac{Q'_1 \times S + S/2 - d_{f1}}{4}$ with Q'_1 provided in Eq.(6.20). It can be concluded from the definition of m_i and Eq.(6.19) that $m_i \in (-\frac{2S}{5}, \frac{2S}{5})$ and the modulated integer quotient as

$$Q'_i = \begin{cases} Q_i, & \text{if } Q_i\%2 = w_i; \\ Q_i - 1, & \text{if } Q_i\%2 = \overline{w}_i \text{ and } R_i < S/2; \\ Q_i + 1, & \text{if } Q_i\%2 = \overline{w}_i \text{ and } R_i \geq S/2. \end{cases} \tag{6.20}$$

Consequently, the resulting d'_{fi} is used to adjust the position of the face centroid. Only one vertex in f_i is selected to move the face centroid to the desired position. Suppose $\boldsymbol{v}_{is}$ is the position of the selected vertex, the adjusted vertex position would be

$$\boldsymbol{v}'_{is} = \left[\boldsymbol{v}_c + (\boldsymbol{v}_{ic} - \boldsymbol{v}_c) \times \frac{d'_{fi}}{d_{fi}} \right] \times N_i - \sum_{j=1, j \neq s}^{N_i} \boldsymbol{v}_{ij}, \tag{6.21}$$

where $\boldsymbol{v}_{ij}$ is the vertex position in f_i with N_i vertices and $\boldsymbol{v}_{ic}$ as the former face centroid. To prevent the embedded watermark bits from being changed by the subsequent encoding operations, all vertices in the face should not be moved any more after the adjustment.

The detailed procedure to reversibly embed the watermark is as follows: At first, the original mesh centroid position is calculated by Eq.(6.13). Then the furthest vertex to the mesh centroid is found out using Eq.(6.15) and the distance D from it to the mesh centroid is obtained. After that, the modulation step S is chosen by specifying the value of N in Eq.(6.16). Using the key *Key*, the sequence of face indices I are scrambled to generate the scrambled version I', which determine the sequence of mesh faces. For a face f_i indexed by I', if there is at least one unvisited vertex, the distance from f_i to the mesh centroid is calculated by Eq.(6.14) and modulated by Eq.(6.19) according to the watermark bit value. Subsequently, the position of the unvisited vertex is modified using Eq.(6.21), whereby the face centroid is moved to the desired position. If there is no unvisited vertex in f_i, the checking mechanism will be skipped to the next face indexed by I' until all watermark bits are embedded.

6.4.3 Decoding Stage

In the decoding process, the original mesh centroid position v_c, the modulation step S, as well as the secret key *Key* and the original watermark are required. The embedded watermark needs to be extracted from the watermarked mesh and compared with the original watermark to detect illegal tampering on the watermarked mesh. The original mesh can be recovered if the watermarked mesh is intact.

The detailed decoding process is conducted as follows: At first, the sequence of face indices I is scrambled using the key *Key* to generate the scrambled version I', which is followed to retrieve the embedded watermark. If there is at least one unvisited vertex in a face f'_i, the modulated distance d'_{fi} from f'_i to the mesh centroid is calculated by Eq.(6.14). With the given S', the modulated integer quotient Q'_i is obtained by

$$Q'_i = \left\lfloor \frac{d'_{fi}}{S'} \right\rfloor. \tag{6.22}$$

And the watermark bit w_i' is extracted by

$$w'_i = Q'_i \% 2. \tag{6.23}$$

If there is no unvisited vertex in f'_i, no information is extracted and the decoding process will be automatically skipped to the next face index by I' until all watermark bits are extracted.

After the watermark extraction, the extracted watermark W' is compared with the original watermark W to detect the modifications that might have been made to the watermarked mesh. Supposing the length of the watermark is K, the normalized cross-correlation value NC between the original and the extracted watermarks is given by

$$NC = \frac{1}{K}\sum_{i=1}^{K} I(w_i', w_i), \tag{6.24}$$

with

$$I(w_i', w_i) = \begin{cases} 1, & \text{if } w_i' = w_i; \\ -1, & \text{otherwise.} \end{cases} \tag{6.25}$$

If the watermarked mesh model is intact, the NC value will be 1; otherwise, it will be less than 1.

To recover the original mesh, the modulation information m_i, needs to be calculated according to d'_{fi}, Q'_i and S'. For $i = 1, 2, \ldots, K$,

$$m_i = d'_{fi} - (Q_i' \times S' + S'/2). \tag{6.26}$$

According to the definition of m_i, for $i = 2, \ldots, K-1$, the original distance $d_{fi} = d'_{fi} - m_{i+1} \times 4$, while $d_{fK} = d'_{fK} - m_1 \times 4$ and $d_{f1} = Q'_1 \times S' + S'/2 - m_2 \times 4$. With the obtained d_{fi}, all the vertices whose positions have been adjusted can be moved back by

$$\boldsymbol{v}_{is} = (\boldsymbol{v}_c + (\boldsymbol{v}_{ic}' - \boldsymbol{v}_c) \times \frac{d_{fi}}{d'_{fi}}) \times N_i - \sum_{j=1, j \neq s}^{N_i} \boldsymbol{v}_{ij}', \tag{6.27}$$

where $\boldsymbol{v}'_{ij}$ is the vertex position in the face f'_i consisting of N_i vertices with $\boldsymbol{v}'_{ic}$ as the adjusted centroid position, $\boldsymbol{v}_{is}$ is the recovered vertex position and $\boldsymbol{v}_c$ is the original mesh centroid position. After the original mesh is recovered from the watermarked mesh, an additional way to detect the modifications on the watermarked mesh is to compare the centroid position of the recovered mesh with that of the original mesh, which should be identical to each other.

6.4.4 Experimental Results and Discussions

The above algorithm is conducted in the spatial domain and applicable to all meshes without any restriction. The modulation step S should be carefully set, providing a trade-off between imperceptibility and false alarm probability. Wu and Yiu [43] have investigated the algorithm on several meshes listed in Table 6.1. A 2D binary image is chosen as the watermark, which can also be a hashed value.

The capacities of the meshes are also listed in Table 6.1, which depends on the vertex number and mesh traversal. Wu and Yiu [43] wished to hide sufficient watermark bits in the mesh so that the modification made to each vertex position can be efficiently detected. Fig. 6.3(a) and Fig. 6.3(b) illustrate the original mesh model "dog" and its watermarked version, while Fig. 6.3(c) shows the recovered one. It can be seen that the watermarking process has not caused noticeable distortion.

Table 6.1 The meshes used in the experiments [43] (©[2005]IEEE)

Models	Vertices	Faces	Capacity (bits)
Dog	7,158	13,176	5,594
Wolf	7,232	13,992	5,953
Raptor	8,171	14,568	7,565
Horse	9,988	18,363	7,650
Cat	10,361	19,098	8,131
Lion	16,652	32,096	14,564

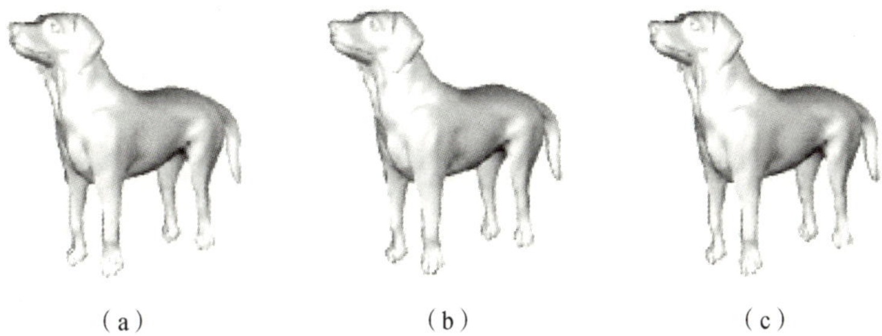

Fig. 6.3. Experimental results on the "dog" mesh with $N = 10000$ [43]. (a) Original mesh; (b) Watermarked mesh; (c) Recovered mesh (©[2005]IEEE)

To evaluate the imperceptibility of the embedded watermark, the normalized Hausdorff distance between two meshes is calculated to measure the introduced distortion, based upon the fact that the mesh topology is unchanged. Fig. 6.4 shows the amount of the distortion subject to the modulation step S. The upper curve denotes the distance between the original and watermarked mesh models, while the distance between the original and recovered meshes is plotted in the lower curve. From Fig. 6.4, it can be seen that the distortion of the watermarked mesh increases as the modulation step S increases. The recovered mesh is nearly the same as the original mesh since the distance between them is very small and nearly unaffected by the modulation step. Given the same modulation step, the difference between the original and recovered meshes is much smaller than the difference between the original and watermarked meshes. In this sense, the mesh distortion introduced by the encoding process has been significantly reduced by performing the reversibility mechanism.

In the experiments, the watermarked mesh models went through translation, rotation and uniform scaling transforms, modifying one vertex position by adding the vector $\{2S, 2S, 2S\}$, reducing one face and adding the noise signal $\{n_x, n_y, n_z\}$

to all the vertex positions with n_x, n_y and n_z uniformly distributed within the interval $[-S, S]$, respectively. The watermarks were extracted from the modified meshes with and without the key *Key*. The centroid positions of the meshes recovered from those modified meshes were compared with the original meshes. The obtained *NC* values are all below 1, and the recovered mesh centroid positions are different from the original one in most of the cases so that modifications on the watermarked mesh can be efficiently detected.

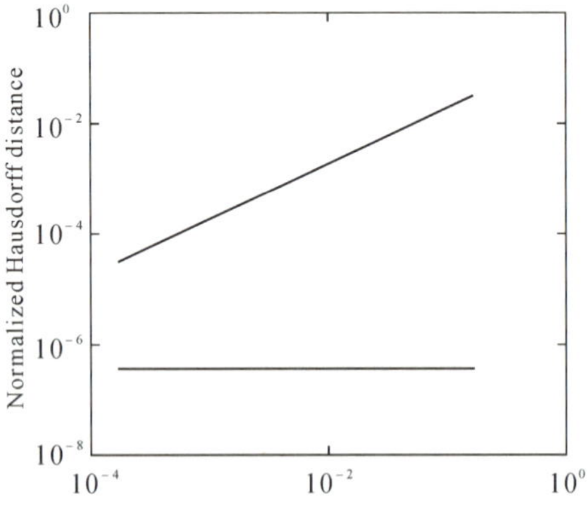

Fig. 6.4. The normalized Hausdorff distance subject to the modulation step *S* [43] (©[2005]IEEE)

6.5 Compressed Domain 3D Model Reversible Data Hiding

Data hiding has become an accepted technology for enforcing multimedia protection schemes. While major efforts concentrate on still images, audio and video clips, recently the research interests in 3D mesh data hiding have been increasing. Reversible data hiding [43, 52, 56-64] has only recently been the subject of focus. It embeds the payload (data to be embedded) into a digital content in a reversible manner. As non-reversible data hiding, the embedding of the payload should not be noticeable. In particular, a reversible data hiding algorithm guarantees that when the payload is removed from the stego content, the cover content can be exactly restored. The first publication on invertible authentication that we are aware of is the patent of Honsinger et al. [56], owned by the Eastman Kodak Company. In 2003, Jana Dittmann and Oliver Benedens [52] first explicitly presented a reversible authentication scheme for 3D meshes. In 2005, Wu and Cheung [43] proposed a reversible data-hiding method to authenticate 3D meshes by modulating the distances from the mesh faces to the mesh center, which has been described in Section 6.4. It is also noticeable that when combining graphics technology with the Internet, the transmission delay for

3D meshes becomes a major performance bottleneck. Consequently, many 3D mesh compression techniques based on vector quantization (VQ) have surged in recent years and thus more and more 3D meshes have been represented in the form of VQ bitstreams. So it is urgent to authenticate the VQ bitstream of a 3D mesh that is equivalent to its counterpart in the original format.

In this section, we introduce a new kind of data hiding method for 3D triangle meshes proposed in [44, 45] by the authors of this book. While most of the existing data hiding schemes introduce some small amount of non-reversible distortion to the cover mesh, the new method is reversible and enables the cover mesh data to be completely restored when the payload is removed from the stego mesh. A noticeable difference between our method and others' is that we embed data in the predictive vector quantization (PVQ) compressed domain by modifying the prediction mechanism during the compression process.

6.5.1 Scheme Overview

A general reversible data embedding diagram [44] is illustrated in Fig. 6.5. First, we compress the original mesh $\boldsymbol{M}_0$ into the cover mesh $\boldsymbol{M}$ that is the object for payload embedding based on the VQ technique. Although the VQ compression technique introduces a small amount of distortion to the mesh, as long as the distortion is small enough, we can ignore it. Besides, VQ technique enables the distortion to be as tiny as possible by simply choosing a higher quality level of codebook. In this sense, $\boldsymbol{M}_0$ as well as $\boldsymbol{M}$ can both be reversibly authenticated as long as they are close enough. Then we embed a payload into $\boldsymbol{M}$ by modifying its prediction mechanisms during the VQ encoding process, and obtain the stego mesh $\boldsymbol{M}'$. Before it is sent to the decoder, $\boldsymbol{M}'$ might or might not have been tampered with by some intentional or unintentional attacks. If the decoder finds that no tampering happened in $\boldsymbol{M}'$, i.e. $\boldsymbol{M}'$ is authentic, then the decoder can remove the embedded payload from $\boldsymbol{M}'$ to restore the cover mesh, which results in a new mesh $\boldsymbol{M}''$. According to the definition of reversible data embedding, the restored mesh $\boldsymbol{M}''$ should be exactly the same as the cover mesh $\boldsymbol{M}$, vertex by vertex and bit by bit.

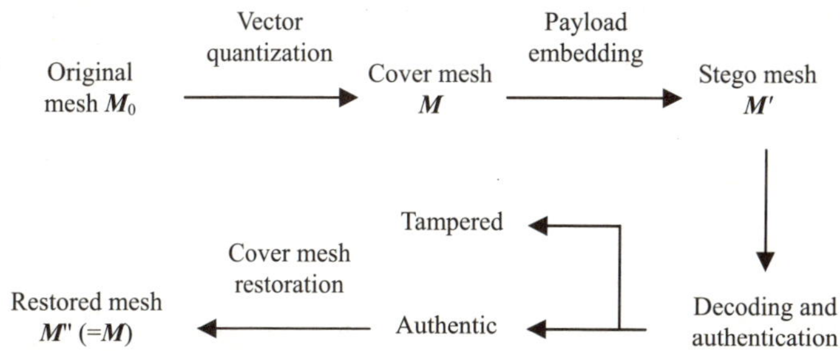

Fig. 6.5. Reversible data hiding diagram

6.5.2 Predictive Vector Quantization

Vector quantization [65] can be defined as a mapping procedure from the k-dimensional Euclidian space to a finite subset, i.e. $Q: R^k \rightarrow C$, where the subset $C = \{\boldsymbol{c}_i | i = 1, 2, \ldots, N\}$ is called a codebook, where $\boldsymbol{c}_i$ is a codevector and N is the codebook size. The best match codevector $\boldsymbol{c}_p = (c_{p0}, c_{p1}, \ldots, c_{p(k-1)})$ for the input vector $\boldsymbol{x} = (x_0, x_1, \ldots, x_{(k-1)})$ is the closest vector to $\boldsymbol{x}$ among all the codevectors in C.

The vertex $\boldsymbol{v}_n$ in a 3D triangle mesh can be predicted by its neighboring quantized vertices $\{\hat{\boldsymbol{v}}_{n-1}, \hat{\boldsymbol{v}}_{n-2}, \hat{\boldsymbol{v}}_{n-3}\}$. The prediction sketch is depicted in Fig. 6.6, where $\hat{}$ denotes the quantized vertex and $\tilde{}$ denotes the predicted vertex. The detailed prediction design is illustrated in [66].

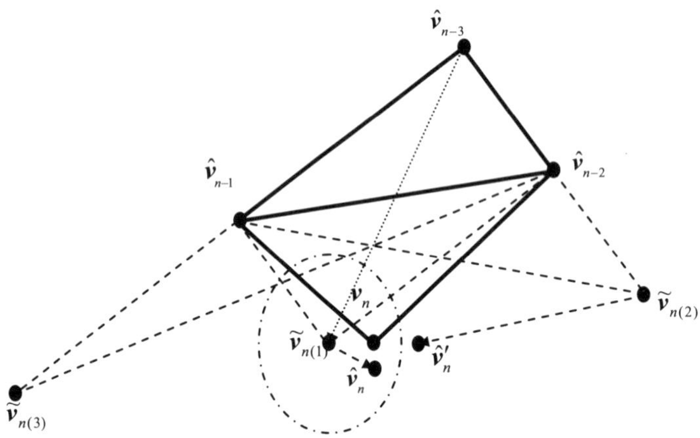

Fig. 6.6. The sketch of mesh vertex prediction

A common prediction mechanism is the parallelogram prediction as follows:

$$\tilde{\boldsymbol{v}}_n = \hat{\boldsymbol{v}}_{n-1} + \hat{\boldsymbol{v}}_{n-2} - \hat{\boldsymbol{v}}_{n-3}, \tag{6.28}$$

which corresponds to the $\tilde{\boldsymbol{v}}_{n(1)}$ in Fig. 6.6. However, there are two less common prediction mechanisms as follows:

$$\tilde{\boldsymbol{v}}_n = 2\hat{\boldsymbol{v}}_{n-2} - \hat{\boldsymbol{v}}_{n-3}, \tag{6.29}$$

and

$$\tilde{\boldsymbol{v}}_n = 2\hat{\boldsymbol{v}}_{n-1} - \hat{\boldsymbol{v}}_{n-3}, \tag{6.30}$$

which correspond to $\tilde{\boldsymbol{v}}_{n(2)}$ and $\tilde{\boldsymbol{v}}_{n(3)}$ in Fig. 6.6, respectively. During the encoding process, we employ the mechanism Eq.(6.28). The residual $\boldsymbol{e}_n = \boldsymbol{v}_n - \tilde{\boldsymbol{v}}_n$

is quantized, resulting in $\hat{\boldsymbol{e}}_n$ and its corresponding codevector index i_n. Consequently, the vertex $\boldsymbol{v}_n$ is approximated by the quantized vertex $\hat{\boldsymbol{v}}_n$ as follows:

$$\hat{\boldsymbol{v}}_n = \tilde{\boldsymbol{v}}_n + \hat{\boldsymbol{e}}_n. \tag{6.31}$$

In this work, 42507 training vectors were randomly selected from the famous Princeton 3D mesh library [67] for training the approximate universal codebook off-line.

6.5.3 Data Embedding

The payload is embedded by modifying the prediction mechanism. In order to ensure reversibility, we should select specific vertices as candidates. Let

$$D = \min\{\|\tilde{\boldsymbol{v}}_{n(2)} - \hat{\boldsymbol{v}}_n\|_2, \|\tilde{\boldsymbol{v}}_{n(3)} - \hat{\boldsymbol{v}}_n\|_2\}. \tag{6.32}$$

Then we select an appropriate parameter α (0< α <1), which is used for payload capacity control. $\hat{\boldsymbol{v}}_n$ can be embedded with a bit of payload when it satisfies the following condition:

$$\|\tilde{\boldsymbol{v}}_{n(1)} - \hat{\boldsymbol{v}}_n\|_2 < \alpha \times D. \tag{6.33}$$

Under the above condition, if the payload bit is "0", we maintain the codeword index unchanged. Otherwise, if the payload bit is "1", we make a further judgment as follows.

Firstly, the nearer vertex to $\hat{\boldsymbol{v}}_n$ out of $\tilde{\boldsymbol{v}}_{n(2)}$ and $\tilde{\boldsymbol{v}}_{n(3)}$ is adopted as the new prediction of $\boldsymbol{v}_n$. For example, in Fig. 6.6, the new prediction of $\boldsymbol{v}_n$ is $\tilde{\boldsymbol{v}}_{n(2)}$, thus we quantize the residual vector $\boldsymbol{e}'_n$ as follows:

$$\boldsymbol{e}'_n = \hat{\boldsymbol{v}}_n - \tilde{\boldsymbol{v}}_{n(2)}. \tag{6.34}$$

The quantized residual vector $\hat{\boldsymbol{e}}'_n$ and its corresponding codeword vector i'_n are acquired by matching the codebook. Thus, the new quantized vector is

$$\hat{\boldsymbol{v}}'_n = \tilde{\boldsymbol{v}}_{n(2)} + \hat{\boldsymbol{e}}'_n. \tag{6.35}$$

Then we compute a temporary vector $\hat{\boldsymbol{v}}''_n$ as follows:

$$\hat{\boldsymbol{v}}_n'' = Q[\hat{\boldsymbol{v}}_n' - \tilde{\boldsymbol{v}}_{n(1)}] + \tilde{\boldsymbol{v}}_{n(1)}, \tag{6.36}$$

where $Q[\cdot]$ is the VQ operation. If the following condition is satisfied

$$\hat{\boldsymbol{v}}_n'' = \hat{\boldsymbol{v}}_n, \tag{6.37}$$

i.e., the reconstructed vector after the change of prediction mechanisms can be exactly restored to the original reconstructed vector before embedding, $\hat{\boldsymbol{v}}_n$ can be embedded with the payload bit "1". In this situation, we replace the codeword index of $\hat{\boldsymbol{e}}_n$ with i_n', while $\hat{\boldsymbol{v}}_n$ remains unchanged.

The payload bit "1" cannot be embedded even when Eq.(6.33) and Eq.(6.37) are satisfied in the unlikely case as follows: the nearest vertex to $\hat{\boldsymbol{v}}_n'$ out of $\tilde{\boldsymbol{v}}_{n(1)}$, $\tilde{\boldsymbol{v}}_{n(2)}$ and $\tilde{\boldsymbol{v}}_{n(3)}$ is not $\tilde{\boldsymbol{v}}_{n(1)}$. This case can be avoided by reducing α or increasing the size of the codebook to achieve a better quantization precision.

One flag bit of the side information is required to indicate whether a vertex is embedded with a payload bit or not. In this work, the bit "1" indicates that the vertex is embedded with a payload bit while "0" indicates not.

6.5.4 *Data Extraction and Mesh Recovery*

When the flag bit is "1", we find the residual vector by table lookup operations in the codebook. Then we compute a temporary vector $\boldsymbol{x}_n$ by subtracting the residual vector from $\tilde{\boldsymbol{v}}_{n(1)}$. It can be easily deduced from the payload embedding process that if the nearest vector to $\boldsymbol{x}_n$ out of $\tilde{\boldsymbol{v}}_{n(1)}$, $\tilde{\boldsymbol{v}}_{n(2)}$ and $\tilde{\boldsymbol{v}}_{n(3)}$ is $\tilde{\boldsymbol{v}}_{n(1)}$, the embedded payload bit is "0"; otherwise, the embedded payload bit is "1". Whenever Eq. (6.37) is not satisfied during the decoding process, we terminate the procedure because the stego mesh must have been tampered with and is certainly unauthorized.

When the flag bit is "1", the nearest vector to $\boldsymbol{x}_n$ out of $\tilde{\boldsymbol{v}}_{n(1)}$, $\tilde{\boldsymbol{v}}_{n(2)}$ and $\tilde{\boldsymbol{v}}_{n(3)}$ is obviously the prediction of $\hat{\boldsymbol{v}}_n'$. $\hat{\boldsymbol{v}}_n'$ is computed by adding its prediction and $\hat{\boldsymbol{e}}_n'$. Then we can easily acquire $\hat{\boldsymbol{v}}_n$ based on Eqs.(6.36) and (6.37). After all vertices have been restored to their original values, the restored mesh is acquired.

6.5.5 *Performance Analysis*

There is a bit-error rate when the VQ-compressed codeword indices are transmitted in a noisy channel, due to malicious attacks or low channel performances. A wrong index results in a distortion of its corresponding reconstructed vector. Because the embedded payload bits are judged by the nearest vector to $\boldsymbol{x}_n$ out of the three predictions, a distortion within a certain range can be tolerated. We use the following model to simulate the channel noise effect on indices:

$$\boldsymbol{e}_i^* = \hat{\boldsymbol{e}}_i + \beta \times \|\hat{\boldsymbol{e}}_i\|_2 \times N_i, \tag{6.38}$$

where $\hat{\boldsymbol{e}}_i$ is the residual vector specified by its index in the VQ bitstream, N_i is i-th value of a zero-mean Gaussian noise sequence with a standard deviation of 1.0, β is the parameter to control the noise intensity and $\boldsymbol{e}_i^*$ is the noise distorted vector. After requantizing $\boldsymbol{e}_i^*$, we can get its newly quantized version $\boldsymbol{e}_i^*$. If βN_i is very small, maybe $\hat{\boldsymbol{e}}_i^* = \hat{\boldsymbol{e}}_i$, then the corresponding index is unchanged; otherwise, $\hat{\boldsymbol{e}}_i^* \neq \hat{\boldsymbol{e}}_i$, the index is changed, but they are close and thus the watermark bit may be also correctly extracted. Based on this, the proposed method is robust to noise attack. Besides, attacks on mesh topology such as mesh simplification, re-sampling or insection are not available because the geometric coordinates and topology of the mesh are unknown before the VQ bitstream is decoded.

6.5.6 Experimental Results

To evaluate the effectiveness of the proposed methods, we first adopt the 3D Shark and Chessman meshes as the experimental objects. The Shark mesh consists of 1,583 vertices and 3,164 faces while the Chessman mesh consists of 802 vertices and 1,600 faces. First, we quantize the original mesh $\boldsymbol{M}_0$ to acquire the cover mesh $\boldsymbol{M}$ with a universal codebook consisting of 8,192 codewords. The PSNR values between $\boldsymbol{M}_0$ and $\boldsymbol{M}$ are 47.90 dB and 47.85 dB, for Shark and Chessman, respectively. Here, PSNR values are computed between the restored meshes $\boldsymbol{M}''$ and the original quantized meshes $\boldsymbol{M}$ as

$$PSNR = 10 \times \log \frac{B}{\sum_{i=1}^{B} \|\boldsymbol{v}_i' - \boldsymbol{v}_i\|_2^2},$$

where B is the number of vertices of $\boldsymbol{M}$, $\boldsymbol{v}'_i$ and $\boldsymbol{v}_i$ are the i-th vertex of $\boldsymbol{M}''$ and $\boldsymbol{M}$, respectively, and all the vertices in $\boldsymbol{M}$ are previously normalized in a zero mean sphere with a radius of 1.0. A higher PSNR value is considered as better quality. The PSNR values can be further improved by many other sophisticated VQ encoding techniques, which are not what we aim at in this work. In fact, when the codebook is generated by the cover mesh itself and the codebook size is the same as the number of VQ quantized vertices of the mesh, the PSNR may be ∞, i.e. in this case the proposed reversible authentication scheme will be perfect.

As shown in Table 6.2 and Table 6.3, with α increasing, the embedding capacities in Shark and Chessman increase while the correlation values between the extracted payloads and the original ones remain as 1.0, with β set to be 0.005. Data in Table 6.4 and Table 6.5 indicate the robustness performances for Shark and Chessman with α set to be 0.8. Here, the capacity is represented by the ratio of payload capacity to the number of mesh vertices. From the above results, we can see that the proposed scheme is effective.

Table 6.2 Capacity and robustness values for Shark with different α (β= 0.005)

α	Capacity	Correlation
0.2	0.004	1.00
0.3	0.016	1.00
0.4	0.027	1.00
0.5	0.049	1.00
0.6	0.078	1.00
0.7	0.110	1.00
0.8	0.134	1.00
0.9	0.166	1.00
1.0	0.209	1.00

Table 6.3 Capacity and robustness values for Chessman with different α (β= 0.005)

α	Capacity	Correlation
0.2	0.002	1.00
0.3	0.022	1.00
0.4	0.060	1.00
0.5	0.091	1.00
0.6	0.145	1.00
0.7	0.171	1.00
0.8	0.198	1.00
0.9	0.219	1.00
1.0	0.243	1.00

Table 6.4 PSNR and robustness values for Shark with different β (α= 0.8)

β	PSNR	Correlation
0.001	∞	1.00
0.002	∞	1.00
0.005	∞	1.00
0.10	48.63	0.99
0.20	25.02	0.84

Table 6.5 PSNR and robustness values for Chessman with different β (α= 0.8)

β	PSNR	Correlation
0.001	∞	1.00
0.002	∞	1.00
0.005	∞	1.00
0.10	42.19	1.00
0.20	23.62	0.94

6.5.7 Capacity Enhancement

Although the data hiding scheme [44] is very robust to zero-mean Gaussian noise in a noise channel, the main drawback of this algorithm, however, is that the capacity for data hiding is not high. To evaluate the capacity enhancement performance, 20 meshes were randomly selected from the famous Princeton 3D mesh library [67] and 42507 training vectors were generated from these meshes for training the approximate universal codebook off-line. The residual vectors are then used to generate the codebook based on the minimax partial distortion competitive learning (MMPDCL) method [68] for optimal codebook design. In this way, we expect the codebook to be suitable for nearly all triangle meshes for VQ compression and can be pre-stored in each terminal in the network [45]. Thus the compressed bitstream can be transmitted alone with convenience. The improvement in [45] over [44] can be illustrated as follows.

6.5.7.1 Data Embedding

The payload is hidden by modifying the prediction mechanism. In order to ensure reversibility, we should select specific vertices as candidates.

Let

$$D = \left\| \tilde{\boldsymbol{v}}_{n(2)} - \hat{\boldsymbol{v}}_n \right\|_2 . \tag{6.39}$$

Then we select an appropriate parameter α ($0 < \alpha < 1$), which is used for payload capacity control. $\hat{\boldsymbol{v}}_n$ can be hidden with a bit of payload when it satisfies the following condition

$$\left\| \tilde{\boldsymbol{v}}_{n(1)} - \hat{\boldsymbol{v}}_n \right\|_2 < \alpha \times D. \tag{6.40}$$

Under the above condition, if the payload bit is "0", we maintain the codeword index unchanged. Otherwise, if the payload bit is "1", we should make a further judgment as follows.

$\tilde{\boldsymbol{v}}_{n(2)}$ is adopted as the new prediction of $\boldsymbol{v}_n$. Thus we quantize the residual vector $\boldsymbol{e}'_n$ as follows:

$$\hat{\boldsymbol{e}}'_n = Q[\boldsymbol{e}'_n] = Q[\hat{\boldsymbol{v}}_n - \tilde{\boldsymbol{v}}_{n(2)}]. \tag{6.41}$$

The quantized residual vector $\hat{\boldsymbol{e}}'_n$ and its corresponding codeword index i'_n are acquired by matching the codebook. Thus, the new quantized vector is

$$\hat{\boldsymbol{v}}'_n = \tilde{\boldsymbol{v}}_{n(2)} + \hat{\boldsymbol{e}}'_n . \tag{6.42}$$

Then we compute a temporary vector $\hat{\boldsymbol{v}}''_n$ as follows:

$$\hat{\boldsymbol{v}}''_n = Q[\hat{\boldsymbol{v}}'_n - \tilde{\boldsymbol{v}}_{n(1)}] + \tilde{\boldsymbol{v}}_{n(1)} . \tag{6.43}$$

If the following condition is satisfied,

$$\hat{\boldsymbol{v}}''_n = \hat{\boldsymbol{v}}_n . \tag{6.44}$$

In other words, the reconstructed vector after the change of prediction mechanisms can be exactly restored to the original reconstructed vector before embedding, $\hat{\boldsymbol{v}}_n$ can be hidden with the payload bit "1". In this situation, we replace the codeword index of $\hat{\boldsymbol{e}}_n$ with i'_n, while $\hat{\boldsymbol{v}}_n$ remains unchanged.

The payload bit "1" cannot be hidden even when Eq.(6.40) and Eq.(6.44) are satisfied in the unlikely case as follows: the nearest vertex to the vector $\hat{\boldsymbol{v}}_n - \hat{\boldsymbol{e}}'_n$ out of $\tilde{\boldsymbol{v}}_{n(1)}$ and $\tilde{\boldsymbol{v}}_{n(2)}$ is not $\tilde{\boldsymbol{v}}_{n(2)}$. This case can be avoided by reducing the value of α or increasing the size of the codebook to achieve a better quantization precision. When the payload bit "1" cannot be hidden, proceed to the next vertex until the bit satisfies the hiding conditions.

One flag bit of the side information is required to indicate if a vertex is hidden with a payload bit. In this work, the bit "1" indicates that the vertex is hidden with a payload bit while "0" indicates not. The vertex order in the payload embedding process is the same as for the VQ quantization process.

6.5.7.2 Data Extraction and Mesh Recovery

When the flag bit is "1", we find the codevector specified by the received index by table lookup operations in the codebook. Then we compute a temporary vector x_n by subtracting the codevector, $\hat{\boldsymbol{e}}_n$ or $\hat{\boldsymbol{e}}'_n$ from $\hat{\boldsymbol{v}}_n$. It can be easily deduced from the payload hiding process that if the nearest vector to $\boldsymbol{x}_n$ out of $\tilde{\boldsymbol{v}}_{n(1)}$ and $\tilde{\boldsymbol{v}}_{n(2)}$ is $\tilde{\boldsymbol{v}}_{n(1)}$, the hidden payload bit is "0"; otherwise, the hidden payload bit is "1".

Whenever Eq.(6.44) is not satisfied during the decoding process, we terminate the procedure because the mesh bitstream must have been tampered with and is certainly unauthorized. Thus, if a mesh bitstream is tampered with, the decoding process cannot be completed in most cases.

When the hidden payload bit is judged to be "1", $\hat{\boldsymbol{v}}'_n$ is computed by adding $\tilde{\boldsymbol{v}}_{n(2)}$ and $\hat{\boldsymbol{e}}'_n$. Then we can easily acquire $\hat{\boldsymbol{v}}_n$ according to Eqs.(6.43) and (6.44). When the hidden payload bit is judged to be "0", no operation is needed.

After all vertices have been restored to their original values, the restored mesh $\boldsymbol{M}''$ in its uncompressed form is acquired. For content authentication, we compare the authentication hash hidden in the bitstream with the hash of $\boldsymbol{M}''$. If they match exactly, then the mesh content is authentic and the restored mesh is exactly the same as the cover mesh $\boldsymbol{M}$. Most likely a tampered mesh will not go through to this step because some decoding error could happen, as mentioned, in the payload extraction process. We reconstruct a restored mesh first, and then authenticate the content of the stego mesh.

The capacity bottleneck is to satisfy Eq.(6.44), which is the same as that in [44]. In [44], two other uncommon prediction rules are used other than the parallelogram prediction. When the payload bit "1" is embedded, one of the two uncommon prediction rules is used, resulting in a large residual vector, so the vector quantization error is large. As a result, Eq.(6.44) is not likely to be satisfied in [44]. In the work [45], both $\hat{e}_n$ and $\hat{e}_n'$ are small, so a small vector quantization error ought to be expected and thus Eq.(6.44) is more likely to be satisfied. As a result, a high capacity of payload hiding can be achieved.

Attacks on mesh topology such as mesh simplification, re-sampling or insection are not available because the geometric coordinates and topology of the mesh are unknown before the VQ bitstream is decoded.

Residual vectors are kept small after the payload hiding process, so the statistical characteristic of the bitstream does not change much. Thus, one cannot judge whether a codeword index corresponds to a payload bit by simply observing it. Instead, the payload can only be extracted by the payload extraction algorithm. The flag bits in the bitstream can be shuffled with a secure key. In this sense, the payload is imperceptible.

Any small change to the authenticated mesh will be detected with a high probability because the chances of obtaining a match between the calculated mesh hash and the extracted hash are equal to finding a collision for the hash.

In addition, in order to reduce the encoding time of VQ, we adopt the mean-distance-ordered partial codebook search (MPS) [69] as an efficient fast codevector search algorithm, which uses the mean of the input vector to dramatically reduce the computational burden of the full search algorithm without sacrificing performance.

To evaluate the effectiveness of the proposed method in [45], we first adopt the 8 meshes as the experimental objects. First, we quantize the original mesh $\boldsymbol{M}_0$ to acquire the cover mesh $\boldsymbol{M}$ with a universal codebook consisting of 8,192 codewords. The PSNR values between $\boldsymbol{M}_0$ and $\boldsymbol{M}$ are 50.99 dB and 56.40 dB, for Stanford Bunny and Dragon meshes, respectively. The PSNR values can be further improved by many other sophisticated VQ encoding techniques that are not what we aim at in this work.

$\boldsymbol{M}_0$, $\boldsymbol{M}$ and the restored meshes $\boldsymbol{M}''$ for Bunny and Dragon are shown in Fig. 6.7. Comparing these meshes visually, we can know that there are no significant differences among the Bunny meshes and the Dragon meshes. Other original meshes used here are depicted in Fig. 6.8.

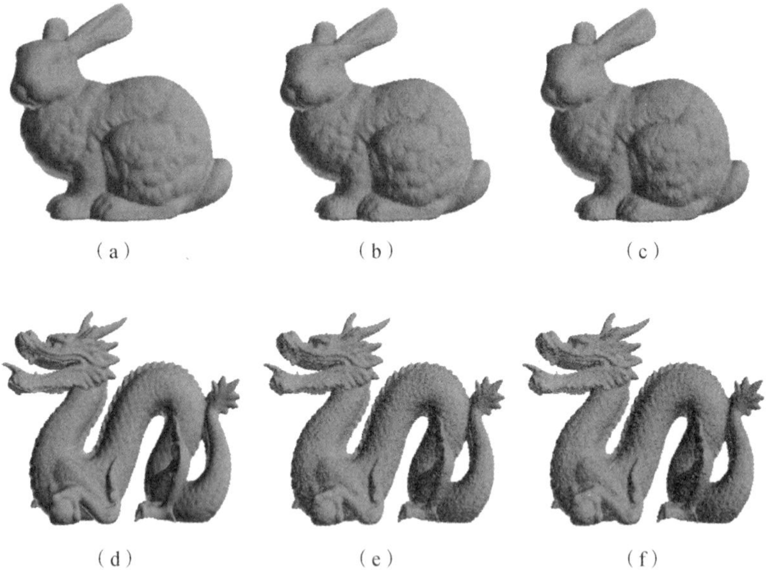

Fig. 6.7. Comparisons of rendered meshes (implemented with OpenGL). (a) Original Bunny mesh; (b) Cover Bunny mesh; (c) Restored Bunny mesh; (d) Original Dragon mesh; (e) Cover Dragon mesh; (f) Restored Dragon mesh

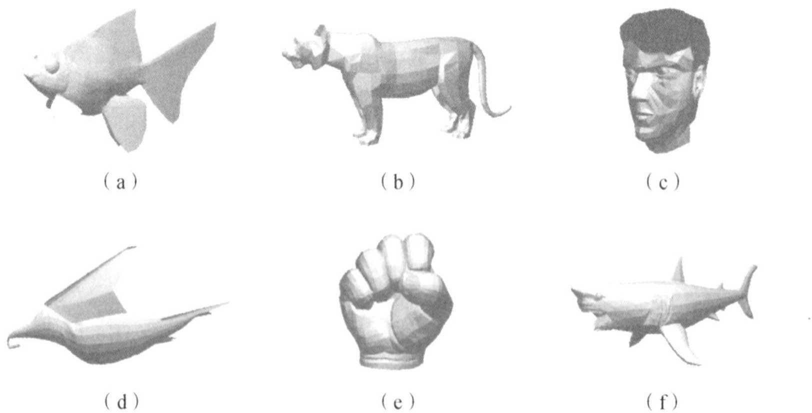

Fig. 6.8. Other original meshes (implemented with OpenGL). (a) Goldfish; (b) Tiger; (c) Head; (d) Dove; (e) Fist; (f) Shark

Table 6.6 lists PSNR values of the vector quantized meshes and numbers of their vertices and faces. As shown in Table 6.7, with α increasing, the embedding capacities for various meshes increase while the correlation values between the extracted payloads and the original ones remain as 1.0. Each capacity in all tables is represented by the ratio of hidden payload bits to the number of mesh vertices. Evident in Table 6.7, the capacity for each mesh is as high as about 0.5, except for the Dragon model. This is because the Dragon model has very high definition and the prediction error vectors are of small norm compared to the codevectors in the universal codebook. Payload in this case can be increased by using a larger codebook that contains enough small codevectors. The payload of the proposed data hiding method is about 2 to 3 times the capacity reported in [44].

Table 6.6 PSNR values of the vector quantized meshes and numbers of their vertices and faces

Mesh	PSNR (dB)	Numbers of vertices	Numbers of faces
Bunny	50.99	8,171	16,301
Dragon	56.40	100,250	202,520
Goldfish	41.15	1,004	1,930
Tiger	44.19	956	1,908
Head	42.31	1,543	2,688
Dove	39.33	649	1,156
Fist	38.82	1,198	2,392
Shark	47.90	1,583	3,164

Table 6.7. Capacity values for various meshes with different α

Mesh	Capacity					
	α=0.5	α=0.6	α=0.7	α=0.8	α=0.9	α=1.0
Bunny	0.06	0.12	0.21	0.30	0.38	0.47
Dragon	0.04	0.06	0.09	0.12	0.15	0.21
Goldfish	0.11	0.17	0.27	0.36	0.42	0.50
Tiger	0.10	0.15	0.24	0.33	0.40	0.48
Head	0.12	0.18	0.25	0.32	0.40	0.51
Dove	0.12	0.15	0.19	0.27	0.33	0.42
Fist	0.12	0.19	0.25	0.32	0.42	0.50
Shark	0.15	0.22	0.30	0.39	0.45	0.51

6.6 Transform Domain Reversible 3D Model Data Hiding

In this section, we introduce a reversible data hiding scheme for a 3D point cloud model proposed in [40] by the authors of this book. This method exploits the high correlation among neighboring vertices to embed data. It starts with creating a set of 8-neighbor vertices clusters with randomly selected seed vertices. Then an 8-point integer DCT is performed on these clusters and an efficient highest frequency coefficient modification technique in the integer DCT domain is employed to modulate the watermark bit. After that, the modified coefficients are inversely transformed into coordinates in the spatial domain. In data extraction, we need to recreate the modified clusters first, and other operations are the inverse process of the data hiding. The original model can be perfectly recovered using the clusters information if it is intact. This technique is suitable for some specific applications where content accuracy of the original model must be guaranteed. Moreover, the method can be easily extended to 3D point cloud model authentication. The following is the detailed description of our scheme.

6.6.1 Introduction

In recent years, 3D point cloud models have gained the status of one of the mainstream 3D shape representations. Point cloud is a set of vertices in a 3D coordinate system. These vertices are usually defined by *X*, *Y* and *Z* coordinates. Compared to a polygonal mesh representation, a point set representation has the advantage of being lightweight to store and transmit, due to its lack of connectivity information. Point clouds are most often created by 3D scanners. These devices measure a large number of points on the surface of an object and output a point cloud as a data file. The point cloud represents the visible surface of the object that has been scanned or digitized. Point clouds are used for many purposes, such as creating 3D CAD models for manufactured parts, metrology/quality inspection, and a multitude of visualization, animation, rendering and mass customization applications. Point clouds themselves are generally not directly usable in most 3D applications, and therefore are usually converted to triangle mesh models, NURBS surface models, or CAD models through a process commonly referred to as reverse engineering, so that they can be used for various purposes. Techniques for converting a point cloud to a polygon mesh include Delaunay triangulation and more recent techniques such as Marching triangles, Marching cubes, and the Ball-Pivoting algorithm. One application in which point clouds are directly usable is industrial metrology or inspection. The point cloud of a manufactured part can be aligned to a CAD model (or even another point cloud) and compared to check for differences. These differences can be displayed as color maps that give a visual indicator of the deviation between the manufactured part and the CAD model. Geometric dimensions and tolerances can also be extracted directly from the point cloud. Point clouds can also be used to represent volumetric data used for example in medical imaging. Using point clouds multi-sampling and data compression are achieved.

Nowadays, most existing data hiding methods are for 3D mesh models. However, fewer approaches for 3D point cloud models have been developed. In [70], Wang et al. proposed two spatial-domain-based methods to hide data in point cloud models. In both schemes, principal component analysis (PCA) is applied to translate the points' coordinates to a new coordinate system. In the first scheme, a list of intervals for each axis is established according to the secret key. Then a secret bit is embedded into each interval by changing the points' position. In the second scheme, a list of macro embedding primitives (MEPs) is located, and then multiple secret bits are embedded in each MEP. Blind extraction is achieved in both of the schemes, and robustness against translation, rotation and scaling is demonstrated. In addition, these schemes are fast and can achieve high data capacity with insignificant visual distortion in the marked models.

A great deal of the existing data hiding process usually introduces irreversible degradation to the original medium. Although slight, it may not be acceptable in some applications where content accuracy of the original model must be guaranteed, e.g. a medical model. Hence there is a need for reversible data hiding.

In our context, reversibility refers to the ability to recover the original model in data extraction. Actually, it is advantageous to recover the original model from its watermarked version for the distortion introduced by the data hiding can be compensated. However, up until now, there has been little attention paid to reversible data-hiding techniques for 3D point cloud models.

The original idea of our method is attributed to the high correlation among neighboring vertices. It is well known that the discrete cosine transform (DCT) exhibits high efficiency in energy compaction of highly correlated data. For high correlated data, higher frequency is associated with smaller amplitude of the coefficient in the statistics. Usually, the first harmonic coefficient is larger than the last one and this fact is the basic principle of our reversible data hiding scheme. However, due to the finite representation of numbers in the computer, floating-point DCT is sometimes not reversible and therefore not able to guarantee the reversibility of the data hiding process.

In this research, we employ an 8-point integer-to-integer DCT, exhibiting similar energy compacting property and ensuring the perfect recovery of the original data in data extraction. First, some vertices clusters are chosen as the entry of integer DCT, then the 8-point integer DCT is performed on these clusters and an efficient highest frequency coefficient modification technique is used to modulate the data bit. After modulation, the inverse integer DCT is used to transform the modified coefficients into spatial coordinates. In data extraction, we need to recreate the modified clusters first, and subsequent procedures are the inverse process of data hiding.

6.6.2 Scheme Overview

Most existing data hiding methods are for 3D polygonal mesh models. 3D polygonal meshes consist of coordinates of vertices and their connectivity information. As we know, these methods can be roughly divided into two categories: spatial domain based and transform domain based. Approaches based on spatial domain directly modify either the vertex coordinates or the connectivity, or both, to embed data. Ohbuchi et al. [71-74] presented a sequence of watermarking algorithms for polygonal meshes. However, their approaches are not robust enough to be used for copyright protection. In [75] Benedens developed a robust watermarking for copyright protection. Nevertheless, this method requires a significant amount of data for decoding and is therefore not suitable for public data hiding. Yeo et al. [76] introduced a fragile watermarking for 3D objects verification. Wagner [77] presented two variations of a robust watermarking method for general polygonal meshes of arbitrary topology. In contrast, relatively fewer techniques based on transform domain have been developed. Praun et al. [78] introduced a watermarking scheme based on wavelet encoding. This method requires a registration procedure for decoding and is also not public. Ohbuchi et al. [79] also developed a frequency-domain approach employing mesh spectral analysis to modify mesh shapes.

However, there exists little work on data-hiding 3D point cloud models. Ohbuchi et al. [80] proposed a method for a 3D point set. In fact, it needs to construct a non-manifold mesh from the point set using mesh spectral analysis. Data is hidden based on the connectivity information. In data extraction, the mesh must also be recreated first. In contrast, our method is a pure data-hiding scheme without using any connectivity information.

Popular 3D models have many kinds of representations such as solid models, polygonal meshes and point clouds. A 3D point cloud model is just a bunch of points sampled on the model surface in the 3D space. Different from the method in [80], our method embeds data in 3D point cloud models by modifying the vertex coordinates without employing connectivity information. This research takes the same 8-point integer DCT to shape modification as the 2D vector data hiding algorithm reported in [81].

The data embedding and extraction can be summarized below. Actually, data extraction is the inverse process of data hiding, with nothing but the clusters' information required.

6.6.2.1 Data Embedding

(1) Use a pseudo-random number generator to attain a set of non-repeating seed vertices.

(2) Create disjoint clusters with the seed vertices. Use a secret key K to permute the clusters information and it is stored.

(3) Perform forward 8-point integer DCT to these clusters.

(4) Modulate the AC_7 coefficients of clusters according to the watermark bit.

(5) Perform inverse 8-point integer DCT to the modified clusters; meanwhile the watermarked model is obtained.

6.6.2.2 Data Extraction

(1) Use the key K to retrieve the clusters information, furthermore the modified clusters.

(2) Perform forward 8-point integer DCT on the modified clusters.

(3) Demodulate the AC_7 coefficients of clusters and extract the embedded data sequence.

(4) Perform inverse 8-point integer DCT on the restored clusters and the recovered model is obtained.

The block diagram of data embedding and the extraction process is as shown in Fig. 6.9. Details are illustrated in the next sections.

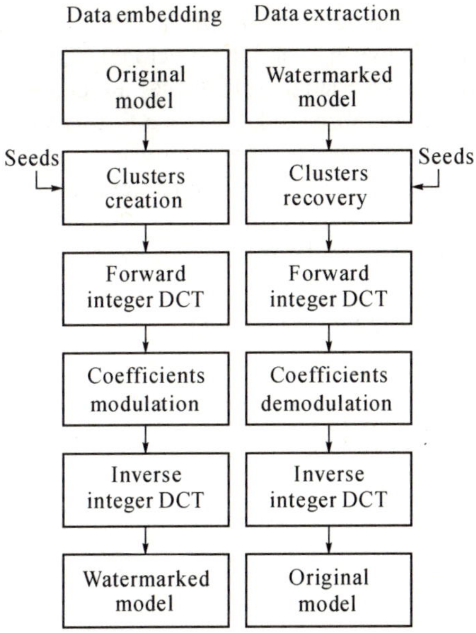

Fig. 6.9. Block diagram of data embedding and extraction

6.6.3 Data Embedding

Suppose the cover model $\boldsymbol{M}$ has n vertices $V = \{\boldsymbol{v}_1, \boldsymbol{v}_2, \ldots, \boldsymbol{v}_n\}$ with 3D space coordinates $\boldsymbol{v}_i = (x_i, y_i, z_i)$ $(1 \le i \le n)$.

6.6.3.1 Selection of Seeds

To a 3D point cloud model, firstly we use a pseudo-random number generator to select a set of non-repeating seed vertices $S = \{s_1, s_2, \ldots, s_m\}$. In our case, each cluster contains 8 vertices and, obviously, the total number of the seeds m must satisfy Eq.(6.45).

$$m \le \left\lfloor \frac{n}{8} \right\rfloor. \tag{6.45}$$

6.6.3.2 Clustering

This step aims to select appropriate point sets as the target of data hiding. As an example shown in Fig. 6.10, a point cluster consists of a given seed s_j $(1 \le j \le m)$ and its 7 nearest neighbor vertices $N_1, N_2, \ldots, N_7$, with distance to s_j ranking in an ascending order. The clustering starts from the first seed s_1, and 3D Euclidian distances are calculated between s_1 and the other $n-1$ vertices. Then the nearest 7 vertices corresponding to the 7 smallest distances are chosen and a cluster with 8 vertices including the seed is formed. Now move to s_2, its nearest 7 points can be chosen according to $n-9$ distances, except the visited points in the first cluster. Such operations are repeated for all seeds and j clusters are created. Generally, suppose d_l denotes the number of distances of s_l needing to be computed. Apparently it can be estimated by Eq.(6.46).

$$d_l = n - 8l + 7 \quad (1 \le l \le j). \tag{6.46}$$

The clusters' information must be saved for data extraction. In our approach, it refers to the indices of the vertices of all clusters. A secret key K is used to permute the index information.

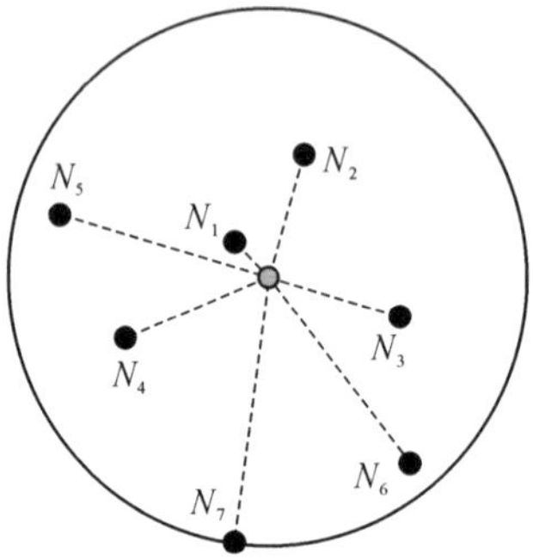

Fig. 6.10. An example of a cluster

6.6.3.3 Forward Integer DCT

To highly correlated data, the DCT's energy compacting property results in large values of the first harmonics. Once the point cloud model is clustered, we apply the 8-point integer-to-integer DCT introduced in [82] to all clusters. To each cluster, coordinates of 8 vertices are input in the following order: The seed is the first entry, and other vertices coordinates are successively input as the distance to the seed grows. Let us take the example in Fig. 6.10, the input sequence is s_j, N_1, N_2, …, N_7. In this way, 8 DCT coefficients, DC and AC_1, AC_2, …, AC_7, can be acquired from a cluster.

6.6.3.4 Modulation of Coefficients

Since a cluster has x, y and z coordinate sets, it has three sets of DCT coefficients. Here we only take the example of coefficients associated with the x-coordinates to demonstrate data embedding and extraction. The operation on the other two sets of coefficients is similar.

It is reasonable to suppose that, in most cases, the magnitude of the highest frequency coefficient AC_7 is quite small, which is smaller than the largest magnitude among $|AC_1|$, $|AC_2|$, …, $|AC_6|$, as long as the 8 neighboring vertices are relatively closely distributed and thus highly correlated. That is to say, in most cases, the results of the 8-point integer DCT should satisfy Eq.(6.47):

$$|AC_7|<|AC_{\max}|, \tag{6.47}$$

where $|AC_{\max}|$ is the maximum magnitude among $|AC_i|$ ($i = 1, 2, …, 6$). All clusters in the DCT domain can be divided into two categories according to Eq.(6.47). If it is satisfied, the cluster is a normal cluster NC, otherwise an exceptional cluster EC. A NC can be used to embed data, while an EC cannot. In data embedding, if the cluster is an EC, then the coefficients are modified as Eq.(6.48). This operation can be regarded as magnitude superposition.

$$AC_7' = \begin{cases} AC_7+|AC_{\max}|, & \text{if } AC_7>0; \\ AC_7-|AC_{\max}|, & \text{if } AC_7<0. \end{cases} \tag{6.48}$$

To an NC, data is hidden by the following rule: when embedding "0", coefficients of the cluster are kept unchanged; when embedding "1", the coefficients are modified in the way described in Eq.(6.49):

$$AC_7' = \begin{cases} AC_7+|AC_{\max}|, & \text{if } AC_7>0; \\ AC_{\max}, & \text{if } AC_7=0; \\ AC_7-|AC_{\max}|, & \text{if } AC_7<0. \end{cases} \tag{6.49}$$

In this way, data is inserted into the clusters, namely the point cloud model. It is clear that Eq.(6.50) is satisfied for all modified clusters:

$$\begin{cases} |AC_7'|=|AC_7|+|AC_{\max}|\geq|AC_{\max}|; \\ AC_7\cdot AC_7'\geq 0. \end{cases} \tag{6.50}$$

In a word, to embed data, we add $AC_{\max}$ to AC_7 to modulate the data bit "1", and keep all coefficients unchanged to modulate the data bit "0". Obviously, the modified AC_7' no longer satisfies Eq.(6.47), and thus a new exceptional cluster occurs. We regard it as an artificial exceptional cluster AE.

6.6.3.5 Inverse Integer DCT

After coefficient modulation, the last step is to perform the inverse 8-point integer DCT on all clusters and the watermarked model is obtained.

6.6.4 Data Extraction

Data extraction includes four steps, i.e. cluster recovery, formard integer DCT, coefficient demodulation and inverse integer DCT.

6.6.4.1 Cluster Recovery

The clusters must be recovered first for further data extraction. The same key K and the cluster information are used to retrieve the index of vertices of all clusters, and the coordinates of these clusters, i.e. the modified clusters, are used as entries of the integer DCT.

6.6.4.2 Forward Integer DCT

This step is to perform forward 8-point integer DCT on the modified clusters. Meanwhile, each cluster is transformed into three sets of DCT coefficients.

6.6.4.3 Coefficient Demodulation

This step is the inverse process of the coefficient modulation in data embedding. We still take the example of coefficients corresponding x-coordinates of a cluster to describe the demodulation operation. After embedding data, clusters can be classified into three kinds of states: NC, EC and AE. These three categories are distinguished according to Eq.(6.51):

$$\begin{cases} \text{NC: } |AC_7'| \le |AC_{\max}|; \\ \text{EC: } |AC_7'| \ge 2|AC_{\max}|; \\ \text{AE: } |AC_{\max}| \le |AC_7'| \le 2|AC_{\max}|. \end{cases} \tag{6.51}$$

No data is inserted into EC. A bit "0" is inserted into an NC, while a bit "1" is inserted into an AE. Data extracted is as shown in Eq.(6.52), where W denotes the extracted watermark bit:

$$W = \begin{cases} 0, & \text{if} \quad \text{NC}; \\ 1, & \text{if} \quad \text{AE}. \end{cases} \tag{6.52}$$

The demodulation operation is as shown in Eq.(6.53):

$$\begin{cases} |AC_7| = |AC_7'|, & \text{if} \quad \text{NC}; \\ |AC_7| = |AC_7'| - |AC_{\max}|, & \text{if} \quad \text{EC or AE}. \end{cases} \tag{6.53}$$

6.6.4.4 Inverse Integer DCT

This step is to perform the inverse 8-point integer DCT on the demodulated coefficients, thus spatial coordinates of vertices are recovered. Namely, the original model is perfectly restored if it is intact.

6.6.5 Experimental Results

To test the performance and effectiveness of our scheme, a point cloud model, the Stanford Bunny with 34,835 vertices is selected as the test model, as shown in Fig. 6.11(c). The original data to be hidden is a 32×32 binary image "KUAS" as shown in Fig. 6.11(a). Experiment results show 502 clusters (i.e. 1,506 sets of coordinates) can be inserted into 1,024 bit data. In other words, in these clusters 1,024 sets of coordinates belong to NC, and the left 482 sets belong to EC. From Figs. 6.11(c) and 6.11(d), slight degradation is introduced to the visual quality of the original model. The recovered model is exactly the same as the original model if the watermarked model suffers no alteration. This can be verified as the Hausdorff distance between the original model and the recovered model is equal to 0. Although the original model is not required, our method is semi-blind, for the clusters information is required for data extraction.

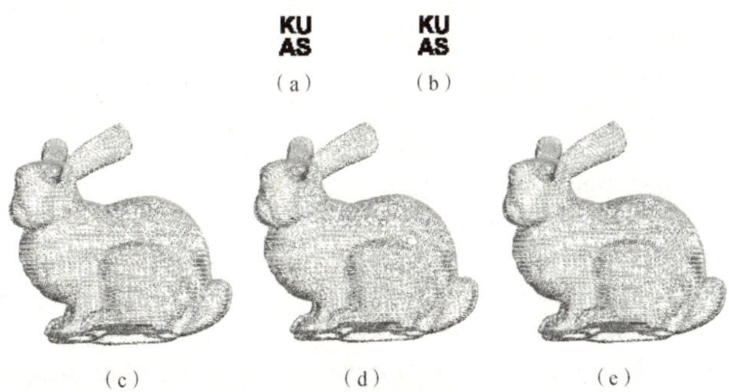

Fig. 6.11. Experimental results. (a) Original watermark; (b) Extracted watermark; (c) Original Bunny; (e) Watermarked Bunny; (e) Recovered Bunny

6.6.6 Bit-Shifting-Based Coefficients Modulation

In Subsection 6.6.3, the coefficients modulation is based on a magnitude superposition strategy. In this subsection, another strategy for coefficients modulation is introduced, i.e., bit shifting.

To 7 AC coefficients of a cluster, two distinct parts P_1 and P_2 are selected, where P_1 is the range when we are looking for the maximum and P_2 is the modification area. The embedding procedure is: As long as a coefficient in P_2 is smaller than the largest coefficient $AC_{\max}$ in P_1, its value is doubled and the watermark bit is embedded. In the case where a coefficient of P_2 is larger than $AC_{\max}$ in P_1, $AC_{\max}$ is added to the coefficient. The embedding process can more formally be written as

$$AC_i' = \begin{cases} 2AC_i + W, & \text{if } AC_i \le AC_{\max} \\ AC_i + AC_{\max}, & \text{if } AC_i > AC_{\max} \end{cases}, \quad i \in P_2, \tag{6.54}$$

where W denotes the watermark bit and $AC_{max} = \max_{j \in P_1} AC_j$.

In the retrieving process we check if a coefficient out of P_2 is larger than $2AC_{\max}$ and, if so, we subtract $AC_{\max}$ from it to get the original coefficient. In the other case we know that a doubling has been performed during embedding and after reading the watermarking bit the coefficient is divided by two to get the original coefficient. Next, an improved scheme is proposed to further increase the capacity. There are also two ranges P_1 and P_2 among AC_1 to AC_6, instead of among AC_1 to AC_7 in the basic scheme. In the embedding procedure we have to first discriminate between a typical and a non-typical distribution of the AC coefficients. A distribution is defined as typical when the highest frequency coefficient AC_7 is lower than the largest component $AC_{\max}$ in P_1 and as a non-typical one if AC_7 is higher than $AC_{\max}$. Depending on the kind of distribution, a modification of the coefficients of region P_2 is performed or not performed. In the case of a typical distribution, the coefficients of region P_2 are shifted by 1 bit or 2 bits, depending on a certain threshold T. That means during embedding all coefficients which are smaller than a certain threshold are shifted by 2 bits, otherwise a 1-bit-shift is performed. In the retrieving process, the three cases (non-typical distribution, typical distribution (1-bit-shift) and typical distribution (2-bit-shift) are distinguished. In other words, we use the highest frequency component AC_7 to discriminate between the three cases. After coefficient modulation, the last step is to perform inversely the 8-point integer DCT on all clusters and the watermarked model is obtained.

In data extraction, the corresponding bit-shifting-based coefficient modulation is adopted. We still take the example of coefficients corresponding to x-coordinates of a cluster to describe the demodulation operation. The retrieving procedure can be clearly arranged as follows: First we find the $AC_{\max}$ in the range P_1 then, if the $AC'_7 > 2AC_{\max}$, an exceptional distribution is detected. If the $AC'_7 \le 2AC_{\max}$, we judge if the $AC'_7 > AC_{\max}$, and if so, a 1-bit-shift is detected, otherwise a 2-bit-shift. If a 1-bit-shift is detected, a 1-bit watermark can be extracted and, for a 2-bit-shift, a 2-bit watermark can be extracted. After demodulation of coefficients, the inverse 8-point integer DCT is performed on the demodulated coefficients, and thus spatial coordinates of vertices are recovered. Namely, the original model is perfectly restored if it is intact.

To test the performance and effectiveness of bit-shifting-based coefficient modulation, the point cloud model Stanford Bunny with 34,835 vertices is selected as the test model. Capacities with a different number of clusters are listed in Table 6.8, where $T = 2{,}000{,}000$.

Table 6.8 Capacities (bits) with different number of clusters

P_1; P_2	1; 2-6	1-2; 3-6	2-3; 4-6	3-4; 5-6	4-5; 6
100	168	223	262	274	308
200	360	487	552	585	636
300	570	748	844	898	985
400	781	1,018	1,152	1,224	1,342
500	982	1,290	1,457	1,537	1,663
600	1,185	1,552	1,739	1,826	1,982
700	1,361	1,819	2,034	2,125	2,318
800	1,537	2,063	2,324	2,429	2,641
900	1,707	2,308	2,598	2,707	2,964
1,000	1,918	2,601	2,905	3,021	3,305

6.7 Summary

First, this chapter is started by introducing the background and performance evaluation metrics of 3D model reversible data hiding. As many available 3D model reversible data hiding techniques come from ideas that complement digital image reversible data hiding schemes, some basic reversible data hiding schemes for digital images are briefly reviewed. With respect to 3D model reversible data hiding techniques, we first introduced a reversible watermarking algorithm for authentication of 3D meshes in the spatial domain. The experimental results have demonstrated that the proposed method is able to embed a considerable amount of information into the mesh. The embedded watermark can be extracted using some a priori knowledge so that the watermarked mesh can be authenticated by comparing the extracted watermark with the original one, additionally the recovered mesh centroid with the original mesh centroid. Therefore, modifications to the watermarked mesh can be efficiently detected. The original mesh model can be recovered by performing the reverse process of the watermark embedding if the watermarked mesh is intact. Future efforts are needed to realize the on-line applications of mesh authentication.

Second, a new invertible authentication scheme was introduced for 3D meshes based on a data hiding technique. The hidden payload has cryptographic strength and is global in the sense that it can detect every modification made to the mesh with a probability that is equivalent to finding a collision for a cryptographically secure hash function. This technique embeds the hash or some invariant features of the whole mesh as a payload. This method can be localized to blocks rather than applied to the whole mesh. In addition, it is argued that all typical meshes can be authenticated and this technique can be further generalized to other data types, e.g. 2D vector maps, arbitrary polygonal 3D meshes and 3D animations.

Third, a reversible data hiding scheme for a 3D point cloud model was presented. Its principle is to employ the high correlation among neighboring vertices to embed data, and an 8-point integer-to-integer DCT is applied to guarantee the reversibility. Two strategies of transform domain coefficient modulation/demodulation are introduced. Low distortion is introduced to the original model and it can be perfectly recovered if intact, using some prior knowledge.

Future work in 3D model reversible data hiding will involve further improving the capacity and robustness of the schemes.

References

[1] R. Ohbuchia, H. Masudab and M. Aonoa. Data embedding algorithms for geometrical and non-geometrical targets in three-dimensional polygonal models. Computer Communications, 1998, 21:1344-1354.

[2] E. E. Abdallah, A. B. Hamza and P. Bhattacharya. Robust 3D watermarking technique using eigendecomposition and nonnegative matrix factorization. Lecture Notes in Computer Science, 2008, Vol. 5112, pp. 253-262.

[3] O. Benedens. Watermarking of 3D polygonal based models with robustness against mesh simplification. In: Proc. SPIE Security and Watermarking of Multimedia, 1999, pp. 329-340.

[4] M. Corsini, F. Uccheddu, F. Bartolini, et al. 3D watermarking technology: visual quality aspects. VSMM, 2003, pp. 1-8.

[5] O. Benedens and C. Busch. Toward blind detection of robust watermarks in polygonal models. In: Proc. EUROGRAPHICS Comput. Graph. Forum, 2000, Vol. 19, pp. C199-C208.

[6] O. Benedens. Two high capacity methods for embedding public watermarks into 3D polygonal models. In: Proc. Multimedia and Security, 1999, pp. 95-99.

[7] R. Ohbuchi, A. Mukaiyama and S. Takahashi. A frequency domain approach to watermarking 3D shapes. Computer Graphics Forum, 2002, 21(3):373-382.

[8] R. Ohbuchi, S. Takahashi, T. Miyazawa, et al. Watermarking 3D polygonal meshes in the mesh spectral domain. In: Proceedings of Graphics Interface, 2001, pp.9-18.

[9] S. Kanai, H. Date and T. Kishinami. Digital watermarking for 3D polygons using multiresolution wavelet decomposition. In: Proceeding of the Sixth International Workshop on Geometric Modeling: Fundamentals and Applications, 1998, pp. 296-307.

[10] I. J. Cox, M. L. Millter, J. A. Bloom, et al. Digital Watermarking and Steganography (2nd ed.). Morgan Kaufmann, 2008.

[11] H. T. Sencar, M. Ramkumar and A. N. Akansu. Data Hiding Fundamentals and

Applications. Elsevier Academic Press, 2004.
[12] M. Wu and B. Liu. Multimedia Data Hiding. Springer-Verlag, 2003.
[13] M. Awrangjeb. An overview of reversible data hiding. In: Proc. 6th Int. Conf. Computer and Information Technology, Jahangirnagar University, Bangladesh, 2003, pp. 75-79.
[14] F. Mintzer, J. Lotspiech and N. Morimoto. Safeguarding digital library contents and users: digital watermarking. D-Lib Magazine, 1997.
[15] S. Lee, C. D. Yoo and T. Kalker. Reversible image watermarking based on integer-to-integer wavelet transform. IEEE Trans. Information Forensics and Security, 2007, 2(3):321-330.
[16] J. Fridrich, J. Goljan and R. Du. Invertible authentication. In: Proc. SPIE, Security and Watermarking of Multimedia Contents, 2001, Vol. 4314, pp. 197-208.
[17] M. U. Celik, G. Sharma, A. M. Tekalp, et al. Lossless generalized-LSB data embedding. IEEE Trans. Image Processing, 2005, 14(2):253-266.
[18] B. Yang, M. Schmucker, C. B. W. Funk, et al. Integer DCT-based reversible watermarking for images using companding technique. In: Proc. SPIE, Security, Steganography, and Watermarking of Multimedia Contents, 2004, Vol. 5306, pp. 405-415.
[19] G. Xuan, Y. Q. Shi, Q. Yao, et al. Lossless data hiding using histogram shifting method based on integer wavelets. In: International Workshop on Digital Watermarking, Lecture Notes in Computer Science, Springer-Verlag, 2006, Vol. 4283, pp. 323-332.
[20] J. Tian. Reversible data embedding using a difference expansion. IEEE Trans. Circuits and Systems for Video Technology, 2003, 13(8):890-896.
[21] A. M. Alattar. Reversible watermark using difference expansion of triplets. In: Proc. IEEE Int. Conf. Image Processing, 2003, Vol. 1, pp. 501-504.
[22] A. M. Alattar. Reversible watermark using difference expansion of quads. In: Proc. IEEE Int. Conf. Acoustics, Speech, and Signal Processing, 2004, Vol. 3, pp. 377-380.
[23] A. M. Alattar. Reversible watermark using the difference expansion of a generalized integer transform. IEEE Trans. Image Processing, 2004, 13(8):1147-1156.
[24] L. Kamstra and H. J. A. M. Heijmans. Reversible data embedding into images using wavelet techniques and sorting. IEEE Trans. Image Processing, 2005, 14(12):2082-2090.
[25] D. M. Thodi and J. J. Rodriguez. Expansion embedding techniques for reversible watermarking. IEEE Trans. Image Processing, 2007, 16(3):721-730.
[26] Z. Ni, Y. Q. Shi, N. Ansari, et al. Reversible data hiding. IEEE Trans. Circuits and Systems for Video Technology, 2006, 16(3):354-362.
[27] E. Varsaki, V. Fotopoulos and A. N. Skodras. A reversible data hiding technique embedding in the image histogram. Technical Report HOU-CS-TR-2006-08-GR, Hellenic Open University, 2006.
[28] J. Hwang, J. W. Kim and J. U. Choi. A reversible watermarking based on histogram shifting. In: International Workshop on Digital Watermarking, Lecture Notes in Computer Science, Springer-Verlag, 2006, Vol. 4283, pp. 348-361.
[29] W. C. Kuo, D. J. Jiang and Y. C. Huang. Reversible data hiding based on histogram. In: Int. Conf. on Intelligent Computing, Lecture Notes in Artificial Intelligence, Springer-Verlag, 2007, Vol. 4682, pp. 1152-1161.

[30] P. Tsai, Y. C. Hu and H. L. Yeh. Reversible image hiding scheme using predictive coding and histogram shifting. Signal Process, 2009.
[31] S. K. Lee, Y. H. Suh and Y. S. Ho. Lossless data hiding based on histogram modification of difference images. In: Pacific Rim Conference on Multimedia, Lecture Notes in Computer Science, Springer-Verlag, 2004, Vol. 3333, pp. 340-347.
[32] C. C. Lin, W. L. Tai and C. C. Chang. Multilevel reversible data hiding based on histogram modification of difference images. Pattern Recognition, 2008, 41(12):3582-3591.
[33] Z. Ni, Y. Shi, N. Ansari, et al. Reversible data hiding. In: IEEE Proceedings of ISCAS'03, 2003, (2):II-912~II-915.
[34] X. Luo, Q. Cheng and J. Tian. A lossless data embedding scheme for medical images in applications of E-Diagnosis. In: Proc. IEEE 25th Annual Int. Conf. Engineering in Medicine and Biology Society, 2003, Vol. 1, pp. 852-855.
[35] P. Ross, M. A. Viegerver, M. C. A. Van Dijke,et al. Reversible infraframe of medical images. IEEE Trans. Medical Image, 1998, 7:328-336.
[36] F. Bartolini, G. Bini, V. Cappellini, et al. Enforcement of copyright laws for multimedia through blind, detectable, reversible watermarking. In: IEEE Int. Conf. Multimedia Computing and Systems, 1999, Vol. 2, pp. 199-203.
[37] M. Barni, F. Bartolini, V. Cappellini, et al. Near-lossless digital watermarking for copyright protection of remote sensing images. In: Proc. IEEE Int. Conf. Geoscience and Remote Sensing Symposium, 2002, Vol. 3, pp. 1447-1449.
[38] D. Vleeschouwer, J. E. Delaigle and B. Macq. Circular interpretation of bijective transformations in lossless watermarking for media asset management. IEEE Trans. Multimedia, 2001, 5(1):97-105.
[39] Chou, C. Y. Jhou and S. C. Chu. Reversible watermark for 3D vertices based on data hiding in mesh formation. International Journal of Innovative Computing, Information and Control, 2009, 5(7):1893-1901.
[40] H. Luo, Z. M. Lu and J. S. Pan. A reversible data hiding scheme for 3D point cloud model. In: IEEE International Symposium on Signal Processing and Information Technology, 2006, pp. 863-867.
[41] H. .Luo, J. S. Pan, Z. M. Lu, et al. Reversible data hiding for 3D point cloud model. In: Proceedings of the International Conference on Intelligent Information Hiding and Multimedia Signal Processing, 2006.
[42] H. T. Wu and J. L. Dugelay. Reversible watermarking of 3D mesh models by prediction-error expansion. MMSP, 2008, pp. 797-802.
[43] H. T. Wu and M. C. Yiu. A reversible data hiding approach to mesh authentication. In: Proceedings of the 2005 IEEE/WIC/ACM International Conference on Web Intelligence, 2005.
[44] Z. Sun, Z. M. Lu and Z. Li. Reversible data hiding for 3D meshes in the PVQ-compressed domain. In: IEEE International Conference on Intelligent Information Hiding and Multimedia Signal Processing, 2006, pp. 593-596.
[45] Z. M. Lu and Z. Li. High capacity reversible data hiding for 3D meshes in the PVQ domain. In: The 6th International Workshop, IWDW, LNCS 5041, 2007, pp. 233-243.
[46] R. Ohbuchi, H. Masuda and M. Aono. Watermarking three-dimensional polygonal models through geometric and topological modifications. IEEE J. Select. Areas Commun., 1998, 16:551-560.
[47] O. Benedens. Geometry-based watermarking of 3-D models. IEEE Comput.

Graph., Special Issue on Image Security, 1999, 1/2:46-55.
[48] E. Praun, H. Hoppe and A. Finkelstein. Robust mesh watermarking. In: Proc. SIGGRAPH, 1999, pp. 69-76.
[49] M. M. Yeung and B. L. Yeo. Fragile watermarking of three dimensional objects. In: Proc. 1998 Int. Conf. Image Processing, ICIP98, 1998, Vol. 2, pp. 442-446.
[50] F. Cayre and B. Macq. Data hiding on 3-D triangle meshes. IEEE Trans. Signal Processing, 2003, 51(4):939-949.
[51] H. Y. S. Lin, H. Y. M. Liao, C. S. Lu, et al. Fragile watermarking for authenticating 3D polygonal meshes. IEEE Transactions on Multimedia, 2005, 7(6):997-1006.
[52] J. Dittmann and O. Benedens. Invertible authentication for 3D meshes. In: Proceedings of SPIE - The International Society for Optical Engineering, 2003, Vol. 5020, pp. 653-664.
[53] X. Mao, M. Shiba and A. Imamiya. Watermarking 3D geometric models through triangle subdivision. In: Proceedings of SPIE, Security and Watermarking of Multimedia Contents III, 2001, Vol. 4314, pp. 253-260.
[54] H. T. Wu and Y. M. Cheung. A new fragile mesh watermarking algorithm for authentication. Paper presented at The IFIP 20th International Information Security Conference, 2005, pp. 509-523.
[55] B. Chen and G. W. Wornell. Dither modulation: a new approach to digital watermarking and information embedding. In: Proc. SPIE: Security and Watermarking of Multimedia Contents, 1999, Vol. 3657, pp. 342-353.
[56] C. W. Honsinger, P. Jones, M. Rabbani, et al. Lossless recovery of an original mesh containing embedded data. US Patent Application, Docket No: 77102/E–D, 1999.
[57] J. Tian. High capacity reversible data embedding and content authentication. In: ICASSP, IEEE International Conference on Acoustics, Speech and Signal Processing – Proceedings, 2003, Vol. 3, pp. 517-520.
[58] G. Xuan, Y. Q. Shi, Z. C. Ni, et al. High capacity lossless data hiding based on integer wavelet transform. In: Proceedings - IEEE International Symposium on Circuits and Systems, 2004, Vol. 2.
[59] Y. Q. Shi, Z. Ni, D. Zou, et al. Lossless data hiding: fundamentals, algorithms and applications. In: Proceedings - IEEE International Symposium on Circuits and Systems, 2004, Vol. 2.
[60] Z. Ni, Y. Q. Shi, A. Nirwan, et al. Reversible data hiding. IEEE Transactions on Circuits and Systems for Video Technology, 2006, 16(3):354-361.
[61] C. Mehmet, U. S. Gaurav, T. A. Murat, et al. Reversible data hiding. Paper presented at The IEEE International Conference on Image Processing, 2002, Vol. 2, pp. II/157-II/160.
[62] R. Xuan, C. Y. Yang, Y. Z. Zhen, et al. Reversible data hiding based on wavelet spread spectrum. In: 2004 IEEE 6th Workshop on Multimedia Signal Processing, 2004, pp. 211-214.
[63] Z. C. Ni, Y. Q. Shi, A. Nirwan, et al. Robust lossless image data hiding. Paper presented at The IEEE International Conference on Multimedia and Expo (ICME), 2004, Vol. 3, pp. 2199-2202.
[64] J. Fridrich, M. Goljan and R. Du. Invertible authentication watermark for JPEG images. In: Proc. IEEE Int. Conf. on Information Technology: Coding and Computing, 2001.
[65] R. Gray and D. Neuhoff. Quantization. IEEE Trans. Information Theory, 1998,

44(10):2325-2384.
[66] P. H. Chou and T. H. Meng. Vertex data compression through vector quantization. IEEE Transactions on Visualization and Computer Graphics, 2002, 8(4):373-382.
[67] Princeton University. 3D Model Search Engine. http://shape.cs.princeton.edu.
[68] C. Zhu and L. M. Po. Minimax partial distortion competitive learning for optimal codebook design. IEEE Trans. on Image Processing, 1998, 7(10):1400-1409.
[69] S. W. Ra and J. K. Kim. Fast mean-distance-ordered partial codebook search algorithm for image vector quantization. IEEE. Trans. on Circuits and Systems-II, 1993, 40(9):576-579.
[70] C. M. Wang and P. C. Wang. Steganography on point-sampled geometry. Computers & Graphics, 2006, 30:244-254.
[71] R. Ohbuchi, H. Masuda and M. Aono. Embedding watermark in 3D models. In: Proceedings of the IDMS'97, Lecture Notes in Computer Science, Springer, 1997, pp. 1-11.
[72] R. Ohbuchi, H. Masuda and M. Aono. Watermarking three-dimensional polygonal models. In: Proceedings of the ACM Multimedia'97, 1997, pp. 261-272.
[73] R. Ohbuchi, H. Masuda and M. Aono. Watermarking three-dimensional polygonal models through geometric and topological modifications. IEEE Journal on Selected Areas in Communications, 1998, 16(4):551-560.
[74] R. Ohbuchi, H. Masuda and M. Aono. Watermark embedding algorithms for geometrical and non-geometrical targets in three-dimensional polygonal models. Computer Communications, 1998.
[75] O. Benedens. Geometry-based watermarking of 3D models. IEEE Computer Graphics and Applications, 1999, 19(1):46-55.
[76] B. L. Yeo and M. M. Yeung. Watermarking 3D Objects for Verification. IEEE Computer Graphics and Applications, 1999, 19(1):36-45.
[77] M. G. Wagner. Robust watermarking of polygonal meshes. In: Proceedings of Geometric Modeling and Processing, 2000, pp. 10-12.
[78] E. Praun, H. Hoppe and A. Finkelstein. Robust mesh watermarking. Microsoft Technical Report TR-99-05, 1999.
[79] R. Ohbuchi, A. Mukaiyama and S. Takahashi. A frequency-domain approach to watermarking 3D shapes. In: Proc. EUROGRAPHICS 2002, 2002.
[80] R. Ohbuchi, A. Mukaiyama and S. Takahashi. Watermarking a 3D shape model defined as a point set. In: Proc. of Cyber Worlds 2004, IEEE Computer Society Press, 2004, pp. 392-399.
[81] M. Voigt, B. Yang and C. Busch. Reversible watermarking of 2D-vector watermark. In: Proceedings of the Multimedia and Security Workshop 2004 (MM&SEC'04), 2004, pp. 160-165.
[82] G. Plonka and M. Tasche. Invertible integer DCT algorithms. Appl. Comput. Harmon. Anal., 2003, 15:70-88.

Index